W9-AMZ-768

THE CHEMISTRY BOOK

BIG IDEAS

THE ART BOOK
THE ASTRONOMY BOOK
THE BIBLE BOOK
THE BIOLOGY BOOK
THE BLACK HISTORY BOOK
THE BUSINESS BOOK
THE CHEMISTRY BOOK
THE CLASSICAL MUSIC BOOK
THE CRIME BOOK
THE ECOLOGY BOOK
THE ECONOMICS BOOK
THE FEMINISM BOOK
THE HISTORY BOOK
THE ISLAM BOOK
THE LAW BOOK
THE LITERATURE BOOK
THE MATH BOOK
THE MEDICINE BOOK
THE MOVIE BOOK
THE MYTHOLOGY BOOK
THE PHILOSOPHY BOOK
THE PHYSICS BOOK
THE POLITICS BOOK
THE PSYCHOLOGY BOOK
THE RELIGIONS BOOK
THE SCIENCE BOOK
THE SHAKESPEARE BOOK
THE SHERLOCK HOLMES BOOK
THE SOCIOLOGY BOOK
THE WORLD WAR II BOOK

SIMPLY EXPLAINED

THE
CHEMISTRY
BOOK
DK

DK LONDON

SENIOR ART EDITOR
Duncan Turner

SENIOR EDITORS
Helen Fewster, Camilla Hallinan

EDITORS
Alethea Doran, Annelise Evans, Becky Gee, Lydia Halliday, Tim Harris, Katie John, Gill Pitts, Jane Simmonds, Jess Unwin

SENIOR US EDITOR
Kayla Dugger

ILLUSTRATIONS
James Graham

ADDITIONAL TEXT
Richard Beatty

JACKET DESIGN DEVELOPMENT MANAGER
Sophia MTT

JACKET DESIGNER
Stephanie Cheng Hui Tan

SENIOR PRODUCTION EDITOR
Andy Hilliard

SENIOR PRODUCTION CONTROLLER
Meskerem Berhane

MANAGING ART EDITOR
Michael Duffy

MANAGING EDITOR
Angeles Gavira Guerrero

ASSOCIATE PUBLISHING DIRECTOR
Liz Wheeler

ART DIRECTOR
Karen Self

DESIGN DIRECTOR
Phil Ormerod

PUBLISHING DIRECTOR
Jonathan Metcalf

DK DELHI

PROJECT ART EDITOR
Anjali Sachar

ART EDITORS
Mridushmita Bose, Debjyoti Mukherjee

ASSISTANT ART EDITOR
Aarushi Dhawan

SENIOR EDITOR
Anita Kakar

SENIOR MANAGING EDITOR
Rohan Sinha

MANAGING ART EDITOR
Sudakshina Basu

DTP DESIGNERS
Rakesh Kumar, Mrinmoy Mazumdar, Vikram Singh

SENIOR JACKETS COORDINATOR
Priyanka Sharma Saddi

SENIOR PICTURE RESEARCHER
Surya Sankash Sarangi

ASSISTANT PICTURE RESEARCHER
Mayank Shankar Choudhary

PICTURE RESEARCH MANAGER
Taiyaba Khatoon

PRE-PRODUCTION MANAGER
Balwant Singh

PRODUCTION MANAGER
Pankaj Sharma

EDITORIAL HEAD
Glenda Fernandes

DESIGN HEAD
Malavika Talukder

SANDS PUBLISHING SOLUTIONS

EDITORIAL PARTNERS
David and Sylvia Tombesi-Walton

DESIGN PARTNER
Simon Murrell

original styling by

First American Edition, 2022
Published in the United States by
DK Publishing, 1450 Broadway, Suite 801, New York, NY 10018

22 23 24 25 26 10 9 8 7 6 5 4 3 2 1
001–325004–Aug/2022

Published in Great Britain by Dorling Kindersley Limited

A catalog record for this book is available from the Library of Congress.
ISBN 978-0-7440-5632-7

DK books are available at special discounts when purchased in bulk for sales promotions, premiums, fund-raising, or educational use. For details, contact: DK Publishing Special Markets, 1450 Broadway, Suite 801, New York, NY 10018
SpecialSales@dk.com

Printed in China

For the curious
www.dk.com

This book was made with Forest Stewardship Council ™ certified paper–one small step in DK's commitment to a sustainable future. For more information go to www.dk.com/our-green-pledge

CONTRIBUTORS

ANDY BRUNNING

Andy Brunning is a former chemistry teacher and University of Bath graduate. He is the creator of the award-winning chemistry infographics website, Compound Interest, and the author of *Why does asparagus make your pee smell?*, a book exploring food chemistry. He has also written for Crash Course Organic Chemistry.

CATHY COBB

Cathy Cobb, adjunct Professor, University of South Carolina Aiken, has written five books on chemical history and chemistry for the nonchemist. In *The Chemistry of Alchemy*, she presents the history of alchemy; the chemistry behind alchemy; and home demonstrations of alchemical practices, such as making fool's gold and the peacock's tail.

ANDY EXTANCE

Before becoming a full-time science writer, Andy Extance worked for six and a half years in early-stage drug discovery research. Today, Andy's science writing explores everything related to chemistry, from Earth's environment to space, from food to fusion, and from solar cells to how we smell.

JOHN FARNDON

Shortlisted five times for the Royal Society's Young People's Science Book Prize, John Farndon has written over 1,000 books on science, nature, and other topics. He has also contributed to many histories of science, including DK's *Smithsonian Science* and *Science Year by Year*.

TIM HARRIS

A widely published author on science and nature for both children and adults, Tim Harris studied geology in college. He has written more than 100 mostly educational reference books and contributed to many others, including *Chemistry Matters!*, *Great Scientists*, *Routes of Science*, *The Physics Book*, and *The Biology Book*.

CHARLOTTE SLEIGH, CONSULTANT

Professor Charlotte Sleigh is at the Department for Science and Technology Studies, University College London. She is the author of several books on the history and culture of science and president of the British Society for the History of Science.

ROBERT SNEDDEN

Robert Snedden has worked in publishing for over 40 years, researching and writing science and technology books on a range of topics—from chemical and environmental engineering to materials science, space exploration, physics, and Albert Einstein.

CONTENTS

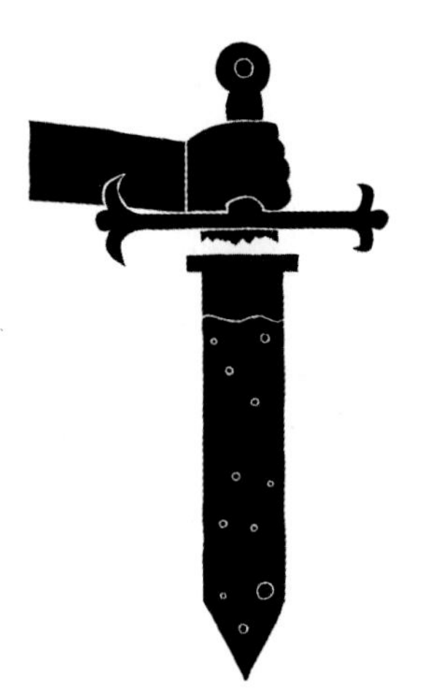

THE CHEMICAL REVOLUTION

THE INDUSTRIAL AGE

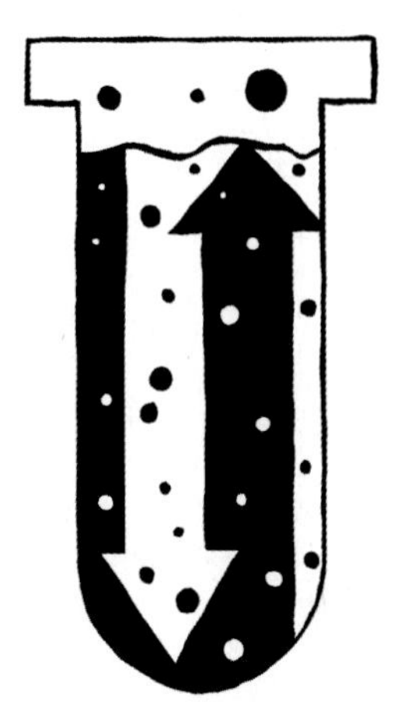

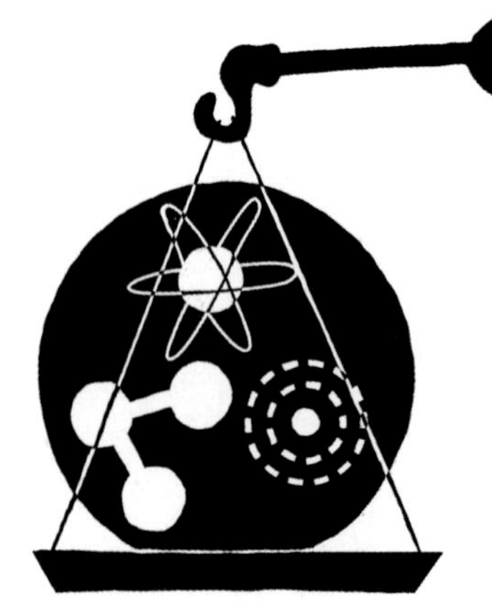

THE MACHINE AGE

THE NUCLEAR AGE

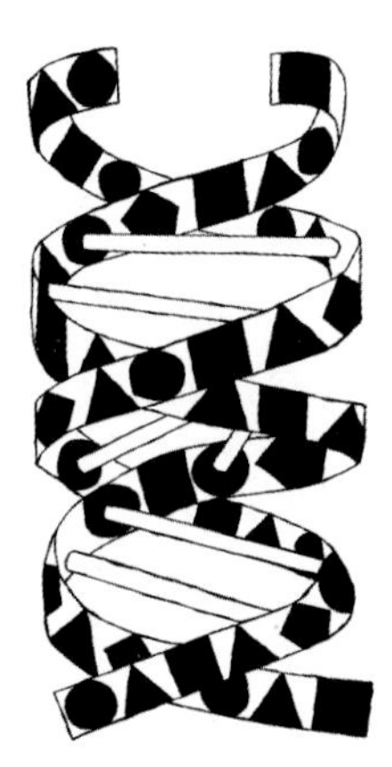

A CHANGING WORLD

INTRODU

CTION

Chemistry can be defined as the study of the elements and compounds that make up ourselves and the world around us, and the reactions that transform their multitude of substances into different ones. But to define it so simply diminishes the mystique and wonder of chemistry, which has repeatedly drawn people to study it over the ages.

Chemistry is the science of flair and spectacle. Two colorless liquids, mixed together, produce a blooming bright yellow cloud of precipitate. A sliver of shimmering metal, dropped into a bowl of water, bubbles and bursts dramatically into an ethereal lilac flame.

> It is the great beauty of our science, chemistry, that advancement in it … opens the door to further and more abundant knowledge.
> **Michael Faraday**

Outwardly, left unexplained, these reactions have the appearance of magic; however, unlike magic, chemistry has surrendered its secrets over the centuries, although some of the tools required to probe them may be complex. And as our knowledge of chemistry has developed, so have our perceptions of this science.

From alchemy to chemistry

In ancient times, the discipline that would become chemistry began as a practical means of separating and refining substances, driven by the recognition that the components making up a mixture could have different properties. Early practitioners of these techniques in Babylon, China, Egypt, and Turkey developed specialized equipment to better refine their processes. Some of these methods, such as those to produce soaps, make glass, and refine metals, are still used in modified forms today.

In medieval times, the practice known as alchemy held a promise of riches and immortality. The alchemists tirelessly sought the legendary philosopher's stone, a mythical object purported to have the ability to turn common metals into gold and allow creation of an elixir that would give the drinker immortality. Though these lofty goals would be unfulfilled, and may raise eyebrows today, the alchemists' work in pursuing them led to the development of experimental chemistry and even to the discoveries of new elements.

By the 18th century, something resembling modern chemistry was beginning to emerge from the increasingly disparaged practice of alchemy. A revolution in chemical thinking led to clearer ideas about proportions in which substances react with each other and combine. The 19th century saw the founding of modern atomic theory, as well as the emergence of chemistry's most recognizable visual representation: the periodic table. It also saw an explosion of industrial applications for chemistry, transforming the science into a technical discipline that made innovations possible.

The 20th century witnessed the realization of these innovations. Plastics, fertilizers, antibiotics, and batteries are essential parts of modern life as we know it, while few inventions can claim to have effected such colossal societal change as the contraceptive pill. But there were also warnings of the power of chemistry and its potential for harm; the extended use of leaded gasoline and its

potential impact on neurological health, the damage done to the ozone layer by ozone-depleting compounds, and the advent of nuclear weapons were all reminders that chemicals can be dangerous as well as beneficial.

Today, our relationship with chemistry is an uneasy one. It continues to provide vital and life-saving innovations that constantly extend the boundaries of our knowledge: most recently, the COVID-19 vaccines are dependent upon the chemistry that underpins them. But there is also continued concern about the impact of chemicals on our health, our climate, and our planet. Ironically, to solve these chemical problems, we will rely on solutions from chemistry in combination with the other sciences.

The divisions of chemistry

It has become customary for modern chemistry to be split into three broad divisions: physical chemistry, organic chemistry, and inorganic chemistry.

Physical chemistry is at the interface between physics and chemistry, commonly involving the application of mathematical concepts to understand chemical phenomena. Its aspects include thermodynamics, which chemists can apply in order to discern the stability of chemical compounds, whether or not certain reactions take place, and the speed with which reactions happen.

> Chemistry provides not only a mental discipline, but an adventure and an aesthetic experience.
> **Sir Cyril Hinshelwood**

Organic chemistry is the study of carbon-based compounds. Carbon is unique in its ability to form large networks of bonds with other carbon atoms and atoms of different elements such as oxygen, hydrogen, and nitrogen. Biological compounds, including our DNA, are organic compounds, as are many of the medicines we use. Organic chemistry is concerned with understanding the structures and reactions of these compounds.

Finally, inorganic chemistry deals with compounds outside organic chemistry, including compounds of metals, determining their structures and how they react. Advances in this area have led to the creation of pigments, new materials, and the lithium-ion batteries that power many of our modern devices.

While textbooks and chemistry classes still commonly organize chemistry into these divisions, increasingly the boundaries between them—and between chemistry, biology, and physics—have become blurred. Many of the biggest scientific advances in recent years—the use of particle accelerators to discover new elements, genome editing, and the COVID-19 vaccines—transcend these simple classifications and require expertise from across the scientific disciplines.

Chemistry has become the central science, intersecting with the other sciences to deliver new and exciting advances. This book charts the course of this evolution, beginning with chemistry's practical roots in ancient times, recounting the emergence of modern chemistry from alchemy, and ultimately uncovering how chemistry's reach has extended to touch almost every aspect of today's world. ■

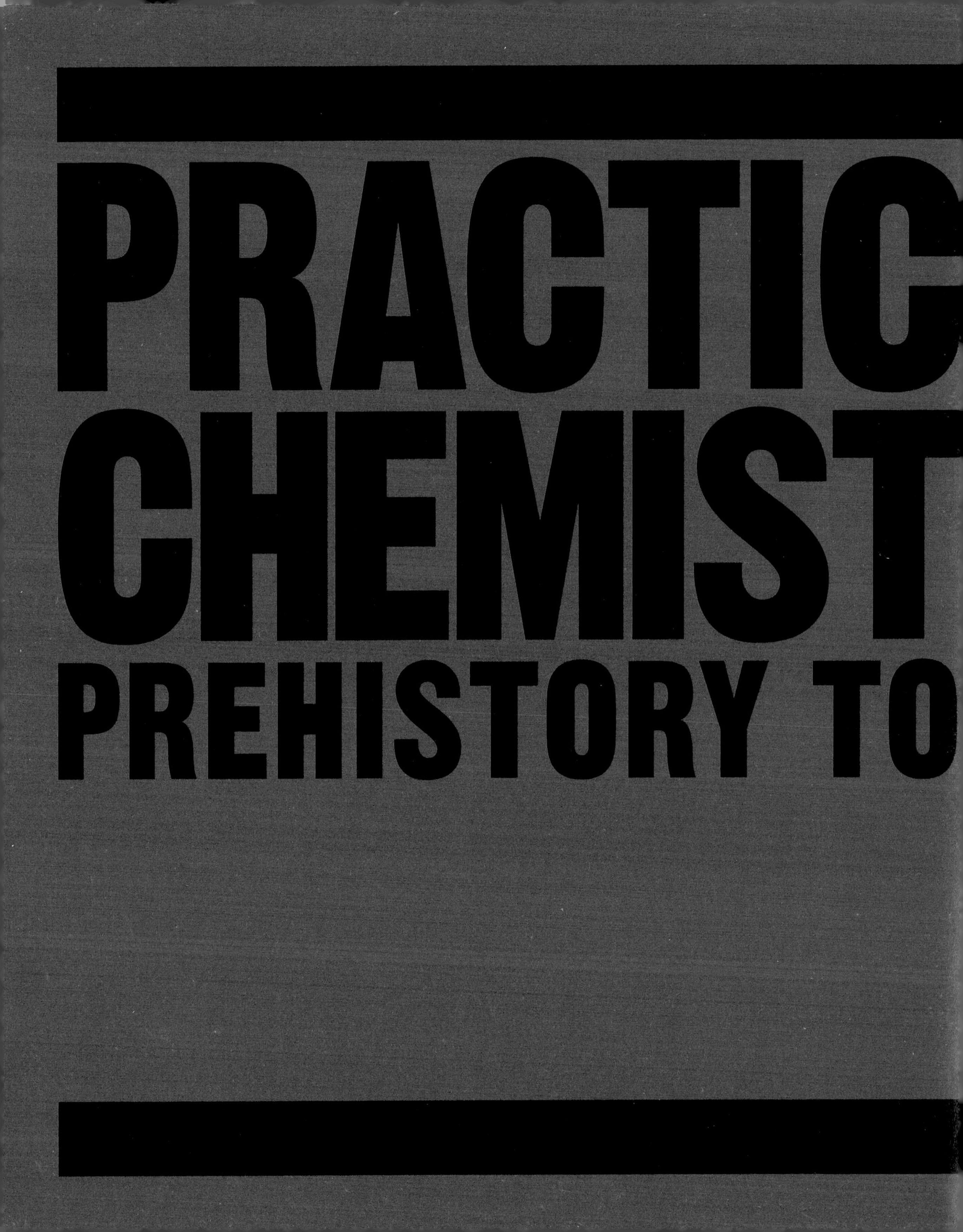
PRACTIC
CHEMIST
PREHISTORY TO

AL
RY
800 CE

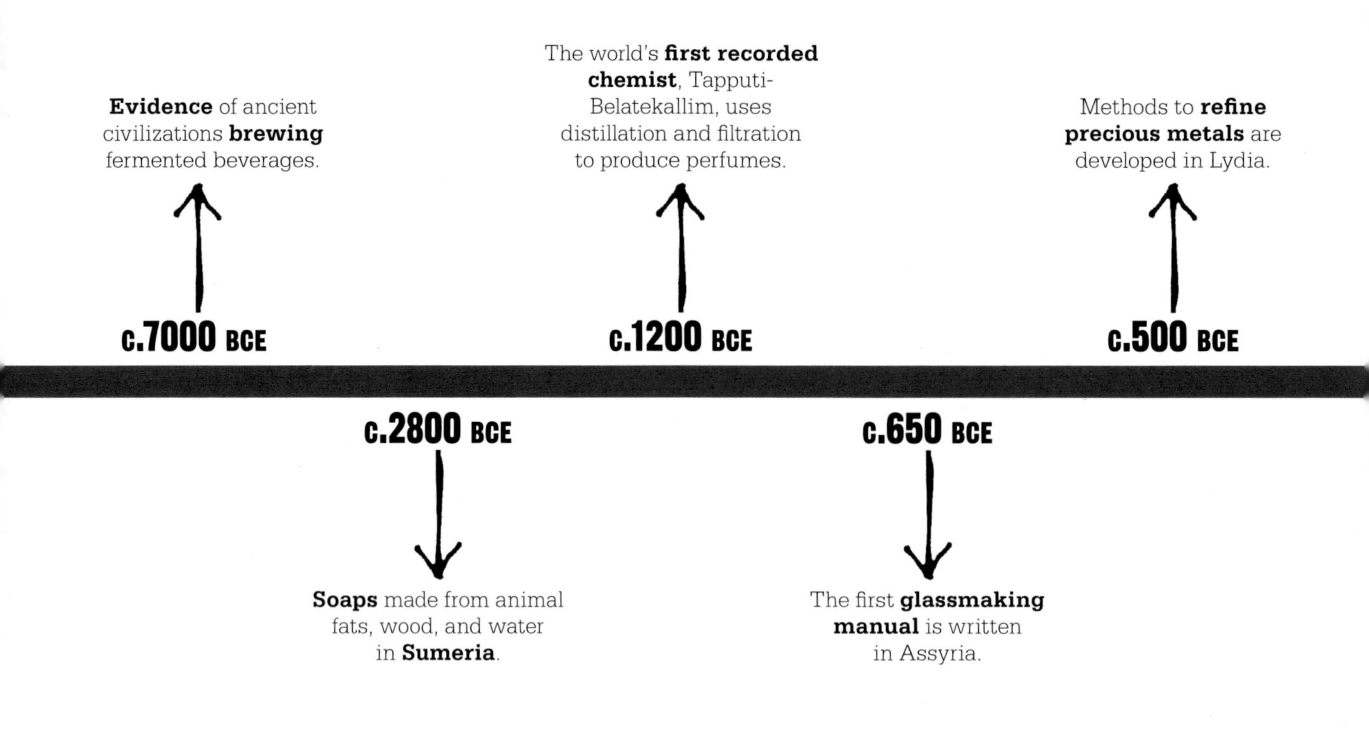

c.7000 BCE
Evidence of ancient civilizations brewing fermented beverages.
c.2800 BCE
Soaps made from animal fats, wood, and water in Sumeria.
c.1200 BCE
The world's first recorded chemist, Tapputi-Belatekallim, uses distillation and filtration to produce perfumes.
c.650 BCE
The first glassmaking manual is written in Assyria.
c.500 BCE
Methods to refine precious metals are developed in Lydia.

Practicality was often the driver for early forays into chemistry. While the western world often focuses on itself as the theater for much of the documented history of chemistry, the foundations of practical chemistry were laid down by ancient empires across the globe.

Initially, they used chemical processes to make items for everyday use or convenience, such as soaps, pottery, fabric dyes, and home building materials.

From archaeological evidence, we know that fermentation was among the first biochemical processes with which our ancestors experimented, creating bread and fermented beverages. In what is now China, early rice wines were produced by fermenting rice, honey, and fruit. While it is likely that the Chinese also developed processes for distillation, that technique is thought to have originated in ancient India. Several indigenous civilizations in the Americas and sub-Saharan Africa are also known to have developed their own alcoholic beverages.

Chemical artistry

Distillation was not used solely for alcohol production. In Babylon (present-day Iraq and Syria), the development of early chemistry apparatuses and techniques allowed for the separation of mixtures by exploiting the properties of the mixtures' components.

These processes were turned to artisanal purposes, including the production of perfumes. Babylon can lay claim to the first documented chemist, Tapputi-Belatekallim, who recorded her work on clay tablets. She detailed her use of extraction, distillation, and filtration to concoct perfumes for medicinal and ritual purposes.

Glassmaking was another chemical process that led to artisanal uses. In Assyria, which covered parts of present-day Iran, Iraq, Syria, and Turkey, the first glassmaking manual was discovered in the library of King Ashurbanipal—but we know from archaeological evidence that other civilizations, including those in Egypt, China, and Ancient Greece, were experimenting with glassmaking before this time. The glasses produced were used to make weapons, decorative objects, and hollow vessels, though the art of glassblowing would not develop until the 1st century CE.

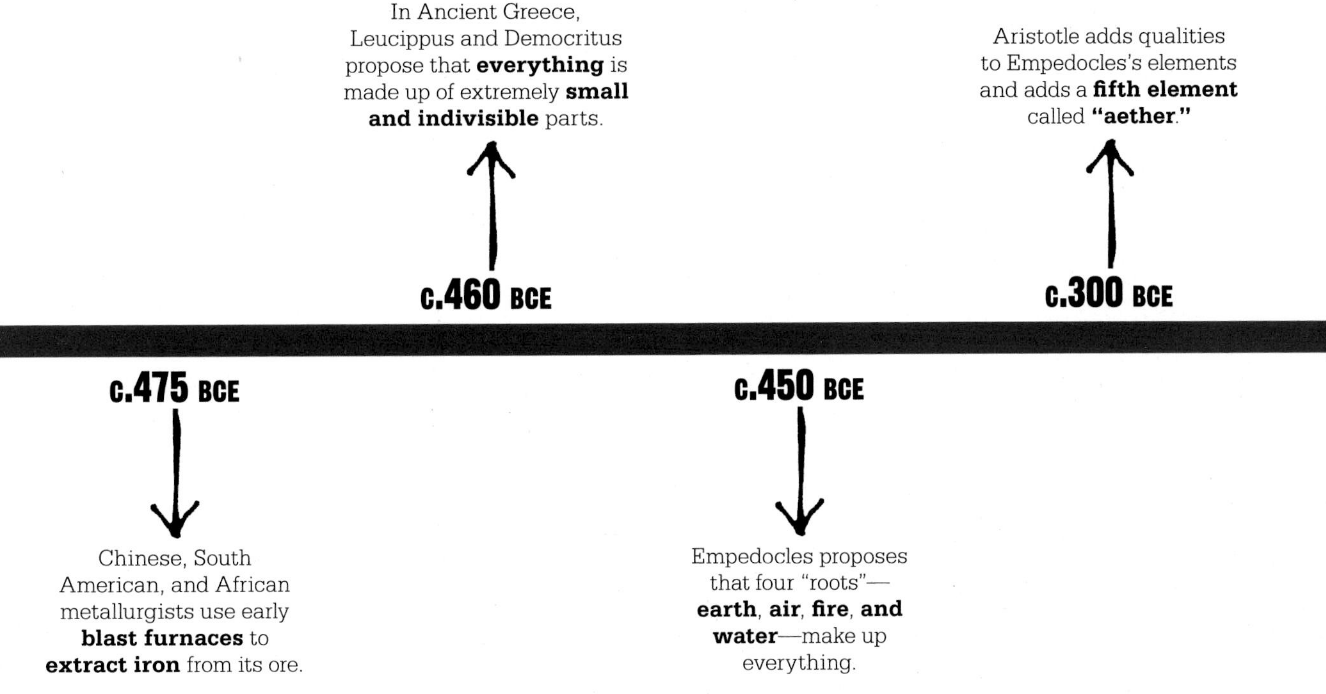

In Ancient Greece, Leucippus and Democritus propose that **everything** is made up of extremely **small and indivisible** parts.
c.460 BCE
Aristotle adds qualities to Empedocles's elements and adds a **fifth element** called **"aether."**
c.300 BCE
c.475 BCE
Chinese, South American, and African metallurgists use early **blast furnaces** to **extract iron** from its ore.
c.450 BCE
Empedocles proposes that four "roots"—**earth**, **air**, **fire**, **and water**—make up everything.

Metallurgy

Early chemistry also allowed metal reserves to be exploited. Precious metals such as gold and silver were less problematic to use than other metals, which often existed in combination with other elements, but techniques developed in Lydia (modern Turkey) for refining gold and silver enabled the creation of standard coinage systems.

More important were the techniques devised to isolate other metals from their ores, where these were found chemically combined with other elements. Early blast furnaces in ancient China were used to extract iron, and there is evidence of copper smelting in some indigenous South American civilizations. Such processes transformed metals from being primarily used to make decorative cultural items to being harnessed in a range of practical applications, including in weaponry.

Elemental foundations

About 2,500 years ago, Ancient Greek thinkers turned their minds to theorizing about what makes up the world around us. Their philosophy laid the foundations of theoretical frameworks that would remain highly significant in the study of the material world for centuries afterward.

The philosophers Leucippus and Democritus introduced the concept of atoms as solid, indivisible pieces of matter that make up everything around us. Democritus also postulated that different forms of atoms made up different substances and that atoms could combine with each other in various ways.

In the same period, Empedocles proposed that every substance is formed from a combination of four basic "roots": earth, air, fire, and water. Plato is thought to have been the first Greek philosopher to refer to these substances as "elements." Aristotle went on to define an element as "not itself divisible into bodies different in form defined." He also gave descriptive qualities to the elements to explain the qualities of substances. His theory persisted until the 17th century, until it began to be superseded by the discovery of physical elements. Conversely, the theory of atoms disappeared, but it began afresh in the 18th century.

These classical ideas, with the techniques and apparatuses created in various ancient cultures, would form the basis of modern chemistry. ■

HE WHO DOES NOT KNOW BEER, DOES NOT KNOW WHAT IS GOOD

BREWING

IN CONTEXT

KEY FIGURES
Unknown brewers
(c. 11,000 BCE)

BEFORE
c. 21,000 BCE Near the Sea of Galilee, Israel, hunter-gatherers construct brush huts where they stockpile seeds and berries. The huts have hearths, sealed floors, and sleeping areas.

AFTER
c. 6000 BCE Chemical evidence of wine production is preserved in jars near present-day Tbilisi, Georgia.

c. 1600 BCE Egyptian texts set out approximately 100 medical prescriptions citing beer as a cure for a variety of conditions.

c. 100 BCE In the southwestern United States, the Papago use wine made from the saguaro cactus in their sacred rituals.

c. 1000 CE Hops are used extensively in the beer brewing process in Germany.

Alcohol has been associated with social activities—both sacred and profane—since before written records began, and its production is among the oldest chemical processes for which we have evidence.

First drafts

We cannot be sure how alcohol was first discovered, but brewing is a crucial early human foray into chemistry. It is likely that humanity's earliest experience of alcohol was a chance occurrence, possibly associated with rotting fruit. There is some evidence that the earliest examples of alcohol production may even predate the first cultivation of crops, some 11,000 years ago.

The Natufians, a Neolithic people who lived around the eastern Mediterranean from c. 15,000 to c. 11,000 BCE, may have been one of the first cultures to brew beer. Archaeologists have analyzed residue found in stone mortars (bowls) dating to c. 11,000 BCE, discovered in a Natufian burial site located near what is now Haifa, Israel. They detected signs that these mortars had been used for the brewing of wild wheat or barley, as well as for storing food. The archaeologists speculate that the Natufians used a three-stage brewing process in which starch from the wheat or barley was first turned into malt by germinating

This Egyptian brewing scene, dating to 2500–2350 BCE, is part of the painted limestone decoration of a funerary chapel in North Abydos, an ancient city in Upper Egypt.

See also: Purifying substances 20–21 ▪ Catalysis 69 ▪ Enzymes 162–163

1. Mashing
Barley malt is combined with hot water. The resulting mash is then filtered, giving a sugar solution called wort.

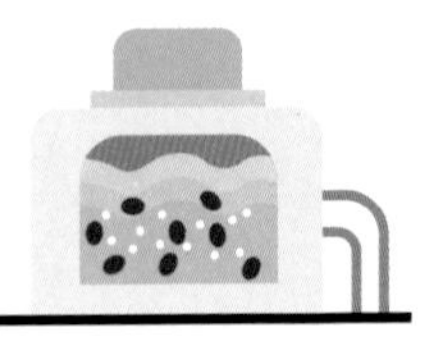

2. Boiling
Hops are added to the wort, which is boiled in a brewing kettle. The wort is then cooled and the hops are filtered.

3. Fermenting
The solution is moved to a fermentation vessel, where yeast is added. The yeast converts sugars into alcohol and CO_2.

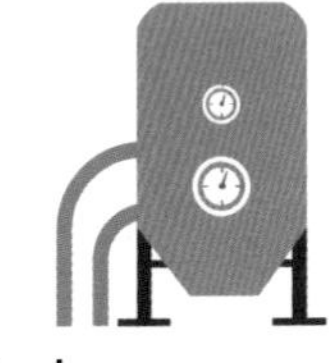

4. Conditioning
Beer is undrinkable until it has been conditioned. During conditioning, yeast breaks down the foul-tasting compounds.

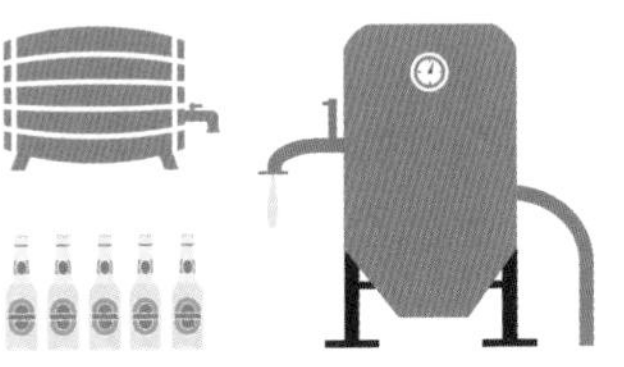

5. Filtering
Finally, the beer is filtered for clarity. Some beer styles are not filtered and retain their "haze."

The beer-making process starts by germinating barley to turn it into barley malt, a process that ensures the presence of sugars and starch, as well as amylase and protease enzymes. There are then five main steps to follow.

the grains in water before it was dried and stored. The malt was then mashed and heated, and finally it was left to be fermented. During the fermentation process, airborne wild yeast, which occurs naturally in the environment, converted the sugars from the barley or wheat into ethanol (alcohol). The results were more like a "beer oatmeal" than the liquid we are used to today.

It is thought that brewing was being carried out by several civilizations by c. 7000 BCE, and chemical evidence of one of the oldest alcoholic beverages dates to this time. Archaeologists analyzed the residue on ceramic pots found in Jiahu, Northeast China, and discovered trace amounts of a fermented drink made from honey, rice, and fruit. Examination of vessels and residue from several archaeological sites suggests that people used a grain-based starter, called *qu*, for making a beerlike drink during the early period of plant domestication in that region, which has also been dated to c. 7000 BCE. Like the Natufian findings, these vessels come from sites that were associated with burials, possibly suggesting that drinking played a role in death rituals.

Bread and beer

The oldest written record of beer production is a 6,000-year-old tablet from ancient Mesopotamia (a historical region between the Tigris and Euphrates Rivers that broadly covered parts of modern-day Syria and Turkey and most of Iraq). It is believed to have been created by the Sumer civilization (in modern-day Iraq), who had a patron goddess of brewing named Ninkasi. The oldest surviving beer recipe, describing the production of beer made from barley bread, was found in a 3,900-year-old poem written in praise of her.

Egypt was one of the ancient world's biggest producers of wine and beer. In fact, the world's oldest known brewery (c. 3400 BCE), in the city of Hierakonpolis, is thought to have produced more than 300 gallons (1,100 liters) of beer a day. Egyptian breweries were often associated with bakeries, with both relying on the activity of yeast to convert sugars from grains such as barley and emmer into ethyl alcohol and carbon dioxide (CO_2). The difference is that the alcohol is the desired product for brewers, whereas bakers look to the CO_2 to leaven the bread. It seems likely that our forebears were brewing beer before they were baking bread. Today, yeast left over from the brewing process is often used for making bread. ■

SWEET OIL, THE FRAGRANCE OF THE GODS

PURIFYING SUBSTANCES

A **mixture of liquids** is placed in a flask.

When the liquids are **heated**, the one with the **lowest boiling point** forms a **vapor first**.

The **vapor is cooled** in a condenser.

The resulting purified liquid is collected as a distillate.

IN CONTEXT

KEY FIGURE
Tapputi-Belatekallim
(c. 1200 BCE)

BEFORE
c. 4000 BCE People in the Tigris Valley make bell-shaped pots that may form part of a distillation apparatus.

c. 3000 BCE A terracotta distillation apparatus in the Indus Valley is most likely used to produce essential oils.

c. 2000 BCE An enormous perfume-making factory operates in Cyprus.

AFTER
c. 9th century CE Arab philosopher al-Kindi's *Book of the Chemistry of Perfume and Distillations* sets out more than 100 recipes and methods.

c. 11th century Persian polymath Ibn Sina invents a process for extracting oils from flowers by distillation to create more delicate perfumes.

Distillation is a process for separating out liquids either from solids, as when extracting alcohol from fermented materials, or from a mixture of liquids with different boiling points, such as the separation of crude oil into its components (including butane and gasoline).

Early technology

One of the first technological discoveries made by early humans was that tar could be distilled from the bark of birch trees. This natural adhesive was key to making compound tools and was used to install stone blades into wooden handles for axes, spears, and hoes. Ancient beads of tar have been uncovered in Middle Paleolithic European sites that predate the arrival of modern *Homo sapiens* in Western Europe by about 150,000 years. These early distillers were Neanderthals, who most likely heated the bark in the embers of a fire to extract the tar.

In more (relatively) recent times, people learned to use distillation to create perfumes. From the evidence of hieroglyphs, it is an art that stretches back at least 5,000 years to the priests of ancient Egypt, who used aromatic resins in their rituals. One of the first stages in the making of a perfume is to extract fragrant essential oils from plants, and the most common way to do this is by distillation.

See also: Brewing 18–19 ▪ Refining precious metals 27 ▪ Attempts to make gold 36–41 ▪ Cracking crude oil 194–195

The still

In Mesopotamia, western Asia, stills were being used as early as 3500 BCE for distilling and filtering liquids. At this time, they consisted of a double-rimmed clay vessel with a lid. Liquid was heated inside the container and condensate (liquid formed by condensation) accumulated inside the lid, which was cooled with water. This condensate ran from the lid into a trough formed from the double rim of the vessel, where it was collected. The processes used were highly inefficient and often distillations had to be repeated several times to achieve the concentrations required.

The first chemist

Clay tablets inscribed with cuneiform text dating to around 1200 BCE describe perfumeries in ancient Babylon (a city in southern Mesopotamia, present-day Iraq) that employed an early form of distillation. A Babylonian perfume maker identified in the tablets as Tapputi-Belatekallim is the first chemist identified by name since records began. "Belatekallim" means "overseer" and Tapputi was the overseer of the royal perfumery. The tablets describe her treatise on perfume making—the first such ever recorded—and how she filtered and distilled perfumes for religious rituals and medicines, as well as for use in the royal household. Although the still greatly predates Tapputi, the tablets provide the first written description of its use.

Perfume makers such as Tapputi also employed a range of other equipment, much of it adapted from domestic utensils. Examples include earthenware and stone pots and beakers, weights and measures, sieves, pestles and mortars, filtering cloths, and furnaces capable of reaching a range of temperatures.

Another surviving clay tablet describes the step-by-step process Tapputi followed to produce an ointment for the royal household containing water, flowers, oil, and calamus (possibly lemongrass). It details the refining of the ingredients in her still and is the oldest recorded reference to this technique. The ingredients were softened first with water and then with oil and boiled to release their essences, which were quickly condensed on the walls of the still. The concentrate collected could then be diluted in a mixture of water and alcohol, just as perfumes are today. ■

The alembic, depicted in this 18th-century Arabic text, is said to have been invented by Egyptian alchemist Maria Hebraea around the 2nd century CE. The condensate flows from the cooling vessel into a collecting flask.

"Women perfumers developed the chemical techniques of distillation, extraction, and sublimation."
Margaret Alic
***Hypatia's Heritage* (1986)**

Distillation and sublimation

Distillation is an effective way of separating a mixture of liquids that boil at different temperatures. The most volatile component vaporizes at the lowest temperature. The vapor is passed through a condenser, where it cools to its liquid state, and is collected as a distillate. Adjusting the temperature enables different components to be separated. Another method of separation is sublimation. This is when a solid turns into a vapor without first becoming a liquid. A modern example would be frozen carbon dioxide (dry ice) becoming a vapor at room temperature. Substances such as iodine, camphor, and naphthalene sublime when heated and can be recovered as a solid deposit, or sublimate, by cooling the vapor in a similar way to collecting a liquid distillate.

FAT FROM THE RAM, ASHES FROM THE FIRE

MAKING SOAP

IN CONTEXT

KEY FIGURES
Soap makers of Sumeria
(c. 2800 BCE)

AFTER
c. 600 BCE Phoenicians make soap using goat tallow and wood ash.

79 CE Evidence of a soap-making factory is buried in the ruins of Pompeii, Italy.

700 Arabian chemists use vegetable oils, such as olive oil, to make the first solid soap bars. They are perfumed and colored using aromatic oils such as thyme oil.

12th century An Islamic document describes the key ingredient of soap as *al-qaly*, or "ashes," from which comes the chemical term "alkali."

1791 French chemist Nicolas Leblanc opens the first factory to produce sodium carbonate (soda ash) from common salt, which reduces the cost of manufacturing soap.

Soap may well have been the first chemical preparation—a deliberate mixture of two or more chemicals—in history. Clay tablets from c. 2500 BCE, found in the Sumerian city of Girsu (in present-day Iraq), record the earliest description of a method of making a soaplike material. Archaeolgists, however, consider it likely that soap had been in use for at least 300 years before this.

The chemistry of soap making is fundamentally the same across all cultures. Girsu was a center of textile production, and the surviving soap recipe concerns the washing and dyeing of wool. The Sumerians used a mix of wood ash and water to remove the natural oiliness from wool, a necessary process if dyes are to hold. It is likely that Sumerian priests used a similar mixture to purify themselves before rituals.

Alkaline ashes

The ashes-and-water mix works because the alkali in the ashes reacts with the oil, converting it into soap. (Alkali in this instance means a base that dissolves in water; a base is the chemical opposite of an acid.) The soap dissolves the remaining oil and dirt. People realized they could make soap products relatively easily and boiled animal fats and oils with the alkaline ash mix to make cleaning solutions for textiles such as wool or cotton.

> This water consecrates the heavens, it purifies the earth.
> **Hymn to Kusu**
> **(3rd millennium BCE)**

On the human body, soap seems to have been used more often as a treatment for skin ailments at this time rather than as a cleanser. A Sumerian text from c. 2200 BCE describes its application on a person with an unidentified skin condition. The ancient Egyptians developed a similar method to the Sumerians for making soap, using it to treat skin

See also: The new chemical medicine 44–45 ▪ Acids and bases 148–149 ▪ Enzymes 162–163 ▪ Cracking crude oil 194–195

diseases and sores, as well as for washing themselves. The Ebers papyrus from c. 1550 BCE, one of the oldest known medical works, records the making of soap by mixing animal and vegetable oils with alkaline salts.

During the Zhou dynasty in China, around 1000 BCE, the Chinese discovered that the ashes of certain plants could be used to remove grease. A document called "The record of trades," created toward the end of the dynasty, records how the cleaning mixture was improved by the addition of crushed seashells to the ash. This produced an alkaline chemical that could remove stains from fabrics.

Soap saga

The ancient Romans and ancient Greeks cleansed their bodies by massaging oil into the skin and then scraping the dirt away using a metal or wooden strigil, the earliest examples of which date to the 5th century BCE.

The first recorded use of the term "soap" is in the 1st century CE, when Roman author and naturalist Pliny the Elder mentions "sapo" in his encyclopedic tome, *Natural History*. In it, he gives recipes for making soap from tallow (derived from beef fat) and ashes, and he describes the resulting product as a means to "disperse scrofulous sores."

In the 2nd century CE, Galen, the influential Greek physician, described making soap with lye (the bases potassium hydroxide and sodium hydroxide derived from wood ash). He prescribed it as an effective means of cleaning both the body and clothing.

Modern soaps

The most common fats and oils used for manufacturing soaps today are coconut oil, sunflower oil, olive oil, palm oil, and tallow. The properties of soaps are determined by the type of fat used: animal fats make very hard, insoluble soaps, whereas coconut oils make more soluble soaps. The type of alkali used is also important: sodium soaps are hard, whereas potassium soaps are softer.

Many modern-day laundry detergents use enzymes, which are biological catalysts, to break down the fats, proteins, and carbohydrates present in food and other stains. ■

Soap chemistry

Oils or fats derived from plants or animals contain triglycerides. These are composed of a glycerol molecule attached to three long chains of fatty acids. When triglycerides are mixed with a strong alkali solution, the fatty acids are separated from the glycerol. This process is known as saponification. The glycerol is converted into an alcohol and the fatty acids form salts—the soap molecules. The head of the fatty acid salt is hydrophilic (attracted to water) and soluble, but its long tail is hydrophobic (repelled by water) and insoluble.

Fatty acid salts are strong surfactants—substances that accumulate at water surfaces. In water, the soap molecules form tiny clusters called micelles. The hydrophilic part of the soap molecule points outward, forming the outer surface of the micelle and the hydrophobic part points inward. Hydrophobic molecules such as fat and oil are trapped within the micelle, which is soluble in water and can easily be washed away.

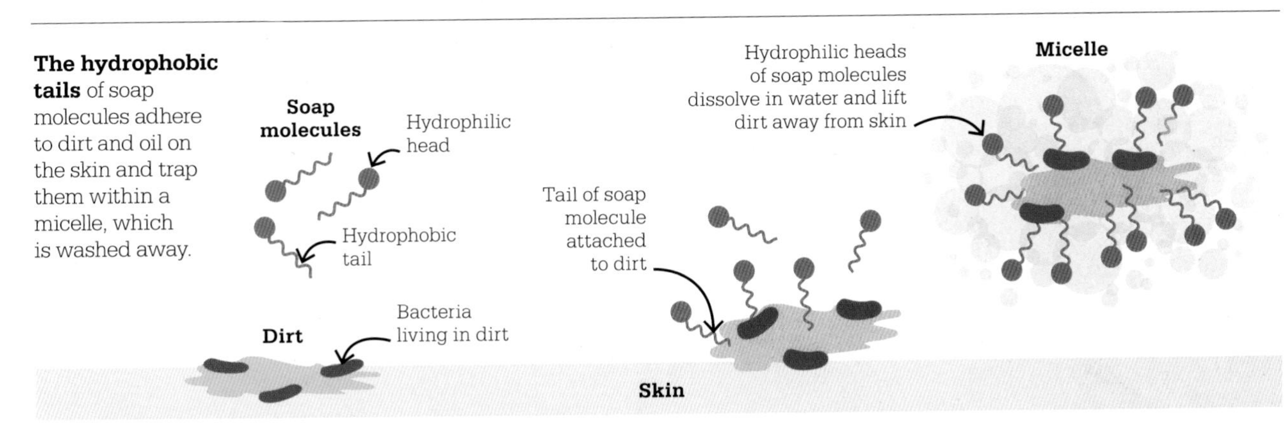

The hydrophobic tails of soap molecules adhere to dirt and oil on the skin and trap them within a micelle, which is washed away.

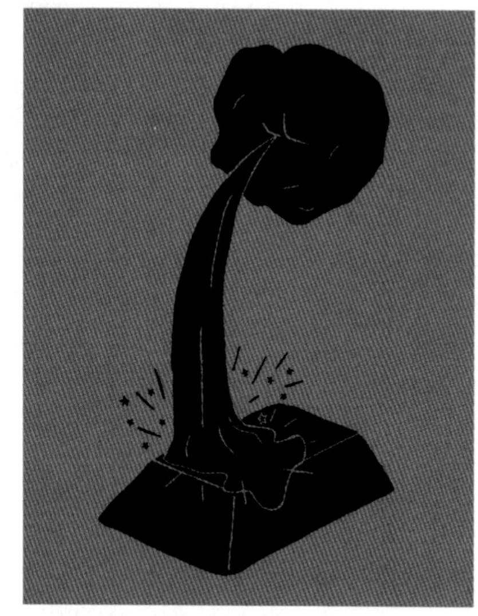

DUSKY IRON SLEEPS IN DARK ABODES

EXTRACTING METALS FROM ORES

IN CONTEXT

KEY FIGURES
Anatolian metalworkers (*c.* 2000 BCE)

BEFORE
***c.*5000 BCE** Evidence of the deliberate extraction of copper from ore from sites in southeast Europe and Iran.

***c.*4000 BCE** Copper axes cast in the Balkans show that the technology is known for metals to be melted and shaped.

AFTER
***c.*400 BCE** Indian metalworkers invent a smelting method that bonds carbon to wrought iron, producing steel.

***c.*12th century CE** The first blast furnaces in the West are built in Durstel, Switzerland.

The discovery of metal extraction was a pivotal advance in technology, enabling tools and other items such as jewelry to be produced by metalworking. The first metals used were copper, silver, and gold, which are found naturally in their metallic, or native, states. Most other metals are found in combination with other materials as part of a rocky ore. Separating metals from their ores, a process known as smelting, requires high temperatures.

Extracting copper

The first people to discover the smelting process were most likely potters experimenting with new techniques for firing their ceramics and observing a shining rivulet of molten metal trickling out from the kiln. Smelting copper requires

In a Bronze Age workshop, copper–tin alloy is shown being poured into a sand mold, having been previously mixed and smelted in a furnace. It will form a bronze object. Another man examines a newly cast sword blade.

See also: Refining precious metals 27 ■ Attempts to make gold 36–41 ■ Oxygen and the demise of phlogiston 58–59 ■ Isolating elements with electricity 76–79

Gold and iron at the present day, as in ancient times, are the rulers of the world.

William Whewell

***Lecture on the Progress of Arts and Science* (1851)**

heating the ore to temperatures of over 1,800°F (980°C), not an easy task with an open wood fire, but achievable in a kiln.

Shafts for excavating copper ore dating back around 6,000 years have been identified in the Balkans and on the Sinai Peninsula in Egypt. A major challenge for early miners was breaking open rocks to get at the ore. One of the first great advances in mining technology was fire setting. This involved first heating the rock to make it expand and then dousing it with cold water, causing it to contract and break open. Crucibles (clay vessels that withstood high temperatures and were used to melt minerals including metals) were found near the mines, indicating that smelting of the ore also took place at the site.

Alloys

Copper itself is a relatively soft metal, with limited usefulness for toolmaking. The discovery that mixing, or alloying, copper with other materials produced a stronger metal came around 5,000 years ago. Many of the early attempts to produce copper involved heating copper sulfide ores in the presence of red-hot charcoal, a process that produced copper alloys. These alloys contained arsenic and were much stronger than pure copper. The first copper–tin alloys probably resulted when a tin-containing ore was present during smelting. The addition of tin to copper made an alloy much harder than either metal alone, which was also easier to cast; this alloy was named bronze. Made from about 3000 BCE onward in the Tigris-Euphrates delta of Mesopotamia, this useful new metal spread widely through trade, heralding the onset of the Bronze Age.

Any old iron

The extraction of iron from its ore was likely first accomplished accidentally in copper-smelting furnaces around 2000 BCE in Anatolia (in present-day Turkey). Smelting iron required the use of charcoal for fuel, which burns at a hotter temperature than wood and reacts chemically to remove some of the impurities from the iron ore. The invention of bellows allowed air, and therefore oxygen, to be pumped into the furnace, making higher temperatures more achievable. These ancient furnaces, known as slag pits or bloomer furnaces, could not achieve the temperatures needed to melt iron. Rather, they produced a bloom, a mix of almost pure iron and other materials that was then refined by repeated heating and hammering. Iron made in this way is known as wrought iron.

Iron is the fourth-most-common element on Earth and easier to obtain in large quantities than copper and tin. Between 1200 and 1000 BCE, knowledge of ironworking and trade in iron objects, particularly agricultural tools and weapons, spread rapidly across the Mediterranean and Near East regions. In China, blast furnaces were developed, which made production more efficient. ■

The blast furnace

Used for smelting metals such as iron, the blast furnace has fuel and ore continually fed into the top of the furnace while air is blown (or blasted) into the bottom of the chamber, ensuring a supply of oxygen. Chemical reactions take place throughout the chamber, resulting in the production of molten metal and slag, which is removed from the bottom, and flue gases that escape from the top.

By the 5th century BCE, cast-iron tools were widespread in China, indicating that blast furnace technology was firmly established throughout the region at this time. These furnaces had clay walls and used phosphorus-rich minerals as a flux to lower the melting point of the metal. In the 1st century CE, Chinese engineer Du Shi developed waterwheels to power piston bellows, saving labor and increasing the effectiveness of these blast furnaces. In Europe at this time, iron production was limited to bloomeries, producing wrought iron.

IF IT WERE NOT SO BREAKABLE, I SHOULD PREFER IT TO GOLD

MAKING GLASS

IN CONTEXT

KEY FIGURES
Mesopotamian glassmakers (c. 2500 BCE)

BEFORE
c. 5000 BCE Paleolithic societies use naturally occurring glass to produce cutting tools.

AFTER
c. 1500 BCE Glassmaking spreads to Egypt and Greece in the late Bronze Age.

c. 7th century BCE Instructions for making various kinds of glass found on a clay tablet in the library of King Ashurbanipal (685–631 BCE) of Assyria (now northern Iraq).

c. 1st century BCE Phoenicians discover glassblowing, using a hollow iron tube to blow air into a blob of molten glass to shape it into a vessel.

c. 1st century CE Romans discover that adding manganese oxide makes glass clearer and use glass in windows for the first time.

Glass is a noncrystalline substance that occurs naturally in Earth's crust, most often as obsidian, a black volcanic glass formed when lava cools rapidly. It is found worldwide and can be flaked to produce razor-sharp edges, making it useful for knives, saws, and spearheads.

Beads from Mesopotamia in 2500 BCE are among the earliest manufactured glass objects to be found. The manufacturing process may have been discovered by potters when adding an impervious glaze to the outside of ceramics at a high temperature. The Mesopotamians made glass from three ingredients: silica (SiO_2, usually sand); soda (sodium hydroxide, NaOH) or potash (potassium hydroxide, KOH), acting as a flux to lower the temperature at which the sand melts; and lime (calcium hydroxide, $Ca(OH)_2$) to stabilize the mixture. Melting the raw materials needs a temperature over 1,800°F (1,000°C), something few furnaces could achieve. Because they were so hard to make, glass objects were highly prized.

Molten glass could be shaped. By the mid-16th century BCE, small glass vessels were made in Mesopotamia by core-forming. A core of clay or animal dung was attached to a metal rod and dipped into molten glass. After the glass had cooled, the core was removed.

Glassmakers had developed the reverberatory furnace by the 5th century BCE, with a combustion chamber at one end and a vent at the other. This allowed several tons of raw materials to be melted at once, greatly increasing productivity. ■

> Glass, like copper, is smelted in a series of furnaces, and dull black lumps are formed.
> **Pliny the Elder**
> ***Natural History*** **(c. 77 CE)**

See also: Borosilicate glass 151

MONEY IS BY NATURE GOLD AND SILVER

REFINING PRECIOUS METALS

IN CONTEXT

KEY FIGURE
Croesus, King of Lydia
(reigned c. 560–546 BCE)

BEFORE
c. 7000 BCE Neolithic communities begin hammering copper into tools. Some of the earliest examples are found in eastern Anatolia (in modern-day Turkey).

c. 3000 BCE The earliest surviving gold jewelry is in use in Egypt.

AFTER
c. 300–500 CE Tumbaga, an alloy of gold and copper, is widely used in Panama and Costa Rica for small ornaments and regalia.

1867 Francis Bowyer Miller invents a process using chlorine gas that refines gold to a purity of 99.5 percent.

The first metals worked by humans were copper and gold; 8,000-year-old copper beads have been found in northern Iraq, and gold may have been used for decoration earlier. By 4000 BCE, seven metals were in use: copper, gold, and silver, all found in their native states and relatively easy to obtain; and lead, iron, tin, and mercury, which were extracted from mineral ores by smelting.

Native metals were not always pure. In the late 7th century BCE, the Lydians in Anatolia recovered electrum—a pale, natural alloy of gold and silver—from river sands and used this to produce coinage. In the 6th century BCE, King Croesus of Lydia introduced the world's first gold coins of a standardized purity.

The Croesus coins were a refined, purified gold, achieved by beating the electrum flat and placing it in earthenware pots between layers of salt. When heated to a temperature below the melting point of gold for several hours, the silver in the electrum reacted with the salt to form silver chloride. This was absorbed by "carrier" clay, such as the furnace bricks and pottery containers, leaving behind almost pure gold.

A gold coin—one of the world's first—dating from the time of King Croesus. It shows a lion and a bull, which were imprinted by hammering their images into the gold.

To recover the silver, the carrier clays were smelted with copper or lead. The silver was then separated from the other metals by cupellation. This involved heating the alloy in cupels (bowls), using bellows for a high temperature. The copper oxide or lead oxide formed was absorbed by the cupel, and the silver was isolated and used for more coins. ■

See also: Extracting metals from ores 24–25 ▪ Isolating elements with electricity 76–79

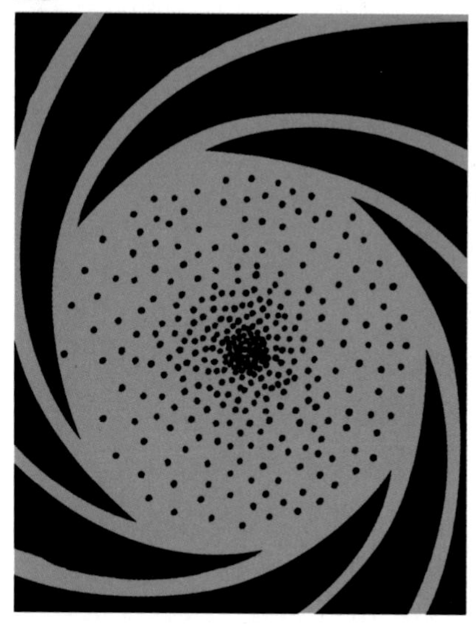

ATOMS AND THE VACUUM WERE THE BEGINNING OF THE UNIVERSE

THE ATOMIC UNIVERSE

IN CONTEXT

KEY FIGURE
Democritus (*c.* 460–*c.* 370 BCE)

BEFORE
c. 475 BCE The Greek philosopher Leucippus develops the first theory of atomism, the idea that everything is composed of indivisible elements.

AFTER
c. 11th century CE The Islamic philosopher al-Ghazali writes of atoms as the only perpetual material things in existence.

1758 Croatian polymath Roger Boscovich (Ruder Boškovic) publishes the first general mathematical theory of atomism.

The idea that all matter is made of atoms has a very long history. It began in the 5th century BCE with the Greek philosopher Democritus. He drew on the work of his near-contemporary Anaxagoras—who believed that matter was infinitely divisible—and of his teacher Leucippus, who suggested that all matter consists of an infinite number of invisibly small, indivisible particles.

The eternal *atomos*

Democritus knew that if you cut a stone in half, each half had the same properties as the original stone. He reasoned that if you kept on dividing the stone, eventually the pieces would be so tiny that it would be physically impossible to divide them any further. He defined these infinitesimally small pieces of matter with the word *atomos*, meaning "indivisible"—from which we derive the word "atoms." He suggested that atoms were eternal and could not be destroyed; however, they were constantly combining and recombining into different substances.

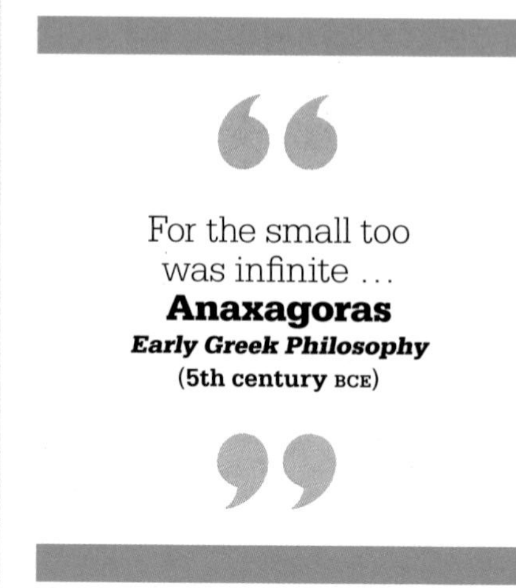

For the small too was infinite …
Anaxagoras
Early Greek Philosophy
(5th century BCE)

These atoms were solid, with no internal structure. They were all the same matter, but of different sizes, weights, and shapes. Each material came from a specific form of atom—the atoms of a stone were unique to it and distinct from the atoms of a feather, for example. The nature of a material resulted from the shape of the atoms from which it was formed and the way those atoms joined together; for example, iron atoms were jagged and locked together, while water atoms were smooth and rolled over each other.

Democritus's universe

In Democritus's thought, the universe had existed and would exist forever. Its structures arose

See also: The four elements 30–31 ▪ Corpuscles 47 ▪ Dalton's atomic theory 80–81

through random movements of atoms, which collided to form larger bodies and worlds. These collisions set up motions, or vortices, that differentiated atoms by mass.

The world was governed by the nature of atoms, their motion, and the way they were packed together. This was an attempt to apply mathematical laws to nature, since the atoms' behavior was governed by mathematics. For Democritus, nature was a machine.

Democritus arrived at his views by deduction rather than experiment. Other philosophers, particularly Aristotle, did not agree with them. Aristotle, following Empedocles, maintained that everything in the universe was made up of fire, air, earth, and water. In addition, Aristotle criticized the idea that atomic motion had always happened and that there was no beginning to it.

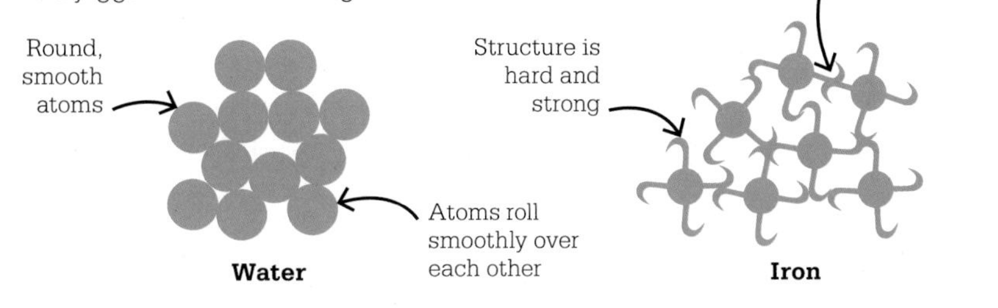

According to Democritus, atoms of different materials had differing shapes. Those of water were smooth and slipped or rolled easily against each other, while those of iron were jagged and hooked together into a solid substance.

Later developments

In the 4th century BCE, the Greek philosopher Epicurus upheld the atomic theory. However, in an attempt to argue against the concept of Democritus's mechanical, deterministic universe and defend the notion of free will, Epicurus maintained that atoms moving through space might occasionally "swerve" from their otherwise predetermined paths, adding an element of chance and thus giving rise to new chains of events. The Roman philosopher Lucretius, in the 1st century BCE, wrote in *On the Nature of Things* that matter was composed of "first beginnings of things": tiny particles perpetually moving at very high speed. The atomic theory, like much of ancient Greek thinking, was forgotten in Europe for centuries until it was rediscovered in Arabic translations of Aristotle, who attacked it. Aristotle's theory of the four elements as eternal principles prevailed over the atomic theory, which Christian scholars regarded as too materialistic and thus contrary to their teachings. The concept of atoms was eventually revisited by Enlightenment philosophers in the 18th century and evolved into the atomic theory of British chemist John Dalton in the early 19th century. ■

Democritus

Known as "the laughing philosopher" for his cheerful, light-hearted outlook on life, Democritus was born around 460 BCE, possibly in Abdera, in the Greek province of Thrace, or perhaps in Miletus, now in western Turkey. Little is known of his life, and none of his writings have survived; his thought has passed down in fragments, mainly through a monograph by Aristotle, and anecdotes related by the Greek biographer Diogenes Laërtius in the 3rd century CE.

Democritus is said to have traveled widely—almost certainly to Egypt and Persia, and possibly to Ethiopia and India as well, meeting scholars in these countries. He also toured Greece to speak with natural philosophers; Leucippus of Miletus became his mentor and had a great influence on his thinking, sharing his theory of atomism with him.

The circumstances of Democritus's death are unclear. He is said to have lived to the age of 90, which puts his death around 370 BCE, although some writers claim that he lived to 109 years of age.

FIRE AND WATER AND EARTH AND THE LIMITLESS VAULT OF AIR

THE FOUR ELEMENTS

IN CONTEXT

KEY FIGURES
Empedocles (492–432 BCE)
Aristotle (384–322 BCE)

BEFORE
c. 6th century BCE The Greek philosopher Thales of Miletus asserts that all phenomena can be understood in natural, rational terms.

AFTER
c. 8th century CE Arabic alchemist Jābir ibn Hayyān expands the four-element hypothesis to include the sulfur-mercury theory of metals.

1661 Anglo-Irish natural philosopher and chemist Robert Boyle rejects the four-element hypothesis in favor of a theory that all matter is made of corpuscles.

The ancient Greeks are credited as being the first people to ask: what is everything made of? Thales of Miletus, as reported in Aristotle's *Metaphysics*, said that water was the "originating principle" (*arche*) of all things. Other philosophers of the time had different views: Heraclitus thought that the *arche* was fire, while Anaximenes of Miletus posited that the *arche* was air.

Primal roots

The Sicilian-born philosopher Empedocles, in the 5th century BCE, declared that all matter, including living things, was composed of four primal "roots" (in Greek, *rhizomata*): air, earth, fire, and water. Materials were rarely pure but instead were formed from combinations of different substances, and the ratio of those roots determined the nature of each substance. In his system, two powers acted on the roots to cause changes: love (*philotes*), which brought different kinds of matter together, and strife (*neikos*), which parted them. Empedocles also believed that all matter, whether alive or not, was in some way conscious.

Empedocles in the *Nuremberg Chronicle*, 1493, an encyclopedia of world history by German humanist Hartmann Schedel. This indicates his importance to medieval scholars.

Empedocles's system was founded on philosophy rather than on experimental evidence. However, he did allegedly demonstrate that air was not just nothingness. Using a clepsydra—a water clock that measured the flow of water through a vessel with holes in the bottom and top—Empedocles observed that if he placed the bottom hole underwater, the vessel filled with water. If he first put his finger over the top hole, the water would not flow into the vessel; however, once he removed his finger, the water flowed in. Empedocles deduced that the air in the container was preventing the water from entering.

See also: The atomic universe 28–29 ▪ Attempts to make gold 36–41 ▪ Gases 46 ▪ Corpuscles 47 ▪ Oxygen and the demise of phlogiston 58–59 ▪ Dalton's atomic theory 80–81 ▪ The ideal gas law 94–97 ▪ The periodic table 130–137

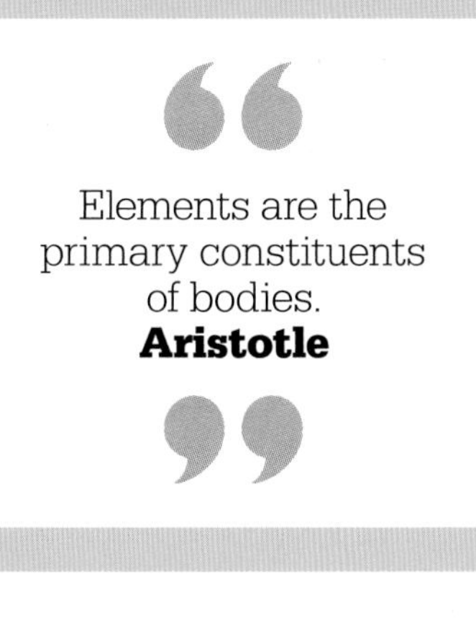

Elements are the primary constituents of bodies.
Aristotle

Complementary qualities

The Athenian philosopher Plato, in his *Timaeus* (c.360 BCE), may have been the first writer to use the name "element" (*stoicheion*; the Greek word for the smallest division of a sundial or letter of the alphabet) for the four basic roots. However, his student Aristotle provided the first definition in his work *On the Heavens*: "An element … is a body into which other bodies may be analyzed … and not itself divisible into bodies different in form."

Aristotle held that all substances were a combination of matter and form. Matter was the material from which substances were made, while form gave a substance its structure and determined its characteristics and functions. He agreed with Empedocles that matter was formed from different ratios of air, earth, fire, and water; however, he believed that these existed only as potentials and not as things in themselves until they acquired form.

Aristotle saw the four elements as having different properties: fire was hot and dry, air was hot and wet, earth was cold and dry, and water was cold and wet. To Empedocles's four elements, he added a fifth, which came to be known as quintessence, or aether—a divine substance that formed the stars and planets.

In Aristotle's Earth-centered cosmos, aether was the lightest element and formed the outermost layer of the cosmos; then, in descending order, came fire, air, water, and earth. Each element would always try to return to its natural level—so rain fell from the air to the earth and returned to the level of water, and flames rose from the earth toward the level of fire.

Enduring influence

The four elements theory became fundamental to alchemy. It also had a major influence on medicine. In the treatise *On the Nature of Man* in the 5th century BCE, Hippocrates, the Greek "father of medicine," associated the elements with four vital fluids, or humors, in the body: blood (air), phlegm (earth), yellow bile (fire), and black bile (water).

The theory of the elements later spread to the Islamic world and from there back to Europe. It dominated thinking into the Middle Ages and beyond. It was not until 17th- and 18th-century scientists such as Galileo and Robert Boyle promoted experimentation and observation over philosophy that Aristotle's four elements were finally superseded. ■

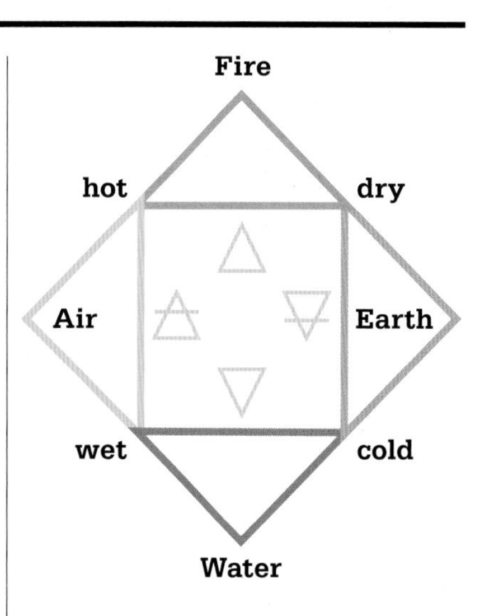

Over the centuries, natural philosophers devised this diagram to illustrate the similar and opposing qualities of the elements. The symbols at the center show the upward or downward movement of energy associated with each element.

Aristotle

Born in 384 BCE in Macedonia, northern Greece, Aristotle became a student at Plato's Academy in 367 BCE, later teaching there. Following Plato's death in 347 BCE, Aristotle founded his own school, the Lyceum, in Athens in 335 BCE. Stories that he tutored the young Alexander the Great are probably a later invention, although he spent some time at the court of Philip of Macedonia, Alexander's father. Aristotle died in 322 BCE at age 62.

Aristotle promoted the concept of natural laws to explain physical phenomena. His writings ranged from philosophy, logic, astronomy, and biology to psychology, economics, poetry, and drama. His ideas dominated western science and philosophy for nearly 2,000 years, until they were challenged by natural philosophers in the 17th century.

Key works

Metaphysics
On Generation and Corruption
c.350 BCE *On the Heavens*

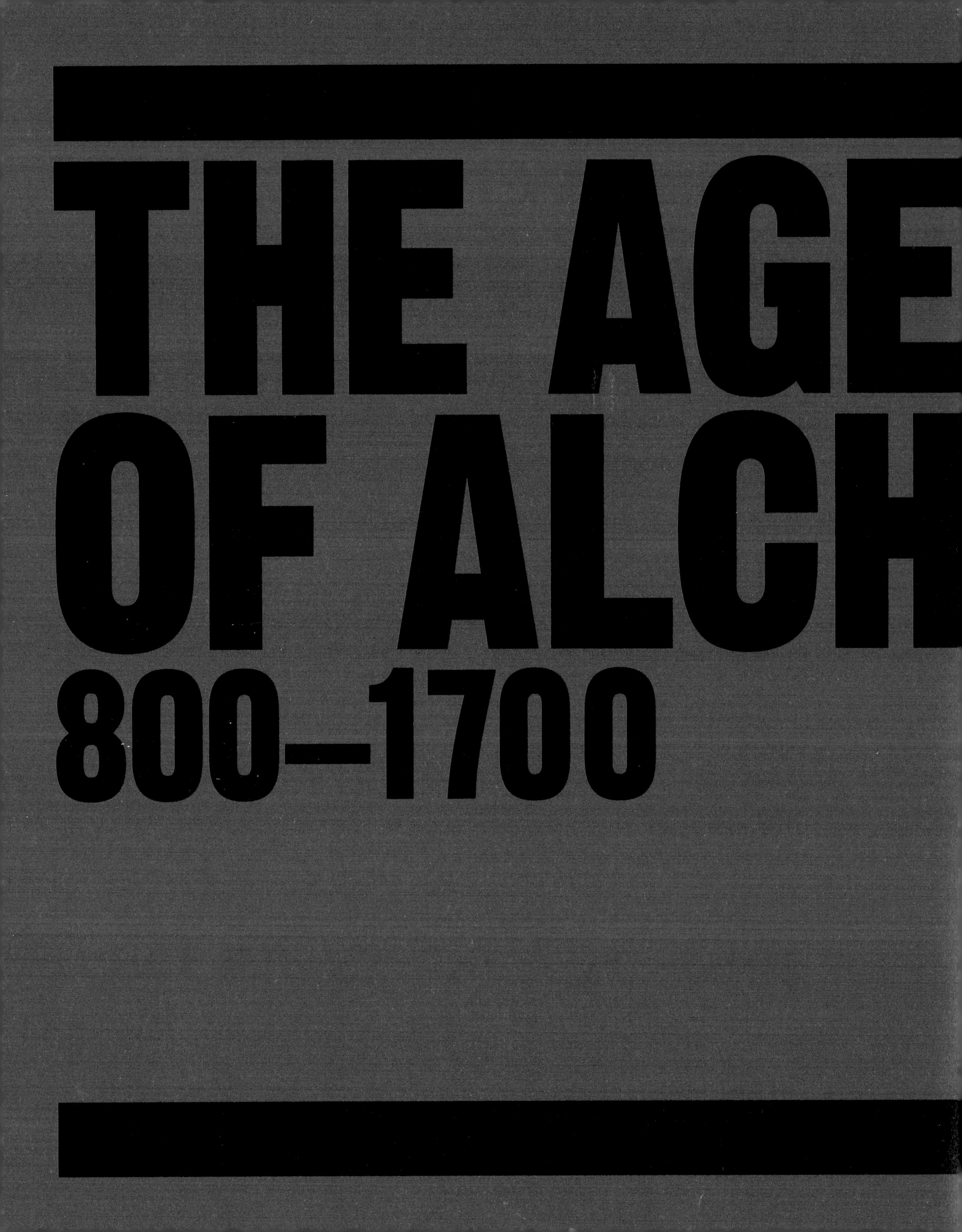
THE AGE
OF ALCH
800–1700

EMY

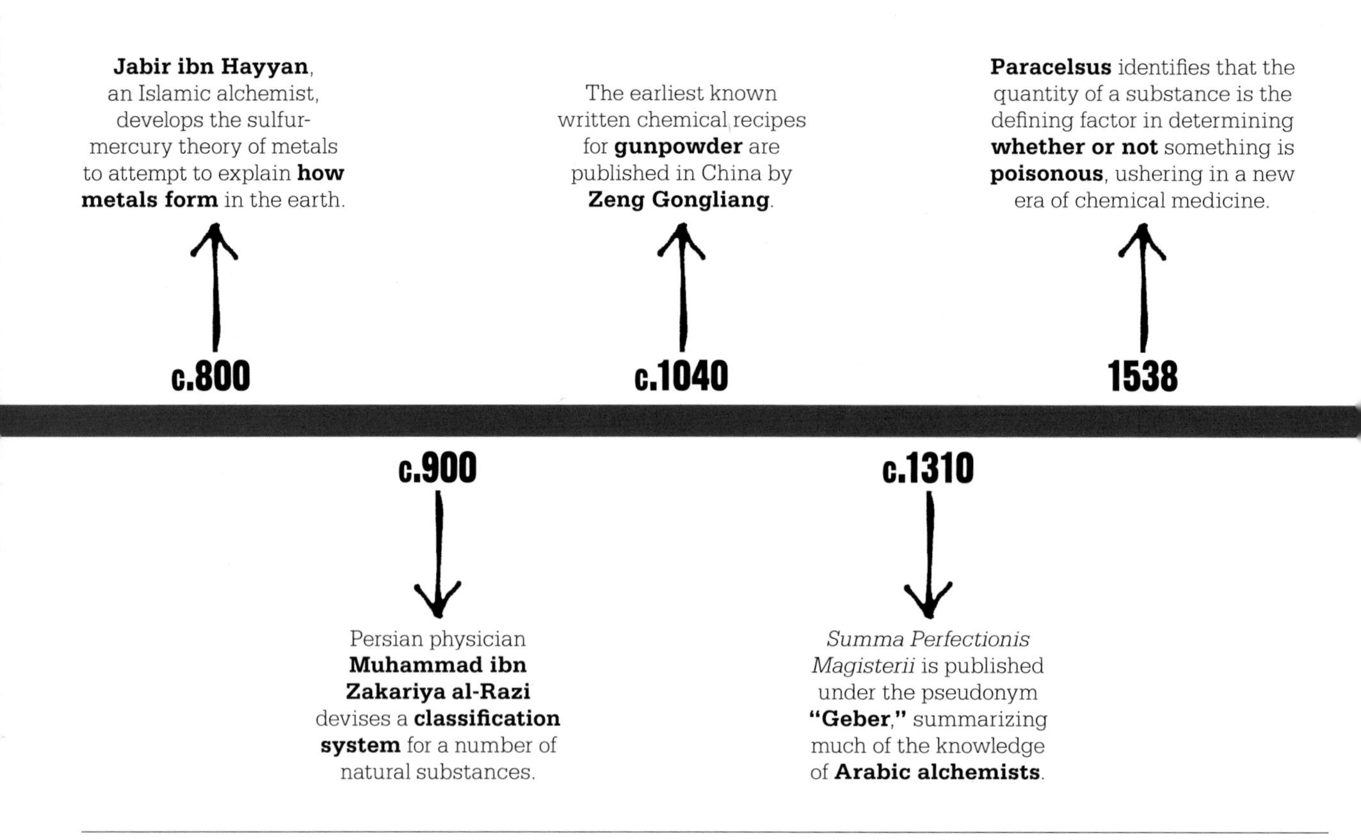

Jabir ibn Hayyan, an Islamic alchemist, develops the sulfur-mercury theory of metals to attempt to explain **how metals form** in the earth.
c.800
The earliest known written chemical recipes for **gunpowder** are published in China by **Zeng Gongliang**.
c.1040
Paracelsus identifies that the quantity of a substance is the defining factor in determining **whether or not** something is **poisonous**, ushering in a new era of chemical medicine.
1538
c.900
Persian physician **Muhammad ibn Zakariya al-Razi** devises a **classification system** for a number of natural substances.
c.1310
Summa Perfectionis Magisterii is published under the pseudonym **"Geber,"** summarizing much of the knowledge of **Arabic alchemists**.

The age of alchemy is sometimes derided as a time when pseudoscience and occultism stalled the progress of chemical thinking. It is true that the later, lofty goals of western alchemists, such as turning base metals into gold and discovering the secret to eternal life, never came to pass. However, the view of alchemists as deluded occultists or even frauds obscures the dedicated experimentalism that was such a large part of alchemy and that underlay the gradual accumulation of knowledge and experience that became modern chemistry.

A mysterious art

Perceptions of alchemy were further clouded by the apparently mystical language that the early alchemists used to describe their procedures and discoveries. Early and medieval alchemical recipes are littered with obscure references to concepts such as "the green lion devouring the sun," "the gray wolf," and "the seed of the dragon." Yet when these sometimes baffling utterances are decoded, they turn out to be descriptions of what we would now recognize as chemical reactions, thus showing that the writers accurately understood some of the processes that they were exploring.

Alchemy's origins

Alchemy's exact point of origin is uncertain; in addition, different traditions arose in different parts of the world. Some aspects, such as pursuing an "elixir of life," can be recognized in ancient Chinese and Indian writings. The roots of what would become western alchemy can be identified in ancient Egypt, at the time when it was ruled by the Greeks. This practice grew from a fusion of Greek thought and Egyptian practices such as those for embalming the dead. It also involved refinements to apparatuses and techniques that had already been in use for centuries for practical processes such as distillation and filtration.

Alchemy faded from view in the late Roman empire, but toward the end of the 1st millennium CE, the alchemists of the Islamic world drove it forward. Practitioners such as the renowned Jabir ibn Hayyan developed classification systems for substances that went beyond the earth, air, fire, and water of the Ancient Greeks and began to make systematic explorations of the properties of various substances.

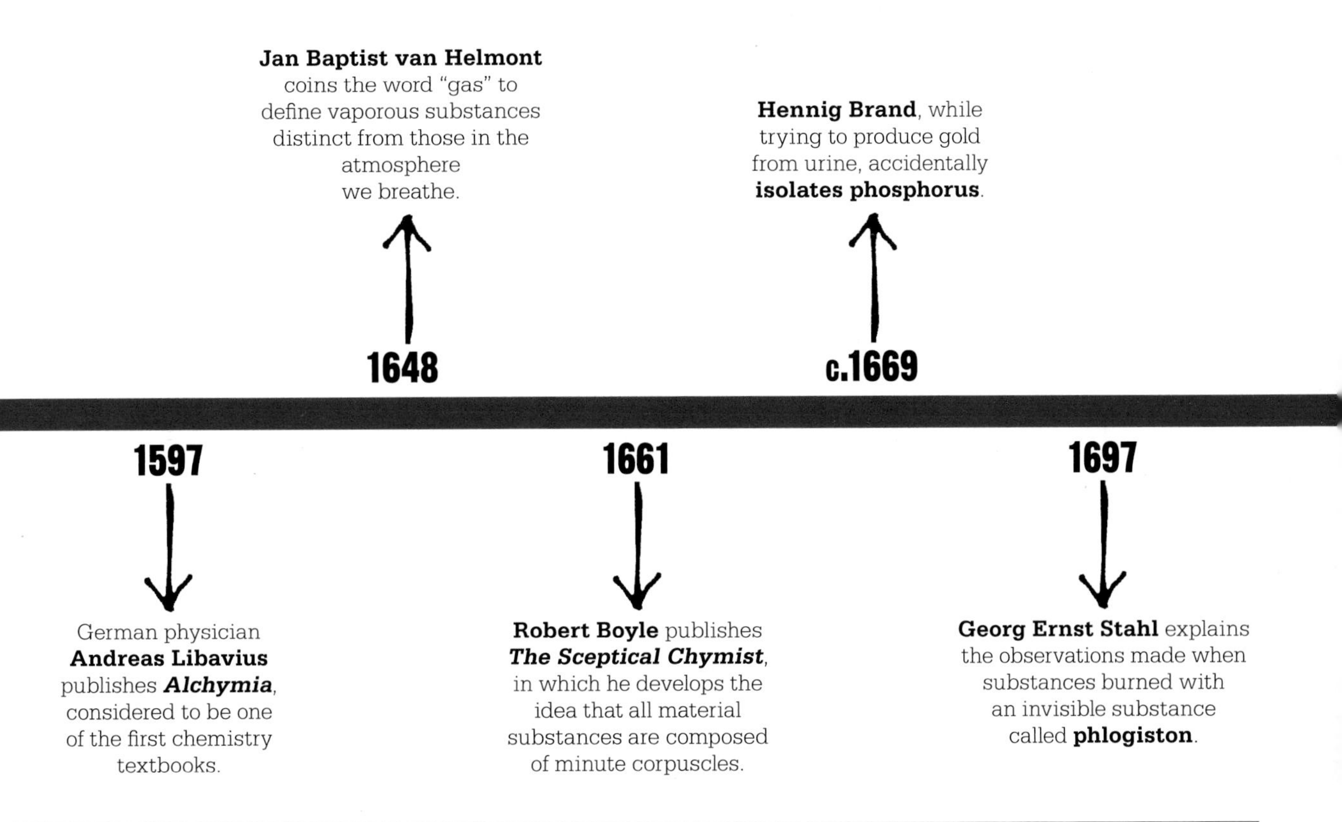

Jan Baptist van Helmont coins the word "gas" to define vaporous substances distinct from those in the atmosphere we breathe.
1648
Hennig Brand, while trying to produce gold from urine, accidentally **isolates phosphorus**.
c.1669
1597
German physician **Andreas Libavius** publishes ***Alchymia***, considered to be one of the first chemistry textbooks.
1661
Robert Boyle publishes ***The Sceptical Chymist***, in which he develops the idea that all material substances are composed of minute corpuscles.
1697
Georg Ernst Stahl explains the observations made when substances burned with an invisible substance called **phlogiston**.

From 1095–1291, when Christianity launched a series of crusades against the Islamic powers, the contact between the two cultures resulted in the reseeding of alchemy in western Europe, as Arabic works on the subject were translated into Latin around the 12th century.

Elements and airs

The European alchemists' determined pursuit of the philosopher's stone indirectly led to significant advances. Both arsenic and phosphorus were first isolated by German alchemists in the 13th and 17th centuries, respectively. The 16th century saw some of alchemy's concepts applied to medicine, ushering in a new understanding of the way in which chemicals affect living organisms. Alchemists started to analyze the chemical complexity of the material world in more detail. In the 17th century, Flemish chemist Jan Baptist van Helmont was among the first chemists to realize that the airlike substances produced by some chemical reactions were not simply different varieties of air, but distinct substances entirely—substances he described as "gas." These were the initial steps toward the more thorough investigations of the air in Earth's atmosphere that would follow in later centuries.

Fire and phlogiston

Toward the end of the period, alchemists turned their minds to a question that had puzzled them for centuries: what makes fire burn? In 1697, Georg Ernst Stahl, a German physician, proposed that a substance he called phlogiston was responsible—a proposal that would spark almost a century of argument. The phlogiston theory persisted until the late 18th century, when the French chemist Antoine Lavoisier put an end to it by isolating oxygen and describing the phlogiston theory as a "gratuitous supposition."

The concept of phlogiston, like the goals of alchemy, is often mocked as "pseudoscience" from a modern perspective. However, like the study of alchemy, the exploration of the supposed substance phlogiston led to more detailed quantitative experiments and the discoveries of other components of air, marking a significant point in the transition from alchemy to chemistry. ■

THE PHILOSOPHER'S STONE

ATTEMPTS TO MAKE GOLD

IN CONTEXT

KEY FIGURE
Jabir ibn Hayyan
(c.721–c.815)

BEFORE
c.3300 BCE Metalworkers in Sumer, the earliest of the Mesopotamian civilizations, discover how to forge bronze from copper and tin.

c.450 BCE The Greek philosopher Empedocles declares that all things are formed from four primal elements: air, earth, fire, and water.

AFTER
1623 English philosopher Francis Bacon publishes *De Augmentis Scientiarum* (*Of The Advancement of Learning*), which contains description of experimental methods.

1661 Anglo-Irish chemist Robert Boyle's *The Sceptical Chymist* draws a line between alchemy and modern chemistry.

'Tis a stone,
And not a stone …
Ben Jonson
English playwright (1572–1637)

The Alchemist, painted c.1650 by Flemish artist David Teniers the Younger, shows an alchemist and his assistant using equipment such as bellows, scales, and retorts.

From ancient times to the 18th century, alchemy was an important and respected branch of inquiry into the way the world worked. Although it is often considered today as pseudoscience, alchemy may be better regarded as a proto-science.

Alchemical practice combined esoteric aspects (spiritual or mystical knowledge restricted to initiates) and exoteric aspects (practical applications). The ultimate goal, or "great work," of the alchemist was "transmutation" of metals, or turning one metal into another—notably, turning base (nonprecious) metals into gold or silver. Alchemists believed that they could achieve this by using a substance known as the "philosopher's stone." They also saw transmutation as having a symbolic counterpart in practices for purifying the soul.

Origins in Egypt

The practices that became Western alchemy appeared in ancient Egypt during the rule of the Greeks (305–30 BCE). In fact, the word "alchemy" is ultimately derived from the Greek word *chémeia* (pouring, or casting, together); it has also been associated with an Egyptian art, *khemeia*, mentioned in hieroglyphs concerned with rituals for the burial of the dead. The practitioners of *khemeia*, skilled in embalming, were seen as magicians. Their arts also extended to processes such as metallurgy and glassmaking.

Late medieval alchemists claimed their practice originated with a figure known as Hermes Trismegistus (Hermes the Thrice-great)—a combination of the Greek god Hermes and the Egyptian god Thoth, who was thought to have been a contemporary of the Jewish prophet Moses. The philosophy attributed to him was known as

See also: Purifying substances 20–21 ▪ Refining precious metals 27 ▪ The four elements 30–31 ▪ The new chemical medicine 44–45 ▪ Corpuscles 47 ▪ Phlogiston 48–49 ▪ Catalysis 69 ▪ The synthesis of urea 88–89

hermeticism. Practices included a procedure for making the philosopher's stone by placing a mixture of materials into a glass vessel, which was then sealed by fusing the neck closed; this seal was known as the Seal of Hermes and has given us the expression "hermetically sealed" to describe something that is airtight.

The search for the philosopher's stone

The Greco-Egyptian alchemist Zosimos of Panopolis, who lived around 300 CE, provided the first recorded mention of the philosopher's stone in the oldest known book on alchemy: the *Cheirokmeta* (*Things Made by Hand*). He describes what we might now understand as a chemical process for turning base metals into gold, involving a catalyst that he refers to as a "tincture."

Zosimos's detailed descriptions of experiments and careful recording of the results can be seen as a precursor to the modern scientific method. Zosimos also describes apparatuses, many of them adapted from workshop tools and cooking utensils, for processes such as distillation and filtration. He acknowledges his debt to the writings of predecessors such as Maria Hebraea (Mary the Jewess), thought to have lived in Alexandria in the 1st century CE. He credits Maria with developing a broad range of apparatuses and techniques. One such technique was gentle, even heating using a bath of hot water rather than an open flame; the bain-marie used by cooks today takes its name from her.

In 296 CE, the Roman Emperor Diocletian banned alchemy

> So the search and endeavors to make gold have brought many useful inventions and instructive experiments to light.
>
> ***De Augmentis Scientiarum***
> **(1623)**

throughout the Roman Empire, fearing that a sudden surplus of alchemical gold would undermine the empire's economy. Western alchemy disappeared from view for several centuries, until it was revived by the Muslims from the 7th century CE; their influence persists in Arabic-derived words such as "alcohol" (*al-kuhl*), "alembic" (*al-inbiq*), and "alkali" (*al-qali*)—as well as "alchemy" itself (*al-kimiya*).

Alchemy in the Muslim world

One of the most renowned Arab alchemists was Jabir ibn Hayyan. He followed the Greek philosopher Empedocles in believing that all matter was composed of four elements: fire, air, earth, and water. He also followed Aristotle in assigning these elements pairs of basic qualities: fire was hot and dry; earth was cold and dry, water was cold and moist, and air was hot and moist. To these elements, Jabir added sulfur, embodying the principle of combustibility, and mercury, defined as the idealized principle of metallic properties.

Jabir believed that metals were formed in the earth from varying combinations of sulfur and mercury and that transmutation of metals could be achieved by adjusting the proportions of mercury and sulfur (see box, overleaf) in a metal. The process would involve applying a catalyst, called *al-iksir* (derived from the Greek word *xerion*, meaning "powder for drying wounds"), from which comes the English "elixir"; this elixir would be obtained from the philosopher's stone. Jabir's elixir came to be seen not just as a means to transmute metals but as a panacea (a medicine that could cure all ills) and even as "the elixir of life," giving immortality and eternal youth. »

This floor mosaic in the Cathedral of Siena, created in 1488, shows Hermes Trismegistus teaching others the "letters and laws of the Egyptians."

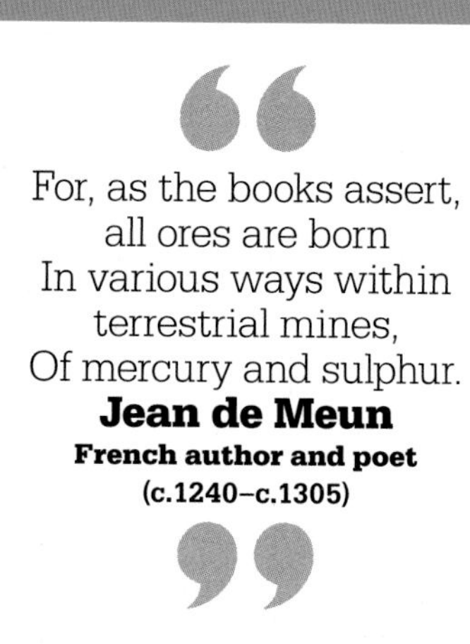

For, as the books assert,
all ores are born
In various ways within
terrestrial mines,
Of mercury and sulphur.

Jean de Meun
French author and poet
(c.1240–c.1305)

The sulfur-mercury theory Exhalations of "earthy smoke" (*dukhan*) and "watery vapor" (*bukhar*) beneath the ground were converted into sulfur and mercury; these then combined in different proportions to form the known metals.

Less positive or less balanced mixtures of mercury and sulfur produce silver and lesser metals

Lead
Tin
Silver
Iron
Copper
Mercury
Gold
Sulfur

Gold = perfect balance of positive mercury and sulfur

Air
Water
Fire
Earth
Sunlight
Vapors (*bukhar*)
Smoke (*dukhan*)
Soil and rock

Although the elixir was never discovered, Jabir systematically explored the properties of substances such as ammonium chloride (NH_4Cl). He distilled acetic acid (CH_3COOH) and prepared weak solutions of nitric acid (HNO_3) from saltpeter (potassium nitrate, KNO_3). He is also credited with inventing aqua regia (HNO_3+3 HCl)—a combination of nitric acid with hydrochloric acid (HCl), and one of the few chemicals that can dissolve gold.

Later Muslim alchemists made further attempts to discover the philosopher's stone, building on classical knowledge. Notably, the 9th-century Persian alchemist Muhammad ibn Zakariya al-Razi devised a classification system for natural substances such as salts, metals, and spirits, as well as defining a range of procedures and equipment that would be used in alchemy for centuries to come.

Further discoveries

During the centuries when "pagan" Greek and Roman learning was suppressed in Christian Europe, alchemy was still being practiced in other parts of the world. In the 4th century CE, Christian heretics fleeing to Persia carried alchemical knowledge with them. Meanwhile, in China, another alchemical tradition had been flourishing since at least the 2nd century BCE; like their Western counterparts, Chinese alchemists were engaged in the search to turn base metals into gold and find the elixir of life.

Knowledge of alchemy returned to western Europe in the 12th century, during the Christian crusades against the Muslims. European natural philosophers studied the works of the Muslim alchemists and the ancient Greeks, above all Aristotle. In the 13th century, the German friar Albertus Magnus combined his study of

Jabir ibn Hayyan

There is some dispute as to whether Jabir (known in Europe as Geber) ever existed. He was said to have been the son of Hayyan al-Azdi, a pharmacist who lived in Kufa in Iraq in the early 8th century CE but fled to Iran to escape the Umayyad caliphs; Jabir was born there around 721 CE in the northeastern city of Tus.

On returning to Iraq, Jabir is said to have studied philosophy, astronomy, alchemy, and medicine with Imam Jafar al-Sadiq. He became court alchemist to Caliph Haroun Al-Rashid and physician to his grand ministers, or viziers. Jabir is also credited as the author of hundreds of books on alchemy and philosophy, but many of these may have been written by his followers. Very few of his works reached medieval Europe. Jabir is thought to have died between 806 and 816 CE.

Key works

Kitab al-Rahma al-kabir (*Great Book of Mercy*)
al-Kutub al-sab un (*The Seventy Books*)

Aristotelian thought with practical experiments and is credited with discovering arsenic. Albertus's English contemporary, the monk Roger Bacon, was influenced by Hermetic philosophy but emphasized the importance of experiment in understanding the material world.

The alchemists, like many craftsmen, hid their practices from laypeople. They used a system of symbols and metaphors to conceal their theoretical and spiritual knowledge, following the ancient Egyptian practices supposedly handed down through Hermeticism.

Many alchemists sought the philosopher's stone. In 14th-century France, the Franciscan monk and alchemist John of Rupescissa produced a distillation of wine that he termed *quinta essentia* ("quintessence"); he claimed that it was a perfect balance of elements and recommended it as a panacea. The 16th-century German Hennig Brand chose a less pleasant method: he left 50 buckets of urine to stand until it "bred worms," then reduced it by boiling and heated it with sand and charcoal. The result was a white, waxy substance that glowed in the dark. Brand called his new material "phosphorus," from the Greek word meaning "light bearer." Phosphorus was the first element to be discovered since ancient times, and Brand the first person known to have discovered a chemical element.

Alchemists protected their knowledge by expressing it in symbolic form, as can be seen in this image, "Allegory of Distillation," by Claudio de Domenico Celentano di Valle Nove, in the *Book of Alchemical Formulas* (1606).

Alchemy persisted into the late 17th century. Isaac Newton, the renowned English mathematician and natural philosopher, was a practitioner and was eager to find the philosopher's stone. The Anglo-Irish natural philosopher Robert Boyle successfully petitioned the English Parliament in 1689 to repeal a law forbidding gold making, as he thought this impeded research into the powers of the stone. However, the alchemists' own increasingly precise methods of experiment led, by the early 18th century, to the discoveries of the Enlightenment period, which brought an end to alchemy as a serious discipline.

The alchemists' beliefs proved to be untrue, but alchemists did contribute to the development of skills and knowledge in many areas, including metallurgy and the production of pigments and dyes. Alchemy also influenced physics and medicine, and it led to the development of processes such as distillation of liquids and chemical alteration of metals, giving rise to the modern science of chemistry. ■

THE WHOLE HOUSE BURNED DOWN

GUNPOWDER

IN CONTEXT

KEY FIGURE
Zeng Gongliang (998–1078)

BEFORE
142 CE Chinese alchemist Wei Boyang describes a substance that may have been a type of gunpowder.

300 CE Ge Hong, a Chinese philosopher, experiments with saltpeter and charcoal while attempting to create gold.

AFTER
1242 English philosopher Roger Bacon writes of an explosive mixture, the first mention of gunpowder in Europe.

15th century Europeans develop mixing techniques and corning to make gunpowder more effective and easier to handle.

Gunpowder—a mixture of saltpeter (potassium nitrate, KNO_3), charcoal (carbon), and sulfur—was the first known chemical explosive. First created in China, it would be used as a weapon across Asia and Europe and later employed in mining.

Fire medicine

Gunpowder is seen as one of the "four great inventions" of ancient China, the others being paper, the compass, and printing. Saltpeter and sulfur had been used in medicine for centuries—ironically, as an elixir to prolong life rather than end it. The earliest confirmed reference to gunpowder dates from the mid-9th century CE, with warnings of dangerous formulas that had caused injury or even set houses on fire. Alchemists called this mixture *huo yao*, or "fire medicine"—a term still used for gunpowder in China today.

The armies of the Tang Dynasty were using gunpowder devices against the Mongols as early as 904 CE. These included "flying fire" (an arrow with a burning tube of gunpowder attached), fire lances (primitive flamethrowers), basic hand grenades, and landmines. But the oldest surviving recipes for

Hand cannon The first true firearm (an explosive weapon operated by one person), the hand cannon could be held in both hands or placed on a rest while the gunner fired.

See also: Oxygen and the demise of phlogiston 58–59 ▪ Explosive chemistry 120 ▪ Why reactions happen 144–147 ▪ Chemical warfare 196–199

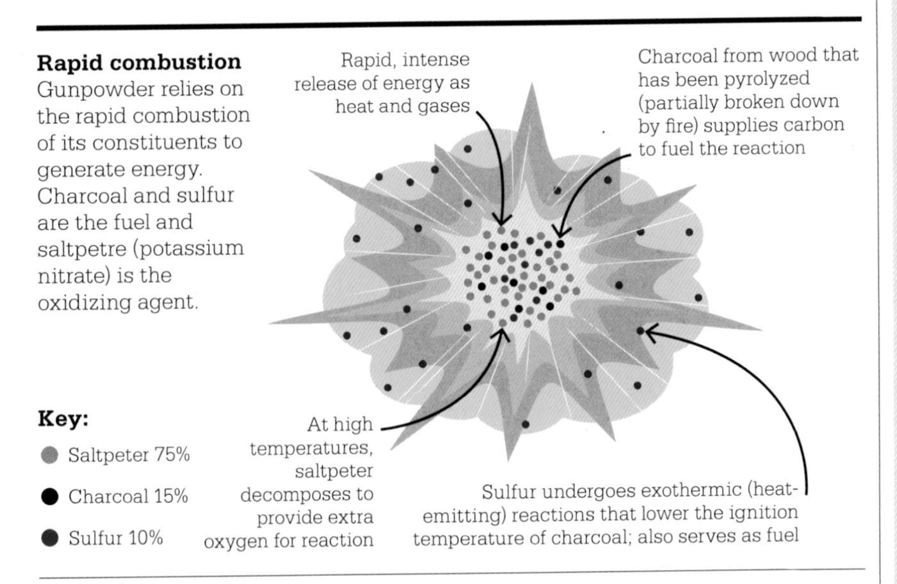

gunpowder date from the Song Dynasty (960–1279) and are found in the *Wujing Zongyao* (*Collection of the Most Important Military Techniques*), a manual from 1044 compiled by military strategist Zeng Gongliang. The manual describes three kinds: two for use in incendiary bombs and one to be used as fuel for bombs emitting poisonous smoke.

Rapid combustion
Gunpowder relies on the rapid combustion of its constituents to generate energy. Charcoal and sulfur are the fuel and saltpetre (potassium nitrate) is the oxidizing agent.
Rapid, intense release of energy as heat and gases
Charcoal from wood that has been pyrolyzed (partially broken down by fire) supplies carbon to fuel the reaction
Key:
Saltpeter 75%
Charcoal 15%
Sulfur 10%
At high temperatures, saltpeter decomposes to provide extra oxygen for reaction
Sulfur undergoes exothermic (heat-emitting) reactions that lower the ignition temperature of charcoal; also serves as fuel

Explosive growth

The Mongols brought gunpowder with them as they invaded Eurasia in the 13th and 14th centuries. In conquered Syria, Arab inventor Hassan al-Rammah described a method for purifying potassium nitrate along with more than 100 recipes for gunpowder. Traders and crusaders learned of the technology when traveling in the Middle East at this time. By 1350, English and French armies were deploying cannons, and by the early 15th century, the first guns had appeared.

Bombs and blasting

The first documented use of explosive weapons was in China during the siege of the Song city of Qizhou in 1221. The attacking Chin forces catapulted iron-cased "iron fire bombs" that showered deadly shards of metal when they exploded and shattered the city walls.

From the 17th century, explosives were used in Europe for quarrying and mining. Blasting rock involved placing gunpowder into a hole, packing it with clay, and laying a trail of gunpowder away from it. This procedure was made safer in 1831 by British inventor William Bickford's safety fuse: two layers of jute yarn woven around a tube of gunpowder, which burned at a steady rate.

Fireworks

Fire medicine was used in festivities before it was used in warfare. In China, bamboo tubes filled with gunpowder were thrown onto fires so that the explosions might scare away evil spirits. In Italy, records show fireworks used in a mystery play in 1377, while in England, the marriage of Henry VII to Elizabeth of York in 1486 was marked by a fireworks display. Modern fireworks were developed by the Italians, who incorporated trace amounts of metals in their gunpowder mix in the 1830s to produce more colorful explosions.

Fireworks also paved the way for rocket science. One 14th-century Chinese military treatise, the *Huolongjing* (*Fire Dragon Manual*), shows a multistage rocket. In the 16th century, Johann Schmidlap, a German firework maker, built a two-stage "step rocket"; when the bigger rocket burned out, the smaller one ignited and continued even higher. This multistage ignition is still used for today's space vehicles. ▪

Gunpowder recipes

Although the standard composition used in fireworks today is 75 percent potassium nitrate, 15 percent charcoal, and 10 percent sulfur, there is no single recipe for gunpowder. Varying the ratio of the ingredients will produce different effects. Gunpowder used in firearms is required to burn at a fast rate to produce the explosive release of gases necessary to accelerate a projectile. In contrast, when used as a rocket propellant, the powder needs to burn more slowly, releasing its energy over a longer period of time.

To ensure that gunpowder burns effectively, the ingredients must be finely ground and thoroughly mixed. In 14th-century Europe, the techniques of "wet grinding" with water to keep the ingredients well mixed and of "corning"—forming the paste into corn-sized grains and drying it—created more durable and reliable explosives with all the ingredients igniting at the same time, which increased the effectiveness of weapons.

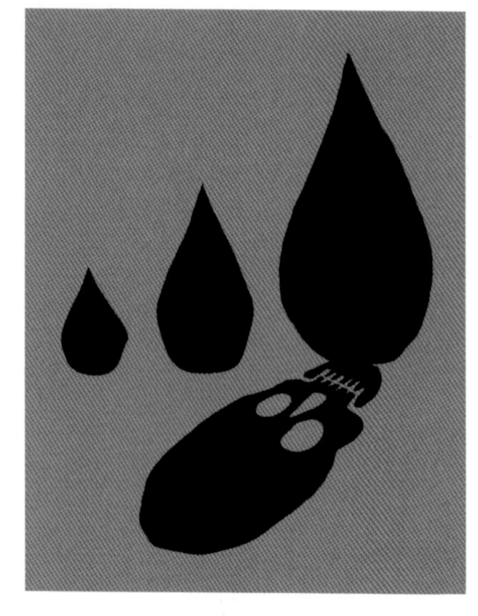

THE DOSE DETERMINES THAT A THING IS NOT A POISON

THE NEW CHEMICAL MEDICINE

IN CONTEXT

KEY FIGURE
Paracelsus (1493–1541)

BEFORE
1st century CE Dioscorides, a Greek physician in the Roman army, compiles *De Materia Medica* (*On Medical Material*); its 600 medicines include minerals, salts, and metals.

10th century CE Persian physician Muhammed ibn Zakariya al-Razi includes small amounts of toxins such as mercury and arsenic in medical treatments.

AFTER
c.1611 Daniel Sennert, Professor of Medicine at the University of Wittenberg in Germany, introduces chemical medicine into the curriculum.

1909 German chemist Paul Ehrlich and Japanese biologist Sahachiro Hata create an arsenic-based treatment for syphilis, which finally supersedes the use of mercury.

In 16th-century Europe, medical thinking underwent radical changes. Philosophers, physicians, and other scholars were rediscovering ideas from ancient Greece and Rome and at the same time challenging orthodoxies that had prevailed for centuries. One of the most influential figures of the time was the Swiss physician and alchemist Paracelsus.

A free thinker, Paracelsus rebelled against the medical authorities of the day. As a lecturer on medicine at the University of Basel, he spoke in German rather than the traditional Latin, so everyone could understand him. He spent years learning from apothecaries, barber-surgeons, bath-house attendants, and others whom he respected for their practical skills in treating sick people.

Philippus Aureolus Theophrastus Bombastus von Hohenheim took the name Paracelsus, Latin for "beyond Celsus," to show he surpassed Celsus, a renowned doctor in the Roman empire.

Paracelsus's theory of medicine, set out in his *Paragranum* of 1529, rested on four foundations: natural philosophy, with an emphasis on physicians learning by observing nature; astrology, detailing cosmic influences on human life; the ethical and religious values that should underpin a physician's work; and alchemy, in particular the art of refining materials to transform their toxic attributes into healing ones.

The alchemy of the body

Up to this time, medicine centered on an idea originating with Hippocrates, the Greek "father of medicine," 2,000 years earlier—namely, that the body contained four fluids, or "humors": blood, phlegm, yellow bile, and black bile. Good health depended on these being in balance. An excess of any humor caused disease; the cure was to rebalance the humors by practices such as blood-letting. Paracelsus contended that these treatments

See also: The four elements 30–31 ▪ Attempts to make gold 36–41 ▪ Anesthetics 106–107 ▪ Antibiotics 222–229 ▪ Chemotherapy 276–277

were useless or even dangerous. He based his approach on alchemy, in which matter, including the human body, was created from three primal principles—sulfur, mercury, and salt—and separation of one principle from the other two led to disease. Physicians had to understand the composition of specific body parts in order to treat them appropriately.

Chemical medicine

Paracelsus reintroduced the practice of using minerals in treatments—known as iatrochemistry (from the Greek *iatrós*, "doctor"). He based his treatments on the principle "like cures like"; in other words, poisoning within the body can be cured by giving a dose of that same poison from an external source.

Paracelsus believed that certain substances were most effective on specific body organs or sites while leaving others unaffected. This idea, now known as targeted organ toxicity, is still important in modern toxicology. His treatments included arsenic, mercury, sulfur, silver, gold, lead, and antimony; for example, he used mercury ointment to treat syphilis and gave antimony to purge the body of poisons. In response to criticism from his peers, he stressed the importance of dose: "All things are poison, and nothing is without poison; only the dose determines that a thing is not a poison."

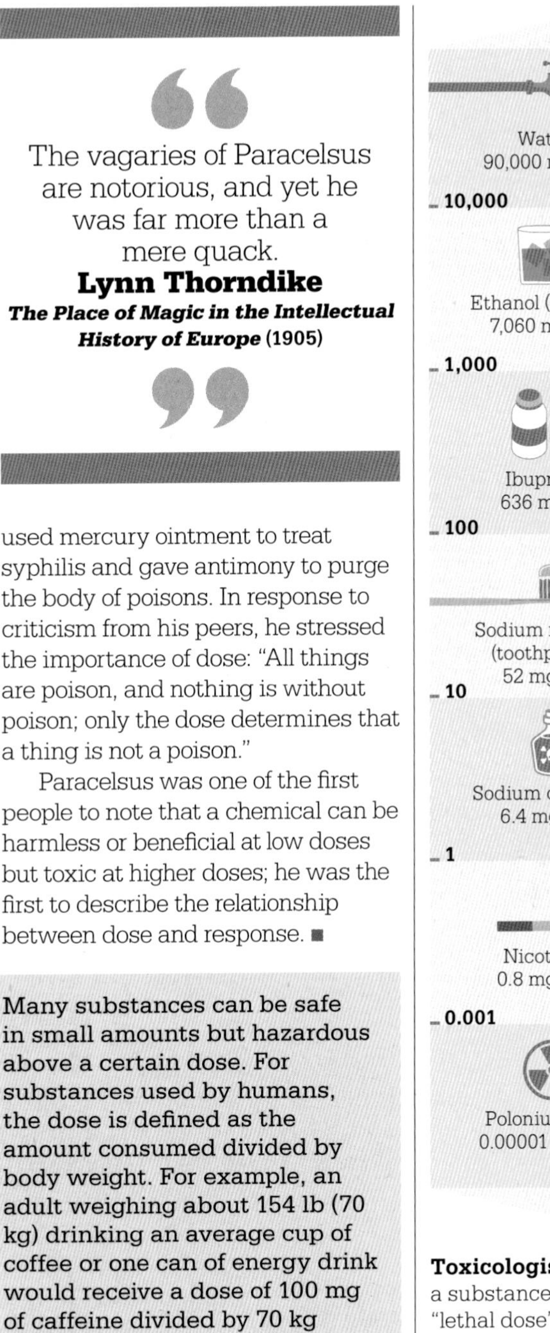

> The vagaries of Paracelsus are notorious, and yet he was far more than a mere quack.
> **Lynn Thorndike**
> ***The Place of Magic in the Intellectual History of Europe*** **(1905)**

Paracelsus was one of the first people to note that a chemical can be harmless or beneficial at low doses but toxic at higher doses; he was the first to describe the relationship between dose and response. ■

Medicines and poisons

The study of the effects that chemicals have on living organisms (including people) is called toxicology. One of the primary considerations in this discipline is the relationship between dose and response. For example, a pharmacist making up medication has to consider the full range of responses, from desirable to undesirable, and determine a dose that will produce benefits without serious adverse effects.

Many substances can be safe in small amounts but hazardous above a certain dose. For substances used by humans, the dose is defined as the amount consumed divided by body weight. For example, an adult weighing about 154 lb (70 kg) drinking an average cup of coffee or one can of energy drink would receive a dose of 100 mg of caffeine divided by 70 kg (weight in kg), or 1.4 mg/kg of caffeine. However, while 100 mg of caffeine may be perfectly safe, 10 grams is potentially lethal.

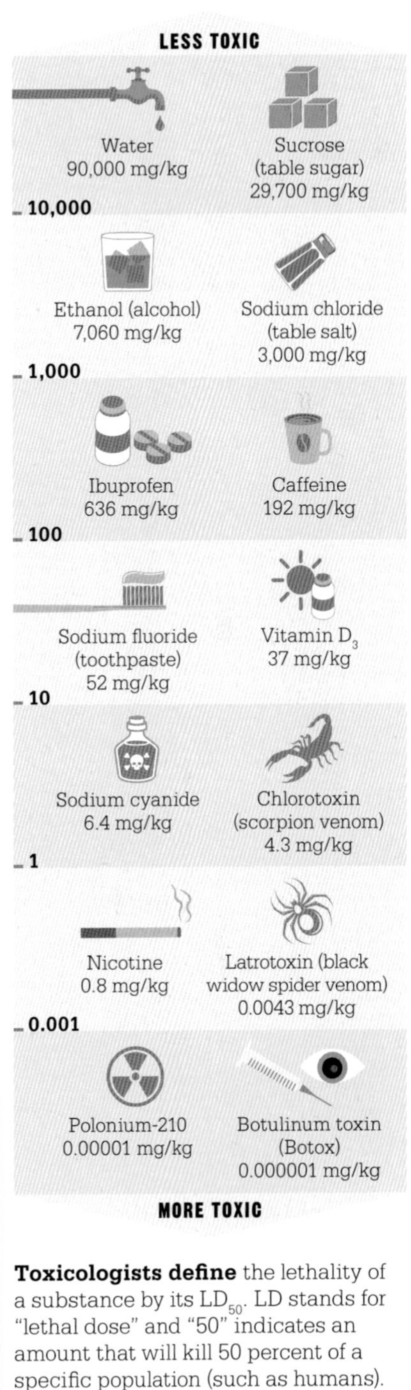

Toxicologists define the lethality of a substance by its LD_{50}. LD stands for "lethal dose" and "50" indicates an amount that will kill 50 percent of a specific population (such as humans). This chart shows the estimated LD_{50} for common substances. The lower the LD_{50}, the more lethal the substance.

A FAR MORE SUBTILE THING THAN A VAPOR

GASES

IN CONTEXT

KEY FIGURE
Jan Baptist van Helmont (1580–1644)

BEFORE
c.450 BCE The Greek philosopher Empedocles declares that air is one of the four primal elements (earth, air, fire, and water), an idea later expanded by Aristotle.

c.1520 The Swiss physician and alchemist Paracelsus discovers a mysterious flammable gas, later to be identified as hydrogen, when experimenting with iron and sulfuric acid.

AFTER
1754 Scottish chemist Joseph Black isolates carbon dioxide, which he calls "fixed air."

1778 French chemist Antoine Lavoisier identifies oxygen and defines its role in combustion.

Until the 17th century, gases were seen as varieties of air. The first person to recognize that gases had distinct properties was Flemish chemist Jan Baptist van Helmont. He may have coined the word "gas" from his Dutch pronunciation of the ancient Greek word *chaos*, which denoted the emptiness of space.

Rejecting Empedocles's concept of four elements (earth, air, fire, and water) and the alchemists' system of three (salt, sulfur, and mercury), van Helmont identified just air and water. He regarded all substances as modified forms of water—except for air, which was the carrier for water vapor and gases.

For truly, Chymistry, hath its principles not gotten by discourses, but those which are known by nature ... and it prepares the understanding to pierce the secrets of nature.
Jan Baptist van Helmont
***Physick Refined* (1638)**

In *Ortus medicinae* (*Origin of Medicine*), published posthumously in 1648, van Helmont was one of the first to investigate how certain chemical reactions liberated gases that were airlike but had properties distinct from those of air.

In one experiment, van Helmont burned 62 lb (28 kg) of charcoal and found that only 1 lb (0.45 kg) of ash remained; he concluded that the rest had escaped as what he called *gas sylvestre* ("wood gas"). He noted that this "air" was given off by fermentation as well as combustion. We now know it as carbon dioxide (CO_2). In another experiment, he heated coal in the absence of air and discovered an inflammable gas that he called *gas pingue*. Now known as coal gas, it is a mixture of methane (CH_4), carbon monoxide (CO), and hydrogen (H_2). ■

See also: Fixed air 54–55 ▪ Inflammable air 56–57 ▪ Oxygen and the demise of phlogiston 58–59 ▪ Conservation of mass 62–63 ▪ The ideal gas law 94–97

I MEAN BY ELEMENTS ... PERFECTLY UNMINGLED BODIES

CORPUSCLES

IN CONTEXT

KEY FIGURE
Robert Boyle (1627–1691)

BEFORE
c.450 BCE The Greek philosopher Democritus proposes the existence of invisible, indivisible atoms from which matter is formed.

c.300 BCE The Greek philosopher Aristotle declares that all matter is formed from just four elements: air, earth, fire, and water.

AFTER
1803 British chemist John Dalton sets out his atomic theory of matter, proposing that the atoms of an element are identical and differ from those of other elements.

1897 British physicist J. J. Thomson discovers the electron, thus demonstrating that the atom is composed of even smaller particles.

The 17th century saw a revival of the classical Greek idea of atomism, proposed by Democritus over 2,000 years earlier. A similar theory was put forward by Anglo-Irish natural philosopher Robert Boyle.

Boyle rejected the Aristotelian belief, still held by some of his peers, that matter was made of four elements (fire, earth, water, air), as well as Paracelsus's theory that matter derived from the "principles" mercury, sulfur, and salt. Instead, he posited that all matter was made from collections of tiny particles called corpuscles ("little bodies") with specific qualities such as shape, size, and motion. Natural phenomena, such as heat, resulted from the collision of corpuscles in motion.

In *The Sceptical Chymist*, published in 1661, Boyle defined the basic particles as "Elements ... certain Primitive and Simple, or perfectly unmingled bodies ... not being made of any other bodies, or of one another."

This painting of Robert Boyle, completed in 1689 by German painter Johann Kerseboom, shows Boyle with a book, suggesting his lifetime of scientific investigation and writing.

A lifelong alchemist, Boyle believed that one element could be transmuted into another by rearranging the corpuscles in each and that this could be proved by experiment. It was his emphasis on testing ideas by experiment that paved the way for the methods used in modern chemistry. ■

See also: The atomic universe 28–29 ▪ The four elements 30–31 ▪ The new chemical medicine 44–45 ▪ Dalton's atomic theory 80–81 ▪ The electron 164–165

AN INSTRUMENT MOST POTENT, FIRE, FLAMING, FERVID, HOT

PHLOGISTON

IN CONTEXT

KEY FIGURE
Georg Ernst Stahl
(1659–1734)

BEFORE
1650 German physicist Otto von Guericke demonstrates that a candle will not burn in a vessel from which the air has been removed.

1665 English scientist Robert Hooke suggests there is an active component in air that combines with combustible substances.

AFTER
1774 British natural philosopher Joseph Priestley isolates an inflammable but breathable gas, which he calls "dephlogisticated air."

1789 French chemist Antoine Lavoisier renames dephlogisticated air as oxygen and lays the phlogiston theory to rest.

For millennia, people have tried to work out what makes fire burn. In the 4th century BCE, Plato suggested that combustible objects contained some inflammable principle, while in the four-element system of Empedocles and Aristotle, when a substance such as wood burned, the flame was the element of fire escaping. The alchemists of the 16th century equated the inflammable principle with sulfur. Robert Boyle challenged this idea with his concept that there were no "principles", only matter. But the questions remained: what is fire, and how does combustion happen? One attempt at an answer was the theory of phlogiston.

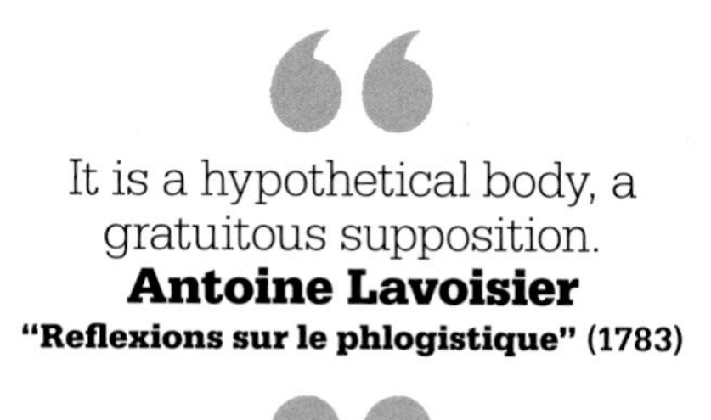

It is a hypothetical body, a gratuitous supposition.
Antoine Lavoisier
"Reflexions sur le phlogistique" (1783)

Fatty earth

In his *Physica Subterranea* of 1667, German physician and alchemist Johann Joachim Becher adapted Paracelsus's system of three principles. He posited that matter was formed from three "earths": *terra pinguis* ("fatty earth") was associated with sulfur and produced combustible, greasy, or fatty properties; *terra fluida* (fluid earth), associated with mercury, contributed fluidity and volatility; and *terra lapidea* (glassy earth), associated with salt, gave solidity. *Terra pinguis* was released when a substance burned.

One of Becher's students, Georg Ernst Stahl, adapted this theory in his 1697 book *Zymotechnia Fundamentalis* (*Foundation of the Fermentative Art*), renaming *terra pinguis* as *phlogiston*. Stahl postulated that sulfur was actually a combination of sulfuric acid and phlogiston; the latter, rather than sulfur itself, was the cause of fire.

Stahl's theory

In Stahl's theory, all flammable substances contain phlogiston (from the Greek *phlogizein*, "to set on fire"); this is released when they burn, the combustion continuing

See also: The four elements 30–31 ▪ Corpuscles 47 ▪ Inflammable air 56–57 ▪ Oxygen and the demise of phlogiston 58–59 ▪ Conservation of mass 62–63

until the phlogiston is exhausted. Flames indicated the rapid release of phlogiston. Air absorbed phlogiston; combustion could not be sustained in a closed container because the air inside became saturated with phlogiston, or "phlogisticated."

Stahl also believed that the corrosion of metals was a form of combustion in which metal lost its phlogiston as it changed to its calx (what is now termed its oxide). He demonstrated his idea by burning mercury to form its calx and then reheating the calx with charcoal to return the metal to its original state.

Positive lightness

There was one major problem with the theory: the calx of a metal was less dense but actually heavier than the metal. To resolve this issue, supporters of the theory described phlogiston as having negative weight or "positive lightness"; this was why phlogiston or flames rose against the pull of gravity.

Accepting this argument made the phlogiston theory difficult to disprove; it held sway until the 1770s, when French chemist Antoine Lavoisier showed that combustion required the presence of "vital air," or as he named it, oxygen. ■

Burning metal

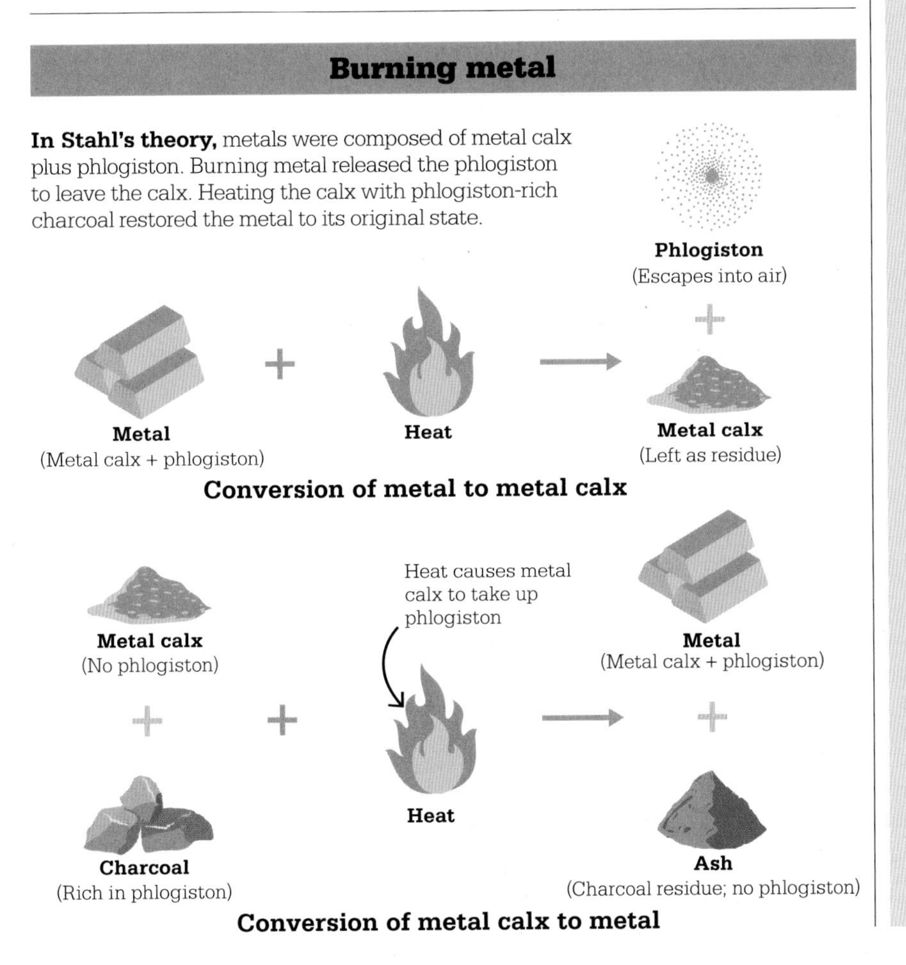

Georg Ernst Stahl

Born in 1659 in Ansbach, Bavaria, Stahl studied medicine at the University of Jena, which was then a center for iatrochemistry (chemical medicine). After graduating in 1684, he taught there until 1687, when he was appointed physician to the duke of Sachsen-Weimar. He then became professor of medicine at the newly established University of Halle in 1694, and became physician to the king of Prussia in 1716, a role he held until his death in 1734.

Although Stahl initially embraced the principles of alchemy, he grew increasingly skeptical in later years. His phlogiston theory is often seen as marking a transition from alchemy to chemistry, and it remained influential among natural philosophers until the late 18th century.

Key works

1697 *The Foundation of the Fermentative Art or the General Theory of Fermentation*
1730 *Philosophical Principles of Universal Chemistry*

ENLIGHT
CHEMIST
1700–1800

ENMENT
RY

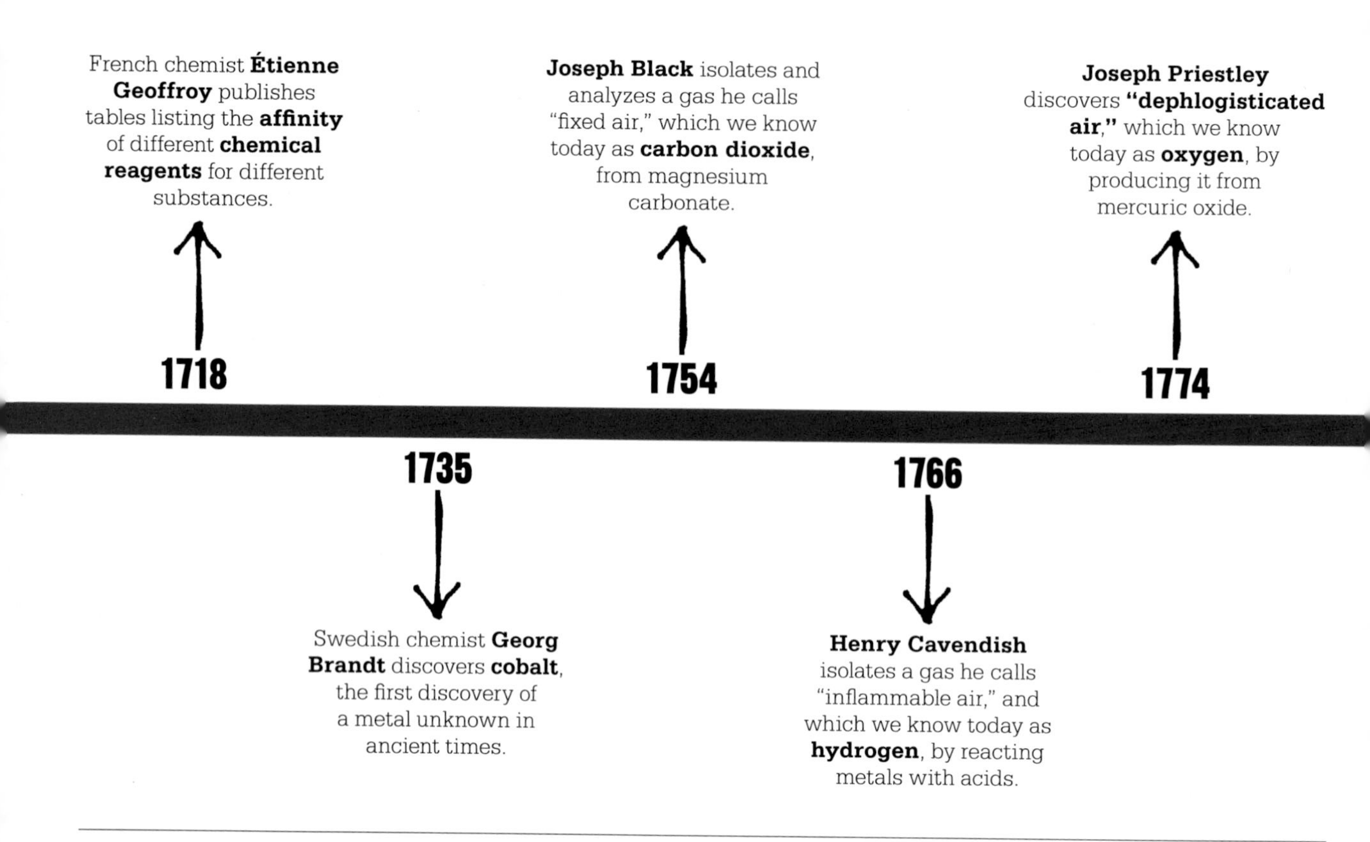

French chemist **Étienne Geoffroy** publishes tables listing the **affinity** of different **chemical reagents** for different substances.
1718
Swedish chemist **Georg Brandt** discovers **cobalt**, the first discovery of a metal unknown in ancient times.
1735
Joseph Black isolates and analyzes a gas he calls "fixed air," which we know today as **carbon dioxide**, from magnesium carbonate.
1754
Henry Cavendish isolates a gas he calls "inflammable air," and which we know today as **hydrogen**, by reacting metals with acids.
1766
Joseph Priestley discovers **"dephlogisticated air,"** which we know today as **oxygen**, by producing it from mercuric oxide.
1774

The 18th century was one of intersecting revolutions. The scientific revolution, which had begun in the previous century, saw the understanding of the material world continue to develop as a discipline distinct from medieval alchemy. This was the age of enlightenment, bringing revolutions in scientific thought that led to a number of pivotal discoveries. Later in the century, the political upheaval of the French Revolution would claim the life of one of the key figures in chemistry's development—but not before his contributions set the stage for a revolution in chemical science.

An explosion of elements

Until the 1700s, only a handful of elements were recognized—mostly those known since antiquity, along with the more recently discovered arsenic and phosphorus. But by the end of the century, more than 20 further elements had been isolated for the first time.

Many of these newly identified elements were metals, including cobalt, platinum, and manganese. Most were discovered as a result of improved mining technologies: platinum was found in the gold mines of modern-day Colombia, while cobalt was discovered in a blue ore from copper mines.

The identification of what would later become known as a whole new family of metallic elements, the rare earths, was started by the discovery of yttrium in an ore from the mine of the Swedish village of Ytterby. This small village would go on to be the site of more element discoveries than anywhere else in the world, with 10 new elements found over the decades in its ores. The names of four of these elements derive directly from their discovery at Ytterby: yttrium, ytterbium, terbium, and erbium. Discoveries of new rare-earth elements continued into the 20th century.

Pneumatic chemistry

Metallic elements were not the only substances being identified for the first time. Building on the initial investigations of combustion reactions at the conclusion of the 17th century, a number of chemists focused on producing, isolating, and identifying new gases.

Crucial to these endeavors was the development of the pneumatic trough, a device for collecting gases. This was not a new concept, but Stephen Hales, an English

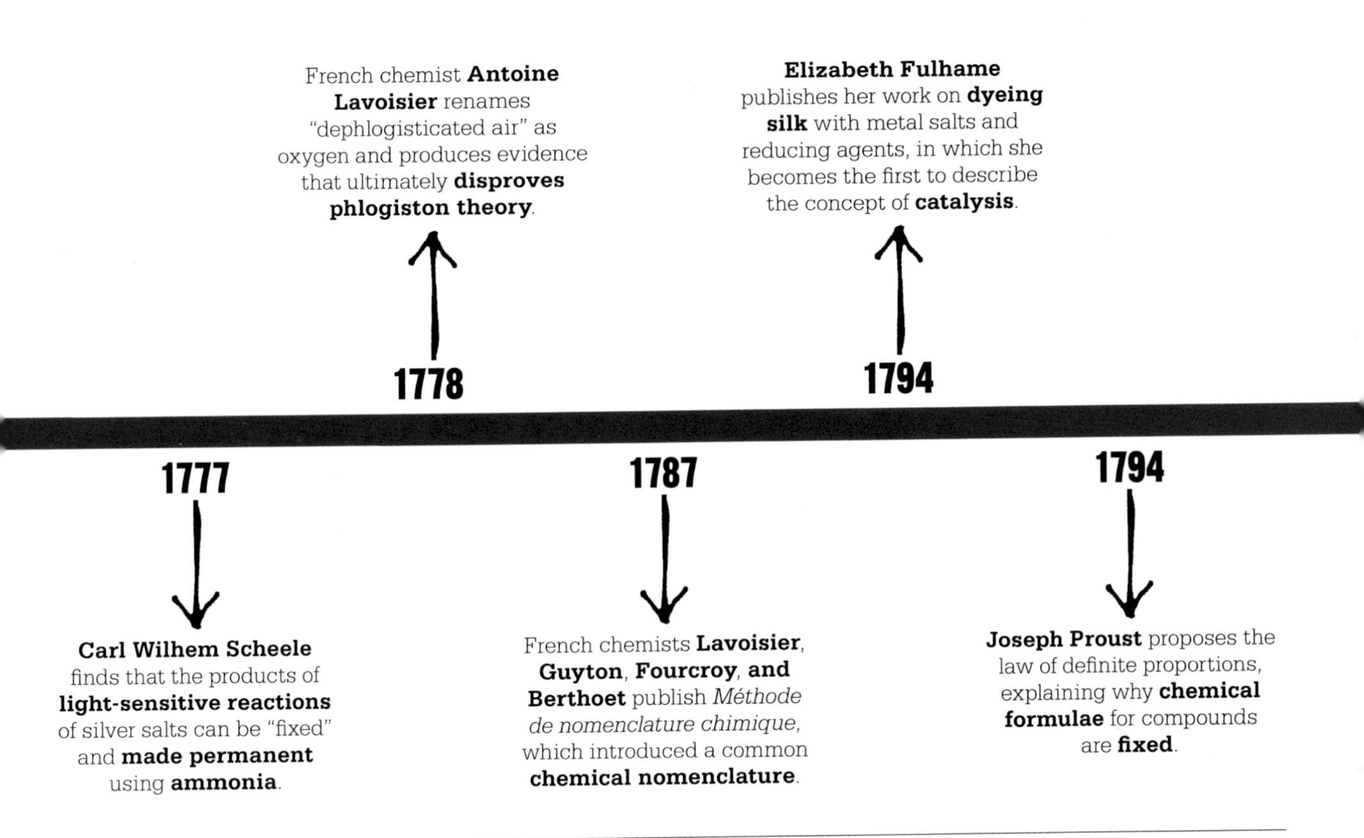

French chemist Antoine Lavoisier renames "dephlogisticated air" as oxygen and produces evidence that ultimately disproves phlogiston theory.
1778
Elizabeth Fulhame publishes her work on dyeing silk with metal salts and reducing agents, in which she becomes the first to describe the concept of catalysis.
1794
1777
Carl Wilhem Scheele finds that the products of light-sensitive reactions of silver salts can be "fixed" and made permanent using ammonia.
1787
French chemists Lavoisier, Guyton, Fourcroy, and Berthoet publish Méthode de nomenclature chimique, which introduced a common chemical nomenclature.
1794
Joseph Proust proposes the law of definite proportions, explaining why chemical formulae for compounds are fixed.

clergyman and chemist, produced a version of the device in 1727 that permitted gases to be collected as they were being produced by chemical reactions. Hale's trough instantly became an essential piece of apparatus for any chemist hunting new gases. The discoveries of carbon dioxide, hydrogen, and oxygen in the following decades all involved the use of the pneumatic trough to gather the gases prior to their identification.

The discovery of oxygen was the death knell for the phlogiston theory (see pp.48–49), which had persisted for nearly a century. The English chemist Joseph Priestley discovered the gas, naming it "dephlogisticated air," but it was French chemist Antoine Lavoisier who recognized it as "the true combustible body" and carried out quantitative experiments that would lead to a new oxygen-based theory of combustion.

The seeds of revolution

Lavoisier's dismissal of phlogiston was not his only contribution to modern chemistry. After decades of criticism concerning the variation in chemical names, many of which were derived from alchemy's cryptic terminology, numerous chemists started to put forward their opinions on how chemical nomenclature should be reformed. This culminated in 1787 with the publication of *Méthode de nomenclature chimique*, a treatise written by four French chemists including Lavoisier, that reformed and standardized the names that chemists used to define elements and compounds.

Two years later, Lavoisier published what is regarded as the first modern chemistry textbook: *Traité Élémentaire de Chimie*. In it, he defined elements as substances that cannot be analyzed any further by chemical means, and lists 33 such substances—23 of which we still consider to be elements today. He also established the principle of mass being conserved in chemical reactions.

Lavoisier would not live to see the full impact of his reforms. A member of France's *ferme générale*, who collected taxes for the king and pocketed significant bonuses for themselves, Lavoisier was accused of tax fraud during the French Revolution and guillotined in 1794. In spite of his death, however, the chemical revolution he had started was just beginning. ■

THIS PARTICULAR KIND OF AIR … IS DEADLY TO ALL ANIMALS

FIXED AIR

IN CONTEXT

KEY FIGURE
Joseph Black (1728–1799)

BEFORE
1630 Jan Baptist van Helmont identifies carbon dioxide as "wood gas," given off by burning wood.

1697 German chemist Georg Ernst Stahl argues that all burning involves a substance he calls "phlogiston."

AFTER
1766 Henry Cavendish, a British chemist, discovers hydrogen.

1774 Joseph Priestley, a British chemist, discovers "dephlogisticated air," oxygen.

1823 British chemists Humphry Davy and Michael Faraday turn carbon dioxide under pressure into liquid.

1835 Adrien-Jean-Pierre Thilorier, a French inventor, makes solid carbon dioxide (dry ice).

In the 1750s, young Scottish student Joseph Black isolated and analyzed carbon dioxide gas for the first time. At the time, physicians in Edinburgh, where Black was studying medicine, were arguing fiercely over the merits of treating kidney stones by dissolving them with a caustic alkali such as limewater (calcium hydroxide, $Ca(OH)_2$). It was a risky procedure, but the alternative—surgical removal without anesthetic—was dangerous and agonizingly painful.

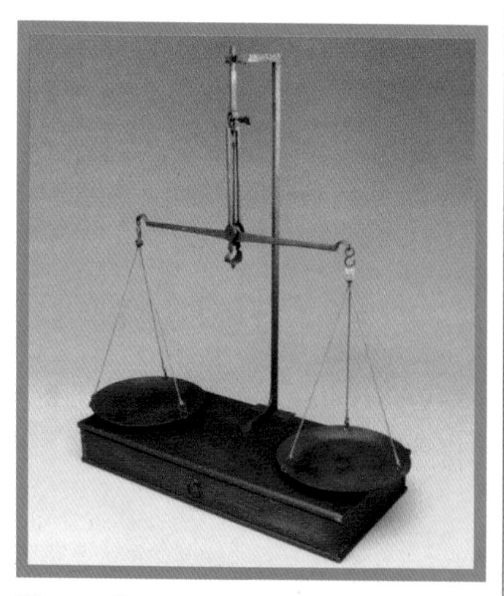

The scales used by Joseph Black in his experiments with alkalis in the 1750s are now exhibited at the National Museum of Scotland, Edinburgh.

To avoid controversy, Black decided to focus his doctoral work on a milder alkali, magnesia alba, which had recently been suggested for the treatment of acid stomach. Magnesia alba is now known as magnesium carbonate ($MgCO_3$).

Methodical approach

What made Black's experiments revolutionary was his painstaking scientific method. As he embarked on his work in 1750, he refined an analytical balance based on a lightweight beam set on a pivot to give precise measurements. He then began to look at different alkali reactions, carefully weighing everything at all stages. Early on, he noticed that, when acid is added to it, magnesia alba effervesces and loses weight. The caustic alkali quicklime (calcium oxide, CaO) does the same. He also observed that when heated in a kiln, magnesia alba turns to "magnesia usta" (magnesium oxide, MgO) and also loses weight. It had previously been assumed that when limestone (calcium carbonate, $CaCO_3$) is cooked in a kiln to make quicklime, the quicklime gets its causticity from a mysterious "fire-stuff," or "phlogiston," added in the kiln.

See also: Brewing 18–19 ▪ Gases 46 ▪ Phlogiston 48–49 ▪ Inflammable air 56–57 ▪ Oxygen and the demise of phlogiston 58–59 ▪ The greenhouse effect 112–115 ▪ The pH scale 184–189 ▪ Carbon capture 294–295

Black's meticulous measurements showed that when either treated with acid or heated, neither mild nor caustic alkalis gained weight or any "fire-stuff"; on the contrary, they lost weight. He then searched for what was lost. There was no liquid, but he was able to collect some gas. He found that this not only snuffed out a candle but was toxic to animals, killing them in seconds, although he did not know why. Bubbled through a pipe into limewater (a solution of quicklime), it left white lime powder. When he blew into the pipe, the result was the same, showing that this gas is also present in the air we breathe out.

Black decided to call the gas he had identified "fixed air," because it could become fixed within a solid—magnesia alba. He soon realized that it was the same *gas sylvestre* ("wood gas") that Flemish scientist Jan Baptist van Helmont had identified being emitted from burning wood a century earlier and that it is ever present in small quantities in the "common air," the air around us that we breathe.

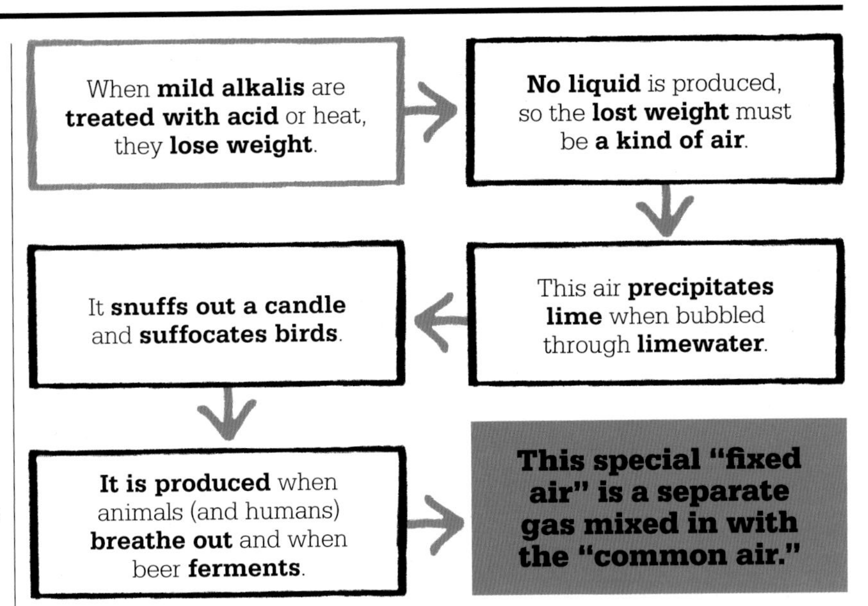
When mild alkalis are treated with acid or heat, they lose weight.
No liquid is produced, so the lost weight must be a kind of air.
This air precipitates lime when bubbled through limewater.
It snuffs out a candle and suffocates birds.
It is produced when animals (and humans) breathe out and when beer ferments.
This special "fixed air" is a separate gas mixed in with the "common air."

When magnesium carbonate ("magnesia alba") is heated, the reaction can be represented by this chemical equation: $MgCO_3$ (magnesium carbonate) → MgO (magnesium oxide, "magnesia usta") + CO_2 (carbon dioxide). And when limestone is heated: $CaCO_3$ (limestone) → CaO (quicklime, calcium oxide) + CO_2.

Black had not only discovered carbon dioxide—although it was not called by that name for many years—but had also established an experimental methodology that was to prove the foundation of modern chemistry. It galvanized the world of chemistry, putting pneumatic chemistry at the forefront of science. ■

Joseph Black

The son of an expatriate Irish wine merchant, Joseph Black was born in Bordeaux, France, in 1728 and lived there for the first 12 years of his life. Later, at Glasgow University, he studied languages and philosophy before switching to medicine. For his doctorate, Black conducted groundbreaking experiments that led to his discovery of carbon dioxide.

Black became a professor at Glasgow at just 28 years of age, and students from Europe and the US came to hear his brilliant lectures. In one, he revealed his discovery that temperature didn't change when ice melted to water, identifying the concept of latent heat. He went on to make the crucial distinction between heat and temperature for the first time, also identifying the concept of specific heat. His work inspired his young friend James Watt to make massive improvements to the steam engine. Black died in 1799.

Key work

1756 *Experiments upon Magnesia Alba, Quicklime, and Some Other Alcaline Substances*

THE GAS WENT OFF WITH A PRETTY LOUD NOISE!

INFLAMMABLE AIR

IN CONTEXT

KEY FIGURE
Henry Cavendish
(1731–1810)

BEFORE
1671 Irish chemist Robert Boyle inadvertently creates hydrogen when experimenting with the effects of dilute acid on iron filings.

1756 Joseph Black identifies "fixed air," carbon dioxide.

AFTER
1772 Scottish chemist Daniel Rutherford discovers nitrogen.

1774 Joseph Priestley discovers "dephlogisticated air," oxygen.

1783 French chemist Antoine Lavoisier reproduces Cavendish's experiment to confirm that water is a compound of hydrogen and oxygen.

The late 18th century was a period of discovery about the nature of gases and the make-up of the atmosphere. In 1766, British scientist Henry Cavendish published three papers, one of which described how he had isolated and identified, for the first time, the gas he called "inflammable air." Antoine Lavoisier later called this gas hydrogen. Cavendish's other papers focused on "factitious airs," his description of gases that combine with other substances.

Cavendish was a somewhat eccentric recluse whose personal wealth allowed him to work in his own well-equipped laboratory.

> By factitious air, I mean … any kind of air which is contained in any other bodies.
> **Henry Cavendish**

What marked him out was his precision. Joseph Black had earlier shown the importance of accurate measurement in experimental chemistry, and Cavendish went even further. Although Robert Boyle had created "inflammable air" a century earlier—without realizing what it was—by pouring acids on iron, Cavendish's meticulousness enabled him to isolate the gas and identify its properties in detail. Using apparatus he designed himself, Cavendish collected the gas that billowed up when he poured acids such as "spirits of salt" (hydrochloric acid) and dilute "oil of vitriol" (sulfuric acid) on metals such as zinc, iron, and tin. We now know that what Cavendish witnessed can be expressed as this chemical reaction: Zn (zinc) + H_2SO_4 (sulfuric acid) $\rightarrow$ $ZnSO_4$ (zinc sulfate) + H_2 (hydrogen).

Explosive reaction

Cavendish found that this gas was different from anything in the air around, then called "common air," and was much less dense. Moreover, he noted that when mixed with ordinary air and ignited, it exploded—hence the name "inflammable air."

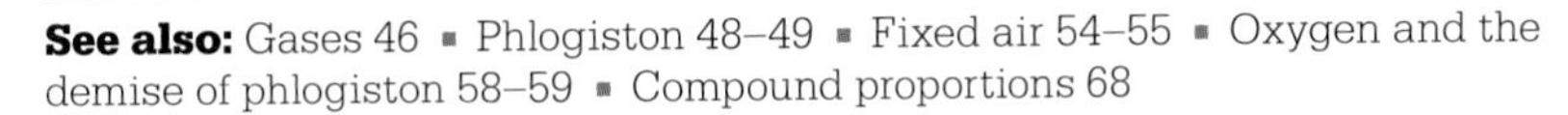
See also: Gases 46 ▪ Phlogiston 48–49 ▪ Fixed air 54–55 ▪ Oxygen and the demise of phlogiston 58–59 ▪ Compound proportions 68

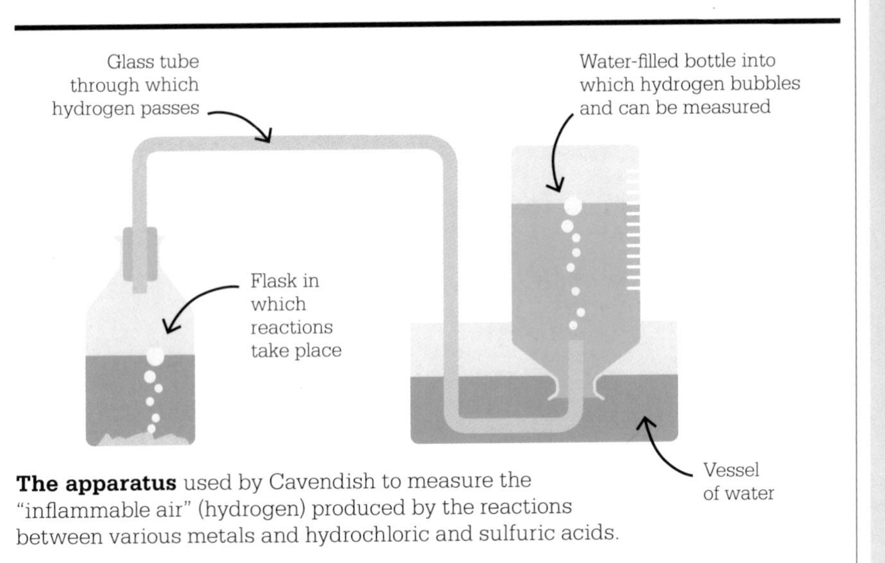

The apparatus used by Cavendish to measure the "inflammable air" (hydrogen) produced by the reactions between various metals and hydrochloric and sulfuric acids.

In his second paper of 1766, Cavendish studied "fixed air" (carbon dioxide) in more detail than Black had done, discovering that it is neither soluble in water nor flammable and that it is much heavier than air. He returned to his investigations of "airs" in the 1780s. He wrongly assumed that what we now know to be hydrogen must be the long-searched-for "phlogiston"—the mysterious ingredient in substances that makes them burn. He believed that burning added something (phlogiston) to the air rather than—as we now know—taking oxygen out.

Composition of water

In 1783, Cavendish conducted an experiment to measure the mystery component. He mixed hydrogen with ordinary air in a sealed flask, then ignited the mixture with an electric spark. The resulting explosion left a residue of water inside the flask. He had shown how hydrogen and oxygen combine to make water, and his measurements revealed that they combine in a ratio of two to one. Never in a hurry to publish, Cavendish delayed his announcement that water is not an element but a compound until the following year, by which time steam engineer James Watt had already presented his own very similar findings.

The debate over who was first to make the discovery about the nature of water continued for years. Nevertheless, Cavendish's role in laying the foundations of modern chemistry was assured. ■

> Inflammable air is either pure phlogiston … or else water united to phlogiston …
> **Henry Cavendish**

Henry Cavendish

Born in 1731 in Nice, France, to one of Britain's richest families, Henry Cavendish was raised by his father after his mother died when he was 2. He was a recluse all his life and painfully shy of women.

The range of Cavendish's scientific investigations is astonishing, but since he rarely published, their full scope is unknown. Beyond discovering hydrogen and the compound nature of water, he analyzed the composition of air, measuring the proportions of oxygen and nitrogen with incredible accuracy. He noted an unaccounted portion, amounting to less than 1 percent, which a century later was found to be the gas argon. In 1798, Cavendish measured the density and mass of Earth in an experiment that demanded mind-blowingly precise measurements with very basic equipment. Cavendish died in 1810.

Key works

1766 "Three Papers Containing Experiments on Factitious Air"
1784 "Experiments on Air"
1798 "Experiments to Determine the Density of Earth"

THIS AIR OF EXALTED NATURE

OXYGEN AND THE DEMISE OF PHLOGISTON

IN CONTEXT

KEY FIGURE
Joseph Priestley (1733–1804)

BEFORE
1674 British physiologist John Mayow theorizes about "nitro-aerial particles" that circulate in the blood after we inhale, predating the discovery of oxygen by 100 years.

1703 German chemist Georg Ernst Stahl proposes phlogiston theory, based partly on the earlier work of Johann Becher.

1754 Joseph Black identifies "fixed air," carbon dioxide.

1766 Henry Cavendish isolates and identifies "inflammable air," hydrogen.

1772 Scottish chemist Daniel Rutherford discovers nitrogen.

AFTER
1783 Antoine Lavoisier reveals that water is not an element but a compound of hydrogen and oxygen.

The discovery of oxygen in the 1770s—which brought about the final demise of the phlogiston theory of how things burn—was a hugely important turning point in chemistry. Historically, this breakthrough has been attributed to Joseph Priestley, but two other chemists—Swede Carl Scheele and French chemist Antoine Lavoisier—also have a strong claim.

In his experiments of 1774, first Priestley heated mercury slowly to create a red "calx" (mercuric oxide). Then, using a magnifying glass to focus sunlight, he heated the calx and collected the emitted gas, which we now know to be oxygen. To his surprise, this gas made a candle in a jar burn vigorously and made hot charcoal glow brilliantly.

> All the facts of combustion … are explained in a much simpler and much easier way without phlogiston than with it.
> **Antoine Lavoisier**

Decoding dephlogistication

The discovery of this gas was, for Priestley, the final proof that air is not an element but a mix of gases. According to phlogiston theory, however, a candle burning in a jar transfers its phlogiston to the air around it—the air in the jar becomes so "phlogisticated" that the candle soon stops burning. Priestley assumed his new gas burned brightly because it was giving up its phlogiston, so he later called it "dephlogisticated air." He also discovered that this gas helped a mouse survive for longer when trapped in a jar—and when Priestley inhaled the gas himself, he gained a sense of health and well-being.

Meanwhile, in Paris, Lavoisier found that substances such as phosphorus and sulfur gain weight when heated. This seemed to contradict the idea that they were losing phlogiston. In October 1774, on a brief tour of Europe, Priestley happened to meet Lavoisier in

See also: Gases 46 ▪ Phlogiston 48–49 ▪ Fixed air 54–55 ▪ Inflammable air 56–57 ▪ Early photochemistry 60–61 ▪ Conservation of mass 62–63

Paris and mentioned his discovery of dephlogisticated air. This inspired Lavoisier to start experimenting with calx. When Lavoisier heated a measured volume of air together with mercury to make mercuric calx, he knew how much air was consumed. When he reheated the calx by itself, it changed back to mercury and produced a gas—equal in volume to that lost earlier.

Lavoisier could see that when something is burned or heated, it did not lose phlogiston at all but combined with something in the air. For him, that could only mean that the old phlogiston theory of combustion no longer made sense. He soon realized, too, that the something in the air was Priestley's dephlogisticated air and that this was an entirely separate element. He dubbed it "oxygen," or acid maker, because he could detect it in most acids.

The dispute over who first discovered oxygen is even more complicated because Scheele, working in Uppsala, Sweden, is thought to have isolated oxygen—which he called "fire gas"—earlier than both Priestley and Lavoisier. However, he did not publish this until 1777, some time after Priestley had gone to print. Scheele apparently wrote a letter to Lavoisier about his discovery just before Priestley and Lavoisier met in Paris in 1774, but Lavoisier claimed he never received it. Despite his part in the oxygen story, Priestley continued to explain its existence and function through a version of phlogiston theory. But the scientific community agreed with Lavoisier, leaving Priestley isolated. Lavoisier went on to realize the full significance of the new element and its role in a new way of thinking about chemistry. ■

Priestley's mouse experiment

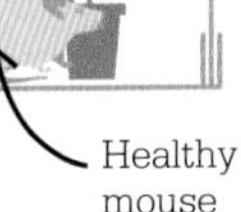

Experiment 1: Priestley placed a burning candle and a healthy mouse in a jar. The candle flame used up the oxygen in the jar, and the mouse died a few seconds later.

Experiment 2: Priestley placed a plant in a jar full of "used-up" air. Seven days later, he put a mouse in the jar and saw that it remained active for "many minutes."

Joseph Priestley

Born in 1733 near Leeds in the UK, Joseph Priestley was precociously talented as a youngster. He became an energetic believer in rational analysis of the natural world and pursued scientific research all his life. He was a member of the Royal Society and the Lunar Society of inventors and thinkers. Priestley wrote an important early book about electricity, invented carbonated water, and discovered several other gases besides oxygen. His unorthodox religious writings and his support for the American and French Revolutions so inflamed some that a mob destroyed his home, and he was forced to flee Britain in 1794. He settled in the US and continued his research until his death in 1804.

Key works

1772 *Directions for Impregnating Water with Fixed Air*
1774–1786 *Experiments and Observations on Air*

I HAVE SEIZED THE LIGHT

EARLY PHOTOCHEMISTRY

IN CONTEXT

KEY FIGURE
Carl Wilhelm Scheele (1742–1786)

BEFORE
1604 Italian alchemist Vincenzo Casciarolo discovers the "Bologna stone," a stone that glows in the dark.

1677 German alchemist Hennig Brand discovers a new element, phosphorus, which glows in the dark—the origin of the term phosphorescence.

AFTER
1822 Joseph Nicéphore Niépce makes the first photograph.

1852 British scientist George Stokes discovers "fluorescence," the way some substances glow under UV light.

1887 German physicist Heinrich Rudolf Hertz discovers the photoelectric effect.

1896 French physicist Henri Becquerel discovers radioactivity.

One of the most remarkable achievements of Swedish chemist Carl Scheele was the key role he played in pioneering photochemistry, which would eventually lead to the invention of photography.

The effect of light on chemicals was first noted by German alchemist Christian Adolf Balduin in 1674. He saw that calcium nitrate exposed to light glows in the dark afterward. This shows phosphorescence is caused by the slow re-emission of light absorbed by atoms.

In 1717, German anatomist Johann Schulze tried to recreate Balduin's results with chalk and nitric acid. To his surprise, his sample turned dark violet when exposed to sunlight—and on investigation, he found that this was due to contamination by traces of silver. Schulze went on to show that silver salts turn black when exposed to light.

> To explain new phenomena, that is my task.
> **Carl Scheele**
> ***Scheeles nachgelassene Briefe und Aufzeichnungen*** **(1892)**

Fixing the image

Six decades later, in 1777, Scheele's experiments also showed that one of the silver salts, silver chloride, turned black in sunlight. Scheele wanted to know why this happened and discovered that light produced a chemical reaction that turned silver chloride back into silver. He then made another crucial discovery: ammonia would dissolve away unexposed silver chloride but not areas of black silver. This "fixed" any exposed part of an image made with silver salts. While Scheele's work gave him all the ingredients to make a photograph, it was left to later inventors to take that extra step.

In the 1790s, British inventor Thomas Wedgwood was intrigued by the camera obscura, a device that used a lens to project an image of a view inside a box. Wedgwood wondered if he could find a way to permanently capture the image.

See also: Attempts to make gold 36–41 ▪ Catalysis 69 ▪ Photography 98–99 ▪ Flame spectroscopy 122–125 ▪ Green fluorescent protein 266

An optical camera obscura was often used as a drawing aid from the 17th century onward. Artists used it to trace out their pictures, enabling accurate portrayals of perspective.

Wedgwood tried silver salts and managed to create silhouettes—in which the outline of an object is created by placing it on silver salts and exposing it to sunlight. He was unaware of Scheele's ammonia fix, so his images vanished when hit by light. The image was also negative, with areas exposed to light turning black and shadows staying light.

Creating a positive

In the 1820s, French inventor Joseph Nicéphore Niépce created the first permanent photographs using light-sensitive lavender oil and bitumen coated on pewter plates rather than silver, but the picture quality was poor. To create the first successful photographic process in 1839, French entrepreneur Louis Daguerre returned to using silver salts. He found that when a metal plate coated with iodized silver is exposed to light, a positive latent image is created, with blacks and whites where they should be. This latent image could then be "developed" by exposure to mercury vapors. However, the process had to be halted at exactly the right moment by quickly rinsing with saltwater; otherwise, the entire image would go black. This technique—which was named the daguerreotype after its inventor—was successful and ushered in the age of photography. ■

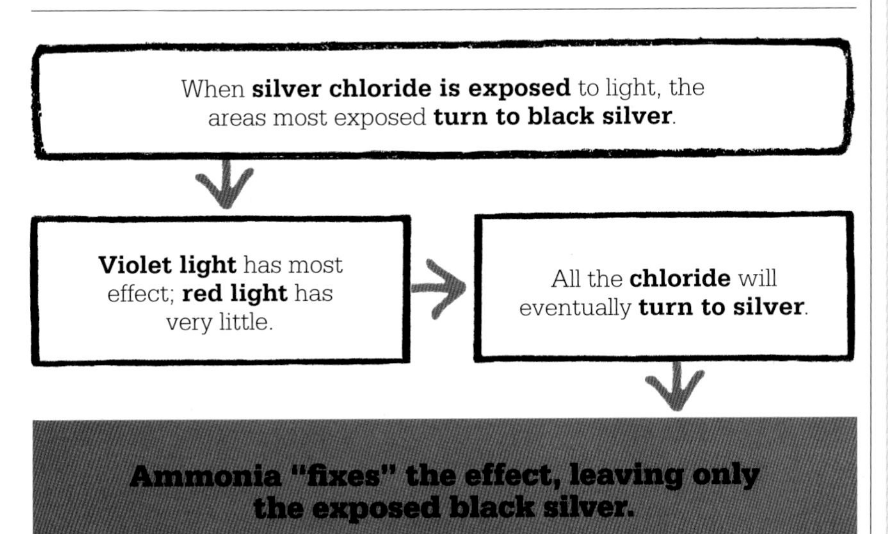

When silver chloride is exposed to light, the areas most exposed turn to black silver.
Violet light has most effect; red light has very little.
All the chloride will eventually turn to silver.
Ammonia "fixes" the effect, leaving only the exposed black silver.

Carl Wilhelm Scheele

Born in Stralsund, western Pomerania (now Germany), in 1742, Carl Scheele relocated to Sweden at age 14 to train as a pharmacist and stayed for the rest of his life. He lived in both the capital Stockholm and Uppsala, all the while pursuing his own research in chemistry. His experiments led to the discovery of chlorine; manganese; and, most famously, oxygen.

Working in many fields of chemistry, Scheele counted among his achievements the discoveries of organic acids tartaric, oxalic, uric, lactic, and citric, as well as hydrofluoric, hydrocyanic, and arsenic acids. He also developed a means of mass-producing phosphorus, which helped Sweden become one of the world's leading producers of matches.

Scheele died in 1786, most likely from the effects of contact with hazardous substances, such as arsenic.

Key work

1777 *Chemical Treatise on Air and Fire*

IN ALL THE OPERATIONS OF ART AND NATURE, NOTHING IS CREATED

CONSERVATION OF MASS

IN CONTEXT

KEY FIGURE
Antoine Lavoisier (1743–1794)

BEFORE
c.450 BCE The Greek thinker Empedocles argues that nothing comes into being out of nothing (translated into Latin as *nihil ex nihilo*) or is destroyed.

1615 French chemist Jean Beguin publishes the first-ever chemical equation.

1754 Scottish chemist Joseph Black discovers carbon dioxide by heating magnesium carbonate and observes that the resulting magnesium oxide weighs less than the original compound.

AFTER
1803 British physicist and chemist John Dalton establishes his atomic theory.

1905 Albert Einstein proposes the theory of mass–energy equivalence.

French chemist Antoine Lavoisier is sometimes referred to as the father of modern chemistry. With his rigorous, systematic approach, he transformed chemistry from a qualitative (descriptive) subject to a quantitative science—founded on precise measurements, with equations at the heart. One key contribution was the principle on which chemical experiments have been based ever since: the conservation of mass.

This principle states that, while matter can take different forms, it cannot be created or destroyed. It can be burned, dissolved, or pulled apart, but the total amount does not change. In a chemical experiment where nothing can enter or escape, the mass of the end products is always the same as the mass of the original reagents (substances added to cause a reaction).

We must always suppose an exact equality between the elements of the body examined and those of the products of its analysis.
Antoine Lavoisier
Elements of Chemistry **(1789)**

The concept was not entirely new: the idea that "nothing comes from nothing" was important in ancient Greek philosophy. By the 18th century, the principle of the conservation of mass was widely assumed by chemists, and in 1756, Russian polymath Mikhail Lomonosov tried to demonstrate it experimentally. But it was Lavoisier who established it as a fundamental truth. He weighed and measured the reactants (consumed substances) and products in his experiments and kept meticulous balance sheets.

Phlogiston and air

Following the discoveries of different types of "air" by British chemists such as Henry Cavendish and Joseph Priestley, Lavoisier wanted to know what was happening as these airs were generated or absorbed. The prevailing phlogiston theory held that when metal burns, rusts, or tarnishes, for instance, it gives up its

See also: The atomic universe 28–29 ▪ Oxygen and the demise of phlogiston 58–59 ▪ Compound proportions 68 ▪ Dalton's atomic theory 80–81 ▪ Atomic weights 121 ▪ The mole 160–161

This solar furnace, made from two huge magnifying lenses to focus the Sun's heat, was designed by Lavoisier to avoid contaminating his experiments with the products of fuel combustion.

phlogiston. If so, it should lose weight. Yet scientists knew that metals gain weight when they rust.

In 1772, Lavoisier performed several experiments using magnified heat from the Sun. In one, he heated a "calx" (what we now know as an oxide) of lead (PbO) with charcoal inside a jar. As the heated calx turned into metal, Lavoisier saw it release a huge amount of air into the jar. If a calx releases air when becoming metal, he wondered, perhaps then, when a metal changes to a calx, it absorbs air. He asked whether this is why it gains weight. He also found that phosphorus and sulfur gained weight when they burned and wanted to know if they, too, could be taking in air.

Lavoisier sent a sealed note to the French Academy of Sciences to establish his claim to this radical new theory. But proving it was more difficult than he expected, and he performed hundreds of carefully quantified experiments, assisted by his wife Marie-Anne.

Proof and explanation

In one key experiment, Lavoisier heated a glass flask containing chunks of tin until the tin turned to a calx. When he broke open the flask and weighed the calx, he found it was a hundredth of an ounce heavier than the tin had been. That tiny extra weight could only have come from the air in the flask.

Lavoisier had confirmed his theory of conservation of mass, but he did not know whether all the air was involved. Then, in late 1774, Priestley visited him and mentioned his discovery of "dephlogisticated air." This was what was combining with or being released from other elements. Lavoisier named this new gas "oxygen" (O_2). ■

Antoine Lavoisier

Born into a wealthy Parisian family in 1743, Antoine Lavoisier studied law at the University of Paris but turned to science after graduating. He published his first scientific paper at the age of 21 and was elected to the elite French Academy of Sciences at just 26 years old. The same year, he bought a share in tax-collecting corporation Ferme Générale. In 1771, he married 13-year-old Marie-Anne Paulze, who became a capable laboratory assistant.

Among his achievements, Lavoisier named the elements oxygen, hydrogen, and carbon, and he identified sulfur. He discovered the role of oxygen in combustion and respiration, disproved the phlogiston theory, and established a system for chemical nomenclature. But his work as a tax collector made him a target during the French Revolution, and he was guillotined on May 8, 1794.

Key works

1787 *Method of Chemical Nomenclature*
1789 *Elements of Chemistry*

I DARE SPEAK OF A NEW EARTH

RARE-EARTH ELEMENTS

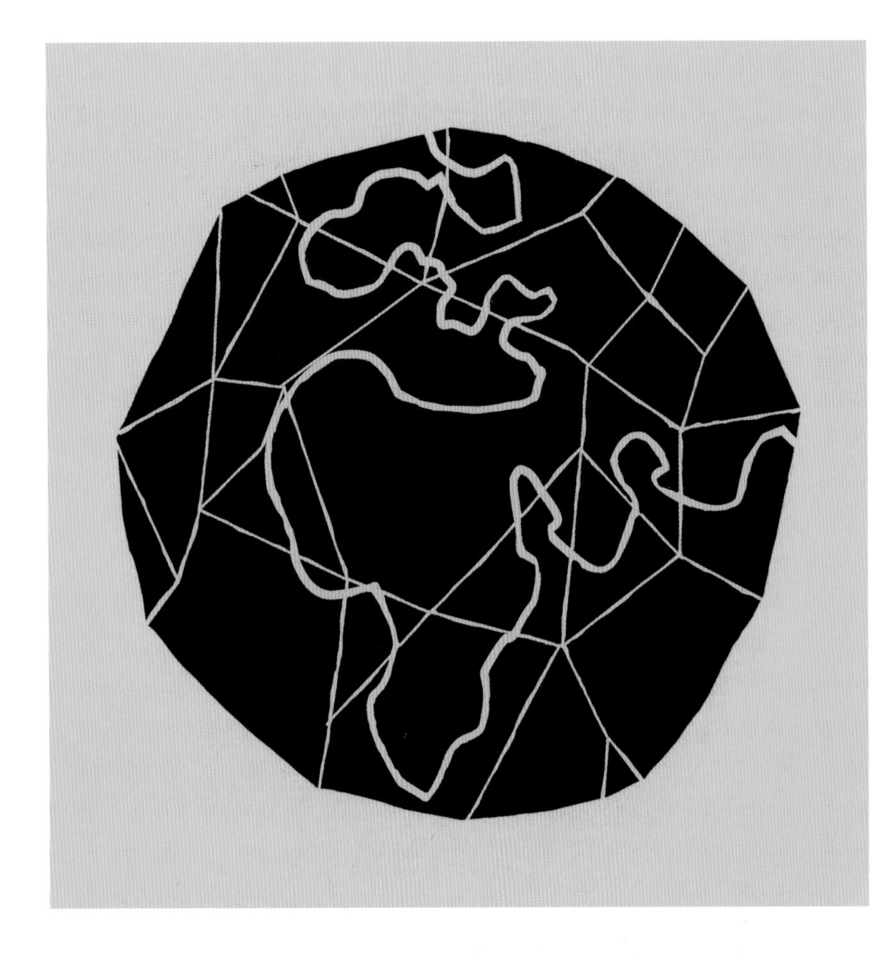

IN CONTEXT

KEY FIGURE
Johan Gadolin (1760–1852)

BEFORE
1735 Swedish chemist Georg Brandt discovers that cobalt is a metal.

1774 Carl Scheele and others isolate the metal manganese.

1783 Spanish chemists and brothers Juan and Fausto Elhuyar discover tungsten.

AFTER
1952 The Mountain Pass rare-earth mine in California begins production, extracting europium for color TV sets.

1984 General Motors and Sumitomo Special Metals simultaneously develop the neodymium magnet, the world's strongest.

1988 China becomes the world's largest producer of rare earths.

The remarkable group of elements called the rare earths are the 15 "lanthanides" plus scandium and yttrium. The first one was discovered by Finnish chemist and mineralogist Johan Gadolin in 1794, but it took almost 150 years to isolate and identify all 17. These silvery metals, which are chemically very alike, are in group three of the periodic table. They have special qualities, including magnetism, conductivity, and luminescence, which make them extremely useful when combined with other metals. For this reason, they have become vital to modern technology—from smartphones to electric cars.

See also: Extracting metals from ores 24–25 ▪ Isolating elements with electricity 76–79 ▪ Electrochemistry 92–93 ▪ Flame spectroscopy 122–25 ▪ The periodic table 130–37 ▪ X-ray crystallography 192–93 ▪ Nuclear fission 234–37

It seems to me rather fatal if each of the new earths should only be found at one site or in one mineral.
Johan Gadolin

The rare earths were initially discovered not as pure metals but as components of oxides, which 18th-century chemists called "earths." They are actually quite geologically abundant—some are as common as lead or copper—but they are never found in high concentrations. Instead, they are so deeply intermingled with other minerals and with other rare earths that they are very hard to find and extract. This is why they acquired the name "rare" and why the journey to their discovery was so long and arduous.

The black rock of Ytterby

In 1787, Swedish army lieutenant Carl Arrhenius, an avid mineralogist, was exploring a feldspar mine on the Swedish island of Resarö when he came across a black lump of rock unlike anything he had ever seen. He wondered if it might contain the dense metal tungsten and passed it on to the inspector of mines in Stockholm, Bengt Geijer. Geijer ran some tests and announced the discovery of a new heavy mineral, calling it "ytterbite" after Ytterby, a village near the mine. He sent it to Johan Gadolin for detailed analysis.

Gadolin dissolved the ground-up stone in various chemicals, including nitric acid (HNO_3) and sodium hydroxide (NaOH), carefully testing and measuring the products. In 1794, he published the results of his analysis, finding the black stone to be 31 parts silica (SiO_2), 19 parts aluminum oxide (Al_2O_3), 12 parts iron oxide (Fe_2O_3), and 38 parts

A sample of ytterbite—the black rock of Ytterby—was analyzed by Johan Gadolin in 1794. It contained the previously unknown oxide "yttria."

an unknown earth. This unknown earth not only was very dense but also had a very high melting point—4,397°F (2,425°C), as we now know. While it dissolved easily in most acids and had similarities to some other metal oxides, the earth was clearly a new substance.

In 1797, Swedish chemical analyst Anders Ekeburg refined the results and named the new earth "yttria." When he heard of French mineral analyst Louis Vauquelin's »

Johan Gadolin

Born in 1760 in Åbo (now Turku) in Finland, Johan Gadolin was the son of a professor of physics. He studied first math and later chemistry at Åbo University, then moved to Uppsala University, Sweden, where he began his work on mineralogy. In 1785, at age 25, he became professor of chemistry at Åbo and, in 1786, went on a grand tour of Europe, visiting mines and meeting a number of prominent chemists.

Gadolin published key studies on specific heat, was an early supporter of Antoine Lavoisier's disproof of the phlogiston theory, and determined the chemistry of the pigment Prussian blue. He is most famous, though, for his analysis of the black rock of Ytterby. Sadly, the great fire of Åbo in 1827 destroyed his laboratory and his incomparable mineral collection, ending his scientific career. He died in 1852, at age 92.

Key works

1794 "Examination of a Black, Dense Mineral from the Ytterby Quarry in Roslagen"
1798 *Introduction to Chemistry*

discovery of beryllium the following year, Ekeburg realized that ytterbite contained beryllium, not aluminum. German analytical chemist Martin Klaproth confirmed Ekeburg's findings and renamed ytterbite "gadolinite," in Gadolin's honor.

The concept of an element was still quite vague at this time. Yttria was an oxide, and it was another 30 years before German chemist Friedrich Wöhler managed to isolate the pure metal yttrium. But Gadolin's analysis marked the discovery of the first rare earth.

Cerium and hidden earths

In 1803, Swedish chemists Jöns Jacob Berzelius and Wilhelm Hisinger and, independently, Martin Klaproth made another key breakthrough with another heavy mineral. This mineral was a reddish-brown lump found half a century earlier by Swedish chemist Axel Cronstedt in the Bastnäs mine in Sweden. Berzelius and Hisinger thought it might contain yttria, but their analysis revealed a new oxide, which Berzelius called "ceria" after the recently discovered asteroid Ceres. The mineral in which it was found was named cerite.

As with yttria, it took decades of hard work to isolate the pure metal cerium. In 1875, American chemists William Hillebrand and Thomas Norton finally achieved this by passing an electric current through molten cerium chloride ($CeCl_3$).

After the discovery of yttria and ceria, chemists gradually realized that other elements were intermingled with these two earths. Separating them out, however, was an exacting task. The development of electrochemistry helped, but it was primarily a matter of painstaking analysis with acids, blowpipes that boosted the heat of a candle flame to furnace temperatures, and fractional crystallization—when a melted mix cools, its components crystallize at different stages, due to the differences in their solubility.

In 1839, Swedish chemist Carl Mosander, a colleague and former student of Berzelius, used nitric acid to separate out a second earth from ceria, which Berzelius called "lanthana." Then, in 1842, Mosander found a third earth in ceria, "didymia," as well as two new earths mingled with yttria: pinkish "terbia" and yellowish "erbia," both also taking their names from Ytterby. This meant six rare earths were now known.

Adding to the list

Confusingly, some chemists running similar analyses believed that they had found elements, not oxides, and this led to many later discredited claims for discoveries of rare-earth elements. Indeed, in the case of didymia, Mosander was mistaken in believing that it was a pure metal oxide. This error came to light decades later, after the technique of flame spectroscopy was pioneered in 1860. Each substance glows with its own

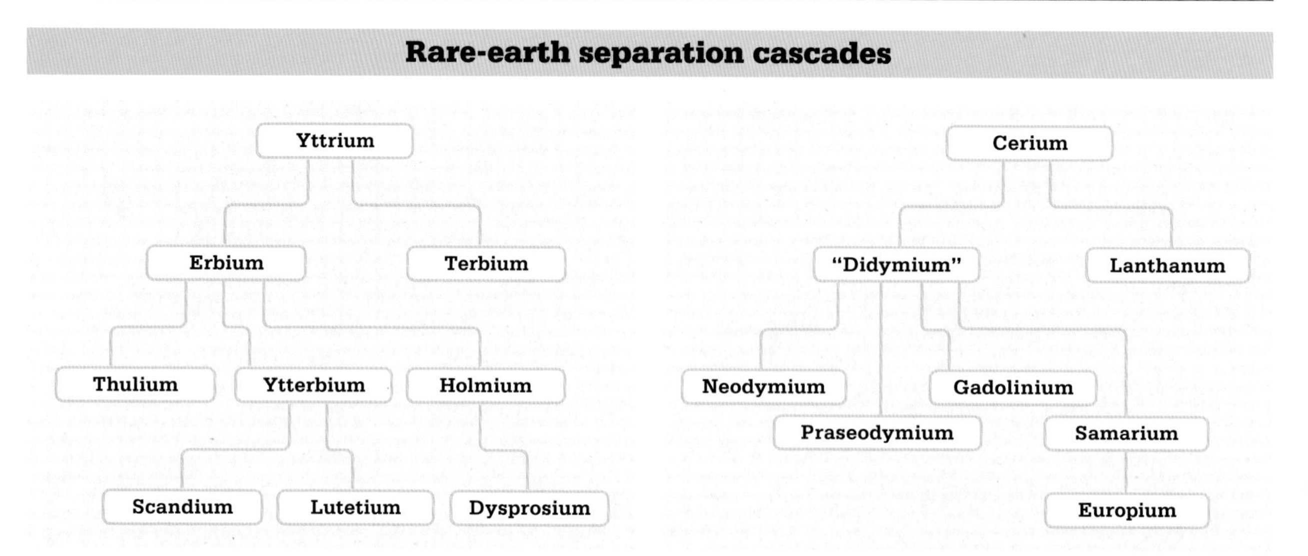

The first 16 rare earths were discovered in two "separation cascades," with each element revealed in the form of its oxide. Yttrium oxide ("yttria") was found to be intermixed with eight other rare-earth oxides.

Cerium oxide ("ceria") turned out to be intermingled with six other rare-earth oxides. The 17th rare earth, promethium, is radioactive and no longer exists naturally; it was created artificially in 1945.

unique spectrum of colors when heated intensely, and this often revealed new substances, even if it took some time to identify them through chemical processes.

Flame spectroscopy raised suspicions in several chemists that what had been thought to be the element "didymium" was actually a mixture of at least two elements. In 1885, Austrian chemist Carl Auer von Welsbach identified neodymium and praseodymium. Then, in 1886, French chemist Paul-Émile Lecoq de Boisbaudran isolated gadolinium.

In 1878, Swiss chemist Jean Charles Galissard de Marignac had separated erbia to find "ytterbia," and Swedish chemist Per Teodor Cleve separated out "holmia" and then, in 1879, "thulia." From holmia, de Boisbaudran extracted "dysprosia" in 1886.

Meanwhile, another two minerals, samarskite and euxenite, had been identified as possible sources of rare earths. In 1879, de Boisbaudran isolated "samaria" from some didymia that had been extracted from samarskite. In the same year, from euxenite, Swedish chemist Lars Fredrik Nilson isolated first erbia, then ytterbia, then a new earth from the ytterbia, "scandia." In 1901, French chemist Eugène-Anatole Demarçay isolated "europia" from samaria. Finally, in 1907, French chemist Georges Urbain was the first to report his extraction of "lutetia," from ytterbia.

Today's smartphones contain several rare earths, many of which are used in the screen to produce color and brightness. Smartphone electronics make use of the high conductivity of rare earths.

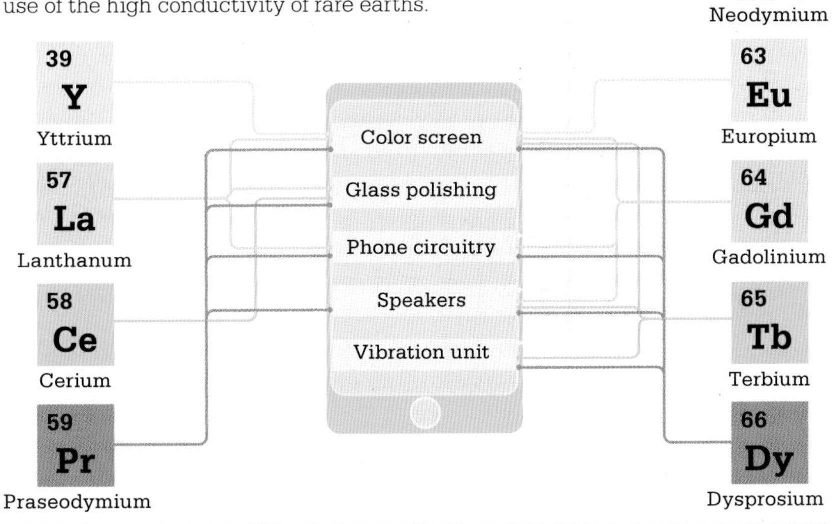

The missing rare earth

There were now 16 rare earths, but just how many more could there be? In 1913, British physicist Henry Moseley used X-ray spectroscopy to pin down atomic numbers for each so they could be slotted into the periodic table. It turned out that there was just a single gap between neodymium and samarium, with atomic number 61. In 1926, scientists in Florence, Italy, claimed to have discovered the missing element and named it florentium, while scientists in Illinois made the same claim, naming it illinium.

Neither team was right. Element 61 is radioactive, decaying far too quickly to have survived since Earth was formed, so it could only occur naturally—extremely rarely—as a product of other radioactive elements. It was finally created artificially in 1945, by scientists who worked on the Manhattan Project to create the atomic bomb during World War II. American chemists Jacob Marinsky, Charles Coryell, and Lawrence Glendenin created the element from the fission products of uranium in a nuclear reactor. They published their results in 1947 and named this final rare earth promethium. ■

The rare-earth revolution

Ion-exchange chromatography (IEC), the technique used to isolate promethium, was developed by American scientists Frank Spedding and Jack Powell while working on the Manhattan Project. They needed to find a way of getting rid of impurities in uranium such as yttrium, which ruined the nuclear chain reaction. While previous separation techniques had produced only tiny samples of rare earths, IEC could separate them en masse. This discovery sparked a rare-earth revolution from the early 1950s. For the first time, these elements were available on an industrial scale.

Compared with other metals, the quantities of rare earths produced are minuscule and the extraction process very expensive, but they have become indispensable to the modern world: neodymium magnets are essential to electric vehicles and wind turbines, while yttrium, erbium, and terbium are used in visual-display devices.

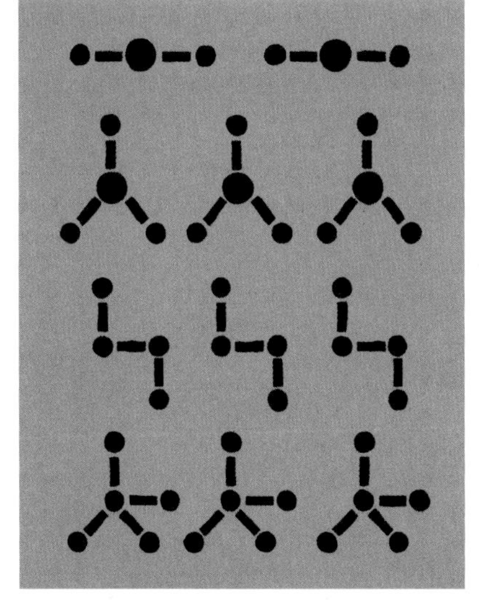

NATURE ASSIGNS FIXED RATIOS

COMPOUND PROPORTIONS

IN CONTEXT

KEY FIGURE
Joseph Proust (1754–1826)

BEFORE
1615 French chemist Jean Beguin writes the first chemical equation.

1661 Robert Boyle identifies differences between mixtures and compounds and shows that a compound can have different properties from those of its constituents.

1718 French chemist Étienne Geoffroy helps codify the idea of a compound in his table of affinities.

AFTER
1808 British chemist John Dalton publishes his atomic theory of elements, explaining how compounds are always made of the same combination of atoms.

1826 Swedish chemist Jöns Jacob Berzelius publishes the first table of atomic weights.

Just as Antoine Lavoisier was pinning down what an element was, another French chemist, Joseph Proust, proposed a fundamental truth about compounds. In 1794, he introduced the law of definite proportions: chemicals only truly combine to form a limited number of substances and always combine in the same proportions by mass. This is what gives us fixed chemical formulas for compounds.

Unique proportions

Proust was interested in the ways metals combined with oxygen, sulfur, and carbon: oxides, sulfides, sulfates, and carbonates. His experiments with oxides showed two distinct proportions, which he called minimum and maximum, and a single proportion with sulfides. In other words, a metal such as iron can combine with sulfur or with oxygen in only one or two ways and always in the same unique proportions.

The concept was attacked by French chemist Claude-Louis Berthollet, who argued that chemicals can come together in a spectrum of different proportions. The issue hinged on how a compound is defined. Berthollet included what we now call mixtures and solutions, which can indeed join in infinitely varying proportions. Proust's "true combination" is our modern definition of a compound: elements bonded chemically and, unlike mixtures and solutions, separable only by a chemical reaction, not physically. ■

Rust is the result of iron joining with oxygen and water in fixed proportions to form hydrous iron oxide. The equation for this reaction is: $4Fe + 3O_2 + 6H_2O = 4Fe(OH)_3$.

See also: Corpuscles 47 ▪ Oxygen and the demise of phlogiston 58–59 ▪ Conservation of mass 62–63 ▪ Dalton's atomic theory 80–81

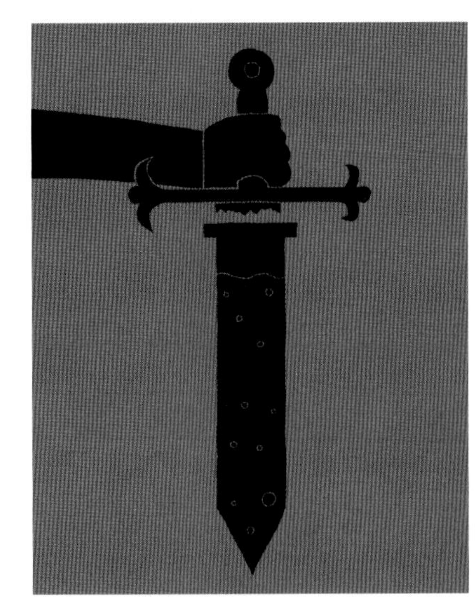

CHEMISTRY WITHOUT CATALYSIS WOULD BE A SWORD WITHOUT A HANDLE

CATALYSIS

IN CONTEXT

KEY FIGURE
Elizabeth Fulhame
(active 1794)

BEFORE
1540 German pharmacologist Valerius Cordus uses sulfuric acid to catalyze the conversion of alcohol to ether.

1781 French pharmacist Antoine Parmentier notes how vinegar stimulates the creation of sugars in potato starch mixed with cream of tartar.

AFTER
1810 Fulhame's book is republished in America, where it finds respect among chemists.

1823 German chemist Johann Wolfgang Döbereiner observes how manganese dioxide speeds up the decomposition of potassium chlorate.

1835 Jöns Jacob Berzelius coins the term catalysis.

Catalysts speed up a chemical reaction by lowering the energy needed to get the reaction going, and they are central to a huge range of everyday processes—from cleaning car exhausts and making plastic to the work of enzymes (biological catalysts) in living organisms. They provide an easier route for the reaction, known as the reaction pathway, without actually becoming involved.

The term catalyst was coined by Jöns Jacob Berzelius in 1835, but one of the first key studies was made by Elizabeth Fulhame in Scotland half a century earlier. The wife of a doctor who studied chemistry with Joseph Black, Fulhame wanted to find how to dye cloth with gold, silver, and other metals. Using different reducing agents, such as hydrogen, Fulhame tested how salts of various metals could be reduced to pure metal. She discovered that many reductions previously thought to need heat can happen at room temperature if water is present to act as the catalyst.

> When combustion occurs, one body, at least, is oxygenated, and another restored … to its combustible state.
> **Elizabeth Fulhame**

Fulhame described the reaction of silver salts when catalyzed by light, contributing to an understanding of chemical processes that would eventually lead to the development of photography. She also observed the catalytic role of oxygen in some reactions, rightly challenging both the old phlogiston theory and Antoine Lavoisier's new alternative. In 1793, Fulhame met Joseph Priestley, who encouraged her to publish her work, which she did the following year. ■

See also: Oxygen and the demise of phlogiston 58–59 ▪ Rare-earth elements 64–67 ▪ Photography 98–99 ▪ Enzymes 162–163

THE CHE
REVOLUT
1800–1850

MICAL
ION

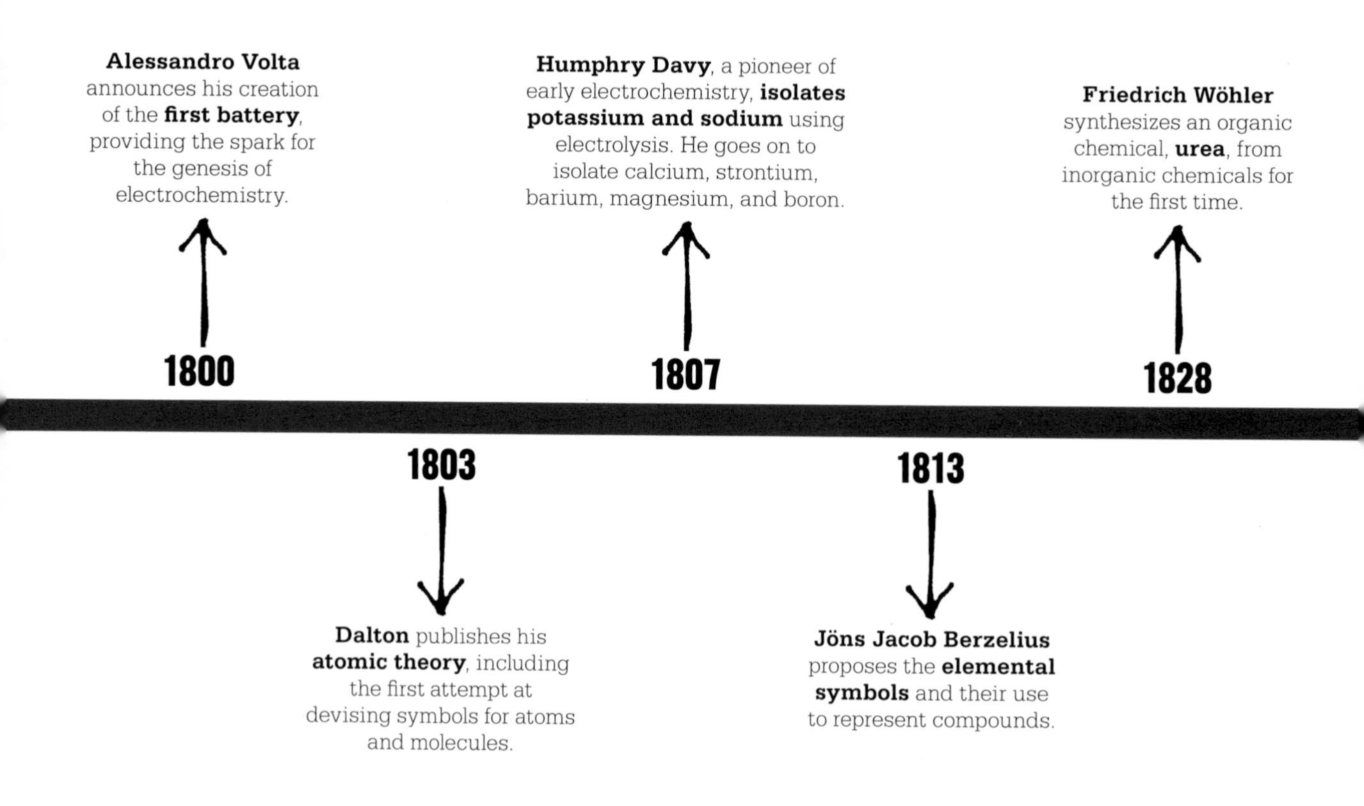

Alessandro Volta announces his creation of the **first battery**, providing the spark for the genesis of electrochemistry.
1800
Humphry Davy, a pioneer of early electrochemistry, **isolates potassium and sodium** using electrolysis. He goes on to isolate calcium, strontium, barium, magnesium, and boron.
1807
Friedrich Wöhler synthesizes an organic chemical, **urea**, from inorganic chemicals for the first time.
1828
1803
Dalton publishes his **atomic theory**, including the first attempt at devising symbols for atoms and molecules.
1813
Jöns Jacob Berzelius proposes the **elemental symbols** and their use to represent compounds.

The establishment of modern chemistry at the end of the 18th century was followed by an explosion in knowledge during the 19th century. The developments included the opening up of whole new fields of study in chemistry.

A common shorthand

The standardization of chemical nomenclature was matched by a revolution in how chemists thought about and represented atoms. The English chemist John Dalton was the first to suggest that atoms of different elements would have different masses and sizes. He also stated that, when elements combine, they do so in whole number ratios. By 1808, Dalton had produced a set of chemical symbols for known elements at the time, considered to be the first attempt to devise such a system.

Just a few years later, Swedish chemist Jöns Jacob Berzelius would suggest his own notation to represent elements with one or two letter symbols, a system still in use today. Now, as well as having a common nomenclature, chemists also had a common shorthand, making the communication of chemical advances much easier—all the better as significant new areas of chemistry were emerging.

The emergence of electrochemistry

Electrochemistry sparked into life at the turn of the 19th century with Alessandro Volta's announcement of the voltaic pile, the first battery. The union of chemistry and electricity would rapidly produce dramatic advances; less than a decade later, English chemist Humphry Davy would discover a number of new metals using electrolysis, splitting common compounds with electricity to isolate the new elements.

In 1813, Davy employed the young Michael Faraday as an assistant. Faraday would go on to become the leading scientist at the Royal Institution and build upon Davy's work with electricity. In the discipline of chemistry, Faraday formulated electrochemical laws that established a direct relationship between the size of an electrical current and the masses of products obtained through electrolysis. Through this work, he also formalized much of the terminology of electrochemistry, which is still used today.

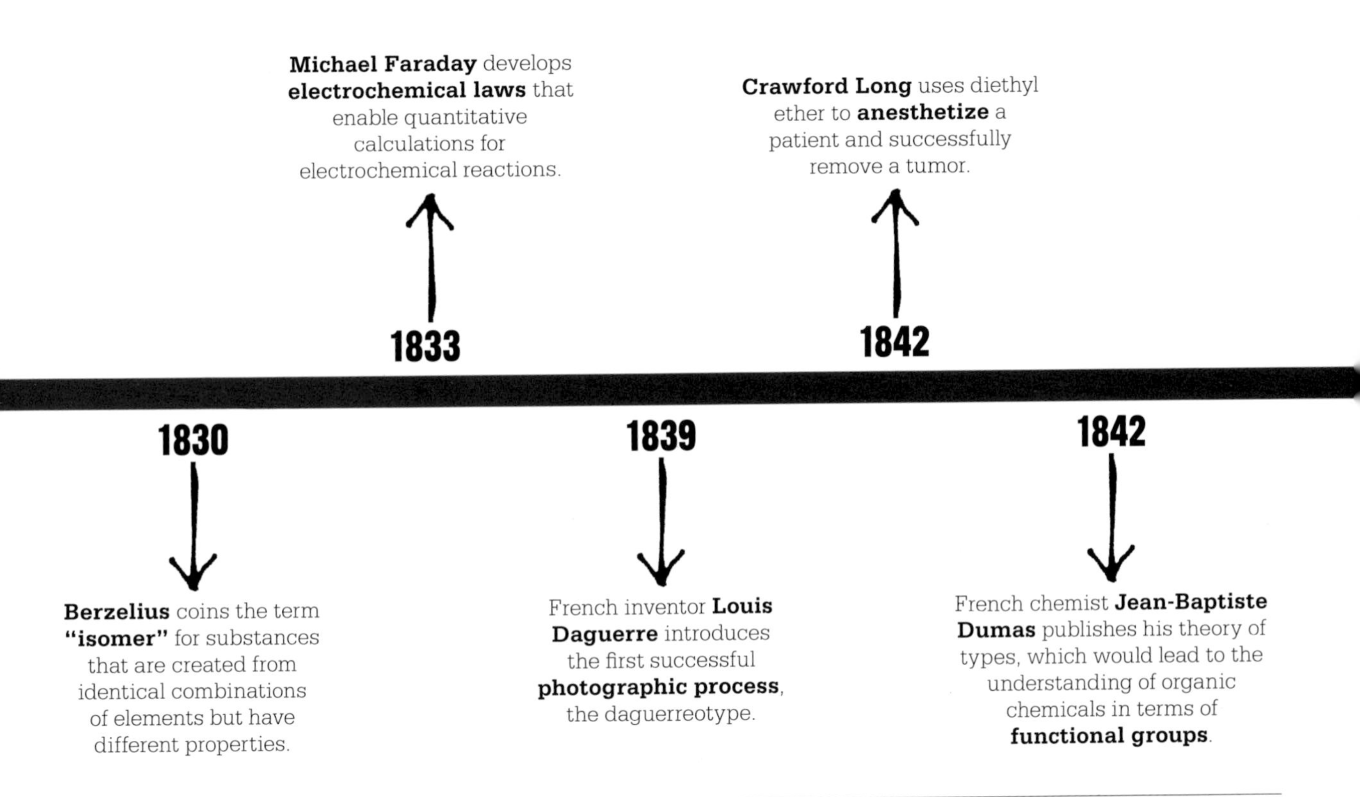

1833
Michael Faraday develops electrochemical laws that enable quantitative calculations for electrochemical reactions.
1842
Crawford Long uses diethyl ether to anesthetize a patient and successfully remove a tumor.
1830
Berzelius coins the term "isomer" for substances that are created from identical combinations of elements but have different properties.
1839
French inventor Louis Daguerre introduces the first successful photographic process, the daguerreotype.
1842
French chemist Jean-Baptiste Dumas publishes his theory of types, which would lead to the understanding of organic chemicals in terms of functional groups.

Beyond vitalism

Faraday's discoveries in chemistry were not limited to those involving electricity. He created the first compounds of carbon and chlorine in 1820, and he was also the first to isolate and identify benzene, in 1825. These were among the earliest forays into organic chemistry, an area that had been classified as distinct from inorganic chemistry by Berzelius in 1806.

Most of the chemists working at this time subscribed to the concept of vitalism. This theory was based on the concept that organic chemicals found in living things could not be synthesized from nonliving (inorganic) compounds. However, in 1828, German chemist Friedrich Wöhler showed that urea, an organic compound, could be synthesized from two inorganic compounds: ammonia and cyanic acid. While the true significance of Wöhler's synthesis was not fully appreciated for several decades afterward, the discovery is often cited as marking the end of the vitalism theory.

Berzelius and Wöhler, together with German chemist Justus von Liebig, also discovered that some inorganic compounds could have identical elemental compositions but have different properties—a phenomenon that they termed isomerism. Later, other chemists would demonstrate that isomerism could be observed in organic compounds as well.

Berzelius, Wöhler, and von Liebig went on to propose the idea that organic compounds were formed from basic substances that they called "radicals." This concept was superseded by Jean-Baptiste Dumas's theory of "types" of organic molecules, which would in turn develop into the concept of functional groups of atoms. These structural patterns are known as "motifs": groups of atoms that give molecules their characteristic properties and reactions. The predictive power of functional groups would become an essential tool for chemists in understanding and predicting the myriad reactions of organic compounds.

In less than half a century, therefore, organic chemistry had gone from being barely recognized as a distinct area of chemistry to one of its fastest-growing fields. And in the decades to follow, it would generate some of chemistry's greatest advances, many of which still touch our lives today. ■

EACH METAL HAS A CERTAIN POWER

THE FIRST BATTERY

IN CONTEXT

KEY FIGURE
Alessandro Volta
(1745–1827)

BEFORE
1729 British dyer and experimenter Stephen Gray shows how electric charge can be transmitted over distance.

1745 German physicist Ewald Georg von Kleist and Dutch scientist Pieter van Musschenbroek independently invent Leyden jars, used for storing an electrical charge.

1752 American inventor and statesman Benjamin Franklin proves that lightning is actually electrical.

AFTER
1808 Humphry Davy invents electrochemistry.

1833 Michael Faraday lays out the laws of electrolysis.

1886 German scientist Carl Gassner invents the dry-cell battery.

During the late 18th century, theaters filled with people eager to see the sparks, flashes, and bangs of electrical effects created using giant static electricity generators or unleashed from Leyden jars that stored static electricity between two electrodes on the inside and outside of a jar. People even wondered whether electricity was the force of life—an idea envisioned in British novelist Mary Shelley's *Frankenstein* (1818).

The link between electricity and life seemed to be confirmed by Italian physicist Luigi Galvani's observations on dismembered frogs' legs in the 1780s. Galvani discovered that the legs twitched not only when the muscles were connected to a static machine or a metal surface during a thunderstorm but also when hung out to dry on a line suspended by a brass hook on an iron fence. Galvani believed the twitching was caused by "animal electricity" inherent in all animal muscles.

Competing experiments

Initially, fellow Italian physicist Alessandro Volta was convinced. He had been experimenting with electric effects since the 1760s and was already a leading authority on electricity. He believed Galvani had shown in his ingenious experiments that animal electricity was "among the demonstrated truths." But Volta gradually developed doubts, and in 1792 and 1793, he argued that the electricity that made the frogs' legs twitch was coming from an outside source—the contact between the brass and iron. Electricity could be created, he said, purely by chemistry and was coming from a chemical reaction between two metals.

This picture of Alessandro Volta by Italian painter Gasparo Martellini shows Volta demonstrating his experiments with the electric battery to Napoleon Bonaparte.

See also: Isolating elements with electricity 76–79 ▪ Electrochemistry 92–93

The apparatus … which will no doubt astonish you, is only the assemblage of a number of conductors of different kinds arranged in a certain manner.
Alessandro Volta

Galvani was not swayed, and both scientists began competing with experiments to settle the issue. After Galvani died in 1798, Volta was determined to prove his theory right but needed to find a way to make electrical charge more detectable.

The next year, Volta stacked alternating disks of copper and zinc and separated each pair with cardboard soaked in saltwater. The result was so dramatic that touching the stack gave Volta a mild shock. Adding more disks gave a bigger charge. Combinations of other metals such as silver and tin also gave different charges.

More importantly, Volta discovered that if he looped a wire from the top of the pile to the bottom, he could create a continuous flow of electricity that could be stopped simply by disconnecting the wire. If the wire was not connected, his pile held its charge. Unlike the Leyden jar, which released its store of electricity in one go, Volta's pile could supply a continuous electric current that could be turned on and off as needed. He had created the first battery, the voltaic pile, which he announced to the world in a letter to London's Royal Society in 1800.

The impact of the voltaic pile on the scientific world was profound. Within a year, chemists had used it to separate water into hydrogen and oxygen, and within a decade British scientist Humphry Davy had created the entirely new science of electrochemistry.

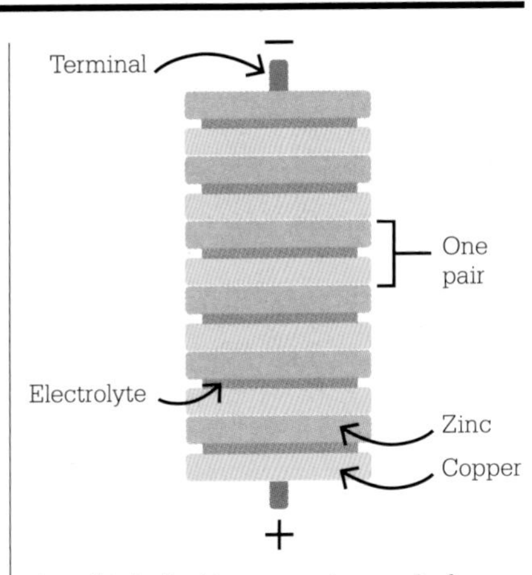

A voltaic battery can be made from a range of components, such as copper and zinc. The disks are stacked in pairs separated by a conductor (electrolyte).

For the next 30 years, until British chemist Michael Faraday and American engineer Joseph Henry discovered how to generate electricity magnetically, the voltaic pile was the main source of electricity, and it remains the forerunner of all the billions of batteries we use today in a range of devices, from mobile phones to electric vehicles. ■

Alessandro Volta

Born in Como, Italy, in 1745, Alessandro Volta was fascinated by electricity as a teenager. He wrote his first scientific paper on electricity in 1768, and in 1776, he discovered methane (natural gas). His experiments showed gases could be ignited in a chamber by an electrical spark, and he invented an electric gun called the voltaic pistol. He became renowned across Europe for his work on electricity. In 1779, he was made professor of physics at the University of Pavia, a post he held for 40 years. He also improved and popularized the electrophorus, a device invented by Swedish physicist Johan Wilcke that produced static electricity. But Volta is best known for his discovery that electricity could be made chemically and his invention of the battery. He was made a count by Napoleon in honor of his work. More than 50 years after his death in 1827, the volt—a unit of electromotive force—was named in his honor.

Key work

1769 "On the Attractive Force of Electric Fire"

ATTRACTIVE AND REPULSIVE FORCES SUSPEND ELECTIVE AFFINITY

ISOLATING ELEMENTS WITH ELECTRICITY

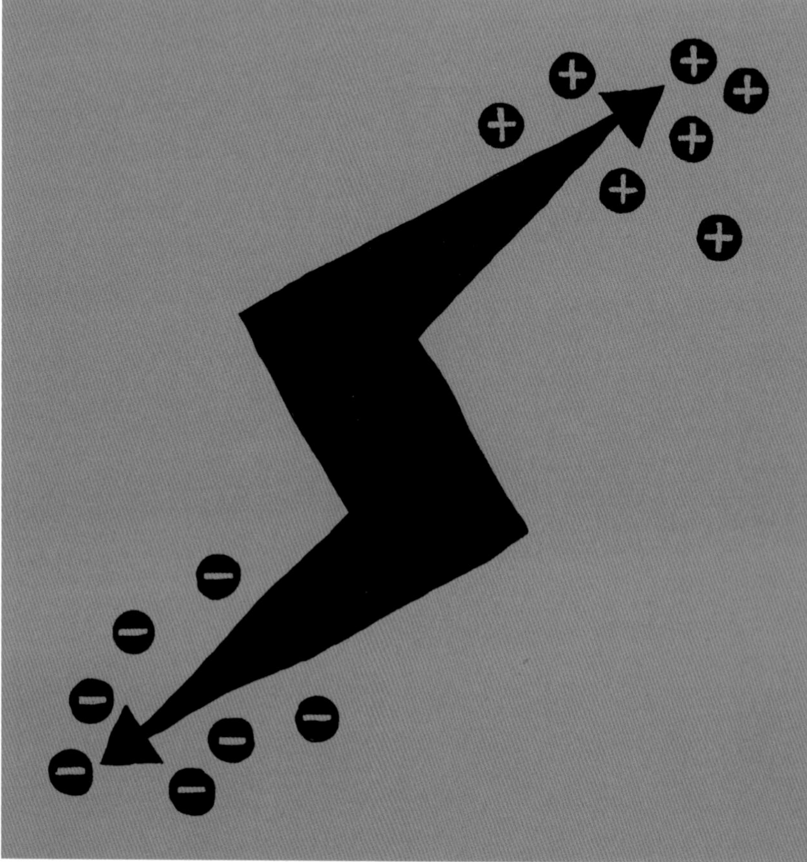

IN CONTEXT

KEY FIGURE
Humphry Davy (1778–1829)

BEFORE
1770s Antoine Lavoisier suggests that oxygen in compounds causes acidity.

1791 Luigi Galvani publishes his idea of "animal electricity."

1800 Alessandro Volta creates the first battery, the voltaic pile.

AFTER
1832 Michael Faraday establishes his two laws of electrochemistry.

1839 British physicist William Grove creates the first fuel cell, combining hydrogen and oxygen to produce water and electricity.

1866 French engineer George Leclanché invents the wet cell, the forerunner of the zinc–carbon battery.

Italian physicist Alessandro Volta's creation of the voltaic pile, the world's first battery, in 1800 had a profound effect on science. Volta's battery gave scientists a controlled current of electricity for experiments for the first time. It also revealed the fundamental link between electricity and chemical reactions and so launched an entire new branch of science: electrochemistry.

Within just a few weeks of Volta announcing his pile, British chemist William Nicholson asked the obvious question: if a chemical reaction can create electricity, can electricity produce a chemical reaction? On May 2, 1800, he and

See also: Extracting metals from ores 24–25 ▪ Oxygen and the demise of phlogiston 58–59 ▪ Conservation of mass 62–63 ▪ Rare-earth elements 64–67 ▪ The first battery 74–75 ▪ Electrochemistry 92–93

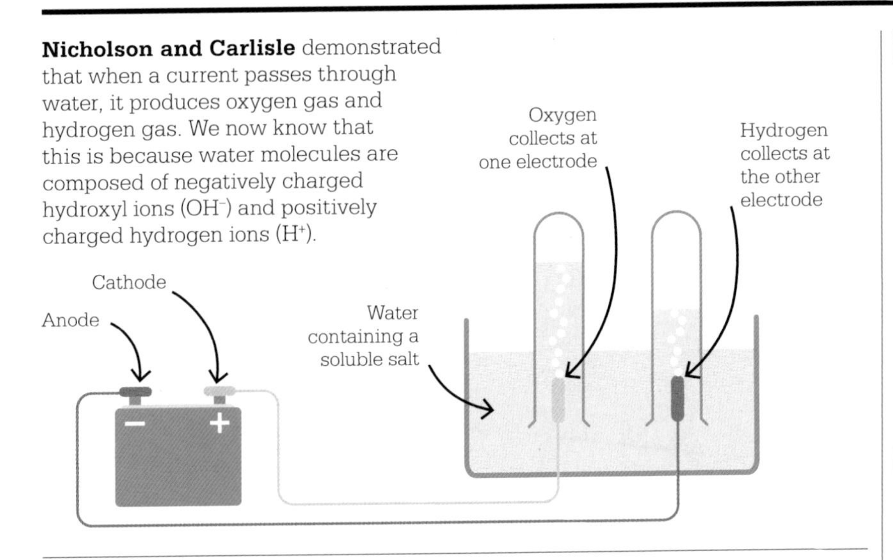

Oxygen collects at one electrode
Hydrogen collects at the other electrode

Water containing a soluble salt

Nicholson and Carlisle demonstrated that when a current passes through water, it produces oxygen gas and hydrogen gas. We now know that this is because water molecules are composed of negatively charged hydroxyl ions (OH^-) and positively charged hydrogen ions (H^+).

surgeon Anthony Carlisle proved that it can—in dramatic fashion. They inserted leads from a voltaic pile into water. Instantly, bubbles appeared at the electrodes (electric terminals) as the water split into hydrogen and oxygen. This was the first real example of electrolysis, the splitting of chemicals electrically.

Just months after Nicholson and Carlisle's breakthrough, German physicist Johann Ritter independently achieved the same result but arranged the electrodes separately so that he could collect and measure precisely the released hydrogen and oxygen. Shortly after, Ritter discovered that he could use electricity to make dissolved metal form a coating on copper. This electroplating technique quickly became a hugely important industrial process.

Increased voltage

Around 1800, British chemist William Cruickshank created the "trough battery," 50 copper–zinc pairs set in a row in cells doused in salt or dilute acid. This battery was much more powerful than Volta's and quickly became the default source of electricity. It was probably with one of these batteries that British chemists William Hyde Wollaston and Smithson Tennant undertook some groundbreaking electrochemical experiments.

The duo discovered no fewer than four new elements as they sought to find a way to purify platinum metal from platinum ore using electrolysis. Tennant worked on the black residue left when platinum ore is treated with *aqua regia* ("royal water," a mixture of hydrochloric and nitric acid) and discovered the new metals iridium and osmium. Wollaston worked on the soluble portion and discovered the elements palladium and rhodium. All four occur naturally in small amounts in platinum ore but were previously unknown. Wollaston also managed to purify platinum by electrolysis—the first time that it had been refined on a commercial basis, making it much sought after for jewelry.

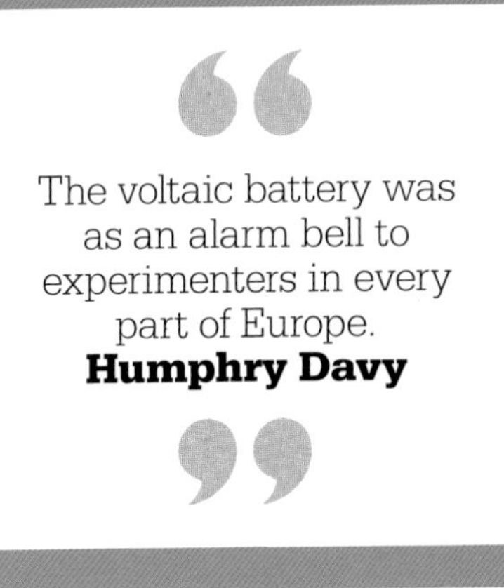

The voltaic battery was as an alarm bell to experimenters in every part of Europe.
Humphry Davy

Science superstar

Another scientist experimenting with electrolysis was British chemist Humphry Davy. Still in his early 20s, Davy was already celebrated for his thrilling science demonstrations. But he was an ingenious practical experimenter, too. Davy wondered whether the bonds between chemicals could be broken down even more effectively with a more powerful current. In 1806, he installed an array of large »

This contemporary etching shows a scientist thought to be Humphry Davy giving a chemistry demonstration at the Surrey Institution, London, in 1809.

Humphry Davy

Born in Penzance, Cornwall, in 1778, Humphry Davy studied science and conducted research at the Pneumatic Institution in Bristol. His experiments with nitrous oxide (laughing gas) made his reputation when he published the results in 1800. In 1801, he was employed by the new Royal Institution in London to give public science lectures.

Many people consider Davy the father of electrochemistry because of his groundbreaking discoveries with electrolysis in 1806–1807. Despite the fact that he discovered neither, Davy was the first to realize that chlorine and iodine were elements rather than compounds. In 1815, with Michael Faraday's help, he created a safety lamp for coal miners to keep the lamp flame separate from flammable gases underground. Davy died in Switzerland in 1829.

Key works

1800 *Researches, Chemical and Philosophical*
1807 "On Some Chemical Agencies of Electricity"
1810 "Historical Sketch of Electrical Discovery"

trough batteries in the basement of the Royal Institution, London, connecting them in series to create a powerful current. With this, he was able to get electrolysis to run for 10 minutes or more.

Davy decided to explore potash (K_2CO_3), the ash collected in a pot from burned wood, which some scientists suspected might be a compound but had never been able to isolate. First, he tried electrolyzing the potash in water, but he only managed to split the water into hydrogen and oxygen. Then he tried electrolyzing dry potash and got nothing at all. Finally, he tried with potash slightly moistened to carry the electric current. The effect was startling. To Davy's huge excitement, a glowing globule of shiny molten metal burst through the crust of potash and caught fire. He had discovered an entirely new metal element: potassium.

The following day, Davy repeated the experiment with moistened caustic soda (lye, NaOH), another apparently inseparable substance. Again, he discovered a new metal element: sodium. These two new metals were unlike any known before. Both are so soft that they can be cut with a knife, and both are so eager to recombine with oxygen that they spit and explode when put in contact with water. Davy's demonstration of this at his lecture made him a scientific superstar, and electrolysis became the scientific sensation of the age.

More elements

In 1808, Davy ran his experiment again with various alkaline earths suspected to contain metal elements. This time, he discovered a further four new metal elements: magnesium, calcium, strontium, and barium. In splitting the alkaline earths, Davy realized that they are

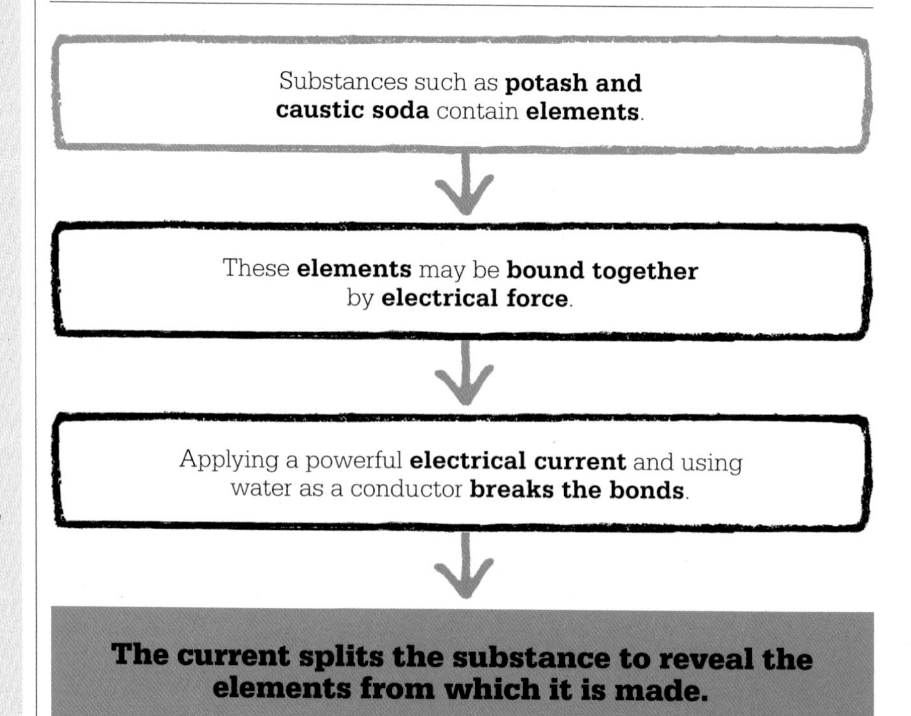

Substances such as **potash and caustic soda** contain **elements**.
These **elements** may be **bound together** by **electrical force**.
Applying a powerful **electrical current** and using water as a conductor **breaks the bonds**.
The current splits the substance to reveal the elements from which it is made.

Sodium reacts explosively with water, producing a bright yellow flame. Its oxidation releases hydrogen gas so rapidly that fragments of sodium seem to "dance" on the water surface.

all metal oxides—combinations of the new metal elements and oxides. If alkaline earths contained oxygen, then how could oxygen be the cause of acidity, as Antoine Lavoisier had claimed? Davy then found that the acid Lavoisier had named oxymuriatic acid—what we now call hydrochloric acid (HCl)—contains no oxygen at all; it is just hydrogen and chlorine. And it was soon confirmed that it is hydrogen, not oxygen, that creates acidity.

With Davy's discovery of six new metal elements, plus Wollaston and Tennant's discovery of two each, the total of new elements revealed in just a few years was now 10. Davy added two more when he showed that iodine and chlorine, previously believed to be compounds, are in fact elements. As other scientists joined in the electrolytic pursuit, more elements were added to the tally, including aluminum, boron, lithium, and silicon.

At the time, chemists believed that certain elements combine with others because they are drawn together by a particular chemical attraction, or "affinity." Davy's experiments led him to believe that this affinity is electrical. Since electrical current overcomes the normal force that hold elements together in compounds, he argued, the force that binds them together must also be electrical, an idea that would in time bear rich fruit.

However, after early successes, Davy gave up research, leaving his young assistant Michael Faraday to continue his work. In 1832, Faraday found that the electrical force was not acting at a distance to split chemicals, like a passing shockwave, as had been thought. Instead, it is the passage of electricity right through the liquid conducting medium that breaks molecules apart. Faraday also discovered that the amount of decomposition depends exactly on the strength of the electrical current. This led Faraday to develop an entirely new theory of electrochemistry that sought to explain how electricity interacts with the forces holding molecules together.

Faraday developed two laws. The first is that the amount of a substance deposited on each electrode of an electrolytic cell is directly proportional to the amount of electricity passing through the cell. The second is that the quantities of different elements deposited by a given amount of electricity are in the ratio of their chemical equivalent weights. French scientist Antoine-César Becquerel soon confirmed Faraday's laws, and he also discovered how to extract metals from sulfide ores using electrolysis.

The potential of electrolysis

By 1840, electrochemistry was well established at the heart of scientific research. It rapidly became a key tool in many industrial processes, unlocking metals and other substances on a huge scale. Today, electrolysis enables us to extract aluminum, sodium, potassium, magnesium, calcium, and other metals from ores. It is hoped that solar-powered electrolysis of water might one day produce hydrogen for automobile fuel cells—but at present, "blue hydrogen" is actually a by-product of fossil fuels. ■

Sodium and potassium

Although sodium and potassium are both hugely important metals, it took Humphry Davy's dramatic experiments to reveal their existence, because on Earth, their pure form is rarely found in nature. Sodium, for example, occurs in abundance in combination with chlorine as common salt (sodium chloride, NaCl)—there are 5.5 quadrillion tons (5 quadrillion metric tons) of it dissolved in the oceans—while potassium, in turn, is a constituent part of various minerals. Sodium and potassium play a key role in the healthy functioning of animals' bodies, including those of humans. They are electrolytes, which means they carry a small electric charge, enabling them to activate various cell and nerve functions. Both are involved in maintaining healthy fluid balance within the body—potassium within cells and sodium in the extracellular fluid. The body cannot survive unless the two are kept in balance.

THE RELATIVE WEIGHTS OF THE ULTIMATE PARTICLES

DALTON'S ATOMIC THEORY

IN CONTEXT

KEY FIGURE
John Dalton (1766–1844)

BEFORE
c.420 BCE The Greek thinker Democritus suggests matter is tiny particles, or atoms, with empty space between.

1630 Jan Baptist van Helmont suggests air is a mixture of gases.

1789 Antoine Lavoisier devises the first list of chemical elements.

AFTER
1890 J. J. Thomson, a British physicist, discovers the existence of the electron.

1911 New Zealand physicist Ernest Rutherford discovers the atomic nucleus.

1913 Danish physicist Niels Bohr creates the planetary model of the atom.

1926 Erwin Schrödinger, a German physicist, proposes the cloud theory of the atom.

One great breakthrough in modern science was the development of the theory of atoms and elements by the British chemist John Dalton in the early 19th century. The idea of atoms was not new. In ancient Greece, Democritus argued that matter is made up of tiny particles separated by empty space, and he coined the term "atoms," from the Greek for uncuttable. But most people could not imagine how air or water could be split like this, and Aristotle's opposing view that matter is continuous and made from just four basic elements—earth, water, air, and fire—prevailed for more than 2,000 years.

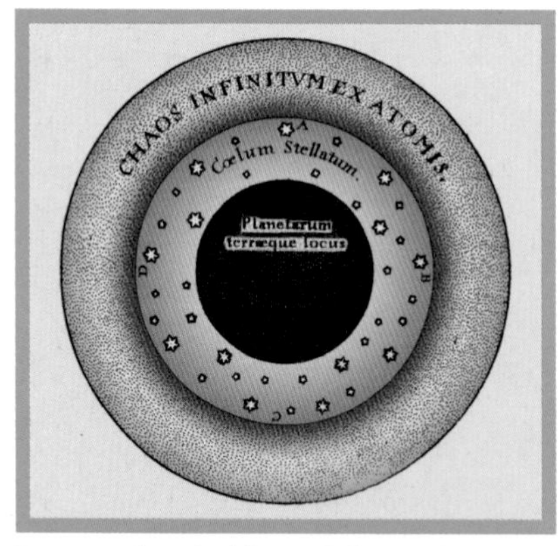

Democritus's idea of the universe placed Earth and the planets at the center. They were surrounded by the starry heavens and an outer ring dubbed the "infinite chaos of atoms."

Questioning Aristotle

Some medieval Islamic scholars had challenged Aristotle's opinion long before. But in 1661, Irish scientist Robert Boyle suggested there were other kinds of "chemical" elements with unique characteristics, and even that matter might consist of atoms. Then, in the 18th century, chemists Antoine Lavoisier and Joseph Priestley finally debunked Aristotle's hypothesis by showing that air and water are combinations of different substances.

No one had clearly defined what an element was, and nobody had linked elements with atoms. It was assumed that if matter is made from atoms, they must all be identical. John Dalton's great insight was to see that the atoms for each of the gases in air might be different and to use this idea as a starting point for a general atomic theory of elements—the broadly correct idea that all the atoms of an element are identical but different from every other element.

Dalton's earliest papers outlining his atomic theory concerned his studies of how the pressure of air

See also: Gases 46 ▪ Oxygen and the demise of phlogiston 58–59 ▪ Conservation of mass 62–63 ▪ The ideal gas law 94–97 ▪ The mole 160–161 ▪ The electron 164–165 ▪ Improved atomic models 216–221

Matter, though divisible in an extreme degree, is nevertheless not infinitely divisible.
John Dalton

All elements are made from **tiny particles called atoms**.

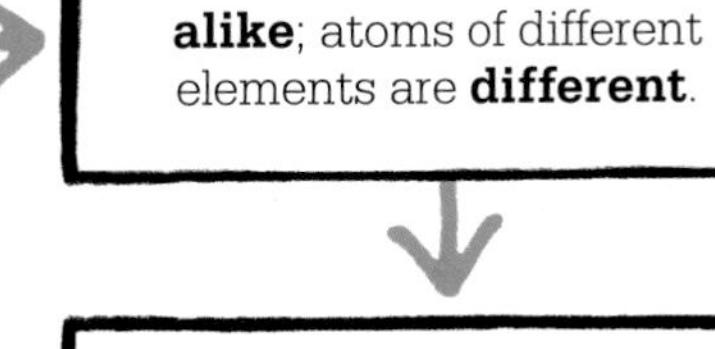

Atoms of each element are **alike**; atoms of different elements are **different**.

Chemical changes happen when different **atoms join or separate**.

Atoms are not created or destroyed by chemical change.

affects the quantity of water it can absorb. In his experiments, he observed that pure oxygen will not absorb as much water vapor as pure nitrogen—and he jumped to the intuitive conclusion that this was because oxygen atoms are bigger and heavier than nitrogen atoms.

Multiple proportions

In a paper read to the Manchester Literary and Philosophical Society on October 21, 1803 (and published in 1806), Dalton told how he had arrived at different weights for the basic units of each elemental gas—in other words, the weight of their atoms, or atomic weight. He argued that the atoms of each element combined to make compounds in simple ratios of whole numbers. So the relative weight of each atom can be worked out from the weight of each element in a compound. This idea later came to be called the law of multiple proportions.

Dalton realized hydrogen is the lightest gas, so he assigned it an atomic weight of 1. Because of the weight of oxygen that combines with hydrogen in water, he gave oxygen an atomic weight of 7. That was a small flaw in Dalton's method because he did not realize that atoms of the same element can combine. He wrongly assumed that a compound made up of atoms—a molecule—had only one atom of each element. That would make water HO, not H_2O. But the basic idea of Dalton's atomic theory—each element has its own unique-sized atoms—proved true and provides the basis for modern chemistry. ■

John Dalton

Born into a Quaker family in the Lake District, UK, in 1766, John Dalton was tutored by a relative, and he began to teach science at a Quaker school at the age of 12. There, he met blind philosopher John Gough, who inspired him to watch the weather. Dalton laid the foundations of modern meteorology. He also identified the hereditary nature of red-green color blindness, a condition he and his brother suffered from. Dalton was elected president of the Manchester Literary and Philosophical Society in 1817, a post he held for the rest of his life.

Dalton's atomic theory made him famous but not wealthy. He turned down membership of the Royal Society in 1810, possibly because he could not afford it, but 12 years later, the Society raised the money for him to be elected. He died in 1844.

Key works

1794 *Extraordinary Facts Relating to the Vision of Colours*
1806 "On the Absorption of Gases by Water and Other Liquids"
1808 *A New System of Chemical Philosophy*

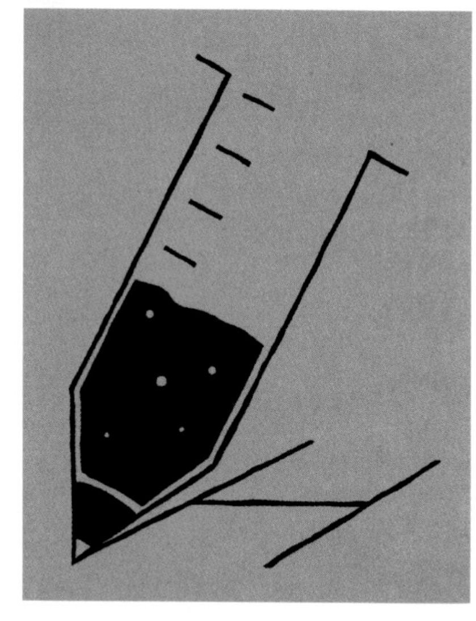

CHEMICAL SIGNS OUGHT TO BE LETTERS

CHEMICAL NOTATION

IN CONTEXT

KEY FIGURE
Jöns Jacob Berzelius (1779–1848)

BEFORE
1775 Swedish chemist Torbern Bergman constructs a table of symbols for alchemicals with "elective affinities."

1789 Antoine Lavoisier introduces the first scientific list of elements.

1808 With his atomic theory, John Dalton introduces a new set of standard symbols for the chemical elements.

AFTER
1869 Russian chemist Dmitri Mendeleev introduces the periodic table, with 63 elements arranged in groups and periods.

1913 British physicist Henry Moseley revises the periodic table to lead with atomic number rather than mass.

The familiar system of notation that we use for chemical elements and compounds today—such as H_2O for water or HCl for hydrochloric acid—was created by Swedish chemist Jöns Jacob Berzelius in 1813. Previously, alchemists had assigned symbols to different substances, but these were not used consistently, and as scientific chemistry developed in the 18th century, chemists began to think of new symbols. In 1789, Antoine Lavoisier introduced the first scientific chemical table, listing 33 "simple substances" (elements), which he divided into gases, metals, nonmetals, and earths (which later turned out to be compounds).

Five years later, another French chemist, Joseph Proust, established that compounds almost always combine in fixed ratios by weight. Soon afterward, John Dalton developed his atomic theory of elements, which introduced both the idea of particular weights for atoms of each element and also his law of multiple proportions. This stated that when one element combines with another in different ways, the ratio of their weights is always a simple one, such as 1:1, 2:1, or 3:1. Dalton created an entirely new set of symbols for the elements, showing pictorially how they combine in compounds.

> I shall take, therefore, for the chemical sign, the initial letter of the Latin name of each elementary substance.
> **Jöns Jacob Berzelius**

A systematic approach

Partly encouraged by Dalton's work, Berzelius began in 1810 to conduct a series of experiments to establish the exact weights of the elements' different compounds. Over the next six years, he analyzed more than 2,000 compounds and produced the most accurate table of atomic weights to date. His work was powerful evidence for Dalton's theory.

See also: Conservation of mass 62–63 ▪ Rare-earth elements 64–67 ▪ Compound proportions 68 ▪ Catalysis 69 ▪ Dalton's atomic theory 80–81 ▪ The periodic table 130–137 ▪ The noble gases 154–159

As he worked, Berzelius realized that the existing array of chemical symbols was a mess, so in 1813, he published his own system. First, he proposed using Latin names for the elements—just as Swedish botanist Carl Linnaeus had for living organisms 80 years earlier—to ensure consistency internationally. Then he decided to use letters instead of the hard-to-draw circles and arrows that Dalton had used.

Letters, not symbols

Berzelius suggested using the first letter of the Latin name as the symbol for each element—for example, carbon (*carbo* in Latin) would be C and oxygen (*oxygenium* in Latin) would be O. With metals that begin with the same letter, such as gold (*aurum*) and silver (*argentum*), he added a second letter, so gold would be Au (the first two letters of the Latin name) and silver Ag (to avoid a clash with arsenic, whose Latin name *arsenicum* has the same first two letters). This system now also extends to nonmetals.

Chemical symbols

Sulfurous oxide (Sulfur dioxide)

⊕○ — **Dalton's style of symbols**

SO^2 — **Berzelius's suggestion**

SO_2 — **Modern use of symbols**

For compounds, the symbol would be the letters for the elements involved, so copper oxide would be CuO. The final ingenious part of the system involved adding small numbers to signify the weight ratios of the different elements in the compound, so carbon dioxide would be CO_2, showing that it is one part of carbon (by weight) and two parts oxygen. Berzelius used superscripts for the numbers (such as CO^2), but they are now always subscripts. It took time for the system to become established, and it has been developed and extended, but its simplicity and effectiveness have ensured it still has a central place in chemistry. ■

The laboratory at the Karolinska Institute, Stockholm, where Berzelius carried out much of his research. He is considered one of the founders of modern chemistry.

Jöns Jacob Berzelius

Born in Linköping, Sweden, in 1779, Jöns Jacob Berzelius began studying medicine at Uppsala University but soon developed a keen interest in experimental chemistry and mineralogy. He proved to be the greatest experimental chemist of his day, producing the first accurate table of atomic weights and establishing the system of chemical notation still used today.

Berzelius also developed a theory of electrical attraction between atoms, discovered the elements cerium (in 1803) and selenium (in 1817), and isolated silicon and thorium for the first time (1824). In 1835, he proposed the term catalysis and described this phenomenon. He published more than 250 papers, covering every aspect of chemistry and influencing many upcoming chemists. Berzelius died in 1848.

Key work

1813 "Essay on the Cause of Chemical Proportions, and on Some Circumstances Relating to Them: Together with a Short and Easy Method of Expressing Them"

THE SAME BUT DIFFERENT
ISOMERISM

IN CONTEXT

KEY FIGURE
Justus von Liebig (1803–1873)

BEFORE
1805 Joseph Gay-Lussac argues that atomic weights in compounds are revealed by gas volumes.

1808 John Dalton introduces his atomic theory.

AFTER
1849 French chemist Louis Pasteur discovers stereoisomers.

1874 Jacob van 't Hoff and French chemist Joseph Le Bel explain stereoisomers.

1922 British scientists James Kenner and George Hallatt Christie discover atropisomers.

2018 A team led by Australian chemists Jeffrey Reimers and Maxwell Crossley discovers akamptisomers.

As chemists became more familiar with chemical compounds in the first decade of the 19th century, they assumed that their properties depended entirely on the combination of elements involved. So when a single atom of sodium joined a single atom of chlorine to form sodium chloride (NaCl), for example, it would always have the same character.

In the 1820s, however, Swedish chemist Jöns Jacob Berzelius and his German protégés Justus von Liebig and Friedrich Wöhler discovered isomers—compounds made from identical combinations of elements but with different

See also: Compound proportions 68 ▪ Dalton's atomic theory 80–81 ▪ Stereoisomerism 140–143

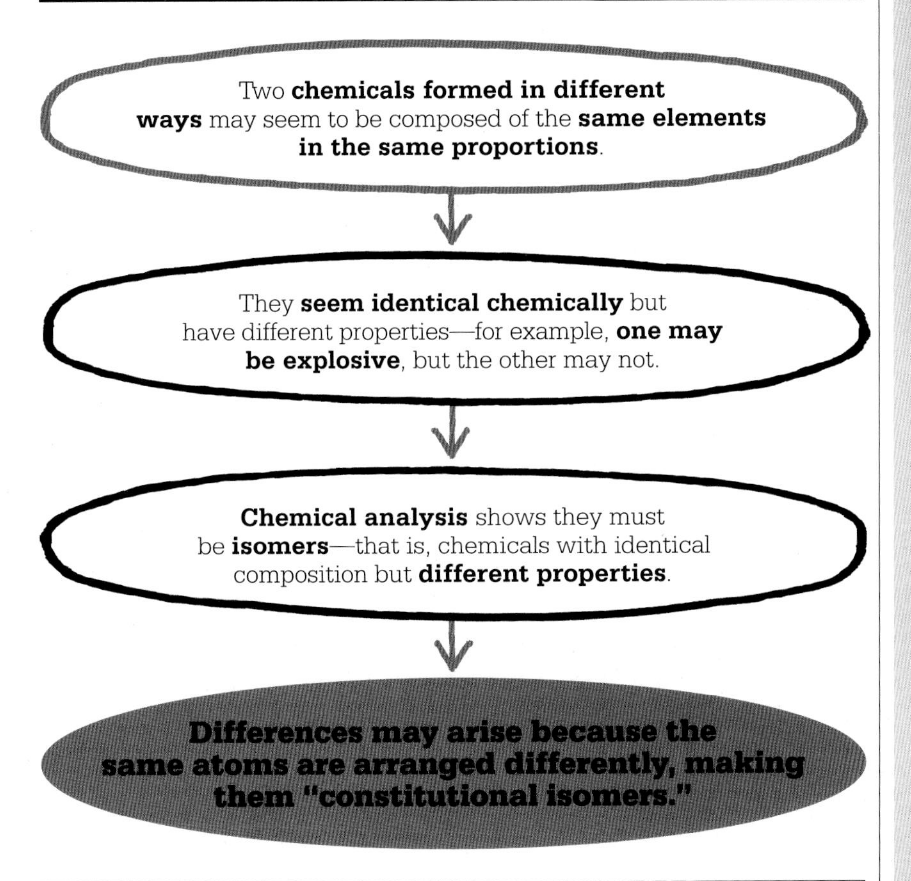

Two chemicals formed in different ways may seem to be composed of the same elements in the same proportions.
They seem identical chemically but have different properties—for example, one may be explosive, but the other may not.
Chemical analysis shows they must be isomers—that is, chemicals with identical composition but different properties.
Differences may arise because the same atoms are arranged differently, making them "constitutional isomers."

properties. Isomers were soon shown to play a key role in the living world and its astonishing array of chemicals, built of carbon and just a few other elements. Von Liebig and Wöhler's work was to lay the foundations of the field of organic chemistry.

Crystal shapes

Important clues about the makeup of compounds emerged in 1819, when German chemist Eilhard Mitscherlich was studying the shapes of crystals.

Mitscherlich found that different compounds can be identically shaped crystals, a phenomenon called isomorphism. They could only be identical shapes if their atoms packed together in the same way. So the way atoms join together must play a part in the chemistry of compounds. By shining light through crystals into a telescope and rotating them, Mitscherlich could reveal very precisely the angles of the faces.

The journey to the discovery of isomers began just a few years later, when Liebig was an ambitious young student in Paris learning the latest techniques in organic analysis from eminent French chemist Joseph Gay-Lussac.

Liebig had been fascinated since childhood with the sheer explosiveness of some derivatives »

Justus von Liebig

Born in Darmstadt, Germany, in 1803, Justus von Liebig became fascinated with chemistry due to his father having a shop selling dyes, drugs, and other chemicals. After studies in Germany and Paris, he was appointed as a professor at the University of Giessen in Germany.

Besides his work with isomerism, Liebig developed a laboratory-oriented approach to instructing his students that would become the model for teaching practical chemistry. He also improved analytic instrumentation by inventing the Kaliapparat.

Liebig laid the foundations of agricultural and nutritional science and created the first nitrogen-based fertilizer before he died in 1873 in Munich.

Key works

1832 *Annalen der Chemie* (a journal founded by Liebig)
1840 *Organic Chemistry in Its Application to Agriculture and Physiology*
1842 *Animal Chemistry; or, Organic Chemistry in Its Applications to Physiology and Pathology*

of fulminic acid (HCNO), and he wanted to discover exactly what was in them.

Liebig was to become one of the greatest chemical analysts ofthe age, and his research into silver fulminate was an early triumph. He showed it to be a compound of silver with carbon, nitrogen, and oxygen—in other words, a silver salt of fulminic acid. He went on to publish his results along with Gay-Lussac in 1824, earning considerable acclaim.

Meanwhile, Wöhler was based in Stockholm, learning chemical analysis from Berzelius. While he was working in Berzelius's laboratory, Wöhler analyzed a silver compound, silver cyanate, which he concluded was the silver salt of a then unknown acid, cyanic acid. His analysis showed that it was a compound of silver with carbon, nitrogen, and oxygen—just like Liebig's silver fulminate—and the quantities involved were identical.

Silver cyanate and silver fulminate both have a single atom of silver combined with a single atom of carbon, nitrogen, and oxygen. The key to their difference lies in the arrangement of those atoms.

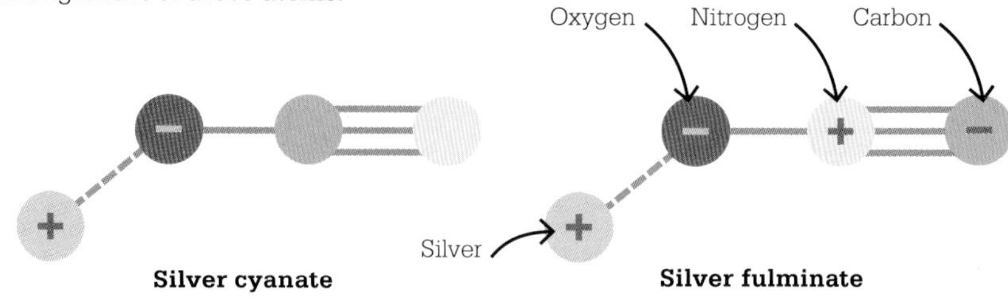

However, while Liebig's silver fulminate was highly explosive, Wöhler's silver cyanate was not.

Hearing of Wöhler's results, Liebig wrote to him suggesting that Wöhler had gotten his analysis wrong. Wöhler then sent Liebig some samples of silver cyanate to analyze, and Liebig's investigation confirmed Wöhler's results.

Liebig and Wöhler—who would go on to become friends—had shown that two compounds with the same composition could have different properties, but neither man knew why. It did not make sense to them because properties were thought to depend solely on composition. Gay-Lussac pondered whether the components might be arranged differently, but nobody really had an answer yet.

Identical compositions

Other similar cases began to emerge. In 1825, Michael Faraday analyzed a gas produced by whale oil and found it to have the same composition as gas coming from marshes (now known as ethylene)—and yet the whale oil gas was much lighter. The same year, Berzelius discovered two different phosphoric acids with the same composition.

In 1828, Wöhler, who was now teaching in Berlin, completed his groundbreaking synthesis

Fireworks sometimes contain silver fulminate, a compound that can only be prepared in small amounts because even the weight of its own crystals can cause self-detonation.

of ammonium cyanate (NH_4OCN). His analysis showed it to have an identical composition to urea, the organic chemical produced in urine, but with very different properties.

While Wöhler was making ammonium cyanate, Gay-Lussac showed that a newly discovered acid that he called racemic acid had the same elemental composition as tartaric acid. Tartaric acid occurs naturally in many fruits, and we now know it to have the formula $C_4H_6O_6$. Racemic acid occurs naturally in grapes, and although it is made from the same combination of elements, it has different properties.

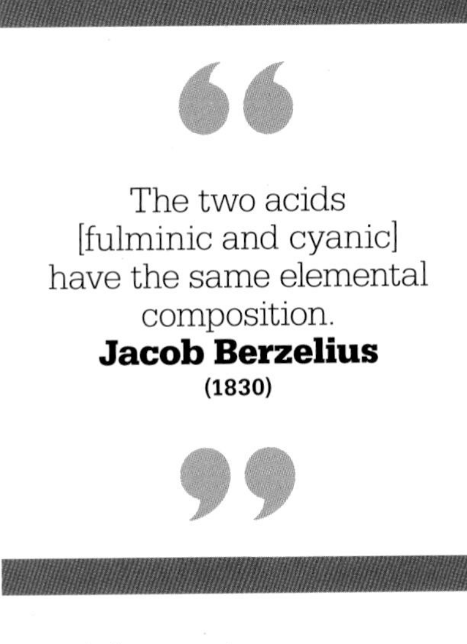

The two acids [fulminic and cyanic] have the same elemental composition.
Jacob Berzelius
(1830)

Experiments on salts

Two years later, Berzelius ran experiments on the lead salts of these two acids and wrote them up in a paper with the very long title "On the Composition of Tartaric Acid and Racemic Acid (John's Acid from the Vosges Mountains), on the Atomic Weight of Lead Oxide, Together with General Remarks on Those Substances which Have the Same Composition but Different Properties." After toying with other possible words, such as homosynthetic, Berzelius coined the term isomers as a way of describing substances that are created from atoms in exactly the same ratio but that have different properties.

In his 1830 book *The History of Chemistry*, British chemist Thomas Thomson responded to Berzelius's work by suggesting that the atoms are simply arranged in different orders in each isomer. So, for instance, in Wöhler's cyanic acid, it could be H-OCN, while in Liebig's fulminic acid, it might be H-CNO. In more complex isomers such as ethanol and dimethyl ether, both have the basic formula C_2H_6O, but the atoms in each are connected in a different order. This can be written as CH_3CH_2OH for ethanol and as CH_3OCH_3 for dimethyl ether.

This concept of atoms linked in different two-dimensional orders came to be known as constitutional isomers. It opened chemists' eyes to the complexity of ways elements combine to form a vast range of substances. This was particularly important in organic chemistry. The variety of materials in living things no longer required special elements unique to life; they could all be the multiple isomers of elemental combinations—just like those in the nonliving world.

We now know two-dimensional constitutional isomers are rare; the truly dizzying variety comes from the three-dimensional arrangement of atoms (stereoisomers), as discovered in 1849. But with the concept of isomers, chemists were at last beginning to understand how a limited range of elements could combine to form every single one of the almost infinite range of substances that make up the universe. ■

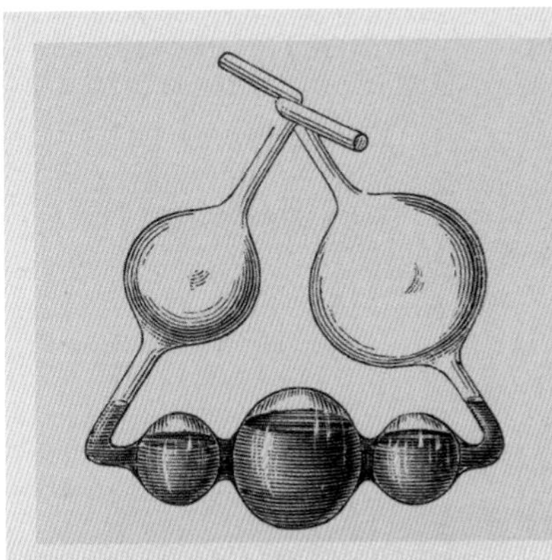

Liebig's Kaliapparat provided an improved method for determining the carbon, hydrogen, and oxygen content of organic substances.

Liebig's five-bulb apparatus

In 1830, Liebig developed a triangular glass device with five bulbs of different sizes that he called the Kaliapparat. This small and simple device was a landmark in organic chemistry, providing chemists with the first precise way of quantifying the carbon levels in a substance.

The gases given off in the combustion of a substance are first passed over calcium chloride to absorb water vapor, removing H_2O, then fed into the Kaliapparat, where three bulbs along the bottom containing potassium hydroxide solution absorb carbon dioxide (CO_2). The difference in weight before and after combustion reveals the amount of CO_2 produced and, therefore, the carbon content of the original substance. The two bubbles on the arm stop gas leaks and prevent the solution from bubbling out.

Liebig also popularized the use of a water-cooling system for distillation that is still referred to as a Liebig condenser.

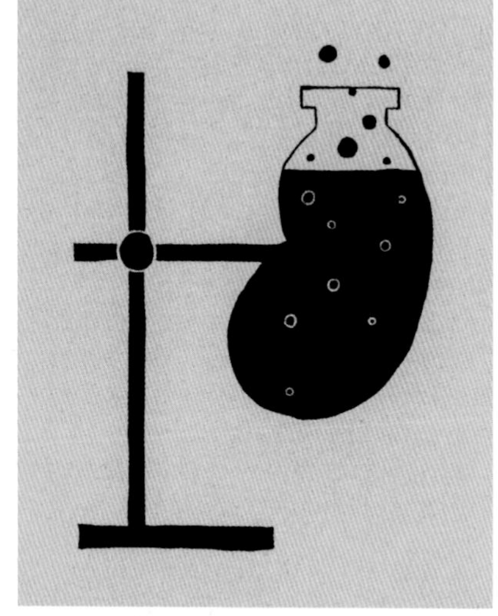

I CAN MAKE UREA WITHOUT KIDNEYS

THE SYNTHESIS OF UREA

IN CONTEXT

KEY FIGURE
Friedrich Wöhler (1800–1882)

BEFORE
1770 Swedish chemist Torbern Bergman recognizes the difference between organic and inorganic chemistry.

1806 The first known amino acid is isolated by French chemists Louis Nicolas Vauquelin and Jean Robiquet.

AFTER
1865 German chemist August Kekulé discovers the benzene ring structure.

1874 Dutch chemist Jacobus Van 't Hoff and French chemist Joseph Le Bel independently explain stereoisomers.

1899 German company Bayer produces aspirin commercially, the first synthetic drug.

1922 British chemist Francis Aston invents the mass spectrometer, a tool for analyzing organic chemicals.

The **reaction** of **ammonia** with **cyanic acid** produces **white crystals**.

These **white crystals** react with **nitric acid** to give brilliant flakes, exactly as **organic urea** does.

The product of **this compound** has precisely the **same composition as urea**.

The reaction of ammonia and cyanic acid synthesizes urea.

Until the early 19th century, it was assumed that the organic chemicals made by living organisms such as juices and dyes had something special about them, linked to the mystery of life, that made them impossible to synthesize (create from inorganic chemicals); this idea is called vitalism. But in 1828, German chemist Friedrich Wöhler became the first person to synthesize one of these chemicals: urea.

By the first decade of the 1800s, chemists understood that all substances are made from basic chemicals or elements that join together in different combinations to form compounds. They had also developed analytic techniques to unlock the combinations of elements that make up compounds and the proportion of atoms involved.

Chemists had shown that organic chemicals were made largely from carbon, hydrogen, oxygen, and nitrogen—different combinations of these few elements created a staggering variety of substances.

Analysis of urea

Urea was one of the first organic chemicals to be fully analyzed. It is produced in the body of all animals to capture toxic amino acids, then excreted through the kidneys as urine. It was first isolated from

See also: Isomerism 84–87 ▪ Sulfuric acid 90–91 ▪ Functional groups 100–105 ▪ Benzene 128–129 ▪ Stereoisomerism 140–143 ▪ Fertilizers 190–191 ▪ Mass spectrometry 202–203 ▪ Retrosynthesis 262–263

urine as white crystals in 1773 by French chemist Hilaire-Marin Rouelle. British chemist William Prout obtained a pure sample in 1818 and was able to establish the exact composition of urea.

The purification of urea was a subject in which Wöhler's mentor, Swedish chemist Jöns Jacob Berzelius, had an interest. Wöhler worked in Berzelius's laboratory in 1823 and may have become acquainted with Prout's analysis.

In 1824, Wöhler was mixing ammonia (a combination of nitrogen and hydrogen, NH_3) with cyanogen (CN_2) when he found they reacted together to make oxalic acid ($C_2H_2O_4$), which is made in rhubarb and other plants. It also produced white crystals, but Wöhler did not know what these were. Returning to the experiment in 1828, he reacted liquid ammonia with cyanic acid, expecting the product to be ammonium cyanate (CH_4N_2O). However, the reaction produced the same white crystals—not the expected appearance of ammonium cyanate. When he treated the crystals with nitric acid, they produced brilliant crystals—just as urea was known to do.

Wöhler compared his analysis of ammonium cyanate with Prout's analysis of urea (both shown here) and found that they had an almost identical chemical makeup. From this, Wöhler concluded that atoms must be able to arrange themselves into molecules in different ways.

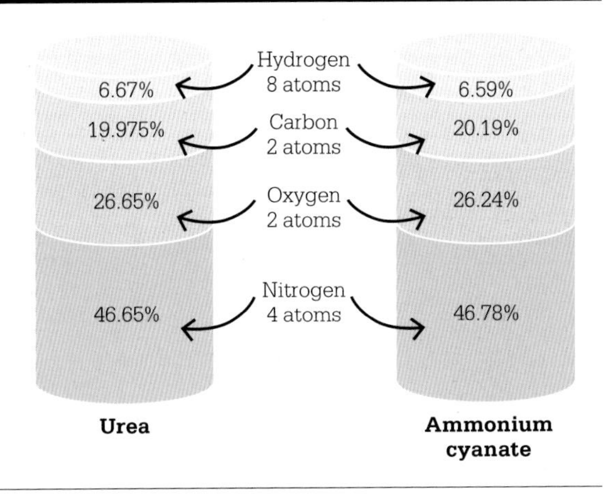

Matching compositions

Wöhler found that the composition of his ammonium cyanate matched Prout's analysis of urea almost perfectly. The ammonia and cyanic acid reaction had created urea—and it was the first time an organic compound had been synthesized. Wöhler and Berzelius were excited, but for them, it was just a curiosity. It did not immediately invalidate the idea of vitalism, and its true significance took a couple of decades to be realized.

The ability to synthesize urea had major applications, including in the production of fertilizers and animal food supplements. Scientists began to understand that organic compounds behave according to the same chemical rules as inorganic chemicals—and that they can also be synthesized through controlled chemical reactions. This helped open up the vast chemical industry we have today. ■

Friedrich Wöhler

One of the great pioneers of organic chemistry, Friedrich Wöhler was born near Frankfurt in Germany in 1800, the son of an agronomist and vet. He graduated from medical school in 1823 but soon switched to chemistry and spent a year studying with Jöns Jacob Berzelius. Over the next few years, Wöhler made the first pure sample of the metal aluminum, as well as succeeding in synthesizing urea.

Working with Justus von Liebig, Wöhler completed the analyses that led to Berzelius identifying isomers. Further collaboration with Liebig led to the discovery of organic radicals, as well as to groundbreaking methods of science education. Wöhler died in Göttingen in 1882.

Key works

1825 *Textbook of Chemistry*
1828 "On the Artificial Production of Urea"
1840 *Outlines of Organic Chemistry*
1854 *Practical Exercises in Chemical Analysis*

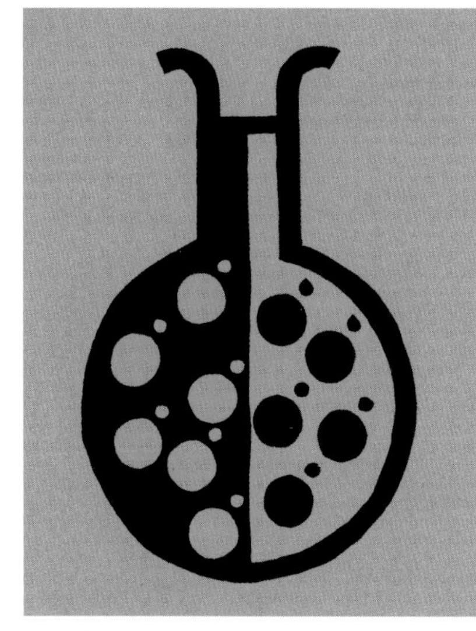

THE INSTANTANEOUS UNION OF SULFUROUS ACID GAS WITH OXYGEN

SULFURIC ACID

IN CONTEXT

KEY FIGURE
Peregrine Phillips (1800–1888)

BEFORE
c.550 The ancient Chinese discover a natural form of sulfur known as shiliuhuang.

1600 Dutch chemist Jan Baptist van Helmont burns sulfur with green vitriol to make sulfuric acid.

1809 Joseph Gay-Lussac and fellow French chemist Louis-Jacques Thenard prove that sulfur is an element.

AFTER
1875 The first industrial plant using the contact process is opened in Freiburg, Germany.

1934 American chemist Arnold O. Beckman develops the acidimeter; it is the forerunner of today's pH meters.

2017 The annual global production of sulfuric acid reaches 275 million tons (250 million metric tons).

Sulfuric acid is one of the most important industrial chemicals, and it is used in making everything from fertilizers to paper. One hundred ten million tons (100 million metric tons) are made worldwide every year, but this is only possible because of the contact process developed by British vinegar maker Peregrine Phillips in 1831.

Dry distillation method

The discovery of sulfuric and other mineral acids is credited to Arabic alchemist Jabir ibn Hayyan around 800 CE. Jabir used what is called the dry distillation method of heating the sulfur salt of copper and iron, which came to be known as blue vitriol and green vitriol respectively. Oil of vitriol—as the alchemists called sulfuric acid—became central to alchemy. It was believed to be key in the search for a way to turn base metal to gold because sulfuric acid, while corroding most metals, leaves gold untouched. In the 15th century, German alchemist Basil Valentinus discovered how to make sulfuric acid by burning sulfur over saltpeter (potassium nitrate, KNO_3). As the saltpeter decomposes, it oxidizes the sulfur to sulfur dioxide (SO_2), then to sulfur trioxide (SO_3), which combines with water to make sulfuric acid (H_2SO_4).

British physician John Roebuck used this method to develop the first industrial-scale process in 1746. Previously, the acid had been made in glass jars, but Roebuck used

The alchemist distilling sulfuric acid in this 1651 woodcut is using a ladle to insert the ingredients that are then heated in a furnace.

See also: Oxygen and the demise of phlogiston 58–59 ▪ Catalysis 69 ▪ Chemical notation 82–83 ▪ Synthetic dyes and pigments 116–119

The production of sulfuric acid is important because this highly corrosive acid has many roles in modern industry. It is used in large quantities by metal manufacturers, in the production of fertilizers, and in oil refining, to name a few.

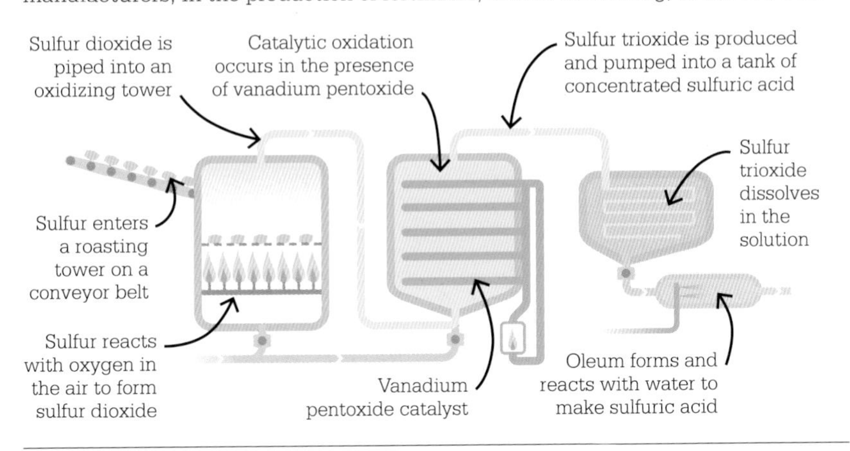
Sulfur dioxide is piped into an oxidizing tower
Catalytic oxidation occurs in the presence of vanadium pentoxide
Sulfur trioxide is produced and pumped into a tank of concentrated sulfuric acid
Sulfur trioxide dissolves in the solution
Sulfur enters a roasting tower on a conveyor belt
Sulfur reacts with oxygen in the air to form sulfur dioxide
Vanadium pentoxide catalyst
Oleum forms and reacts with water to make sulfuric acid

It is … the usual fate of the inventor of a novel process of chemical manufacture to find that his monetary reward is little or nothing …

Ernest Cook
Biographer of Peregrine Phillips (1926)

huge vats made of lead, one of the few metals to resist the acid. Roebuck's lead chamber process was soon feeding a demand for the acid in the bleaching of cotton in the burgeoning textile industry.

The lead chamber process was limited, in that it only gave very dilute acid, about 35–40 percent of pure sulfuric acid. This was far too weak for making dye, so the costly dry distillation method continued to be used. In the 1820s, chemists Joseph Gay-Lussac and John Glover were able to achieve 78 percent, passing the gases through a reaction tower to recover oxides of nitrogen.

A major breakthrough

In 1831, Peregrine Phillips made a breakthrough by patenting a much more efficient process for making sulfur trioxide, which could be mixed with water to create concentrated acid. Phillips passed the sulfur dioxide gas through a tube lined with platinum, which hugely accelerates the conversion into sulfur trioxide. The platinum acts as a catalyst and is directly involved in the reaction; the process came to be known as the contact process because the sulfur trioxide combines with steam.

The contact process was used on a small scale by German chemist Eugen de Haën in the 1870s, in the synthetic dyestuffs industry. But it did not take off until 1915, when the German company BASF, a leader in dye making, substituted vanadium oxide for platinum as the catalyst.

From the 1920s on, almost a century after Phillips's patent, the contact process, using the cheaper vanadium oxide catalyst, began to replace the lead chamber process. The huge increase in concentrated sulfuric acid being produced gave it a key place in countless industrial processes, such as making soap and paint, producing car body panels, and refining petroleum. ■

Peregrine Phillips

Born in 1800, Peregrine Phillips is thought to be the son of a tailor, also named Peregrine Phillips, who opened a shop on Milk Street in Bristol, UK, around 1803.

By the 1820s, Phillips senior had set up a sizable plant for making vinegar, which was widely used as an acid at the time in medicine and in food preservation, among its other uses. The younger Phillips devised his groundbreaking contact process, or wet gas process, for making concentrated sulfuric acid and filed a patent for it in 1831; however, it failed to find support. The process was an expensive one, and consequently it was not used on a large scale until much later, when a cheaper catalyst had been discovered.

The patent, number 6096, still survives and demonstrates how much detail Phillips had developed, as well as his understanding of the chemistry involved in the process. Despite this, Phillips then faded into obscurity and died in 1888.

THE QUANTITY OF MATTER DECOMPOSED IS PROPORTIONAL TO THE QUANTITY OF ELECTRICITY

ELECTROCHEMISTRY

IN CONTEXT

KEY FIGURE
Michael Faraday (1791–1867)

BEFORE
1800 Alessandro Volta creates the first battery, the voltaic pile.

1800 British chemist William Nicholson and surgeon Anthony Carlisle and, independently, German physicist Johann Ritter split water into hydrogen and oxygen using electricity.

1803 John Dalton develops his atomic theory, showing that compounds are comprised of elements in simple whole-number ratios.

AFTER
1859 French physicist Gaston Planté invents the lead–acid storage battery—the first rechargeable battery.

1897 British scientist J. J. Thomson discovers the electron—the unit of negative charge and the first subatomic particle to be identified.

During **electrolysis**, substances in solution **decompose**. Positively charged ions move to the **negative electrode**; negatively charged ions move to the **positive electrode**.

As the amount of **electrical current** increases, so does the **mass of substances** collecting on the electrodes.

Law 1: The mass of substances deposited is in proportion to the amount of electricity.

Law 2: The masses of each substance deposited by a given amount of electricity are in the ratio of their chemical equivalent weights.

In the early 19th century, following the invention of the voltaic pile by Italian physicist and chemist Alessandro Volta in 1800, groundbreaking experiments in electrochemistry generated tremendous excitement among scientists. Within weeks of Volta's discovery, scientists in different countries had used electricity to split water into its component elements. Over the following years, chemists such as the British Humphry Davy and Swedish Jöns Jacob Berzelius isolated entirely new elements by splitting compounds with electricity.

In 1807, Davy suggested that some elements have a negative chemical affinity and others a positive affinity. Berzelius went further, proposing that the attraction of electrical opposites is what binds elements together in compounds. This idea, called dualism, was eventually superseded,

See also: The first battery 74–75 ▪ Isolating elements with electricity 76–79 ▪ Dalton's atomic theory 80–81 ▪ Atomic weights 121 ▪ The mole 160–161 ▪ The electron 164–165

but it led to the idea that natural chemicals might be analyzed quantitatively for their positive and negatively charged constituents. This drew attention to the exact proportions in which elements bind together, which is indicated by the quantities of products and reactants before, during, and after a chemical reaction. These relationships are described by the term "stoichiometry."

Principles of electrolysis

Davy's brilliant protégé Michael Faraday believed there must be a direct quantitative relationship between an electric current and its chemical effect. In 1832 and 1833, he conducted hundreds of electrochemical experiments, measuring how much electricity it takes to decompose various compounds—a process he termed "electrolysis." Working with British polymath William Whewell, he formalized other terminology relating to the process, such as "anode" and "cathode" for the positive and negative terminals in an electrolytic cell (collectively, "electrodes") and "ion" for the charged particles.

In one set of experiments, Faraday placed two foil electrodes on a glass disk and connected them electrically via a disk of filter paper soaked in a chemical solution. By changing the filter for others soaked in different solutions, he could test many different electrolytic reactions quickly. In other experiments, Faraday connected V-shaped glass tubes into his electrolysis circuit to collect and measure the hydrogen and oxygen generated when water was split by the circuit. This allowed him to demonstrate the strength of the electrical current.

Two new laws

Faraday deduced two truths, which he published in 1834 as the laws of electrolysis. The first is that the amount of substance deposited on the electrode of an electrolytic cell depends on the amount of electricity passing through the cell. The second law states that, for a given amount of electric current, the quantity of each substance made at the electrodes depends on its chemical equivalent weight (the amount combining with or displacing a fixed quantity of another substance). Faraday did not link this second idea to atoms, but it was the first concrete evidence of a direct relationship between a unit of electricity and atomic weight, and it became invaluable in determining the relative atomic weights of the elements. ■

Faraday's electrolysis experiments involved meticulous measurement of chemical products. Here, electrolysis of tin chloride produced tin, chlorine, hydrogen, and oxygen.

Michael Faraday

Born in 1791 in London, UK, Michael Faraday was apprenticed to a bookbinder at the age of 14. After he wrote to Humphry Davy with his thoughts on one of Davy's science lectures, Davy took him on as his assistant. Faraday became the Royal Society's leading scientist, famous for his public lectures at the Royal Institution, and one of the greatest practical and theoretical scientists of the age.

Following the discovery of the link between electricity and magnetism in 1820, Faraday revealed first the principle of the electric motor and then how electricity can be generated. He went on to discover the laws of electrolysis. Due to his religious beliefs, Faraday was sure that all forms of electricity were one and the same fundamental force. He died in 1867.

Key works

1834 *On Electro-Chemical Decomposition*
1839 *Experimental Researches in Electricity (Volumes I and II)*
1873 *On the Various Forces of Nature*

AIR REDUC'D TO HALF ITS WONTED EXTENT, OBTAINED TWICE AS FORCIBLE A SPRING

THE IDEAL GAS LAW

IN CONTEXT

KEY FIGURE
Émile Clapeyron (1799–1864)

BEFORE
1650 French mathematician Blaise Pascal coins the term "pressure" to describe the weight of the air.

1662 Robert Boyle formulates a law stating that, in a gas at constant temperature, pressure is inversely proportional to volume.

AFTER
1873 Dutch physicist Johannes van der Waals modifies the ideal gas law to account for the size of molecules, intermolecular forces, and the volume of real gases.

1948 Two chemists, Austrian Otto Redlich and Chinese-American Joseph Kwong, propose the Redlich–Kwong equation of state, a refinement of the ideal gas law equation.

Scientists once doubted that gases have any physical properties at all, but from the 17th century, improved analysis revealed the mutual relationship between their temperature, volume, and pressure. Then, in 1834, French engineer Émile Clapeyron summed up this relationship in the "ideal gas equation."

In 1614, young Dutch scientist Isaac Beeckman had suggested that air, like water, has weight and exerts pressure. The great Italian scientist Galileo disagreed. But some younger scientists concurred with Beeckman, including Galileo's fellow Italians Evangelista Torricelli and Gasparo Berti.

See also: Gases 46 ▪ Oxygen and the demise of phlogiston 58–59 ▪ Intermolecular forces 138–139 ▪ The noble gases 154–159 ▪ The mole 160–161 ▪ Atomic force microscopy 300–301

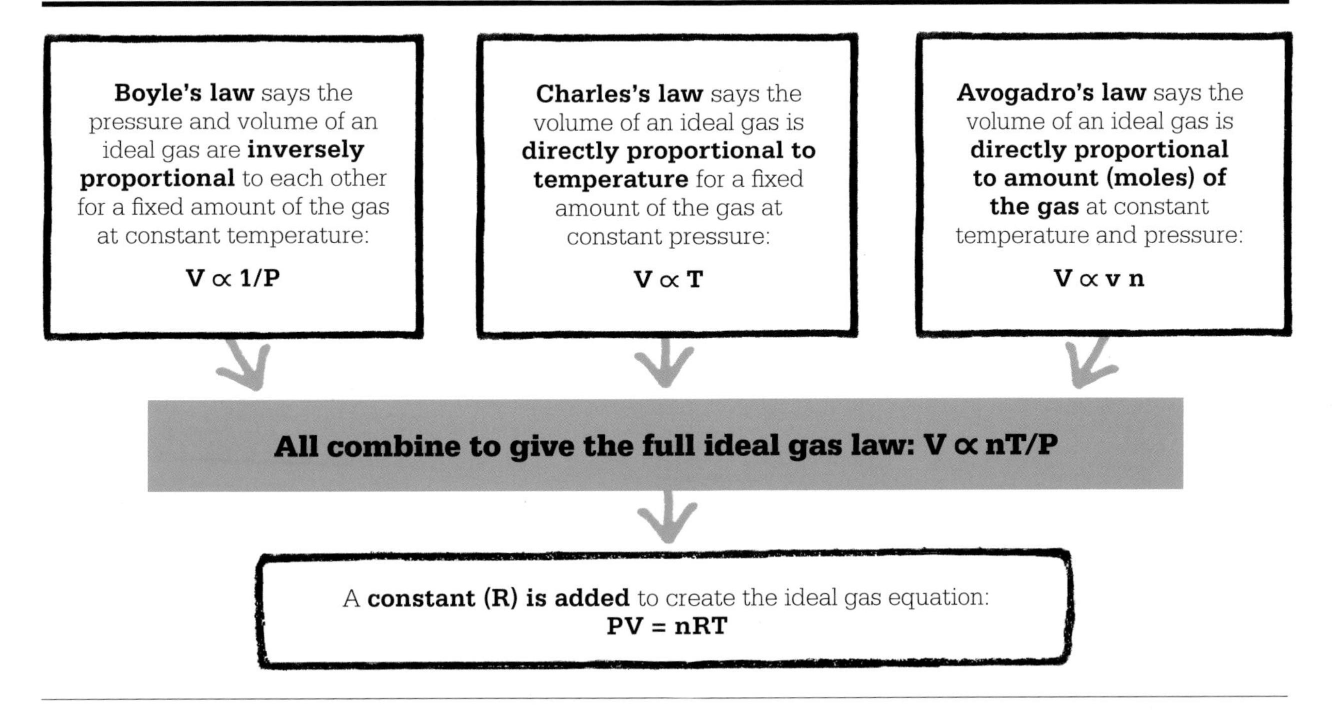

Boyle's law says the pressure and volume of an ideal gas are inversely proportional to each other for a fixed amount of the gas at constant temperature:
V ∝ 1/P
Charles's law says the volume of an ideal gas is directly proportional to temperature for a fixed amount of the gas at constant pressure:
V ∝ T
Avogadro's law says the volume of an ideal gas is directly proportional to amount (moles) of the gas at constant temperature and pressure:
V ∝ v n
All combine to give the full ideal gas law: V ∝ nT/P
A constant (R) is added to create the ideal gas equation:
PV = nRT

Between 1640 and 1643, while he was investigating the existence of vacuums (then a contentious subject), Berti filled a pipe with water and sealed it with a glass jar at the top. When he opened taps at the bottom of the pipe, the water ran out part of the way, creating an empty space at the top. This, he contended, was a vacuum. In 1643, Torricelli investigated the phenomenon further, this time using mercury instead of water. Under his instructions, Vincenzo Viviano filled a glass tube with mercury. He sealed the tube at one end and held a finger over the open end while upturning it into a bowl of mercury. The mercury in the tube dropped to a height of 30 in (76 cm), then stopped, leaving an empty space above.

Torricelli explained that the mercury did not empty out of the tube completely because the weight of air pressing down on the mercury in the bowl was forcing it up. He had refuted the idea that air is weightless. His point was proven when the test was done at higher altitudes, where the air must be lighter, and the mercury in the tube fell to a lower level.

Boyle's law

Inspired by Torricelli, Anglo-Irish chemist Robert Boyle fashioned his own experiments with a J-shaped glass tube also filled with mercury and sealed at the lower tip. Boyle saw that the small space above the mercury in the tip shrank as he added mercury to the tube and expanded when he poured mercury out. He concluded that the pressure of air in the space must be rising or falling as it is squeezed or released.

Boyle likened pressure in the air to springlike particles pushing back as they are squeezed and proposed »

Torricelli conducted his experiment with a glass tube about 3 ft (1 m) long. He argued that day-to-day shifts in the height of the mercury reflected changing atmospheric pressure.

a simple law, now known as Boyle's law: if its temperature remains the same, the volume of a gas and its pressure vary in inverse proportion.

Gas temperature

Over the next century, the advent of steam engines focused attention on the role of heat in expanding air. In 1787, French scientist and balloon pioneer Jacques Charles added temperature to the relationship between volume and pressure. In Charles's law, he showed that, provided the pressure stays steady, gas volume varies with temperature. In fact, the relationship is a straight line, with volume expanding at a steady rate for each degree of temperature rise (from absolute zero) and shrinking equally steadily for each degree drop.

The Montgolfier balloon flights of the 1780s demonstrated Charles's law: heating the air in the balloon caused it to expand, becoming lighter than the surrounding air—and so rising.

In 1802, French scientist Joseph Gay-Lussac completed the equation, linking pressure to temperature. Gay-Lussac's law showed that if gas volume is held constant, then pressure rises with temperature. British chemist John Dalton soon showed this applied to every kind of gas, and Gay-Lussac realized that when different gases combine, they combine in simple proportions by volume: Gay-Lussac's law of combining volumes. This triangular relationship suggested the cause was mechanical. But just what the cause was remained unclear.

Bringing it together

In 1811, Italian scientist Amedeo Avogadro added a crucial ingredient to the equation—the gas particles themselves. Avogadro used the term molecule to describe all particles, including both molecules and atoms, and the crux of Avogadro's hypothesis was that equal volumes of gas at a given temperature and pressure always contain the same

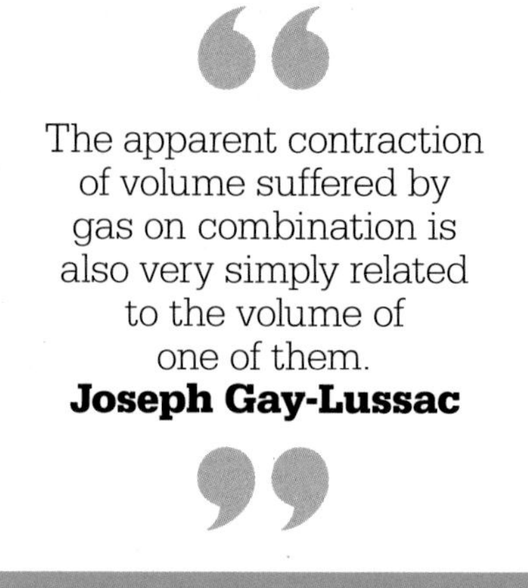

The apparent contraction of volume suffered by gas on combination is also very simply related to the volume of one of them.
Joseph Gay-Lussac

number of particles. In 1865, Austrian high school student Josef Loschmidt calculated this number to be $6.02214076 \times 10^{23}$. This "Avogadro number" of all particles (atoms, molecules, ions, or electrons) in a substance later became known as 1 mole.

All the ingredients added by various scientists over two centuries came together in 1834 in the work of French engineer Émile Clapeyron when he created the ideal gas law. This simple equation brings together all the qualities needed to predict how a gas will behave in changing circumstances based on volume, pressure, temperature, and the number of particles:

$$PV = nRT$$

In the equation, P is pressure, V is volume, T is temperature, and n is the number of moles. R is the constant Clapeyron needed to combine the proportional relationship of pressure, volume, and temperature into a single equation.

The equation is hypothetical because ideal gases rarely exist in the real world except at extremely high temperatures and low pressures. The atoms and molecules in an ideal gas have no size or dimensions; they never interact with each other except to collide occasionally—and when they do collide, they simply bounce off with no loss of momentum. This is not the situation in real gases, but the equation approximates how real gases behave. It includes only the factors a chemist needs to know when making calculations about pressure, temperature, volume, and the number of particles. And that is precisely what makes the equation so effective.

Real-world applications

In the real world, the ideal gas law explains why a bicycle pump gets hot as pressure rises and volume shrinks. It explains, too, why compressed air cools when it is released and expands and is key to the working of a refrigerator, which keeps cool by compressing gas, then allowing it to expand.

Clapeyron's ideal gas equation helped stimulate the burgeoning science of thermodynamics. In the 1850s, German scientists Rudolf Clausius and August Krönig independently developed the kinetic theory of gases, focusing on the energy of their moving particles. To get a snapshot of the mechanical movement of trillions of particles in an ideal gas, they used the statistical distribution of the speeds of particles in an ideal gas to explain the relationship between pressure, volume, and temperature.

Imagine a box filled with billions of particles of an ideal gas. The nonstop movement of the particles in straight lines—their "mean free path"—is what creates the relationship between temperature, volume, and pressure. Temperature is proportional to the average kinetic energy or speed of the particles. Pressure is the statistical effect of all the collisions with the side of the box.

Today, the ideal gas equation is used in various forms. When working with thermodynamics, for instance, physicists add in to the equation the constant derived in 1877 by Austrian physicist Ludwig Boltzmann to factor in the kinetic energy of particles. In all its forms, the equation remains at the heart of our practical understanding of how gases behave. ■

According to Avogadro's hypothesis, at the same temperature and pressure, balloons of equal volume have equal numbers of molecules, regardless of the gas inside them, but the mass of each will differ from the others.

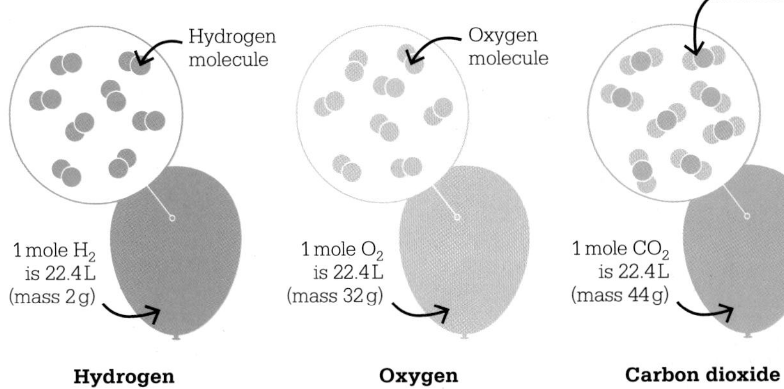

Émile Clapeyron

Born in Paris, France, in 1799, Benoît Paul Émile Clapeyron trained as an engineer at the city's École des Mines. In 1820, he went with friend Gabriel Lamé to teach math and engineering to teams building bridges and roads in Russia. The pair returned to France 10 years later and focused their efforts on railway construction. Clapeyron became a leading steam locomotive designer while also pursuing a more theoretical study of steam engines. His 1834 summary of French mechanical engineer Sadi Carnot's ideas about heat engines, along with graphical representations, marks the birth of thermodynamics, and his synthesis of the gas laws into one equation established him as a leading theorist.

Clapeyron was elected to the Paris Academy of Sciences in 1848, and he died in 1864. Named partly after him, the Clausius–Clapeyron equation determines the heat of vaporization of a liquid.

Key work

1834 "Mémoire sur la Puissance Motrice de la Chaleur" ("Memoir on the Motive Power of Heat")

ANY OBJECT MAY BE COPIED BY IT

PHOTOGRAPHY

IN CONTEXT

KEY FIGURE
Louis Daguerre (1787–1851)

BEFORE
1777 Carl Scheele confirms the reaction of silver compounds to light.

c. 1790 British inventor Thomas Wedgwood captures the first photographic silhouette image.

1822 Joseph Nicéphore Niépce makes the world's first permanent photograph.

AFTER
1850 British photographer Frederick Scott Archer introduces the wet-plate collodion process, which gives greater detail and clarity.

1890 In the US, the Eastman Kodak Company launches rolls of celluloid photographic film.

1975 Kodak introduce the first prototype digital camera.

2000 The first mobile phones with built-in cameras go on sale to the public.

From the 17th century, artists used the camera obscura as an aid, tracing their pictures from the temporary image that it projected. In 1777, Swedish-German chemist Carl Scheele provided the means to permanently capture the image when he demonstrated how to both start and stop silver salts reacting to light. This was finally realized in 1839, when French inventor Louis Daguerre created the first successful photographic process: the daguerreotype.

Daguerre's late colleague, French inventor Joseph Nicéphore Niépce, had made photographs up to 20 years earlier, but the reactions of the chemicals involved were so slow that it took many hours to make a picture, and the image was very blurry. So, in the 1830s, Daguerre experimented with different combinations of chemicals.

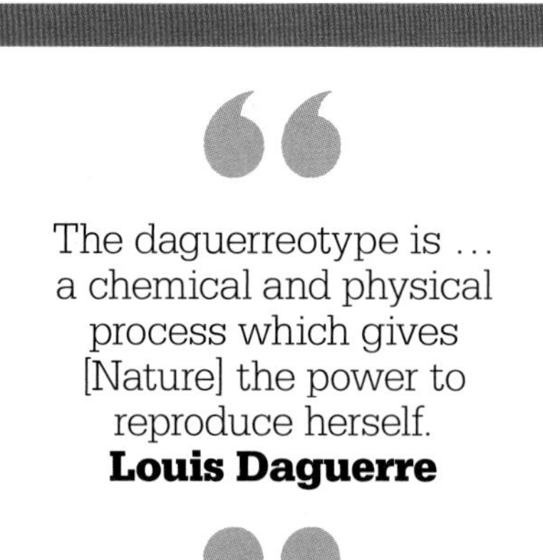

The daguerreotype is … a chemical and physical process which gives [Nature] the power to reproduce herself.
Louis Daguerre

Daguerre divided the process into two stages. In the first, the photograph could be taken quickly to capture a faint "latent image" in the chemicals on the photographic plate. In the second, the latent image was revealed by "developing" it in the photographer's laboratory using mercury fumes.

Other pioneers such as French inventor Hipppolyte Bayard and British inventor William Henry Fox Talbot were trying out similar ideas, but it was Daguerre's breakthrough in 1839 that captured the headlines.

Daguerre's invention

Daguerreotypes were copper plates thinly coated with silver. To make a photograph, the plate was polished to a mirror finish, then cleaned with nitric acid. In darkness, the silver surface was then turned into light-sensitive silver iodine by exposure to iodine fumes before being boosted in sensitivity through use of bromine or chlorine fumes.

See also: Early photochemistry 60–61 ▪ Catalysis 69 ▪ Why reactions happen 144–147 ▪ Infrared spectroscopy 182

The Open Door was taken by Fox Talbot in 1844. He believed photography could be the equal of the old masters—and more faithful to reality.

The sensitized plate was then taken in a lightproof holder into the camera. The holder was opened and the cap on the camera's lens removed briefly to take the picture.

Afterward, the plate was exposed to fumes of heated mercury inside a special box to bring out the latent image. But to "fix" the image, the chemical reaction had to be halted before it went completely black by removing the silver salts unaffected by light with a solution of "hypo" (sodium thiosulfate). Finally, the developed image was encased in protective glass.

The daguerreotype image was actually a negative, dense and dark in the highlights and thinner in the shadows. But the shiny silver surface of the plate reflected light to show the image as a positive. The results were so dazzlingly clear and sharp that the public was immediately smitten, and photography took off. That same year, 1839, French art restorer Alphonse Giroux made the first commercial camera.

The calotype

In 1841, Fox Talbot introduced his own process, the calotype, which had a distinct advantage over the daguerreotype. The daguerreotype image was a one-off that could not be copied. With the calotype, the photo was a negative image on waxed paper. This could be copied many times by shining light through it onto another sheet of light-sensitive paper. This negative–positive method went on to become the standard way of creating photographs until the advent of digital photography in the 1990s. ■

Louis Daguerre

Born in 1787 in Cormeilles-en-Parisis, France, Louis Daguerre became an apprentice landscape painter but also learned about architecture and theater design. After working as a tax officer, he became a scene painter for the opera and a master of illusion. He created a diorama—a mobile device for showing giant scenes—in Paris in 1822.

Daguerre worked with Nicéphore Niépce, the creator of the world's first photograph, to create a practical photographic process. After Niépce died in 1833, Daguerre continued work on the process and eventually unveiled the world's first successful photographic process: the daguerreotype. It was launched to the public by the French Academy of Sciences in 1839, after the government made a deal with Daguerre, acquiring the rights to the process in exchange for a lifetime of pensions. Daguerre died in 1851.

Key work

1839 *History and Practice of Photogenic Drawing on the True Principles of the Daguerréotype*

NATURE HAS MADE COMPOUNDS WHICH BEHAVE LIKE ELEMENTS THEMSELVES

FUNCTIONAL GROUPS

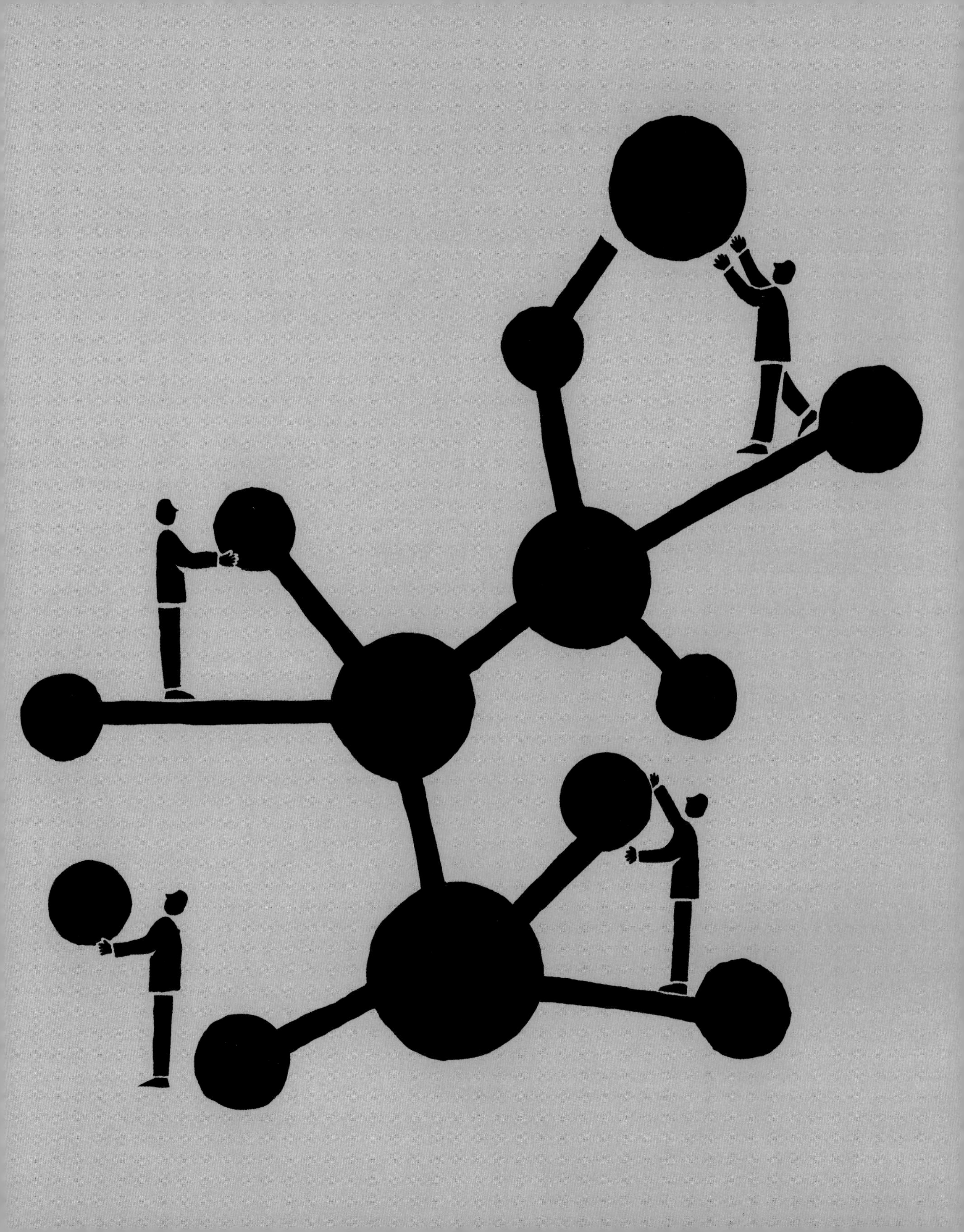

IN CONTEXT

KEY FIGURE
Jean-Baptiste Dumas
(1800–1884)

BEFORE
1782 French chemist Baron de Morveau introduces the word "radical" to describe combinations of chemicals that persist through reactions.

1815–1817 Joseph Gay-Lussac and Jöns Jacob Berzelius develop the idea of inorganic compound radicals. These combinations of elements seem to behave as single units in various combinations.

AFTER
1852 British chemist Edward Frankland discovers the crucial concept of valence, overturning the idea of radicals and types.

1865 German chemist August Kekulé unlocks the ring-shaped molecular structure of benzene.

A serious question facing chemists in the 1830s was just how organic chemicals fit into the picture of chemicals and compounds. The table of elements was increasing in size, and it had become clear that compounds were made by elements joining together. One issue was how the myriad substances produced by living things could be created from just three—or occasionally four—elements: carbon, hydrogen, oxygen, and (sometimes) nitrogen. In 1840, while trying to solve this mystery, French chemist Jean-Baptiste Dumas published his theory of types, which helped explain how organic chemicals and their characteristics can be understood in terms of functional groups.

Properties of compounds

Dumas's theory followed important work by other chemists. In 1828, Friedrich Wöhler's synthesis of an organic chemical, urea, and his discovery with Justus von Liebig of isomers (molecules that share the same chemical formula but differ in the arrangement of atoms) were both landmark moments. But what

We ... seize the key to all changes of matter, so sudden, so swift, so singular, that occur in animals and plants.
Jean-Baptiste Dumas

really interested both young scientists, and their mentor Jöns Jacob Berzelius, were the insights these discoveries provided into the structure of chemicals. It was becoming clear that the properties of a compound are due not just to the combination of chemicals, but also to the way they are joined.

In around 1820, German chemist Eilhard Mitscherlich had discovered isomorphism in crystals. By measuring crystal angles precisely, he found that different compounds, such as arsenates and phosphates, could produce crystals of exactly

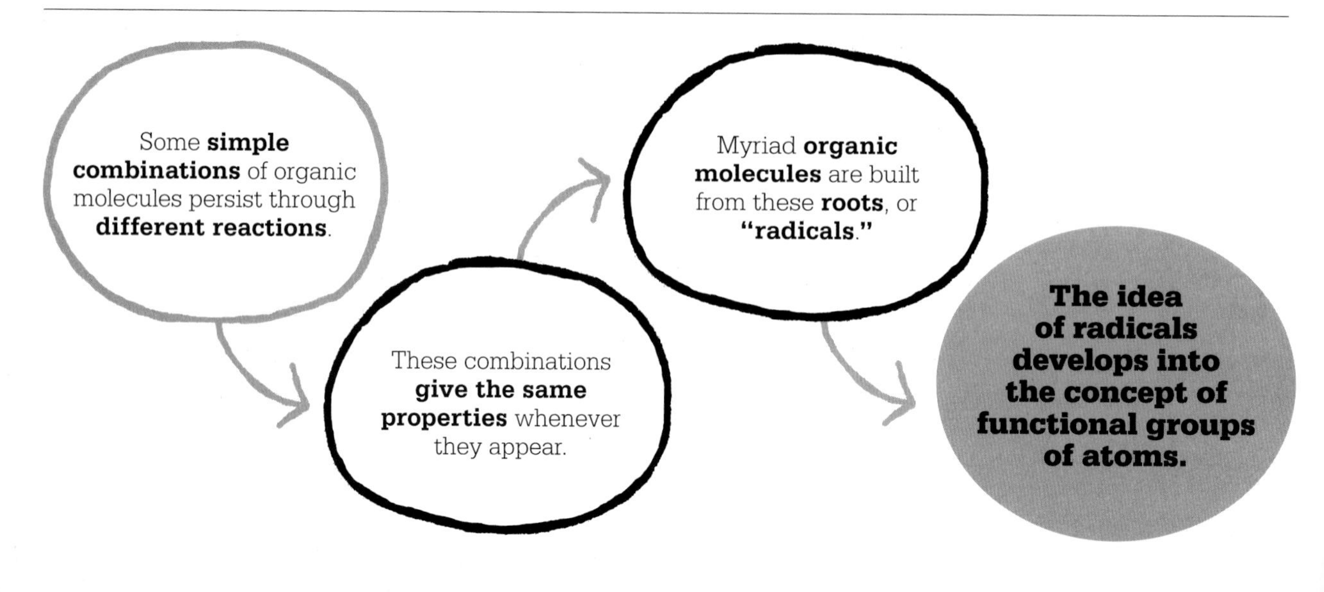

Some simple combinations of organic molecules persist through different reactions.
These combinations give the same properties whenever they appear.
Myriad organic molecules are built from these roots, or "radicals."
The idea of radicals develops into the concept of functional groups of atoms.

See also: The atomic universe 28–29 ▪ Isomerism 84–87 ▪ The synthesis of urea 88–89 ▪ Structural formulae 126–127 ▪ Benzene 128–129 ▪ The electron 164–165 ▪ The hole in the ozone layer 272–273

the same shape. This discovery suggested that some compounds combine in particular arrangements that define their properties.

Liebig and Wöhler began to research organic compounds, starting with oil of bitter almonds, a substance made by the steam distillation of plum and cherry stones. They reacted the oil, which is mostly benzaldehyde (C_7H_6O), with various chemicals, such as oxygen, bromine, and chlorine, then analyzed the compounds made. Fascinatingly, every new compound formed contained C_7H_5O—seven atoms of carbon, five of hydrogen, and one of oxygen—so it seemed as though C_7H_5O was a core group of atoms, persisting through these reactions and not itself reacting. Liebig and Wöhler called it a "radical," or root, after a term introduced many years earlier by French chemist Louis-Bernard Guyton de Morveau for inorganic chemicals. They named it benzoyl.

Together with Berzelius, Liebig and Wöhler developed the idea that clusters of organic chemicals linked to make various radicals, which build into compounds. Berzelius was an advocate of the dualistic idea that compounds represent the coming together of negative and positive elements. He argued that negative and positive radicals combined in the same way.

Oil of bitter almonds, which is used to treat a variety of ailments, is mostly benzaldehyde. This molecule consists of two functional groups—a benzene ring connected to an aldehyde group.

The hunt for radicals

Chemists across Europe eagerly joined the hunt for more radicals. Liebig found acetyl; Berzelius found ethyl; German chemist Robert Bunsen found cacodyl; Italian chemist Raffaele Piria found salicyl; and Dumas found methyl, cinnamyl, and cetyl. It seemed, though, that these radicals could sometimes lose or gain atoms to create other radicals. Dumas discovered that he could make methyl (CH_3), for instance, by adding water to a simpler group called methylene (CH_2), the *-ene* suffix being Greek for daughter.

As chemists discovered other groups of atoms that persisted in different molecules, a naming system developed to reflect the number of carbon, hydrogen, or oxygen atoms. For example, propyl is a group with three carbon atoms, and butyl is a group with four carbon atoms. Propyl has three carbon atoms and seven hydrogen, and propylene (the "daughter") has »

Jean-Baptiste Dumas

Born in 1800 in Alès, France, Jean-Baptiste Dumas moved to Geneva, Switzerland, at age 16 to study pharmacy, chemistry, and botany. He published several papers while still a teenager, and during his 20s, he worked as a professor of chemistry at some of the leading French academies.

One of the pioneers of organic chemistry, Dumas first espoused Berzelius's theory of radicals and discovered several himself, then rejected the idea for his own rival theory of substitutions. From the mid-1840s, Dumas concentrated on teaching, mentoring rising stars such as Louis Pasteur. In 1859, he became president of Paris's municipal council, where he oversaw the city's lighting, drainage, and water supply system. Dumas died in 1884. He is one of the group of French scientists whose names were engraved on the Eiffel Tower.

Key works

1837 "Note on the Present State of Organic Chemistry"
1840 "On the Law of Substitution and the Theory of Types"

three carbon and six hydrogen. Butyl has four carbon and nine hydrogen, and butylene has four carbon and eight hydrogen.

By 1837, Dumas was convinced that radicals were as fundamental a turning point for organic chemistry as Antoine Lavoisier's classification of the elements. Now, he believed, organic chemists would simply have to identify the different radicals, just as inorganic chemistry was a question of identifying elements.

Theory of substitutions

Things turned out to be considerably more complex than Dumas thought, however. One flaw came from his own work. One story tells how guests at a Paris ball were thrown into violent coughing fits by acrid candle fumes, and Dumas was asked to investigate. He discovered that the candles had been bleached with chlorine, and the wax (a fatty ester) had reacted in such a way that the chlorine replaced the hydrogen in the wax to produce hydrogen chloride fumes. Intrigued, Dumas went on to find that chlorine can also replace hydrogen in other organic

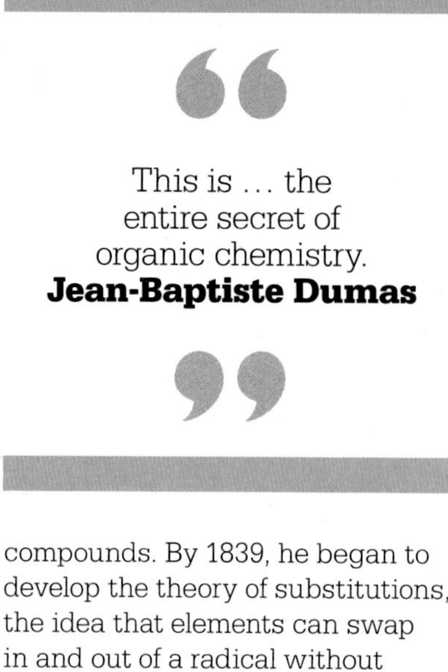

This is ... the entire secret of organic chemistry.
Jean-Baptiste Dumas

compounds. By 1839, he began to develop the theory of substitutions, the idea that elements can swap in and out of a radical without drastically affecting its properties. Berzelius was incensed because this idea seemed to challenge his dualistic plus–minus theory. He reasoned that if chlorine combines with hydrogen, then it must have an opposite charge—so how could the two swap places?

In 1840, Dumas presented his key paper on substitution in the leading chemical journal *Annalen die Chemie und Pharmacie,* edited by Liebig—but Liebig disowned the theory in his editorial and went so far as to publish a spoof letter from Wöhler, posing as the mysterious chemist S. C. H. Windler ("swindler"), in which the author claimed to have discovered new substitutions of chlorine. However, the dualistic theory of plus and minus was on its way out. We now know that substitutions can indeed happen, and this depends on the orientation of the electron orbits. (Electrons were not discovered until 1897.)

Throughout the 1840s, Dumas and other French chemists developed their theory and talked of "types" of molecules rather than radicals. They identified at least four types—water, hydrogen, hydrochloric acid, and ammonia—around which others could build.

Molecular arrangement

Young French chemist Auguste Laurent went further with his nuclear theory, in which compounds are built up from simple clusters of atoms' nuclei. He argued that elements can be substituted without changing most properties of a compound.

The difference between the competing theories can seem impenetrable. The confusion came from the way chemists at the time saw molecules as little more than formulae, hypothetical combinations of atoms. When Dumas suggested that molecules orbit atoms similar to the way in which planets orbit the Sun, he was still talking in the abstract. But around 1850, British chemist Alexander Williamson; Laurent;

Plastic drink bottles are often made from polyethylene terephthalate (PET). This compound contains two hydroxyl groups—one hydrogen atom bonded to one oxygen atom.

and another French chemist, Charles Gerhard, realized that the arrangement of atoms within molecules matters, and they began to present formulae in entirely different ways. In these new formulae, types were presented by arrangements of the chemical symbols, linked by bracket lines. Williamson showed that alcohol and ether, for example, belong to the "water type" by presenting them like this:

$\left.\begin{matrix}H\\H\end{matrix}\right\}O$	$\left.\begin{matrix}C_2H_5\\H\end{matrix}\right\}O$	$\left.\begin{matrix}C_2H_5\\C_2H_5\end{matrix}\right\}O$
Water	**Alcohol**	**Ether**

Williamson argued that these new formulae represented the type, just as orreries (mechanical models of the Solar System) represent the arrangement of the planets. Chemists came to realize that the properties and reactions of molecules depend on their three-dimensional arrangements and links—and crucially, also on their valency (the combining power of an element), which made radicals and types obsolete.

The features of functional groups

Functional groups are groups of atoms within organic molecules that have their own observed characteristic properties. These properties appear no matter which other atoms they are teamed with. More complex molecules may contain more than one functional group. Common examples are alcohols, amines, ketones, and ethers. They might be identified by, for instance, double carbon bonds (C=C), alcohol groups (–OH), carboxylic acid groups (–COOH), and esters (–COO). Molecules with the same functional group behave in similar ways—maybe having a higher or lower boiling point or reacting with certain chemicals. Essentially, all organic chemicals are inert hydrocarbons with one or more functional groups attached, which determine how the chemicals behave. This concept is useful for predicting how organic chemicals will react and also in the synthesis of new molecules with particular properties.

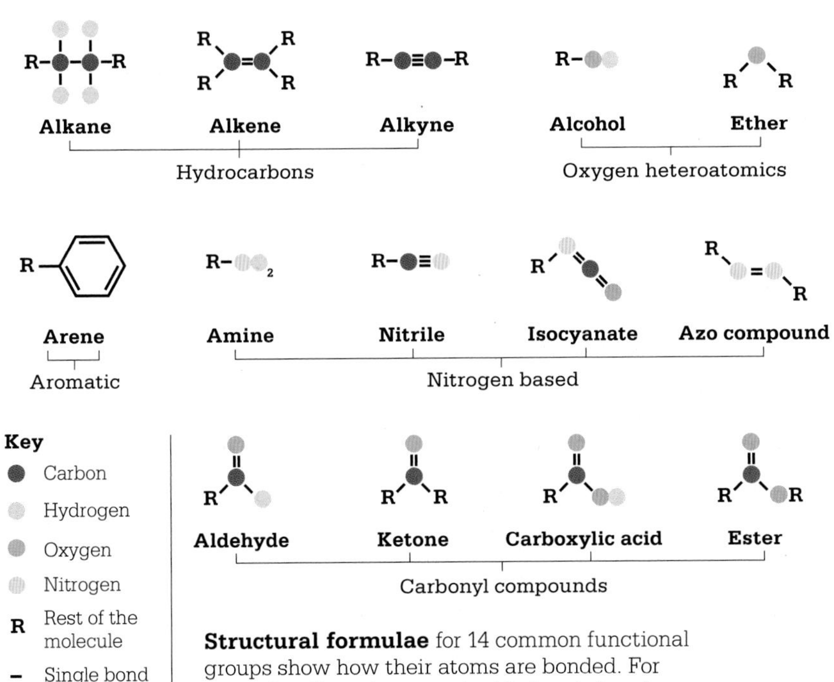

Structural formulae for 14 common functional groups show how their atoms are bonded. For example, alkene has two double-bonded carbon atoms (depicted by a double line), and each of its carbon atoms is bonded to two hydrogen atoms.

Nevertheless, the idea of basic atom combinations persisted and developed into an understanding of functional groups, a key organizing principle in organic chemistry. Functional groups are specific combinations of atoms or bonds within a compound that determine its characteristics and have the same effect regardless of which compound they are components. There are 14 common functional groups and 26 less common ones.

The rise and fall of radicals

The term "radical" persists in medicine, where free radicals are specific chemical entities, very different from Liebig's idea. Discovered by Moses Gomberg in 1900, they are combinations of atoms that roam "free"—so they are not "roots" but molecules with an uneven number of electrons that rob other molecules of electrons.

Although the idea of radicals and types was overturned long ago, it represented the first big step in opening up the entire field of organic chemistry, which today delivers everything from plastics to pharmaceuticals and aids our understanding of how life works. ■

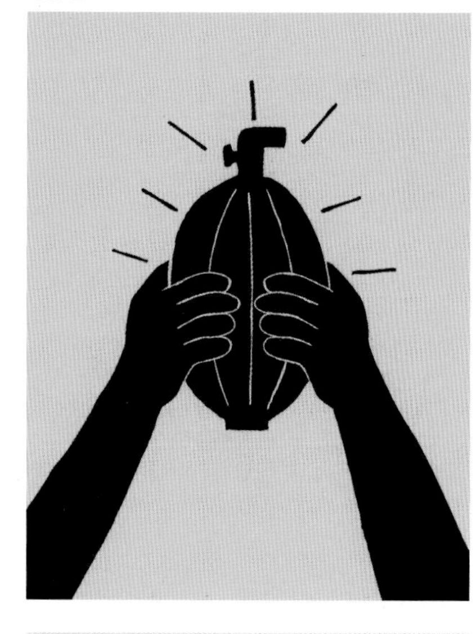

O, EXCELLENT AIR-BAG!

ANESTHETICS

IN CONTEXT

KEY FIGURE
Crawford Long (1815–1878)

BEFORE
6th century BCE The Indian physician Sushruta suggests sedating patients with wine and cannabis oil.

1275 CE Spanish physician Raymundus Lullius discovers ether, calling it "sweet vitriol."

1772 Joseph Priestley discovers nitrous oxide, or "laughing gas."

AFTER
1934 Anesthesiologists in the US use thiopentone, the first intravenous anesthetic.

1962 American chemist Calvin Phillips creates Ketamine, an anesthetic with limited effects on breathing and blood pressure.

1990s Sevoflurane, an inhaled general anesthetic, becomes widely used because it acts quickly and recovery is rapid.

Before anesthetics, a surgeon's only balm for a patient undergoing an amputation was a bottle of rum and speed with the saw. The first true anesthetics emerged in the early 19th century, when some newly discovered gases were found to render patients unconscious.

In 1799, Humphry Davy made up a large batch of nitrous oxide (N_2O) by heating crystals of ammonium nitrate (NH_4NO_3) and bubbling it through water. He piped the gas into a sealed box specially made for inhaling gases and sat in it for over an hour. When he emerged, he was overwhelmed with both the heightening of his senses and his desire to burst out laughing.

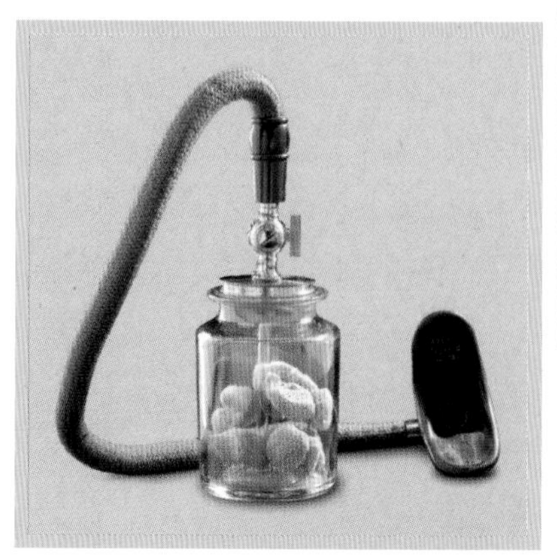

The sensation caught on, and "laughing gas" parties became fashionable. Yet the results were too unpredictable to use for surgery. In 1818, Michael Faraday noticed that ether ($C_4H_{10}O$) had similar effects, and this soon became the party gas of choice because it was easier to make. In the US in 1842, surgeon Crawford Long made a huge breakthrough, using ether to painlessly remove a patient's abscess. American dentist Horace Wells then began to pull teeth using the gas, and in 1846, Wells's former partner, surgeon Robert Morton, used it to put a patient to sleep while he excised a tumor.

General and local

Ether and laughing gas worked well for quick operations, but something else was needed for long surgery. In 1847, Scottish surgeon James Simpson suggested vapor from a few drops of chloroform ($CHCl_3$) sprinkled on a cloth. This worked, and chloroform became the first choice for anesthesia.

Patients used this inhaler in the mid-19th century to breathe in fumes from sponges in the vessel that had been soaked in ether.

See also: The new chemical medicine 44–45 ▪ Gases 46 ▪ The chemicals of life 256–257

When are you going to begin?
Frederick Churchill
Following his leg amputation under ether, London (1846)

The next big breakthrough came almost a century later, when Canadian anesthesiologist Harold Griffith realized that it was not always necessary to send patients to sleep with "general" anesthetics. He knew that some indigenous South American tribes tipped their hunting arrows with the poison curare, which paralyzes muscles. Griffiths developed a safe version of curare called Intracostin. In 1942, he successfully used this on an appendectomy patient. Similar muscle relaxants are now used for a great many major operations, in conjunction with general anesthetics. Initially sold in 1948, lidocaine was the first local anesthetic to work by blocking nerve signals. Since then, many other anesthetic chemicals have been introduced.

Nerve signal interruption

General anesthetics work by interrupting the transmission of nerve signals between the brain and body at the synapses, but it is not known how. It seems likely that they disrupt proteins in the nerve cell membranes. The patient loses all awareness, yet breathing and blood circulation function normally.

Local anesthetics such as lidocaine and novocaine bind to and inhibit the sodium ion channel in the membrane of nerve cells, so blocking nerve transmission to pain centers in the central nervous system. Only the area immediately around the injection is affected. ■

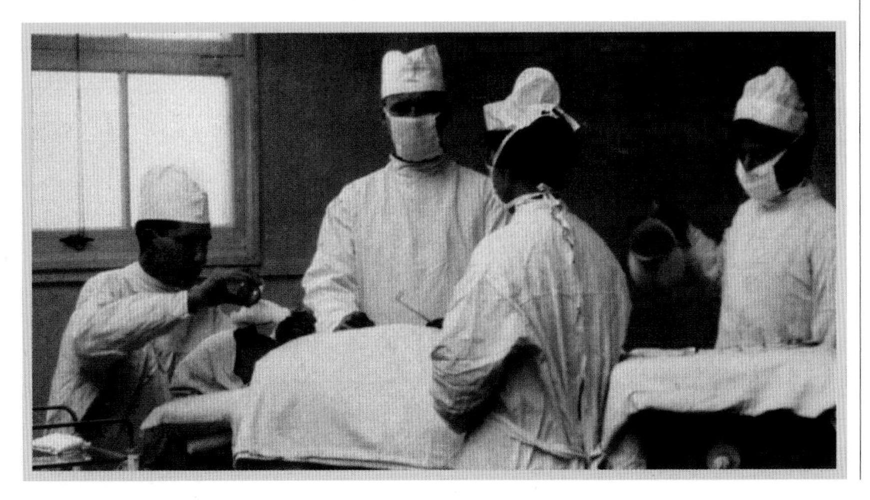

A US Army physician places a chloroform-soaked cloth over a patient's nose to administer anesthesia in Britain during World War I.

Crawford Long

Born in Georgia in 1815, Crawford Long was the son of a senator and plantation owner. He studied medicine, and while training as a surgeon, he saw the painful effects of operating without anesthesia. After completing his training, he worked as an intern in New York City for 18 months before setting up a rural medical practice in Jefferson, Georgia.

While in New York, Long had witnessed "ether frolics"—where partygoers got high on the gas and appeared to feel no pain. Qualified as a pharmacist, he used ether to anesthetize a patient in 1842 and successfully removed a tumor. Long went on to perform dozens more operations under anesthetic. Unaware of this, Robert Morton made a splash with high-profile public operations using ether. It was not until after his death in 1878 that the modest Long was recognized as the true pioneer of ether anesthetics for surgery.

Key work

1849 "The First Use of Sulfuric Ether by Inhalation as an Anesthetic"

THE INDUSTR
1850–1900

IAL AGE

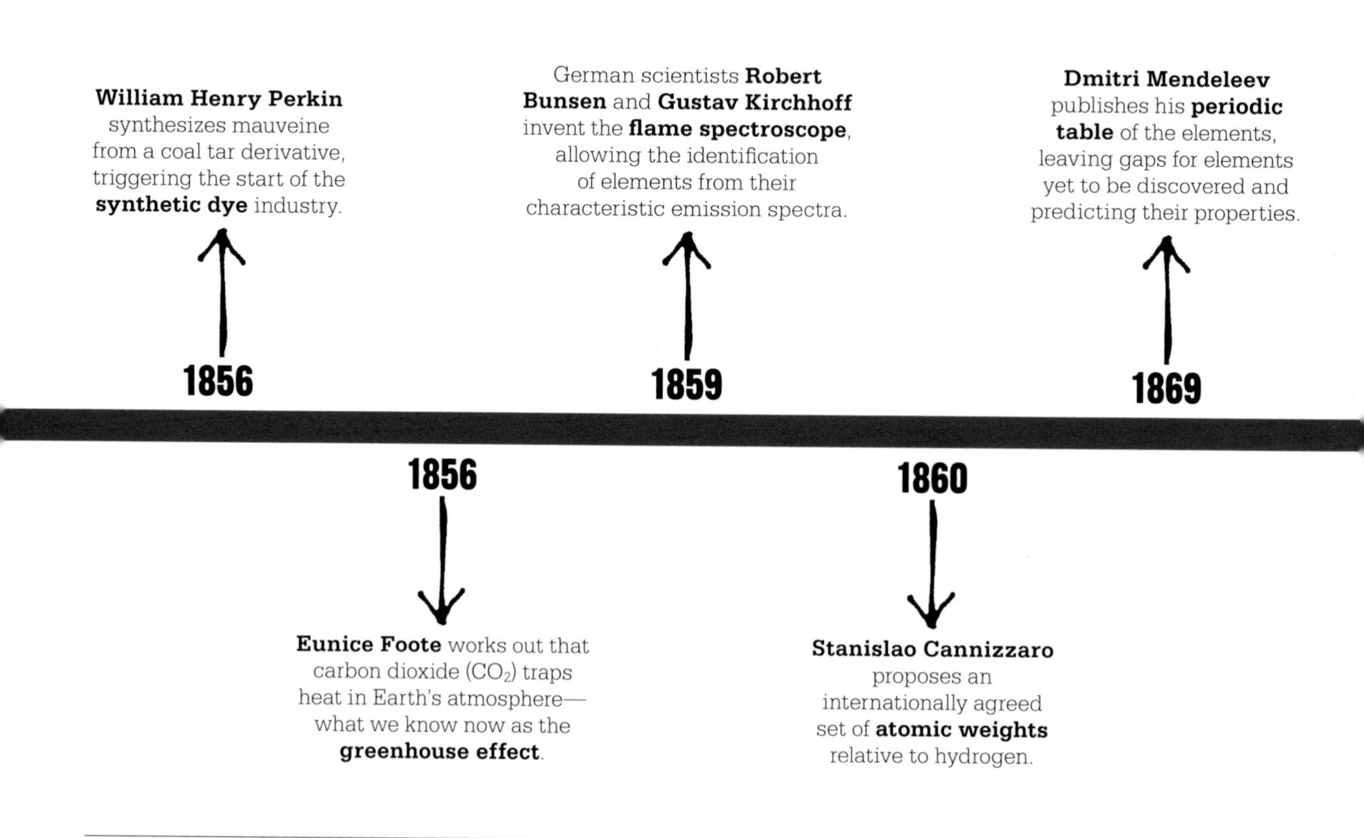

1856
William Henry Perkin synthesizes mauveine from a coal tar derivative, triggering the start of the **synthetic dye** industry.
1856
Eunice Foote works out that carbon dioxide (CO_2) traps heat in Earth's atmosphere—what we know now as the **greenhouse effect**.
1859
German scientists **Robert Bunsen** and **Gustav Kirchhoff** invent the **flame spectroscope**, allowing the identification of elements from their characteristic emission spectra.
1860
Stanislao Cannizzaro proposes an internationally agreed set of **atomic weights** relative to hydrogen.
1869
Dmitri Mendeleev publishes his **periodic table** of the elements, leaving gaps for elements yet to be discovered and predicting their properties.

The industrial advances of the 19th century brought large-scale change in many areas of life, from mechanization to technological advances and rapid industrialization. In the field of chemistry there were a range of significant advances. Notably, coal, the fossil fuel that powered the revolution, offered new raw materials to exploit. The synthetic dyes industry grew from the creation of a dye from aniline, a compound derived from coal tar (a by-product of coal extraction). This industry, in turn, prompted the development of chemical processes generating myriad substances, from fuel to fertilizers to medicines.

Even in the earliest years of their use, however, concerns about the effect of fossil fuels on the atmosphere were present. In the 1850s, American scientist Eunice Foote's research into the thermal properties of gases led her to conclude that an atmosphere with a greater proportion of carbon dioxide (CO_2) would be warmer than one with a lower proportion—the first statement of what we now know as the greenhouse effect. Fifty years later, Svante Arrhenius identified that it was human industrial activity that was increasing the proportion of carbon dioxide in Earth's atmosphere.

Away from industry, scientists were making important advances in their understanding of chemistry at an atomic level, leading to the discovery of new elements, new concepts of atoms and molecules, and—most significantly—the creation of the periodic table of the elements.

Organizing elements

Trying to define and classify the elements had been an obsession of scientists since the time of the Ancient Greeks. In later centuries, as alchemy gave way to chemistry, this organization had been hindered first by a lack of clarity on what constituted an element and later by inaccurate data. However, French chemist Antoine Lavoisier's listing of elements and German chemist Johann Döbereiner's concept of triads, together with international agreement on a standard set of atomic weights for elements in 1860, set the stage for a more logical arrangement.

The periodic table published by Dmitri Mendeleev in 1869 organized the 56 elements known at the time into a system of rows and columns that made the

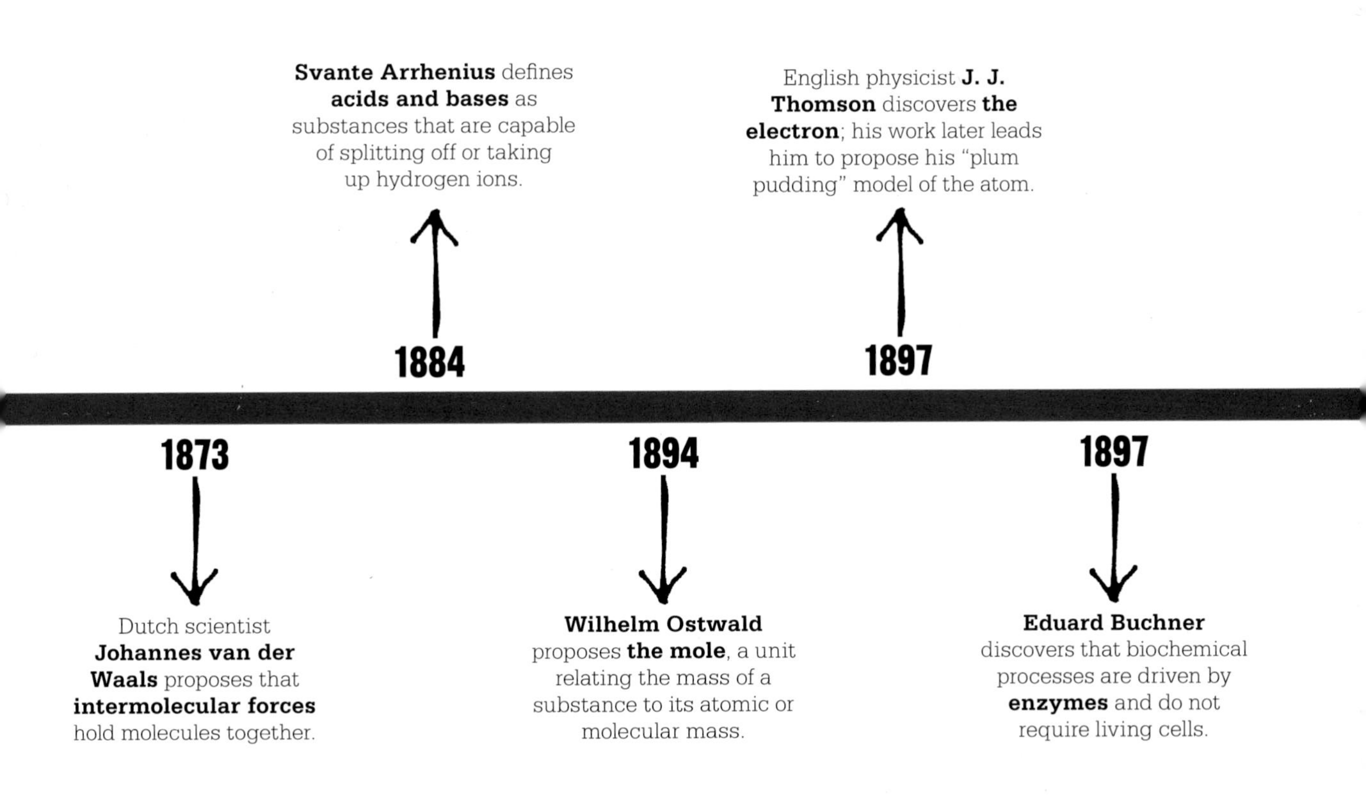

relationships between elements' properties clear—and, crucially, allowed missing elements and their properties to be predicted. The discovery of further elements would prove Mendeleev's predictions to be accurate. His periodic table has undergone further revision and the addition of new elements to take the form that hangs in chemistry laboratories worldwide today.

1884
Svante Arrhenius defines **acids and bases** as substances that are capable of splitting off or taking up hydrogen ions.
1897
English physicist **J. J. Thomson** discovers **the electron**; his work later leads him to propose his "plum pudding" model of the atom.
1873
Dutch scientist **Johannes van der Waals** proposes that **intermolecular forces** hold molecules together.
1894
Wilhelm Ostwald proposes **the mole**, a unit relating the mass of a substance to its atomic or molecular mass.
1897
Eduard Buchner discovers that biochemical processes are driven by **enzymes** and do not require living cells.

Chemical fundamentals

Another area of innovation was the understanding of the behavior of atoms and molecules in general and in chemical reactions. While it had become established knowledge that many substances were made of molecules, what held molecules together within substances was still unclear. A Dutch scientist, Johannes van der Waals, was the first to suggest the concept of weak forces between molecules as the molecular "glue." When these forces were specifically characterized in the 1930s, van der Waals's name was given to describe them.

The introduction of the concept of the mole, a measurement of the quantities of atoms or molecules, made calculations for chemical reactions much easier. It simplified the expression of the huge numbers of chemical entities involved in reactions and was easy to relate to the newly agreed relative atomic masses of elements.

The key question of why some reactions take place with ease, while others do not, was addressed by applying thermodynamics to chemistry. American physicist and mathematician Josiah Gibbs used thermodynamic principles to relate energy and entropy changes in reactions, allowing chemists to calculate the possibility of a reaction taking place.

The effect of varying conditions on a chemical reaction was explored by French chemist Henri-Louis Le Châtelier; this principle would play a vital part in the later synthesis of fertilizers.

Finally, the turn of the century saw the discovery of a fundamental particle; J. J. Thomson carried out a series of experiments with cathode rays that allowed him to identify the electron. We now know that chemical reactions are often, at their simplest level, an interchange of electrons, so this was a vital part of the chemical puzzle—and one that would also lead to rapidly improving models of the atom in the decades to come. ■

THAT GAS WOULD GIVE TO OUR EARTH A HIGH TEMPERATURE

THE GREENHOUSE EFFECT

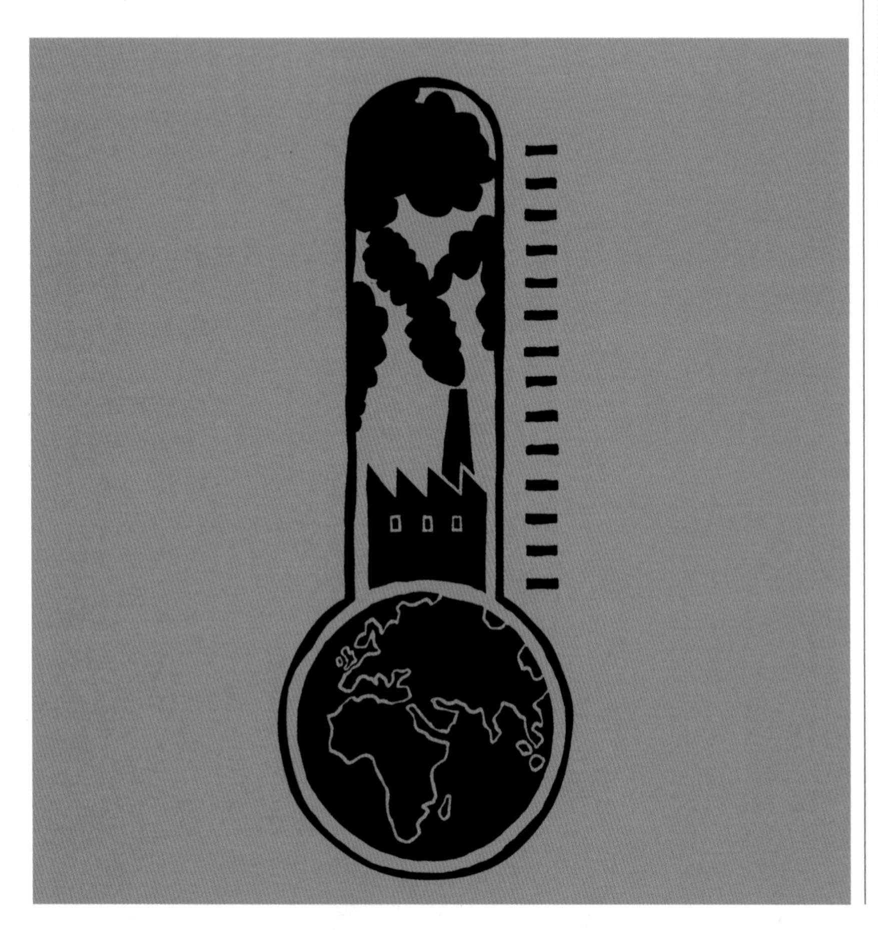

IN CONTEXT

KEY FIGURE
Eunice Newton Foote
(1819–1888)

BEFORE
1824 French physicist Joseph Fourier suggests that Earth's atmosphere insulates it.

1840 Swiss geologist Jean Louis Agassiz posits that Earth experienced a past ice age.

AFTER
1938 British technologist Guy Callendar quantifies the amount of CO_2 emitted by human activity and the rise in global temperature over the previous 50 years.

2019 A record 42 billion tons (38 billion metric tons) of CO_2 are emitted.

2021 NASA announces that the previous seven years have been the warmest seven years on record.

In 1856, American scientist Eunice Newton Foote published a highly significant discovery in the field of climate science. She was the first person to observe that carbon dioxide (CO_2) and water vapor absorb heat and to conclude that an increase in atmospheric CO_2 would cause what we now know as the greenhouse effect. Her paper was read at the annual meeting of the American Association for the Advancement of Science, but its full implications were not appreciated, and it was largely forgotten for more than 150 years.

Foote used glass cylinders containing atmospheric air, oxygen, hydrogen, and carbon dioxide, and

See also: Gases 46 ▪ Fixed air 54–55 ▪ Carbon capture 294–295

she experimented with different concentrations of the gases and different degrees of moisture content. With thermometers in the cylinders, she placed some in sunlight and some in shade, then measured changes in temperature over time. Her results showed that, when heated by the Sun, condensed air warmed more than rarefied air (air with a lower oxygen content), and moist air more than dry air. The cylinders containing CO_2 warmed more than any of the other cylinders, and this convinced Foote that an atmosphere with a higher proportion of CO_2 would be warmer. She presented this as a perspective on past conditions on Earth: "If as some suppose, at one period of its history the air had mixed with it a larger proportion than at present, an increased temperature … must have necessarily resulted."

Quantifying change

Unaware of Foote's work on the other side of the Atlantic, Irish physicist John Tyndall made similar discoveries just a few years later, which he published in 1861. With the benefit of more sophisticated heat apparatus, he measured the capacities of different atmospheric gases to absorb radiant heat (infrared radiation). He found water vapor was by far the strongest absorber of infrared and concluded that, of all the atmospheric gases, it would have the most influence on climate. But he also remarked that changes in the atmospheric content of CO_2 and other hydrocarbons would have climatic effects.

In 1896, Swedish chemist and physicist Svante Arrhenius investigated the link between atmospheric CO_2 and Earth's regular glaciations. After extensive calculations, he concluded that doubling or halving the amount of CO_2 in the atmosphere would cause global temperature to rise or fall by 9–10.8°F (5–6°C). Arrhenius also identified human industrial activity as the main source of new CO_2. But his estimate of the rate of future change was to prove extremely conservative. He suggested that a 100 percent increase in CO_2 levels would take 3,000 years. Today, it »

The Industrial Revolution marked the shift to mechanized manufacturing, powered by fossil fuels such as coal and, later, oil and natural gas.

Eunice Newton Foote

One of 12 children, Eunice Newton was born in Connecticut in 1819. She grew up in Ontario County, New York, attending the Troy Female Seminary and a nearby science college. In 1841, she married judge and inventor Elisha Foote. Eunice was a signatory to the Declaration of Sentiments of the 1848 Seneca Falls Convention, the first women's rights convention. On August 23, 1856, at the eighth annual meeting of the American Association for the Advancement of Science, her paper on greenhouse gases was presented by John Henry of the Smithsonian Institution.

In addition to her pioneering work on the thermal properties of gases, Foote also studied their electrical excitation. She was a keen botanist and an accomplished artist, as well as an inventor: she received a patent for vulcanized-rubber shoe soles and created a new type of paper-making machine. She died in Lenox, Massachusetts, in 1888.

Key work

1856 "Circumstances Affecting the Heat of the Sun's Rays" is estimated that, if current trends continue, this increase will occur by the end of the 21st century.

The Keeling curve

By the early 20th century, many scientists suspected that the level of CO_2 in the atmosphere was rising. But there was no hard data to prove it until March 1958, when American geochemist Charles David Keeling installed an infrared gas analyzer at the Mauna Loa Observatory in Hawaii and recorded a CO_2 concentration of 316 parts per million (ppm). Keeling continued to take readings and made two discoveries: that CO_2 concentrations underwent seasonal variations, with a peak in May and a trough in September; and that there was a year-on-year increase.

Keeling explained the first trend as the result of plants absorbing CO_2 from the atmosphere for growth during summer in the northern hemisphere, which contains most of the planet's land. He attributed the second trend to the burning of fossil fuels, such as coal, oil, and natural gas. The Mauna Loa data set is now known as the Keeling curve, and it shows trends identified by Keeling continuing to the present day.

Greenhouse effect

Approximately half of the Sun's energy that arrives on Earth is short-wave ultraviolet radiation (light), while the rest is long-wave infrared radiation (heat). Clouds and ice reflect some of this radiation directly back out into space, and the rest is absorbed by Earth's surface and atmosphere. Most of the absorbed short-wave energy is reradiated at the surface as heat. If all the heat emitted by Earth's surface passed directly into space, the average temperature on the surface would be around 0°F (-18°C). In fact, the average temperature is about 59°F (15°C) because there are greenhouse gases—water vapor, CO_2, methane (CH_4), nitrous oxide (N_2O), and halocarbons—in the atmosphere. Water vapor is the most abundant by far, followed by CO_2. Greenhouse gases absorb the heat radiated from Earth's surface and then slowly reradiate it in all directions. Life on Earth needs these gases because they make the atmosphere behave like an insulating blanket.

The downside of the greenhouse effect is that, since industrialization began in the mid-18th century, the levels of greenhouse gases (apart from water vapor) in the atmosphere have been greatly increased by human activity. In this period, atmospheric CO_2 has increased by about 47 percent, largely as a result of industrial burning of fossil fuels—especially coal and oil—and deforestation, which reduces the biomass of vegetation available to absorb the gas. Since 1958, CO_2 concentrations have risen by as much as 33 percent: the highest reading, recorded in April 2021, was 421 ppm. More than half the increase has occurred since 1980.

The concentration of methane in the atmosphere has risen by over 150 percent since preindustrial

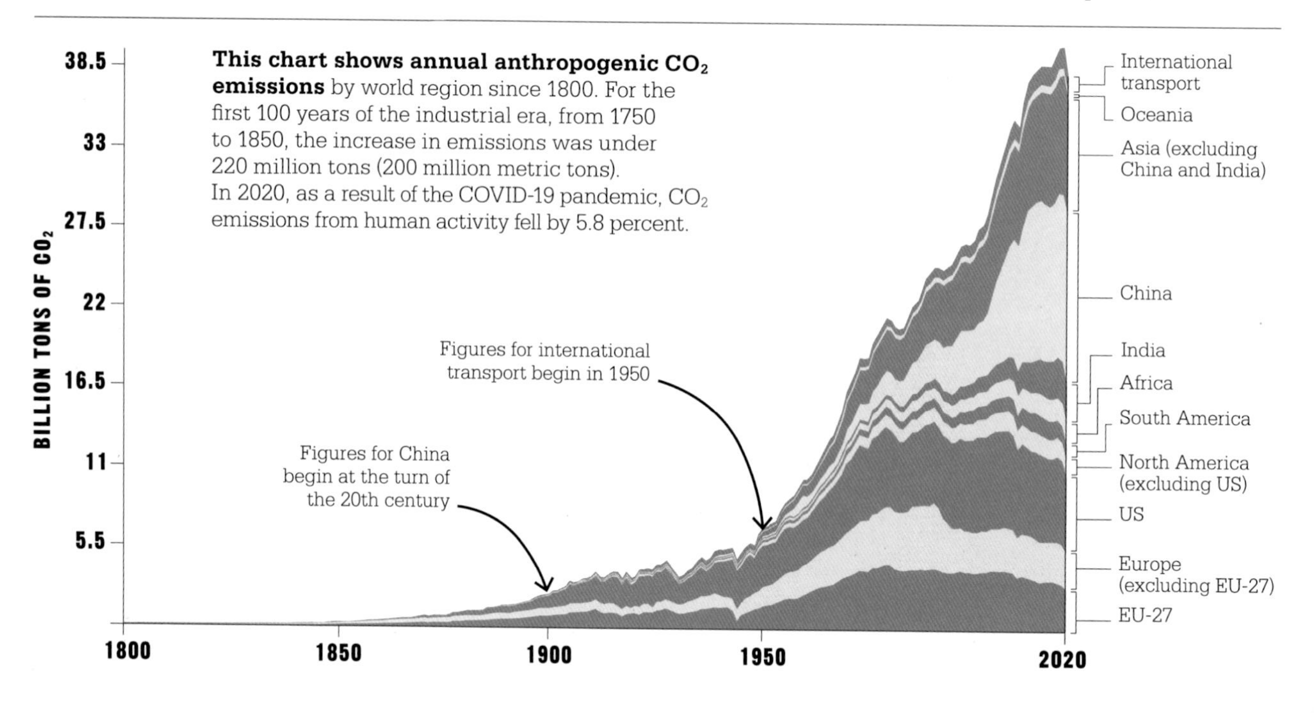

This chart shows annual anthropogenic CO_2 emissions by world region since 1800. For the first 100 years of the industrial era, from 1750 to 1850, the increase in emissions was under 220 million tons (200 million metric tons). In 2020, as a result of the COVID-19 pandemic, CO_2 emissions from human activity fell by 5.8 percent.

times. This is largely as a result of the huge expansion in livestock farming (ruminants, such as cattle and sheep, produce methane when they digest food), in addition to oil and gas production processes. The preindustrial concentration of CH_4 was about 700 parts per billion (ppb); by 2021, it was 1,891 ppb, with most of the increase occurring since 1960. This is of particular concern because, although there is much less CH_4 in the atmosphere than CO_2, methane traps more than 20 times more heat per mass unit than carbon dioxide. (However, while CO_2 can remain in the atmosphere for hundreds of years, atmospheric CH_4 oxidizes within a decade.)

The atmospheric concentration of nitrous oxide has also increased since preindustrial times by nearly 18 percent. Although N_2O represents a fraction of the total volume of greenhouse gases, its insulating effect is 300 times that of CO_2, and it persists in the atmosphere for more than a century.

Global heating acceleration

A warming atmosphere generates positive-feedback effects—processes that accelerate the effect. For example, when permafrost thaws as a result of warmer temperatures, formerly frozen peat bogs release more methane. This is a particular concern in areas that have vast peatlands, such as northern Siberia. When atmospheric warming leads to bigger and more frequent forest fires, the combustion releases even more CO_2. And as ocean water warms, its capacity for absorbing CO_2 from the atmosphere is reduced.

> The eyes of all future generations are upon you. And if you choose to fail us, I say: we will never forgive you.
>
> **Greta Thunberg**
>
> **UN Climate Summit, New York (2019)**

The risk and severity of wildfires is increasing as a result of global heating. Higher temperatures lead to drier, more combustible vegetation and longer fire seasons.

Impacts and response

Although the large number of variables make precise forecasts impossible, climate scientists know that an increased greenhouse effect produces a warmer atmosphere. This in turn melts ice, adding to the volume of ocean water and raising sea levels. A warmer atmosphere is more energetic, resulting in more violent storms and greater extremes of rainfall, wind, and temperature. In August 2021, the Intergovernmental Panel on Climate Change (IPCC) reported that climate changes are being observed in every region on Earth and across the whole climate system. These changes disrupt ecosystems and impact on food production and human health (due to malnutrition, disease, and heat stress), while low-lying coastal regions become uninhabitable as a result of flooding. Poor countries are worst affected because they lack the infrastructure and funds to respond to these threats.

These effects are irreversible in the time frame of hundreds or even thousands of years. To limit the extent of further change, the production of greenhouse gases must be greatly reduced and large-scale deforestation must be halted. Without rapid and substantial reduction in emissions, it will not be possible to limit global heating to 2.7°F (1.5°C) above preindustrial levels. An increase of 3.6°F (2°C) is estimated to be the critical tolerance threshold for agriculture and health. ■

COAL-DERIVED BLUES

SYNTHETIC DYES AND PIGMENTS

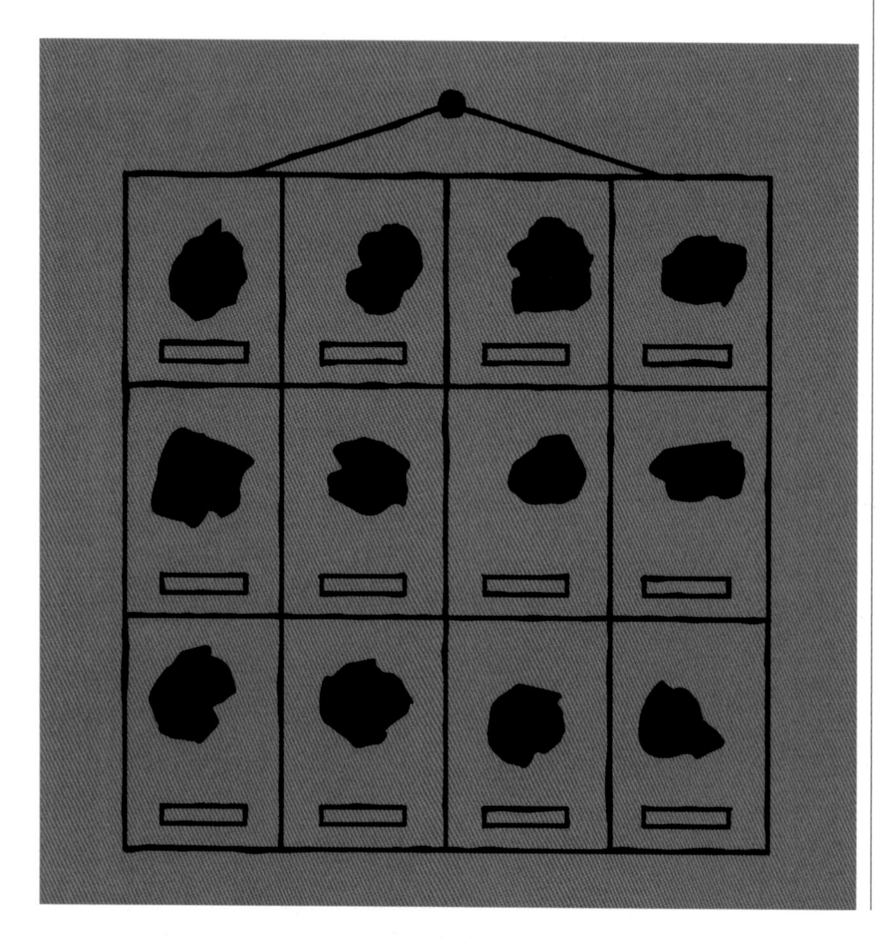

IN CONTEXT

KEY FIGURE
William Henry Perkin
(1838–1907)

BEFORE
***c.*3250 BCE** Earliest known use of Egyptian blue, considered to be the first synthetic pigment.

***c.*1706** Swiss paint-maker Johann Jacob Diesbach makes Prussian blue, the first globally used synthetic pigment.

1834 German chemist Friedlieb Ferdinand Runge isolates aniline from coal tar.

AFTER
1884 Danish microbiologist Hans Christian Gram finds that gentian violet dye stains certain bacteria species but not others—an identification technique still used today.

1932 German chemists Josef Klarer and Fritz Mietzsch, with physician Gerhard Domagk, create Prontosil, the first sulfonamide antibacterial drug, from a red azo dye.

Since prehistoric times, people have used pigments (colored substances) to decorate themselves and their surroundings. Iron oxide (Fe_2O_3) to make reds and minerals to make other colors have been found with pigment-grinding equipment in a cave at Twin Rivers, in Zambia, and dated at 350,000–400,000 years old. The first man-made pigment, Egyptian blue, was created in the 4th millennium BCE by heating a mixture of quartz sand, crushed limestone, soda or potash, and a copper source, such as malachite.

See also: Functional groups 100–105 ▪ Benzene 128–129 ▪ Antibiotics 222–229 ▪ Rational drug design 270–271 ▪ Chemotherapy 276–277

Until the 19th century, dyes for cloth were derived exclusively from plants and other natural sources. Some were expensive to produce. The ultimate example was Tyrian purple, created by the Phoenicians in the 16th century BCE from the mucus of sea snails. It took more than 10,000 snails to make just 1 gram of dye, so the pigment was reserved for the richest and most powerful people; it became known as "imperial purple."

In 1771, Irish chemist Peter Woulfe reported that he had treated the natural dye indigo with nitric acid and produced a yellow dye. This substance, later identified as picric acid, could stain various materials, but it was not used as a dye until the late 1840s, and even then only on a small scale, as it was not particularly colorfast.

Perkin's discovery

A young chemist's accidental discovery in 1856 marked the real birth of the synthetic dye industry. In that year, 18-year-old William Henry Perkin was a research assistant working with August Wilhelm von Hofmann, a pioneering organic chemist, at the Royal College of Chemistry in London. One of Perkin's tasks was to develop a synthetic form of quinine, which was needed as a treatment for malaria in tropical parts of the British Empire but was difficult to extract from its natural source, the bark of the cinchona tree.

In his modest home laboratory, Perkin was experimenting with aniline, a colorless oil derived from coal tar—itself a by-product formed by the production of gas from coal. Attempting to oxidize aniline using potassium dichromate, in the hope of obtaining the alkaloid quinine, Perkin was left with a black residue in his flask. When he tried to wash out the residue with alcohol, he was left with a purple solution. Further experiments showed that the solution would readily dye silk.

Perkin's dye was more resistant to fading than the natural purple dyes in use at the time; in fact, some of his original samples are still noticeably purple today. Modern analysis of historical samples has revealed that, rather than being a single chemical, the dye was a blend of more than 13 different compounds.

In this painting by John Philip, which was commissioned by Queen Victoria, the queen is shown in a mauve gown at the marriage of her eldest daughter on January 25, 1858.

By August 1856, Perkin had taken out a patent for his new synthetic dye. He established a factory and advised the dyeing industry on how best to use the pigment. By 1860, he was rich and famous, largely due to the »

William Henry Perkin

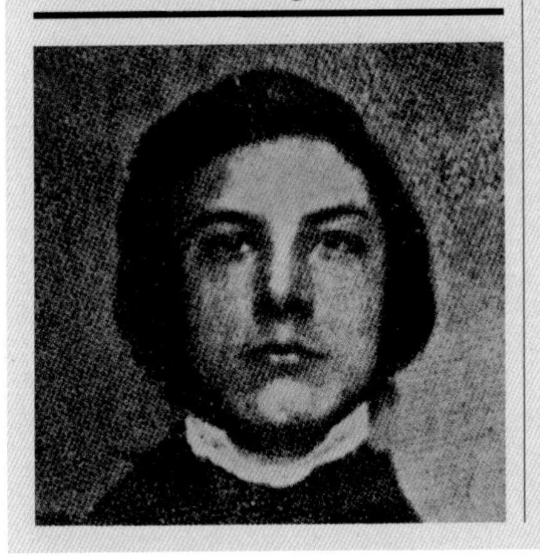

Born in London in 1838, Perkin was the youngest of seven children. His father had hoped he might become an architect; however, after a friend showed him how soda and alum crystallize, the boy developed a strong interest in chemistry.

At 14, Perkin attended the City of London School, one of the first in England to teach chemistry; he had chemistry classes there twice a week. His teacher encouraged him to attend the Saturday afternoon lectures by renowned physicist and chemist Michael Faraday at the Royal Institution.

In 1853, at age 16, Perkin enrolled at the Royal College of Chemistry. August Wilhelm von Hofmann, the director of the college, engaged Perkin in the search for synthetic quinine, which would lead to the discovery of mauveine.

After making a fortune from his discovery of the dye, Perkin sold his factory and retired from industry in 1874, at age 36. In recognition of his achievements, he was elected to the Royal Society in 1866 and knighted in 1906. He died in 1907.

enthusiasm of notables such as Queen Victoria and the Empress Eugénie of France for the new color. He originally named his dye Tyrian purple, after the ancient imperial color, but following its success in France, Perkin renamed it mauveine—and the color it produced mauve—after the French word for the purple mallow flower.

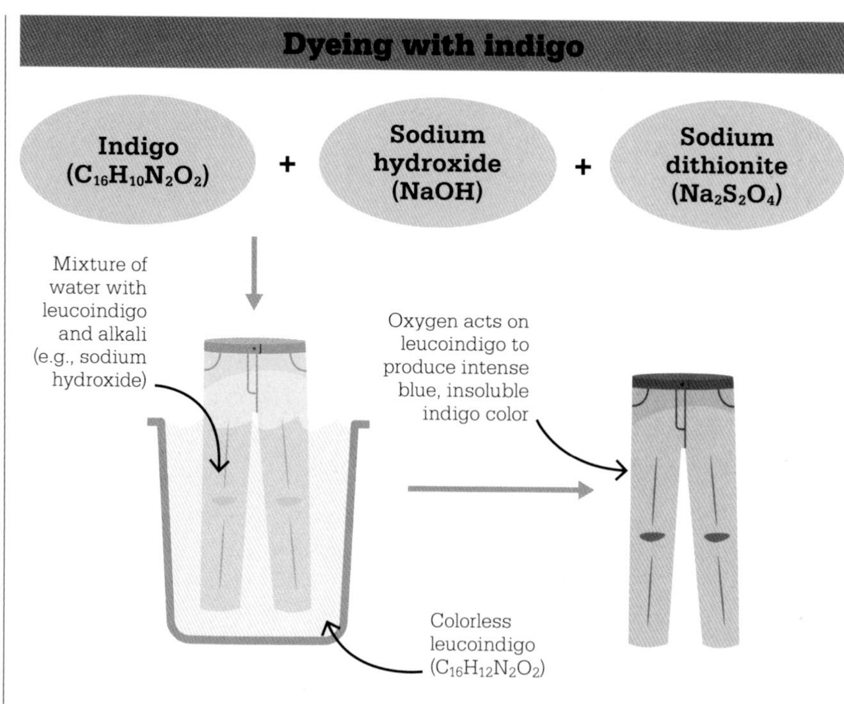

Natural or synthetic indigo is not soluble in water; it must first be mixed with a reducing agent, such as sodium dithionite, to make it soluble. This form, called leucoindigo, is colorless. The fabric or yarn is immersed in a mixture of leucoindigo and alkali. It is then exposed to air; oxygen converts the leucoindigo back to indigo.

Reds, violets, and magenta

One problem that Perkin faced with mauveine manufacturing was the low yield of dye—often as little as 5 percent of the source material. His discovery prompted a push to produce further so-called aniline dyes, and higher yields of dye, from the huge variety of chemicals found in coal tar.

In 1859, François-Emmanuel Verguin, an industrial chemist in Lyon, France, used stannic chloride as an alternative oxidizing agent to potassium dichromate and created a red dye, with a much higher yield than Perkin had managed. Because stannic chloride was expensive, however, cheaper oxidants, such as mercury nitrate and arsenic acid, were used to manufacture the dye. Verguin named his dye fuchsine, possibly after the fuchsia plant; it was later renamed magenta, to mark the French-Sardinian victory over the Austrians at the Battle of Magenta (northern Italy) in 1859.

Hofmann made further studies of amines, the class of compounds that includes aniline, and tried reacting aniline with a variety of organic chemicals. In 1858, he reported that he had obtained "a magnificent crimson color." Then, in 1862, Edward Nicholson, a former student, asked him to determine the chemical composition of magenta. This work led Hofmann to add different functional groups—the atoms producing the characteristic reactions of a compound—to the magenta molecule to create new shades. One reaction produced brilliant violets, which were named Hofmann's violets and were soon being manufactured by Nicholson. These were the first dyes created by deliberate scientific research rather than through trial and error.

> Chemists have always been desirous of producing natural organic bodies artificially.
> **William Henry Perkin**
> **Journal of the Society of Art (1869)**

Turkey red and indigo

Between 1869 and 1870, Perkin and, independently, German chemists Carl Graebe, Carl Liebermann, and Heinrich Caro discovered a way to synthesize alizarin (also known as Turkey red), a commercially important red dye derived from the root of the madder plant. This pigment had been in use for centuries, notably to dye military uniforms. The German chemists patented their process one day before Perkin, but Perkin's factories alone were soon producing

An array of distillation flasks stand ready for use in this illustration of a 19th-century French aniline production workshop. It is taken from Julien Turgan's eight-volume survey of French industry, *Les Grandes Usines en France*, published between 1860 and 1868.

more than 441 tons (400 metric tons) of synthetic alizarin a year at half the cost of the natural product. Following this success, the large-scale cultivation of madder soon ceased.

Another important natural dye was indigo, derived from plants in the genus *Indigofera*. Highly prized for millennia, indigo was referred to as "blue gold." German chemist Adolf von Baeyer determined its chemical structure in 1865, and by 1870, he had succeeded in preparing a sample of synthetic indigo from the chemical isatin. However, this process was too expensive to be scaled up for industry. It was not until 1890 that Karl Heumann discovered a way to manufacture indigo using anilin, a chemical that was readily available. In 1897, the first synthetic indigo was sold by BASF in Germany; by the start of World War I, the Germans were responsible for more than 80 percent of global dye production, eclipsing Britain and other countries.

Dangers and benefits

From the beginning, synthetic pigments could be hazardous to workers and even customers. For example, Scheele's Green, a vivid green pigment invented in 1775 by Swedish chemist Carl Wilhelm Scheele, was a highly toxic arsenic compound that was nevertheless used for wallpaper, paints, and even children's toys. In the 1860s, the arsenic acid-based process for synthesizing fuchsine produced fuchsine arsenite, which could contain up to 6 percent arsenic. Factory workers suffered ulcerations of the nose, lips, and lungs. Newspapers reported that women developed skin rashes after their dresses were exposed to rain or perspiration. In 1864, Swiss dye firm J.J. Muller-Pack was forced to close after people living near its Basel factory fell ill because of arsenic contaminating their well water, and the Renard Frères factory in Lyon, France, ceased magenta production after poisoned wells caused fatalities. Even today, the use of toxic chemicals in the dye industry is a significant health and environmental concern.

By contrast, some synthetic dyes have proved to have beneficial uses in medicine. In particular, synthetic dyes have been used to stain cell samples to reveal disease microorganisms and cellular structures and even to produce drug treatments—including antimalarial drugs, in a surprising reversal of William Perkin's original work. ■

Drugs from dyes

Synthetic dyes have proven useful in medicine as well as the textile industry. German doctor Paul Ehrlich was one of the first scientists to use aniline dyes as biological stains. During the 1880s, he found that certain cells took up certain dyes. In particular, methylene blue, a dye made by the German chemical company BASF, stained live neurons (nerve cells) and plasmodia (malaria parasites) bright blue. In 1891, he began working with German physician Robert Koch, whose use of dyes to stain cells had led to the discovery of the tuberculosis bacillus, to find substances that would attack disease-causing organisms directly while leaving healthy cells unaffected. Ehrlich tested hundreds of chemicals. One such chemical was the dye trypan red, which was effective against trypanosomes—the microorganisms responsible for trypanosomiasis, or sleeping sickness. Ehrlich thus proved that dyes could be used as antibacterial agents and started a pharmaceutical revolution.

POWERFUL EXPLOSIVES HAVE ENABLED WONDERFUL WORK

EXPLOSIVE CHEMISTRY

IN CONTEXT

KEY FIGURE
Alfred Nobel (1833–1896)

BEFORE
c. 9th century Chinese alchemists discover gunpowder, the first explosive.

1679 Gunpowder is used for the first time in civil engineering, in the construction of the Malpas Tunnel in France.

AFTER
1870 Dynamite is first used in a bomb during the Franco–German war.

1891 German chemist Carl Häussermann discovers that TNT (trinitrotoluene), invented as a yellow dye, has explosive properties.

1940–45 Allied forces drop 2.7 million tonnes of high explosive bombs in Europe.

The first modern explosive, nitroglycerine, was invented by Italian chemist Ascanio Sobrero in 1846. Formed by adding glycerine to a mix of nitric and sulfuric acids, it was much more powerful than gunpowder—the only explosive previously available—but too unstable to be used safely.

To make handling nitroglycerine much safer, Swedish chemist and businessman Alfred Nobel invented the blasting cap in 1865. This small wooden detonator held a black powder charge, placed in a metal container of nitroglycerine. The blasting cap could be detonated by lighting a fuse or by an electrical spark to set off the nitroglycerine.

Nobel then mixed the oily nitroglycerine with *kieselguhr* clay; this turned it into a paste, which could be shaped into rods that were easier to handle. In 1867, he patented this idea as dynamite. Then Nobel mixed nitroglycerine with gun cotton (nitrocellulose). Accidentally discovered in 1832 by German-Swiss chemist Christian Friedrich Schönbein, gun cotton was produced by dipping cotton in nitric acid and sulfuric acid, creating a flammable substance that explodes on impact. Patented in 1875, Nobel's new invention was called gelignite. It was more stable and as effective as nitroglycerine and could be used underwater.

Dynamite and gelignite became the standard explosives used in construction, mining, and drilling. ■

In his will, Alfred Nobel stipulated that his fortune was to be used for prizes in Physics, Chemistry, Physiology or Medicine, Literature, and Peace; these became known as the Nobel Prizes.

See also: Gunpowder 42–43 ▪ Why reactions happen 144–147

TO DEDUCE THE WEIGHT OF ATOMS

ATOMIC WEIGHTS

IN CONTEXT

KEY FIGURE
Stanislao Cannizzaro
(1826–1910)

BEFORE
1789 Antoine Lavoisier sets out the principle that mass is neither created nor destroyed in a chemical reaction.

1808 John Dalton introduces his atomic theory, suggesting that atoms of different elements have different masses.

1826 Jöns Jacob Berzelius publishes a table of atomic weights based on experimental analysis that is very close to modern values.

AFTER
1865 Austrian scientist Johann Josef Loschmidt determines the number of molecules in a mole, later termed Avogadro's number.

1869 Dmitri Mendeleev produces his periodic table of the elements based on atomic weights.

In 1811, Amadeo Avogadro had hypothesized that equal volumes of gases at the same temperature and pressure contain equal numbers of molecules. He further suggested that simple gases were not formed from solitary atoms but were compound molecules of two or more atoms. Few scientists accepted these ideas.

In "Sketch of a course of chemical philosophy," an 1858 paper, Italian chemist Stanislao Cannizzaro tried to demonstrate how Avogadro's hypothesis would allow chemists to measure atomic weights. Determining atomic weights was a matter of dispute, because atoms and molecules were often treated interchangeably. In 1860, at the first-ever international chemical congress, Cannizzaro made a compelling case for his ideas. More than 140 of the world's foremost chemists attended the congress. Cannizzaro emphasized that since all atomic weights are relative, one standard weight should be chosen against which all other values could be set. He chose hydrogen, but since he knew it to be diatomic (made of two atoms), as Swedish chemist Jöns Jacob Berzelius had shown, Cannizzaro took half a molecule of hydrogen as his unit value.

> The scales fell from my eyes and my doubts disappeared.
>
> **Julius Lothar Meyer**
> **On reading Cannizzaro's "Sketch"**

Copies of Cannizzaro's "Sketch" were distributed to the congress participants, who were persuaded by his (and Avogadro's) case. Among them were young Russian chemists Julius Lothar Meyer and Dmitri Mendeleev, who at once began to use the recalculated atomic weights in constructing a periodic table of the elements. ■

See also: Conservation of mass 62–63 ▪ Rare earth elements 64–67 ▪ Dalton's atomic theory 80–81 ▪ The ideal gas law 94–97 ▪ The periodic table 130–137

BRIGHT LINES WHEN BROUGHT INTO THE FLAME

FLAME SPECTROSCOPY

IN CONTEXT

KEY FIGURES
Robert Bunsen (1811–1899), **Gustav Kirchhoff** (1824–1871)

BEFORE
1666 Isaac Newton experiments with prisms to produce spectra but believes that light is a stream of particles, not waves.

1835 British inventor Charles Wheatstone reports that metals can be distinguished by the emission spectra of their sparks.

AFTER
1860s British astronomers William and Margaret Huggins use spectroscopy to show that the stars are composed of the same elements as on Earth.

1885 Swiss mathematician Johann Balmer shows that the wavelengths of hydrogen's spectral lines can be represented by a simple mathematical formula.

English physicist Isaac Newton introduced the term "spectrum" in 1666 to describe the dispersion of white light into a rainbow of colors when it passed through a prism. Shortcomings in the equipment he used meant that his spectrum was lacking in detail, with overlapping colors. In 1802, British chemist William Hyde Wollaston described producing a spectrum in which he observed a number of dark lines, but he believed that they were simply gaps between the colors and attached no significance to them. Nonetheless, he was the first to observe what would later be recognized as absorption lines

See also: Photography 98–99 ▪ The periodic table 130–137 ▪ The noble gases 154–159 ▪ Infrared spectroscopy 182 ▪ Mass spectrometry 202–203

in the solar spectrum. The bright lines also observed are known as emission lines.

Fraunhofer lines

German lens maker Joseph von Fraunhofer investigated further and, in 1814, made a careful study of the solar spectrum, mapping several hundreds of the lines and labeling the strongest as A, B, C, D, and so on. These are now known as Fraunhofer lines. He also studied the spectra of stars and planets, using a telescope to collect the light, and noted that the planetary spectra were similar to the solar spectrum.

The color of the flame

Alchemists of the 15th century knew that salts could produce different colors in flames. The earliest reports of investigations into the spectra produced by flames date back to the 18th century. In 1752, Scottish physicist Thomas Melvill used a prism to examine the spectrum produced by burning spirits, to which he introduced various chemicals, such as potash and sea salt. He observed a specific shade of yellow that always took the same position in his spectra, but he did not think this significant. Melvill died in 1753, and it was not until five years later that German chemist Andreas Marggraf reported that he could distinguish between sodium and potassium compounds by their different-colored flames: sodium compounds produced a yellow flame, while potassium salts emitted violet flames.

Flame spectra

The analysis of flame spectra was properly taken up by British scientist John Herschel, who wrote in 1823 that the colors that were produced could provide a way of detecting "extremely minute quantities" of a chemical. His investigations were hampered by the presence of sodium impurities in his samples, which meant that a bright orange-yellow line, identical to that observed by Melvill, always appeared in his spectra. This made it impossible for Herschel to demonstrate that each substance produced a unique spectrum. »

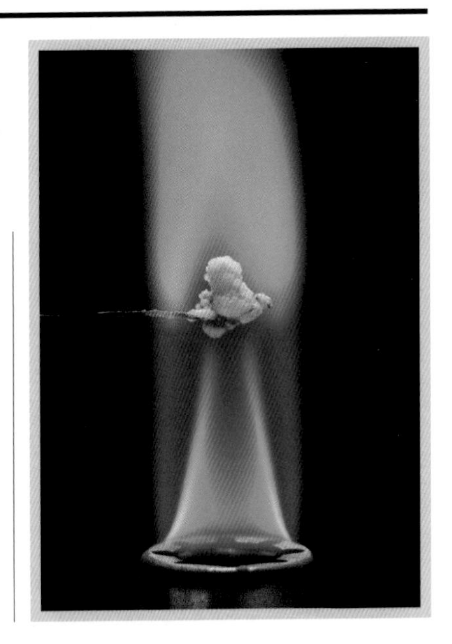

When barium burns, it produces a green flame as a result of the electrons being excited to a higher energy state. On returning to their ground state, they emit energy in the form of light.

A light source, such as **the Sun**, produces a **continuous spectrum**.

↓

As **light** passes through a **cold gas**, elements in the gas absorb characteristic wavelengths, creating **dark absorption lines**.

↓

A hot gas produces **emission lines**—bright lines where light of a **particular wavelength** is emitted.

↓

The absorption lines of an element correspond to its emission lines.

Meanwhile, British photography pioneer William Fox Talbot was undertaking investigations of his own into flame spectra. In 1826, he observed "a red ray" that was "characteristic of the salts of potash" (potassium), in the same way that the yellow ray was for "salts of soda" (sodium), although the red ray could only be seen with the aid of a prism. He believed that the distinctive line patterns seen in flame spectra could be used in chemical analysis to detect the presence of chemicals that would otherwise take hours of work to identify.

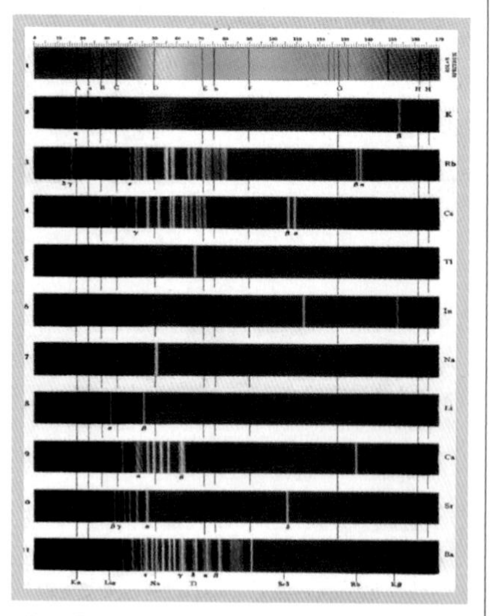

The flame emission spectra of alkali metals show that when elements are heated, they emit a characteristic light pattern. The vertical Fraunhofer lines are seen here in black.

Scientists had suspected for some time that there was a connection between the bright emission lines of the flame spectra and the dark absorption lines of the solar spectrum. Fraunhofer himself had noted that his solar D lines coincided with a bright yellow double line seen in the spectra of flames. In 1849, to illustrate the coincidence of the lines, French physicist Léon Foucault passed sunlight through the light of an arc lamp in order to superimpose the two spectra. He was surprised to find that the D lines were stronger than in the solar spectrum without the arc light. This would prove key to a later discovery.

The flame spectroscope

Ten years later, in 1859, German physicist Gustav Kirchhoff and German chemist Robert Bunsen together invented the flame spectroscope. They used Bunsen's eponymous burner to produce a colorless flame to avoid masking the emission colors of the material under analysis. Their spectroscope consisted of three parts: a flame and collimator, which would narrow the light from the sample into a beam; a prism to disperse the light; and a telescope, used to observe the colors being emitted. This allowed for the precise measuring of the wavelengths of light emitted by the sample in the flame. Repeating Foucault's work, Kirchhoff and Bunsen brought the light from both the Sun and a flame to the slit at the front of their spectroscope and first introduced sodium salt into the

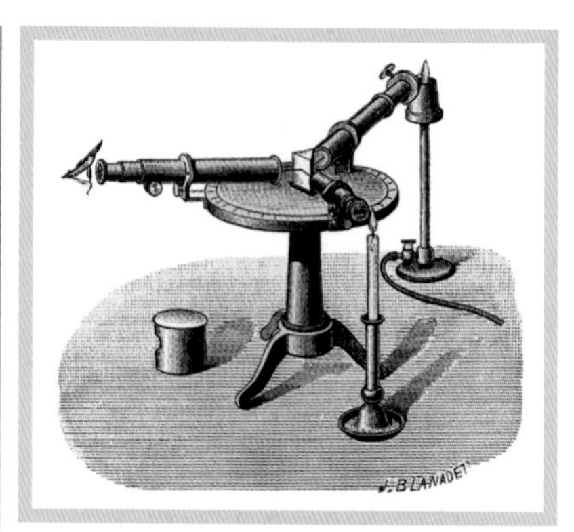

A flame spectroscope, as invented by Kirchhoff and Bunsen, can be used to identify elements and to discover the composition of any chemical based on flame testing.

Robert Bunsen

Born in 1811 in Göttingen, Westphalia (now Germany), Robert Bunsen was the youngest of four brothers. After receiving his doctorate in 1830, Bunsen joined French chemist Joseph Gay-Lussac at his laboratory in Paris in 1832 and returned to Germany the next year to become a lecturer at the University of Göttingen.

Bunsen was gifted as both an experimentalist and an inventor. Working with German physician Arnold Berthold, he discovered an antidote to arsenic poisoning in 1834. He also developed new techniques to analyze gases produced by industries and recommended that the exhaust gases from burning charcoal be recycled to generate more energy. In 1841, he invented the zinc-carbon cell, or Bunsen cell. Combining these cells in large batteries allowed Bunsen to separate metals from their ores by electrolysis.

In 1864, with his research student, British chemist Henry Roscoe, Bunsen invented flash photography when they used burning magnesium as a light source. Bunsen died in Heidelberg in 1899.

flame—then other salts, such as calcium and strontium, to produce different spectra. The bright spectral lines produced aligned exactly with the dark lines of the solar spectrum, showing emission and absorption to be connected processes.

Kirchhoff concluded that when light passes through a gas, the light wavelengths that are absorbed by the gas coincide with the wavelengths that the gas emits when incandescent. A substance that strongly emits light of a certain wavelength will also strongly absorb light of that wavelength.

It quickly became apparent to both Kirchhoff and Bunsen that each chemical element produced its own characteristic pattern of colored lines when heated to incandescence (the point of the emission of light). Each element and compound produced a spectrum, in some ways comparable to a chemical barcode that uniquely identified it. Spectroscopy could therefore be used to determine the composition of any chemical.

Spectroscopy in space

Sodium vapor produced a double yellow line, corresponding with Fraunhofer's D line. The existence of the dark D lines in the solar spectrum indicated that sunlight had passed through sodium vapor on its way to Earth. This, Kirchhoff concluded, was indisputable proof that sodium was present in the solar atmosphere and that it had absorbed light at the characteristic wavelength. By comparing the dark Fraunhofer lines of the solar spectrum with the spectral lines of metals, Kirchhoff established that—in addition to sodium—magnesium, iron, copper, zinc, barium, and nickel were also all present in the solar atmosphere. Bunsen and Kirchhoff's discovery had not only brought a

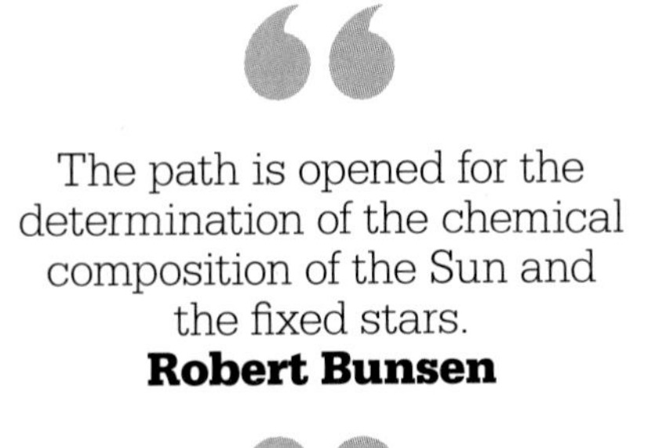

> The path is opened for the determination of the chemical composition of the Sun and the fixed stars.
> **Robert Bunsen**

revolution to chemical analysis, it also provided astronomers with a powerful new tool for their exploration of the cosmos.

New elements

In May 1860, while analyzing the spectral emissions from spring waters known to be rich in lithium compounds, Bunsen spotted a new sky-blue signature in the spectra. He and Kirchhoff realized that it belonged to a new element, which he named cesium (from the Latin for "sky blue"). As an indication of the power of spectral analysis, Bunsen had to evaporate 9,900 gallons (45,000 liters) of the spring water to obtain a large enough sample of cesium salts to determine its properties. He was, however, not able to isolate pure metallic cesium; that feat was achieved in 1881 by Swedish chemist Carl Setterberg.

The following year, in 1861, Bunsen and Kirchhoff discovered yet another new chemical, which produced a dark red spectrum. It was the metal rubidium (from the Latin for "dark red"), and this time, Bunsen succeeded in isolating the element. Bunsen and Kirchhoff's discoveries inaugurated a new era in the way in which unknown elements were found. ■

Helios and helium

On August 18, 1868, during a total eclipse in India, French astronomer Pierre Janssen saw something surprising when he examined the Sun's corona (outermost layer) with his spectroscope: a bright yellow line that failed to match up with any known element. Two months later, unaware of Janssen's work, British astronomer Norman Lockyer also discovered the yellow line. Lockyer quickly claimed to have found a new element and called it helium, after Helios, the Greek god of the Sun.

There were dissenting voices, including Russian chemist Dmitri Mendeleev, who had no space for helium on his periodic table. The discovery of the remaining inert or noble gases, from 1894 to 1900, prompted a revision, and helium was added to the table, along with the five other noble gases, in 1902.

We now know that helium is the second-lightest and second most abundant element in the observable universe after hydrogen.

A total solar eclipse occurs when the moon passes between the Sun and Earth and blocks out the light of the Sun.

NOTATION TO INDICATE THE CHEMICAL POSITION OF THE ATOMS

STRUCTURAL FORMULAE

IN CONTEXT

KEY FIGURE
Alexander Crum Brown (1838–1922)

BEFORE
1808 John Dalton theorizes that atoms hook together and creates diagrams showing combined atoms.

1858 Archibald Couper draws diagrams with lines depicting the bonds between atoms.

AFTER
1865 German chemist August Wilhelm von Hofmann makes the first stick and ball molecular models of molecules.

1916 Gilbert Lewis, an American chemist, introduces "Lewis structures" to represent atoms and molecules, with dots for electrons and lines for covalent bonds.

1931 American chemist Linus Pauling uses quantum mechanics to calculate the properties and structures of molecules.

It is difficult to ascertain what a molecule looks like because most are less than a nanometer (one-billionth of a meter) across. But the structure of a molecule gives important information about its properties and how it reacts with other molecules, so being able to represent it in some way is vital.

Early diagrams

In the 5th century BCE, the Ancient Greeks initiated the idea that atoms have hooks and holes with which they connect to each other, but it was not until the 19th century that serious efforts to depict atom combinations began. In 1803, British chemist John Dalton produced a number of symbols for the elements known at the time, such as hydrogen, oxygen, carbon, and sulfur, and created diagrams to illustrate common molecules. He thought that elements combined mostly in binary forms, so the compounds would only have one atom of each element. This led him to believe the formula for water was OH; we now know that it is H_2O.

Later developments in the depiction of molecules were often tied to progress in the understanding of chemical structure, particularly involving organic molecules. In 1858, Scottish chemist Archibald Couper proposed a theory of molecular structure, emphasizing that carbon formed four bonds and bonded to other carbon atoms to form chains. August Kekulé, a German chemist, put forward the same idea almost simultaneously and saw his work published first in the same year.

Reception

Kekulé accompanied his theory with diagrams that depicted organic compounds using "sausage

> There is … opposition to your formulae here, but I am convinced that they are destined to introduce much more precision into our notions of chemical compounds.
> **Edward Frankland**

See also: Compound proportions 68 ▪ Chemical notation 82–83 ▪ Benzene 128–129 ▪ Coordination chemistry 152–153 ▪ X-ray crystallography 192–193 ▪ Nuclear magnetic resonance spectroscopy 254–255

Evolution of structural formulae

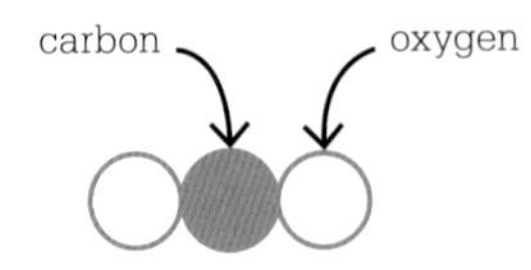

1808: Dalton's atomic symbols, here for carbon dioxide, use differently colored circles to represent the various elements.

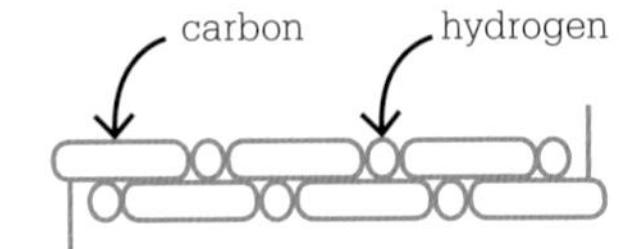

1858: Kekulé's sausage formula, here for benzene (discovered in 1865), has varying-sized sausages to represent valencies.

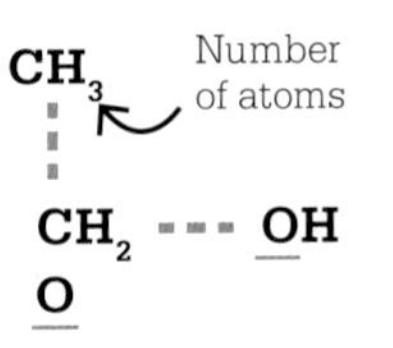

1858: Couper's formula, here for the structure of ethanol, shows elemental symbols for atoms and dashed lines for bonds.

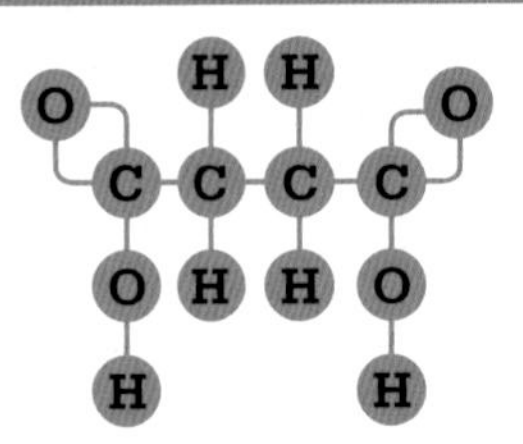

1861: Crum Brown's constitutional formula, here for succunic acid, uses elemental symbols for atoms and solid lines for bonds.

formulae" (elongated ovals and circles). Couper also included diagrams, but his used elemental symbols with lines to represent the bonds between atoms—not unlike the structural representations used today. They both got short shrift from other chemists at the time. Russian chemist Alexander Butlerov, known for his contributions to the theory of chemical structure, dismissed Couper's theory and structures as "too absolute" and "neither clearly enough perceived nor expressed." Whereas Kekulé was renowned for his later proposal of the structure of benzene, Couper was largely forgotten.

Scottish chemist Alexander Crum Brown was apparently unacquainted with Couper's work, but he made equally important contributions to depictions of molecules. His thesis of 1861 contained representations similar to Couper's, using elemental symbols within circles to depict atoms, joined by lines to represent bonds. Crum Brown's formulae were met with opposition in some quarters, but British chemist Edward Frankland, who had carried out work on atomic valence in the early 1850s, used them in his lectures and later included them in a textbook published in 1866, which popularized the formulae further. The second edition of this textbook dispensed with the circles around the atoms that Crum Brown had drawn, leaving depictions near-identical to those used today. Somewhat unfairly, these diagrams became known as "Frankland notation." ■

Alexander Crum Brown

Born in Edinburgh in 1838, Alexander Crum Brown first studied the arts and then medicine. He received a medical doctorate from the University of Edinburgh and a science doctorate from the University of London. From 1869 until his retirement in 1908, he was professor of chemistry at the University of Edinburgh. In addition to his chemical diagrams, Crum Brown was the first to propose that ethene (C_2H_4) contains a carbon-carbon double bond. He also demonstrated that a molecule's structure influences how it acts in the body. He used a range of materials—such as leather, papier-mâché, and wool—to construct 3D mathematical models of interlocking surfaces, and built an impressively accurate model of sodium chloride (NaCl) years before its structure was experimentally determined. He died in 1922.

Key works

1861 "On the theory of chemical combination"
1864 "On the theory of isomeric compounds"

ONE OF THE SNAKES HAD SEIZED HOLD OF ITS OWN TAIL

BENZENE

IN CONTEXT

KEY FIGURES
August Kekulé (1829–1896), **Kathleen Lonsdale** (1903–1971)

BEFORE
1825 Michael Faraday isolates the first sample of benzene from the oily residue of gas lamps.

1861 Johann Loschmidt depicts benzene with a cyclic structure. At the time, the idea of a cyclic compound was novel.

AFTER
1866 Pierre Marcellin Berthelot, a French organic and physical chemist, completes the first synthesis of benzene in the laboratory by heating acetylene in a glass tube.

1928 Kathleen Lonsdale uses X-ray crystallography to confirm that the benzene molecule is a flat hexagon.

In 1865, German chemist August Kekulé proposed that benzene formed a hexagonal ring of six carbon atoms, with each carbon forming four bonds—one to a neighboring hydrogen atom and three further bonds to the neighboring carbon atoms. He claimed that a daytime reverie of a snake turning on itself and biting its own tail inspired the idea of a cyclical structure.

Kekulé's theory

In the decades since Kekulé published his theory, science historians have debated whether he was the first to represent benzene as a flat, cyclic, hexagonal structure. Some believe that Johann Loschmidt, an Austrian chemist, suggested a cyclical structure for benzene four years before Kekulé. Others have argued that Loschmidt's proposal was a lucky coincidence, the result of him choosing a circle to denote that benzene's structure was still unknown.

Controversies aside, Kekulé's initial visual interpretation of the benzene molecule showed the hydrogen atoms as circles and the carbon atoms as elongated ovals. His contemporaries unflatteringly dubbed these diagrams "sausage formulae," which prompted Kekulé to refine his drawings. Later in 1865, he switched his representation to a simple hexagon. In 1866, he added the alternating double and single bonds between the carbon atoms to produce a more accurate depiction, which still bears his name today.

Kekulé's structure correctly predicted the products of some substitution reactions with benzene, where one of its hydrogens swaps out for another atom or group of atoms. However, it did not fully explain observations of benzene's

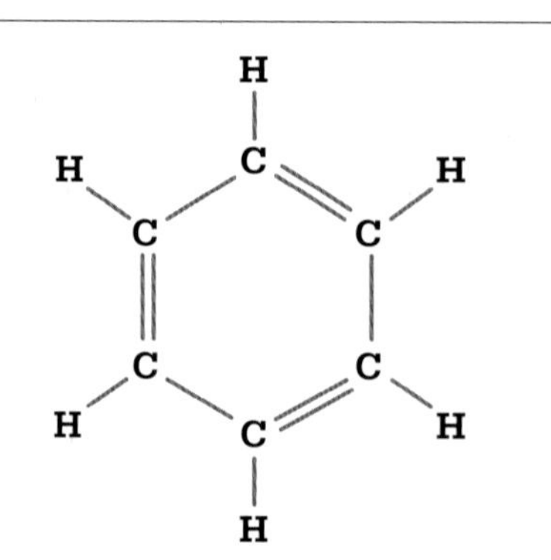

Kekulé's structure of benzene proposes that each carbon atom forms one double and one single bond with a neighboring carbon atom and one single bond with a hydrogen atom.

See also: Structural formulae 126–127 ▪ X-ray crystallography 192–193 ▪ Depicting reaction mechanisms 214–215 ▪ Protein crystallography 268–269

various reaction products, even after Kekulé refined his model further in 1872 to explain that the double and single bonds were constantly swapping places with one another within the molecule. Despite this, his predictions of benzene's structure led to further interest and advances in the composition and structures of aromatic compounds. Today, aromatic compounds are used to make medicines, plastics, dyes, and many other products we use on a regular basis.

X-ray crystallography

Kekulé did not live to see the final details of benzene's structure confirmed—this required X-ray crystallography, a technology that was not developed until more than a decade after his death. In 1928, Kathleen Lonsdale, an Irish crystallographer, used this technique to confirm that the flat, cyclical structure proposed by Kekulé was in fact correct. She also explained some of the oddities of benzene's structure that Kekulé had not been able to fathom.

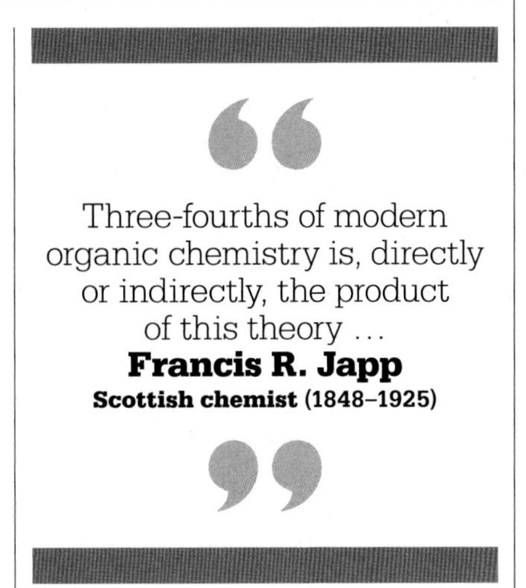

Three-fourths of modern organic chemistry is, directly or indirectly, the product of this theory …
Francis R. Japp
Scottish chemist (1848–1925)

Lonsdale's measurements showed that all the carbon-carbon bonds in a benzene ring are the same length. This is because the electrons in the bonds are delocalized, or spread out, over the entire ring. This delocalized structure also explains benzene's additional stability, in contrast to Kekulé's theory of alternating double and single bonds. Chemists commonly use a hexagon with a circle in it to illustrate the delocalized structure.

Kekulé's depiction was not completely superseded by Lonsdale's. His version is still employed when drawing organic compound structures, partly because it makes it easier to illustrate the electron movements that take place during reaction mechanisms. Somewhat unfairly, though, Kekulé's structure bears his name, but Lonsdale's is frequently cited without mention of hers. ▪

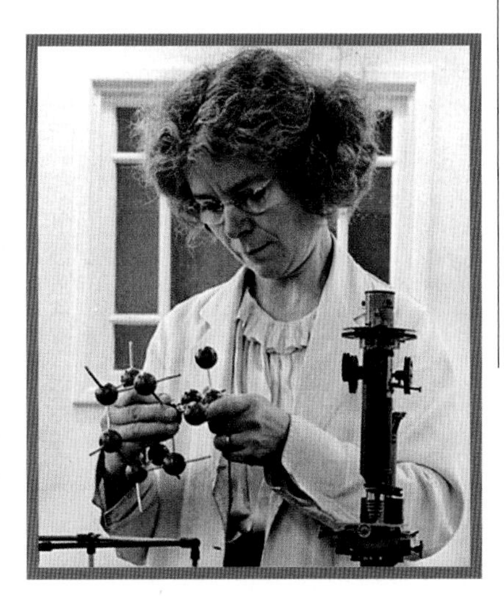

Kathleen Lonsdale in the laboratory in 1948 at University College London, where she was reader in crystallography from 1946 and professor of chemistry from 1949.

August Kekulé

Born in Darmstadt, Germany, in 1829, Friedrich August Kekulé initially intended to study architecture at the University of Giessen. However, he became fascinated by the lectures of Justus von Liebig, one of the founders of organic chemistry, and decided to study chemistry instead. Throughout his career, Kekulé made key contributions to our understanding of chemical structure, particularly in relation to carbon-based compounds. He built on the work of his predecessors to explain the linking of atoms in terms of their valency, with particular emphasis on the idea that a carbon atom forms four bonds to other atoms. Kekulé died in Bonn in 1896. After his death, three of his former students from the University of Bonn each won a Nobel Prize in Chemistry.

Key works

1858 "On the constitution and metamorphoses of chemical compounds and on the chemical nature of carbon"
1859–1887 *Textbook of Organic Chemistry*, Editions 1 to 7

A PERIODIC REPETITION OF PROPERTIES

THE PERIODIC TABLE

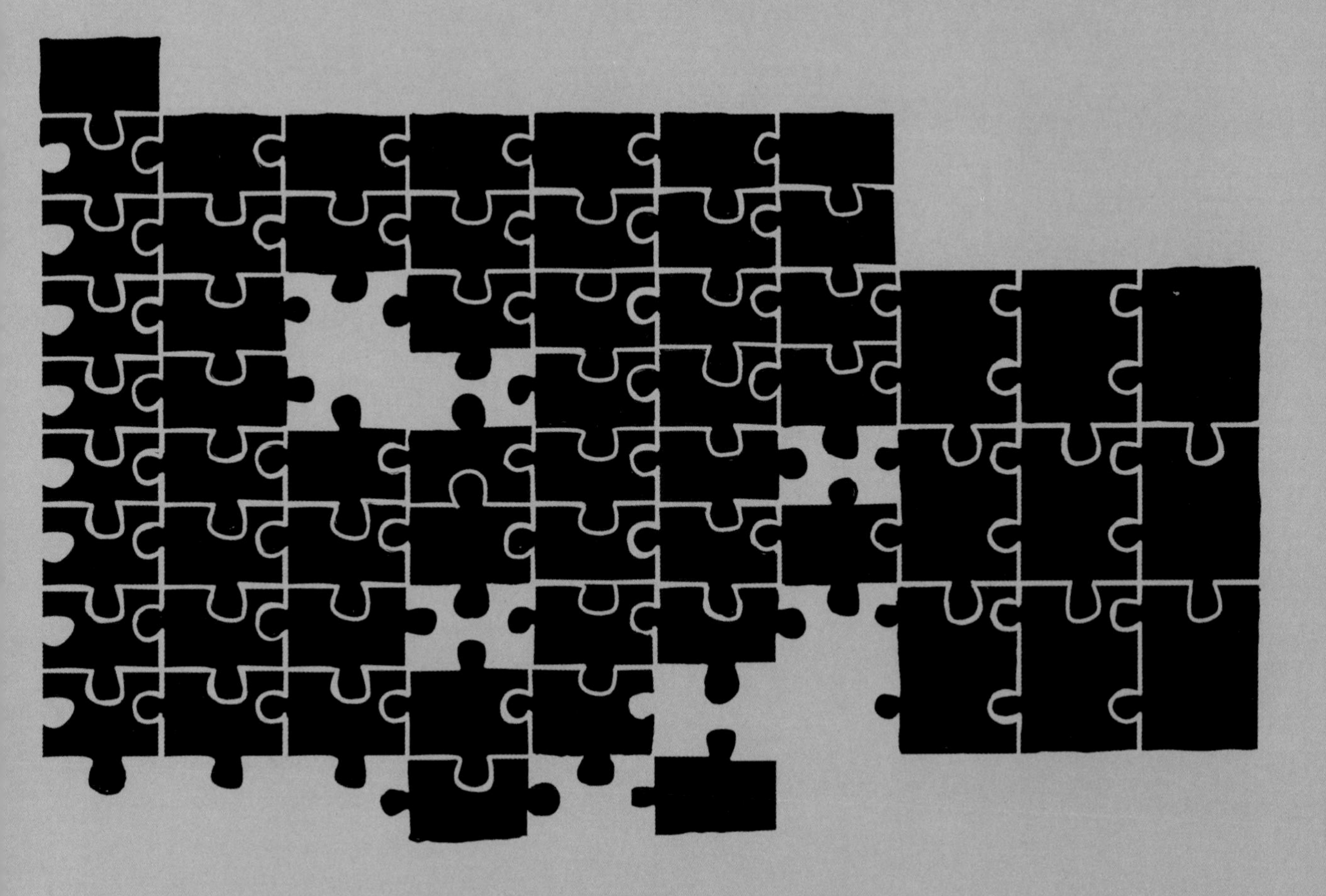

IN CONTEXT

KEY FIGURE
Dmitri Mendeleev
(1834–1907)

BEFORE
1789 Antoine Lavoisier attempts to group the elements into metals and nonmetals.

1803 John Dalton introduces his atomic theory, the first truly scientific theory of the atom, arrived at through experimentation.

AFTER
1904 British physicist J. J. Thomson develops his model of the atom based on the periodic table.

1913 Niels Bohr reasons that elements in the same periodic group have identical configurations of electrons in their outermost shell.

2002 A team of Russian chemists discovers oganesson (Og), element 118.

The periodic table of the elements is one of the most readily recognizable distillations of scientific knowledge ever created. Attached to the walls of laboratories and classrooms around the world, it brings together a storehouse of information on the properties of the chemical elements and their relationships with each other. Although Russian chemist Dmitri Mendeleev is often credited as its inventor, many scientists contributed to achieving this organization of the elements. The periodic table continues to evolve as new elements are discovered and our understanding grows.

> We must expect the discovery of many yet unknown elements … whose atomic weight would be between 65 and 75.
> **Dmitri Mendeleev**
> **(1869)**

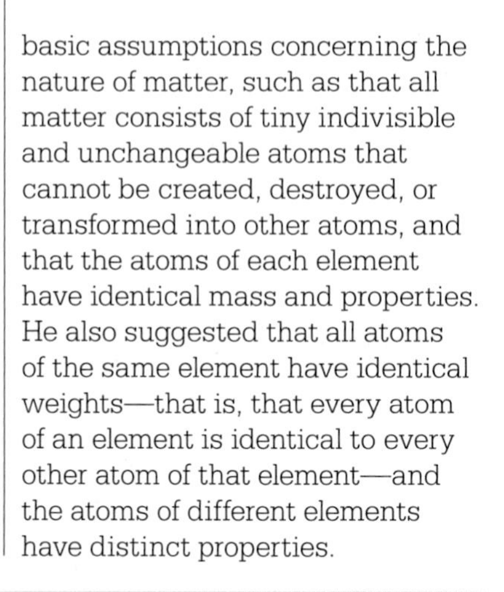

Atomic theory

Eighteenth-century chemists such as Joseph Priestley and Antoine Lavoisier had demonstrated through experiment that some substances could combine to form new materials, that some could be broken down into other materials, and that a few appeared to be "pure" and could not be broken down any further. John Dalton's atomic theory of 1803 drew previous findings together in a coherent whole. He made some basic assumptions concerning the nature of matter, such as that all matter consists of tiny indivisible and unchangeable atoms that cannot be created, destroyed, or transformed into other atoms, and that the atoms of each element have identical mass and properties. He also suggested that all atoms of the same element have identical weights—that is, that every atom of an element is identical to every other atom of that element—and the atoms of different elements have distinct properties.

Dmitri Mendeleev

Born in 1834 in Tobolsk, Siberia, in what was then the Russian Empire, Dmitri Mendeleev was the youngest member of a large family. Following his father's death, his mother took the family to St. Petersburg in 1848. After obtaining his Master's degree in chemistry in 1856, Mendeleev went to Heidelberg, Germany, where he set up his own lab.

In 1860, at the Karlsruhe Conference, he made contact with many of Europe's leading chemists. He became professor of chemical technology at the University of St. Petersburg in 1865. In addition to his more theoretical work, Mendeleev was involved in research into agricultural yields, petroleum production, and the coal industry. In 1893, he was named director of Russia's new Central Board of Weights and Measures. Mendeleev was recognized internationally for his achievements in chemistry, and at his funeral in 1907, his students carried a copy of the periodic table in tribute.

Key work

1871 *The Principles of Chemistry*

See also: Dalton's atomic theory 80–81 ▪ Atomic weights 121 ▪ The noble gases 154–159 ▪ The electron 164–165 ▪ Improved atomic models 216–221 ▪ Nuclear fission 234–237 ▪ The transuranic elements 250–253 ▪ Completing the periodic table? 304–311

Dalton's 1808 table of atomic weights shows 20 "elements," although it is now known that some—such as lime and potash—are compounds. By 1827, his list had grown to 36 elements.

Starting with hydrogen as 1, Dalton assigned atomic weights to the elements known at the time based on the mass ratios in which they combined with hydrogen. The flaw in Dalton's methodology was that he believed the simplest compound of two elements had to be one atom of each. For example, believing the formula of water to be HO rather than H_2O, he assigned oxygen an atomic weight that was half of what it actually is. Nonetheless, Dalton's table of atomic weights was a first step toward the development of the periodic table of the elements.

In 1811, Italian chemist Amedeo Avogadro made the assertion that simple gases, rather than being made up of single atoms, were instead formed from compound molecules of two or more linked atoms. At this time, the words "atom" and "molecule" were used more or less interchangeably. Avogadro referred to an "elementary molecule," what today we would call an atom; essentially, what he was doing was defining this as the smallest part of a substance. It would be several more years before Avogadro's proposals were accepted, mainly because chemists believed that two atoms of the same element could not combine. For example, oxygen was still considered to be a single atom rather than a diatomic molecule.

Karlsruhe Conference

Between 1817 and 1829, German chemist Johann Döbereiner investigated the finding that certain elements could be placed in groups of three, or triads, based on both physical and chemical properties. When the atomic weights of lithium and potassium were averaged, for example, the result approximated the value for sodium, the third member of their triad. The elements in a triad also reacted in a similar way chemically, so the properties of the middle element could be predicted based on the properties of the other two. Although Döbereiner's system worked for some elements, it did not work for all, and progress was hindered by inaccurate measurements. What was needed was an accurate list of the atomic weights of the elements. »

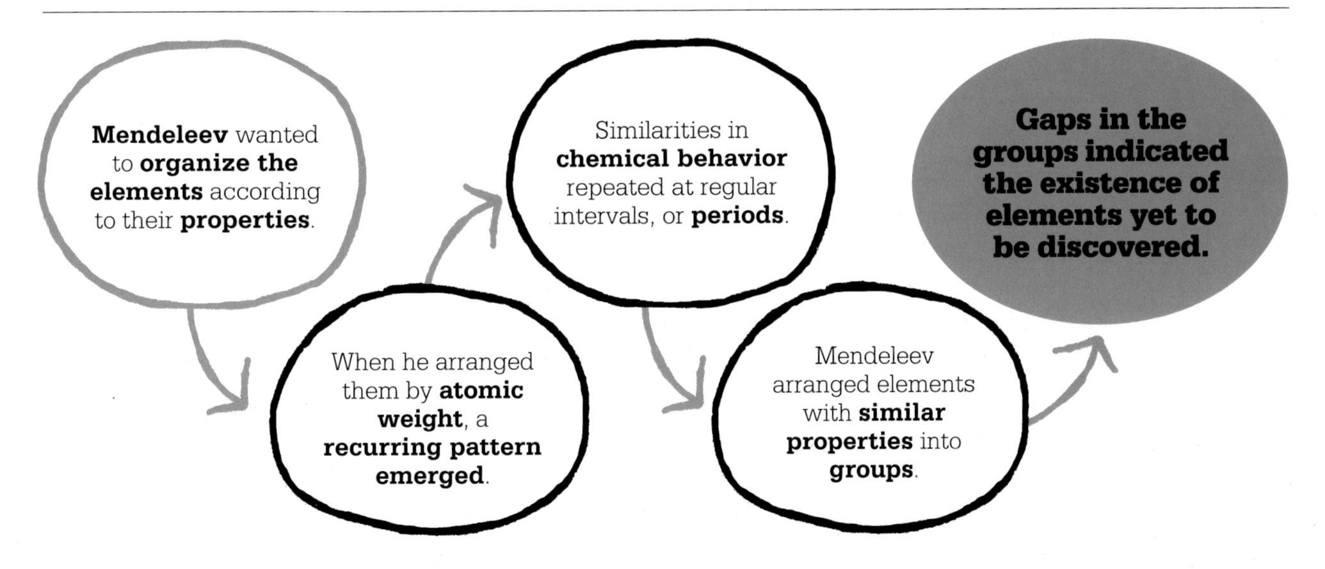

By the middle of the 19th century, several competing ideas on how best to calculate atomic weights and molecular formulas were causing confusion in scientific circles. These were largely resolved at a conference held in Karlsruhe, Germany, in 1860, during which Italian chemist Stanislao Cannizzaro argued forcefully for the acceptance of Avogadro's hypothesis, declaring that it was "invariably in accordance with all physical and chemical laws hitherto discovered." The conference also saw the publication of a revised list of elements and their atomic weights, with hydrogen assigned the atomic weight of 1 and the atomic weight of other elements decided by comparison to it. Armed with a new understanding of atoms and molecules, scientists began working on the task of organizing the elements.

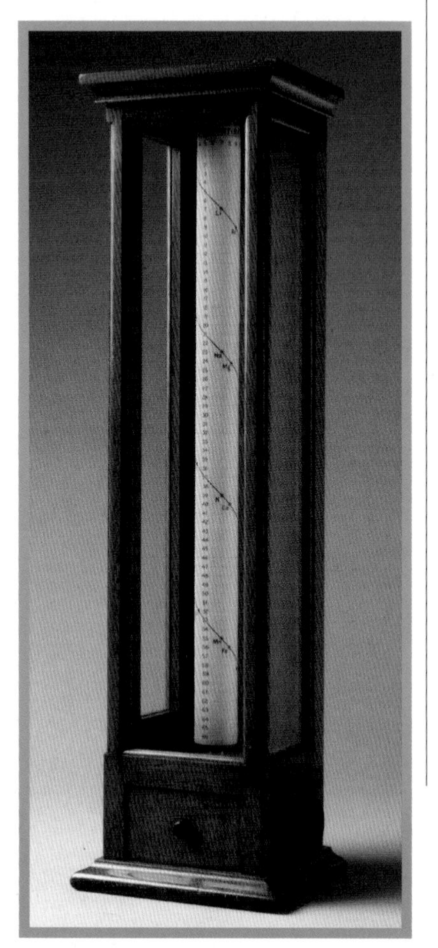

The telluric screw was able to show the periodicity of elements' properties. It got its name from the fact that tellurium was placed at the center.

Turning the screw

One of the first to make a serious attempt at a periodic arrangement of all the known elements was French geologist Alexandre Béguyer de Chancourtois. In 1862, he presented his periodic ordering to the Académie des Sciences in Paris. This took the form of a three-dimensional arrangement of the elements on a "telluric screw," which plotted atomic weights on the outside of a cylinder. One complete turn of this corresponded to an atomic weight increase of 16. As the screw was turned, elements with similar properties aligned vertically—lithium in line with sodium and potassium, for example—to provide a visual demonstration of the periodicity of chemical properties in the same groupings that Döbereiner had discovered. However, some elements did not line up as expected—bromine, for instance, lined up with the chemically very different copper and phosphorus.

The following year, 1863, British chemist John Newlands noted that if the elements were arranged in rows of seven according to their atomic weights, they formed columns of elements with similar chemical properties. He found that every element had similar chemical properties to the one eight places ahead of it. He called this periodicity the law of octaves, a nod to its resemblance to a musical scale.

The eighth element, starting from a given one, is a kind of repetition of the first, like the eighth note of an octave in music.
John Newlands

Newlands's table wasn't without its drawbacks. He sometimes had to double up elements to keep the pattern, and it allowed no space for undiscovered elements.

Meyer and Mendeleev

At around the same time, in Germany, chemist Julius Lothar Meyer, doubtless inspired by his attendance at the Karlsruhe Conference, put together his first periodic table, which contained just 28 elements. The difference was that he arranged them according to their valency, a recently discovered property that measured an element's power to combine with other elements. Four years later, in 1868, he produced a more sophisticated table incorporating more elements, listing them in order of atomic weight, with elements of the same valency arranged in columns. It bore a striking similarity to a table that was soon to be published by Mendeleev, another attendee at the Karlsruhe Conference.

Lacking a textbook to cover the subject of inorganic chemistry, which he taught at the University of St. Petersburg, Mendeleev decided to write his own, *The Principles of Chemistry*. While he was in the

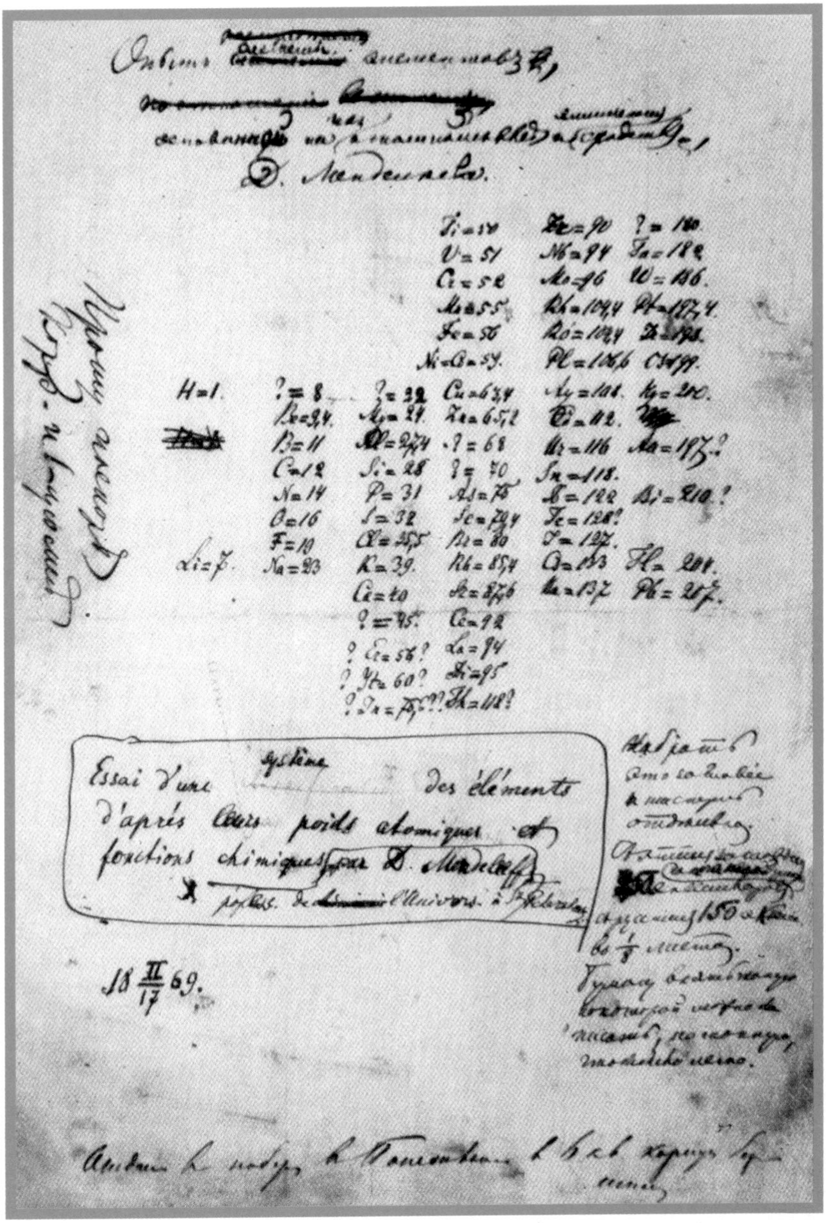

In Mendeleev's first attempt at a periodic system, he placed the elements with ascending atomic weights going down the table rather than across.

process of researching the book in the 1860s, he noticed that there were recurring patterns between different groups of elements, and in 1869, he assembled his version of the periodic table. Although he is reputed to have done this by arranging cards labeled with the various elements and their properties, setting them out like a game of chemical solitaire, no such cards were found in his archive.

However he arrived at his table, Mendeleev printed 200 copies of it, presented it to the Russian Chemical Society, and distributed it to colleagues throughout Europe. Meyer didn't publish until 1870, a year after Mendeleev. Although Meyer and Mendeleev undoubtedly knew each other—both had trained at the University of Heidelberg under Robert Bunsen—they were initially unaware of each other's work. Meyer later readily admitted that Mendeleev had published his version first.

Periodic predictions

Mendeleev's table, which he called a "periodic system," included all 56 elements known at that time. He arranged the elements in order of increasing weight, divided into rows, called periods, so that the elements in each column shared properties such as valence. One decision both Mendeleev and Meyer made was that if an element's atomic weight appeared to put it in the wrong place, they moved it to where it fit in with the patterns they had discovered—for example, reversing the positions of tellurium and iodine. In order of increasing atomic weight, iodine should come before tellurium, but this made no sense in terms of their chemical properties. Later discoveries explained this.

Both Mendeleev and Meyer left blank spaces in their tables, but it was Mendeleev alone who predicted that elements would be discovered to fill those gaps, and he even predicted the atomic weights and properties of five of these elements and their compounds. He wasn't always completely correct in his predictions, but he was close enough that when elements such as gallium and scandium were discovered, they could be inserted into the appropriate empty spaces in his table. »

The discovery of argon, the first of the noble gases, by British physicist Lord Rayleigh (John Strutt) and chemist William Ramsay in 1894, seemed at first to be a challenge to the periodic table. Mendeleev and others argued that rather than being a new element, it was actually a previously unknown form of molecular nitrogen, N_3. Following the subsequent discoveries of helium, krypton, neon, and xenon, however, the noble gases could not be explained away so easily, and some chemists even suggested that they did not belong in the periodic table at all. In 1900, Ramsay suggested the new elements be given their own group, between the halogens and the alkali metals. Mendeleev declared that this was "a glorious confirmation of the general applicability of the periodic law."

Mendeleev's periodic table was arranged by atomic weight, and although this worked very well, there were some anomalies that required tweaking (such as iodine and tellurium).

In 1913, just six years after Mendeleev's death, a new discovery made everything fall into place.

> There is in the atom a fundamental quantity, which increases by regular steps as we pass from one element to the next.
> **Henry Moseley**

At the University of Manchester, UK, physicist Henry Moseley fired beams of electrons at different metals and examined the spectrum of X-rays produced. He found that the frequencies of the X-rays emitted could be used to identify the positive charge on the element's atomic nucleus. This positive charge, equivalent to the number of protons in the nucleus, became known as the element's atomic number.

Moseley concluded that it was the atomic number of the element, not the atomic weight, that decided what the characteristics of the element were. Only whole atomic numbers fit the pattern; there were no elements with fractions of an atomic number. Moseley was able to reorganize the periodic table, with the chemical elements now arranged by atomic number—from hydrogen, atomic number 1, to uranium, 92—rather than atomic weight. Tellurium (atomic number 52) and iodine (atomic number 53) could now be assigned their rightful positions in the periodic table without any fudging. From gaps in his table of X-ray frequencies, Moseley predicted the existence of three unknown elements: rhenium, which was discovered in 1925; technetium discovered in 1937; and finally, promethium, discovered in 1945.

The modern table

In the 1940s, American chemist Glenn Seaborg drew up the version of the periodic table that is most familiar to chemistry students today. When Seaborg and his team discovered plutonium in 1940, they considered naming it ultimium, believing it would be the last element in the periodic table. During research as part of the Manhattan Project to develop the atomic bomb, Seaborg and his team began to suspect that still heavier elements were being formed in the nuclear reactors. The best way to identify and isolate these new elements was by predicting their chemistry according to where their positions should be on the periodic table. Seaborg's efforts to isolate elements 95 and 96, now known as americium and curium, were unsuccessful because the assumption that the chemical properties of those elements would resemble those of iridium and platinum (as suggested by their

Periodic redesigns

Several attempts have been made over the years to redesign the periodic table. In 1928, for instance, French inventor Charles Janet produced his "left-step" periodic table based on the way the electron shells in atoms are filled up. German chemist Otto Theodor Benfey redrew the periodic table in a two-dimensional shape reminiscent of a snail in 1964. Beginning with hydrogen at the center, it spirals out, folding around two outcrops formed from the transition metals and the lanthanides and actinides. Benfey also added an area for superactinides (a largely theoretical grouping), ranging from elements 121 to 157. In 1969, Glenn Seaborg suggested an extended periodic table that would include as yet unknown elements up to atomic number 168, and in 2010, Finnish chemist Pekka Pyykkö proposed an extension to atomic number 172, taking into account quantum mechanical effects.

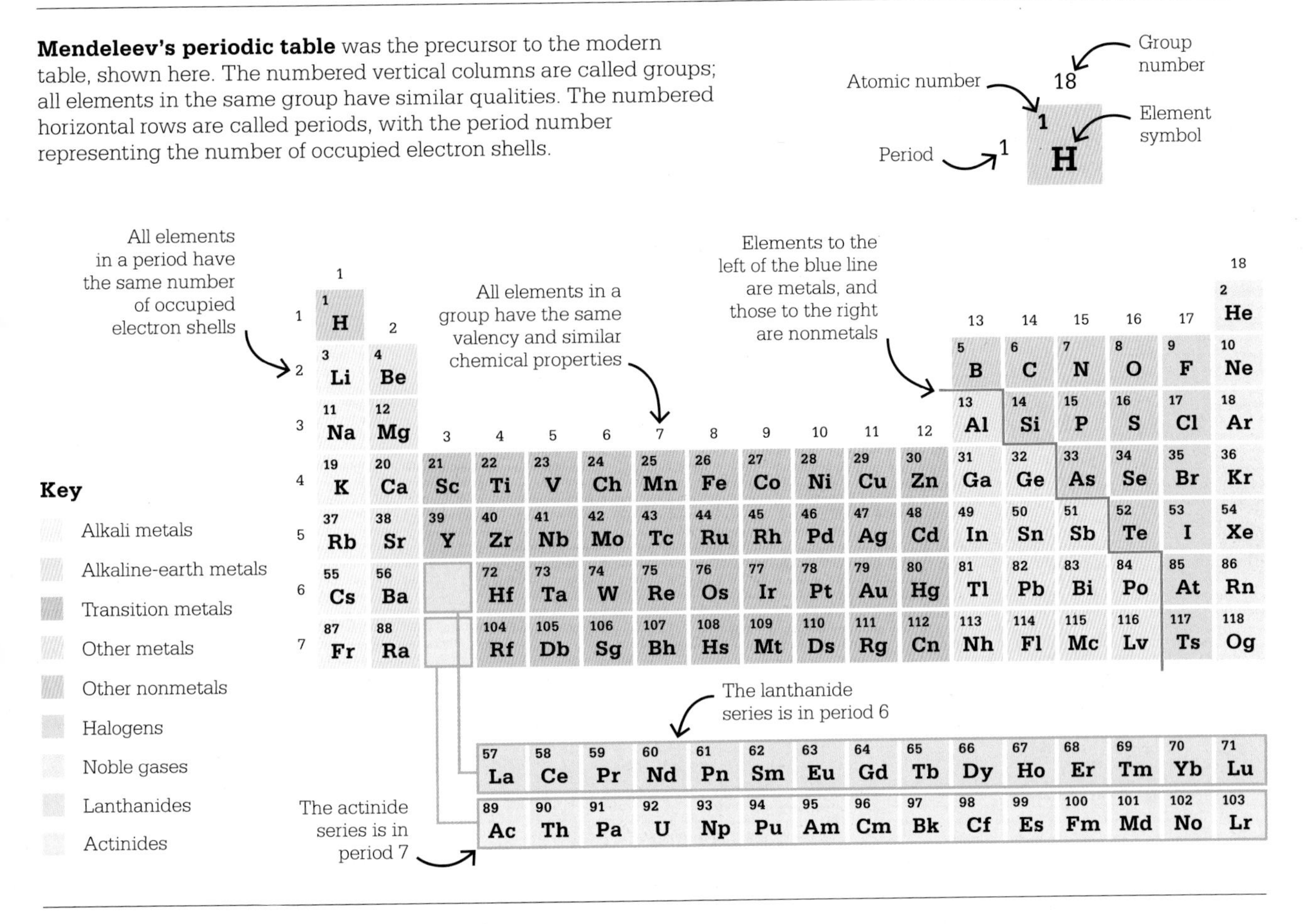

Mendeleev's periodic table was the precursor to the modern table, shown here. The numbered vertical columns are called groups; all elements in the same group have similar qualities. The numbered horizontal rows are called periods, with the period number representing the number of occupied electron shells.

position in the periodic table) turned out to be wrong. Seaborg concluded that their lack of success indicated a flaw in the periodic table itself.

Another group that had proven problematic was the lanthanides, or rare-earth elements. Until Danish physicist Niels Bohr developed his electron shell model of the atom in 1913, the 14 elements following lanthanum had been difficult to place. Bohr proposed that these elements formed an "inner transition" series, in which the number of valence electrons remained constant at three, so they all shared similar properties. Bohr also suggested another inner transition series that was formed by elements beyond actinium.

In 1945, Seaborg published a restructured periodic table with an actinide series located directly under the lanthanides. Although many chemists were initially skeptical of it, Seaborg's table would soon become accepted as the standard version. The actinide series included uranium and the new transuranium elements, such as americium, curium, and fermium, as well as mendelevium (element 101), which was discovered by Seaborg and his colleagues in 1955. Seaborg's contributions to chemistry earned him the rare honor of having an element named after him—seaborgium, number 106—the first element to be named after someone who was still living. ■

The cyclotron particle accelerator at the University of California, which Glenn Seaborg and Edwin McMillan used to discover plutonium, neptunium, and other transuranic elements.

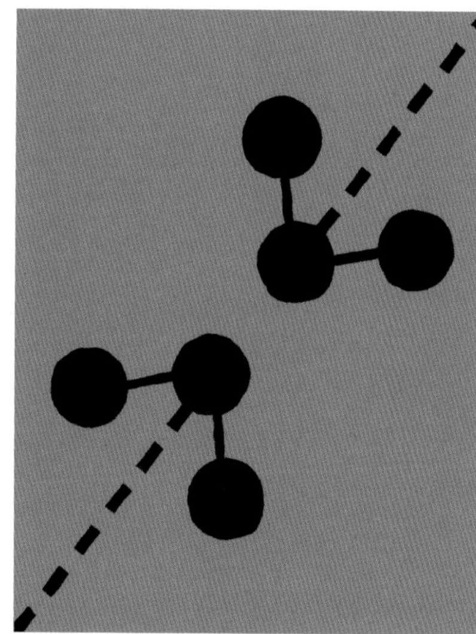

THE MUTUAL ATTRACTION OF THE MOLECULES

INTERMOLECULAR FORCES

IN CONTEXT

KEY FIGURE
Johannes van der Waals (1837–1923)

BEFORE
75 BCE The Roman natural philosopher Lucretius suggests liquids are made from smooth round atoms but that solids are made up of hooked atoms.

1704 British physicist Isaac Newton theorizes that atoms are held together by an invisible force of attraction.

1805 Thomas Young posits surface tension in liquids.

AFTER
1912 Dutch chemist Willem Keesom describes dipole–dipole forces for the first time.

1920 American chemists Wendell Latimer and Worth Rodebush propose the concept of hydrogen bonds in water.

1930 German physicist Fritz London discovers the quantum bonding in noble gases.

In a liquid, **molecules** are **bound together** tightly by **intermolecular forces**.

Heat gives the molecules **more energy**, making them **move**.

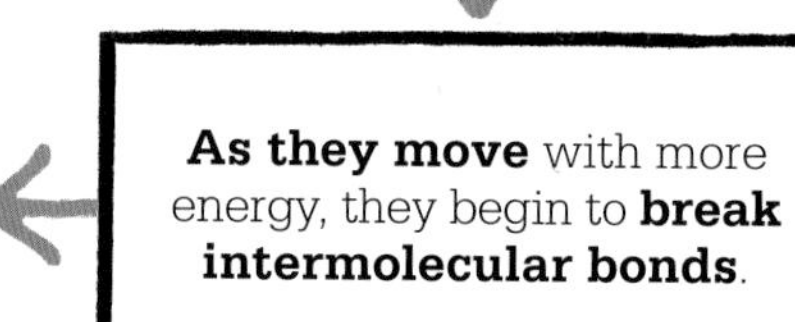

As they move with more energy, they begin to break intermolecular bonds.

As more bonds are broken, the liquid turns to a gas.

By the mid-19th century, chemists had a good idea of what held atoms together and were beginning to understand that matter—solids, liquids, and gases—is made of molecules. But what holds molecules together and makes a gas condense into a liquid? The answer was subtle forces between molecules, an idea developed by Dutch physics teacher Johannes van der Waals in 1873.

Seventy years earlier, British physicist Thomas Young was considering why water formed round drops and why it swells slightly at the top of a glass—in what is called a meniscus. He suggested that there was a pull between molecules at the surface, now called surface tension. Van der Waals's work developed this idea.

Bonds between molecules

Van der Waals studied the moments when a liquid changes to a gas (evaporation) and when a gas becomes a liquid (condensation). Recent advances had focused on the kinetic theory of gases, in which the link between pressure, temperature, and volume in gases is explained by a continuous movement of particles. This theory assumes particles have no attraction between them.

Van der Waals believed that molecules had bonds between them and that condensation and

See also: Dalton's atomic theory 80–81 ▪ The ideal gas law 94–97 ▪ Functional groups 100–105 ▪ Coordination chemistry 152–153

evaporation were not sudden jumps but transitions in which increasing numbers of molecules gain energy to break away from their bonds and become gases or lose energy and get pulled together to become liquids. He proposed a transition layer at the surface of a liquid in which it is neither liquid nor gas.

Van der Waals could not identify the nature of these intermolecular forces holding molecules together, but by 1930, scientists had confirmed their existence and how they work.

Identifying the forces

Three key forces have now been identified: dipole–dipole, hydrogen bonds, and dispersion or London forces. Dipole–dipole forces occur in "polar" molecules, in which electrons are shared unequally between the atoms of the molecule. This makes one side of the molecule more negatively charged, and these sides are attracted to the positive sides of others, binding them together.

Hydrogen bonds, such as those in water, are extreme dipole–dipole bonds that occur when hydrogen meets oxygen, fluorine, or nitrogen. These three atoms attract electrons, while hydrogen is prone to losing them. When they combine with hydrogen, they make a highly polarized molecule that binds strongly, which is why water (H_2O) has such a high boiling point for a substance made of two light gases.

Dispersion or London forces are very weak and occur between nonpolar molecules. In a nonpolar molecule, no part is permanently more positive or negative. However, the continuous movement of individual electrons on the molecule's atoms is enough to create fleeting attractions. ■

Johannes van der Waals

Born in 1837 in Leiden, the Netherlands, Johannes van der Waals was the son of a carpenter. He became a primary school teacher before studying mathematics and physics part time, and in 1873, he finally achieved his doctorate on molecular attraction. He was made a professor in 1877, by which time he was well advanced with his studies of thermodynamics and, in particular, phase changes between liquids and gases.

Van der Waals's work led to the understanding that there is a critical temperature for a gas, above which it is impossible for it to condense into a liquid, and he proved the continuity between liquid and gaseous states. It was for this work that he received the 1910 Nobel Prize in Physics. Van der Waals died in 1923.

Key works

1873 "On the Continuity of the Gaseous and Liquid States"
1880 "Law of Corresponding States"
1890 "Theory of Binary Solutions"

The van der Waals forces

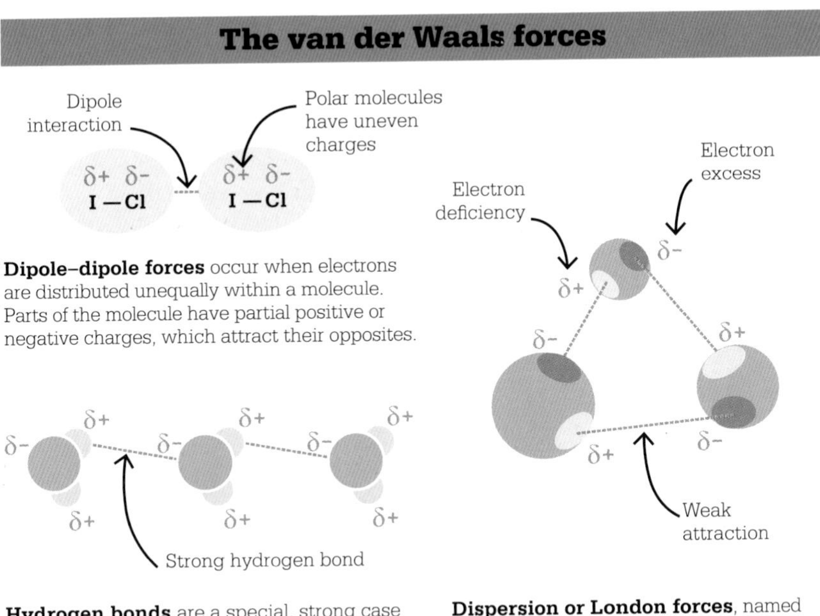

Dipole–dipole forces occur when electrons are distributed unequally within a molecule. Parts of the molecule have partial positive or negative charges, which attract their opposites.

Hydrogen bonds are a special, strong case of dipole–dipole forces, present in molecules where hydrogen is bonded to atoms that strongly attract electrons such as oxygen.

Dispersion or London forces, named after physicist Fritz London, are temporary attractions between adjacent atoms caused by the movement of electrons in molecules.

LEFT- AND RIGHT-HANDED MOLECULES

STEREOISOMERISM

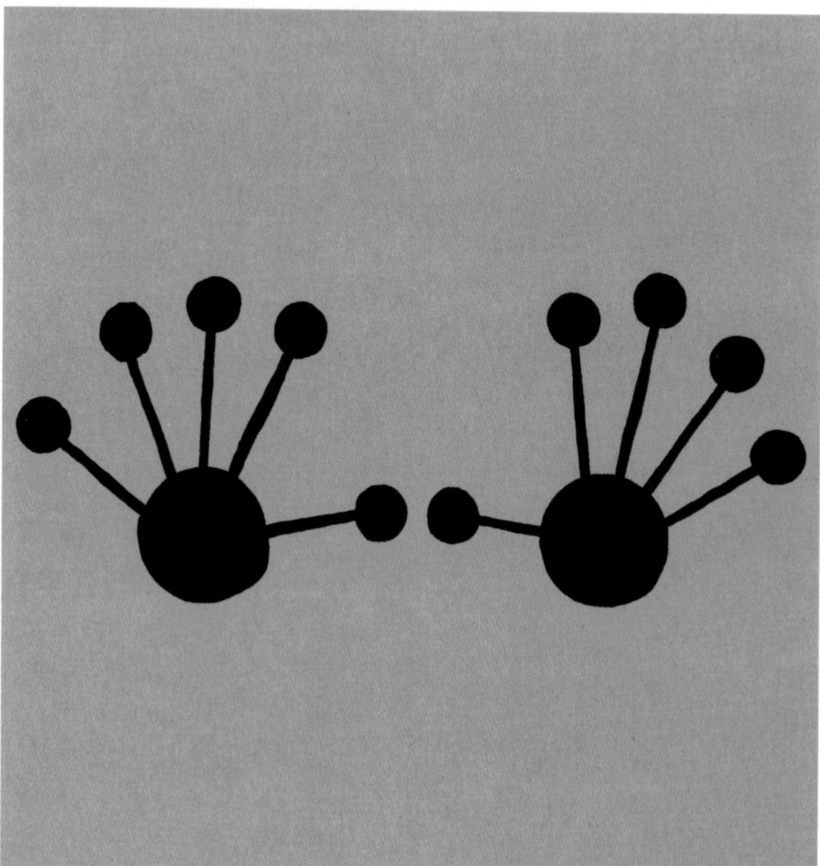

IN CONTEXT

KEY FIGURE
Jacobus van 't Hoff (1852–1911)

BEFORE
1669 Rasmus Bartholin discovers the polarization of light in Iceland spar crystals.

1815 Jean-Baptiste Biot discovers optical activity in tartaric acid salts.

1820 Philippe Kestner discovers racemic acid, a twin of tartaric acid that is not optically active.

1848 Louis Pasteur discovers that tartaric and racemic salt crystals are stereoisomers.

AFTER
1904 Lord Kelvin coins the term "chiral" to define stereoisomers.

1908 Arthur Robertson Cushny notes differences in biological activity between adrenaline and its stereoisomers.

Models and 3D graphics of molecular structures are now so familiar that we take them for granted. But until the work of Dutch chemist Jacobus van 't Hoff and French chemist Joseph Achille Le Bel in the 1870s, chemists did not imagine molecules as having any shape at all.

Compounds had a particular chemical composition and were known to be made of combinations of atoms, but it only occurred to a few chemists that the atoms join to create a real physical object with an actual shape. Van 't Hoff and Le Bel's breakthrough opened up a new field of chemistry, now known as stereochemistry—the study of

See also: Isomerism 84–87 ▪ Structural formulae 126–127 ▪ Intermolecular forces 138–139 ▪ Why reactions happen 144–147 ▪ Rational drug design 270–271 ▪ Atomic force microscopy 300–301

Tartaric acid crystals produced by fruit were key to the discovery of chirality. They were found to have a mirror-image twin, racemic acid, that polarizes light in a different way.

molecules in three dimensions—which plays a key role in many of our pharmaceutical drugs.

Tartaric and racemic acid

The roots of the discovery that molecules have a shape lay in an acid produced naturally on the inside of wine vats, known as tartaric acid. Chemists had long been familiar with it and the salts it created. So it was of great interest when, in 1820, German chemist Philippe Kestner discovered that a similar acid, not quite identical, was produced instead of tartaric acid when some vats got too hot. Jöns Jacob Berzelius called it paratartaric acid, and French chemist Joseph Gay-Lussac gave it the name racemic acid. Berzelius was soon to describe chemicals that have the same composition but different properties as isomers.

The most intriguing difference between tartaric and racemic acids is how they respond to light. Back in 1669, Danish mathematician Rasmus Bartholin had observed how the calcite crystals known as Iceland spar seem to split light in different planes, and by the early 19th century, scientists had confirmed the phenomenon. Light normally vibrates in all directions, but sometimes it is polarized—it vibrates in a single plane because the other directions are filtered out. French physicist Jean-Baptiste Biot was particularly fascinated by the way some crystals and liquids polarize light passing through them, rotating it clockwise or counterclockwise so that it exits on a different plane from the one on which it enters. This is described as optical activity.

In 1815, Biot confirmed that tartaric acid solution showed optical activity—but tests by German chemist Eilhard Mitscherlich revealed that the newly discovered racemic acid did not. In 1848, young French chemist Louis Pasteur investigated in one of his doctoral theses. Using a powerful magnifying glass, he studied crystals of the acid salts (racemic sodium ammonium tartrate) and saw that the crystals were not identical but were instead twinned like pairs of shoes, with one particular crystal facing left and the other right.

With just a pair of tweezers, Pasteur separated the pairs into two piles: left- and right-facing crystals. He made another solution from the left-facers and found that it polarized light in one direction. He made another of the right-facers and discovered that it polarized light in the opposite direction. Mixed equally, they had no effect on light at all. Such equal mixes are now described as racemic. Pasteur »

> The right hemihedral crystals deviate to the right [and] the left hemihedral crystals deviate the plane of polarization to the left.
> **Louis Pasteur (1848)**

had discovered "stereoisomers." He demonstrated it to Biot, who was delighted, saying, "I have loved science so much during my life that this touches my very heart."

In his breakthrough, Pasteur had revealed that chemicals can show "handedness," but he had not explained why. He had an inkling that it might be something to do with the shapes of matter, but no more. The answer arrived in 1874 from two chemists who had met in the laboratories of eminent French chemist Charles-Adolphe Wurtz: van 't Hoff and Le Bel. They worked on the problem independently but came up with the same answer.

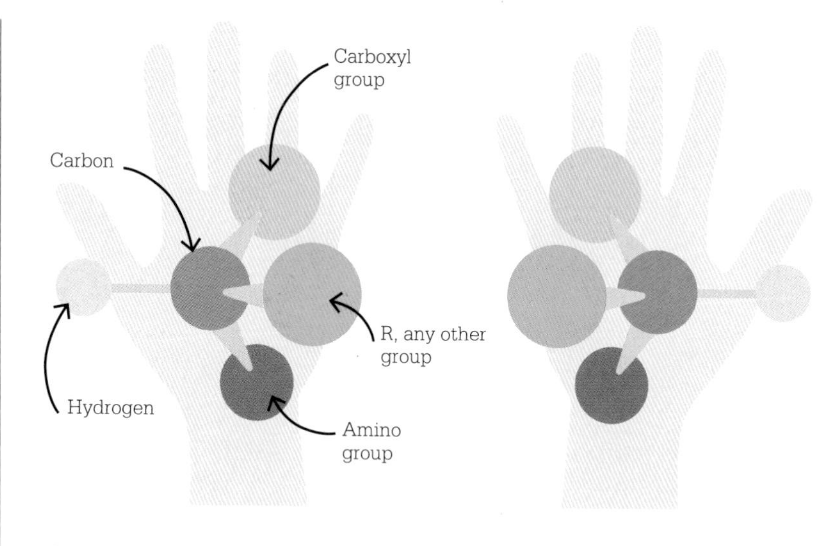

A chiral molecule is one that cannot be superimposed on its mirror image, just as the left hand will not fit into a right-hand glove. However, chiral molecules do share many identical properties.

Shape and structure

The clue, for van 't Hoff and Le Bel, was that most of the known chemicals showing optical activity were compounds of carbon. They argued that all the optically active compounds they focused on could be explained by two possible arrangements of four different groups around a central carbon atom. They suggested that the four atoms to which carbon is bonded sit at the corners of a regular tetrahedron, with carbon in the center.

Van 't Hoff not only visualized this arrangement for the first time through the use of little drawings but actually made up cardboard models of the carbon pyramid or tetrahedron. These models—among the oldest examples of molecular models—still exist. The idea that molecules were real, physical entities with a three-dimensional structure, not just a shapeless combination of elements, seems obvious now, but it outraged some scientists at the time. Ironically, German chemist Hermann Kolbe, one of the pioneers of organic chemistry, turned out to be one of the fiercest critics, saying of van 't Hoff's paper, "To criticize this paper in any detail is impossible because the play of imagination completely forsakes the solid ground."

Jacobus van 't Hoff

Born in Rotterdam, the Netherlands, in 1852, Jacobus van 't Hoff was one of seven children and did not have the funds to work full time in science. However, in 1872, he went to Bonn, Germany, to study math and chemistry for a year, the latter under German chemist August Kekulé. After a period studying in Paris, France, with Charles-Adolphe Wurtz, van 't Hoff returned to the Netherlands and began his groundbreaking work on stereochemistry; he published his ideas in 1874. Following a doctorate and the publication of *Chemistry in Space*, the book that made his ideas known, van 't Hoff became a lecturer in chemistry at the University of Amsterdam, where he stayed for 20 years. There, he studied reaction rates, chemical affinity, chemical equilibrium, and more. In 1896, he moved to the University of Berlin, Germany, and in 1901 became the first Nobel laureate in chemistry for his work in physical chemistry. Van 't Hoff died in 1911.

Key work

1875 *Chemistry in Space*

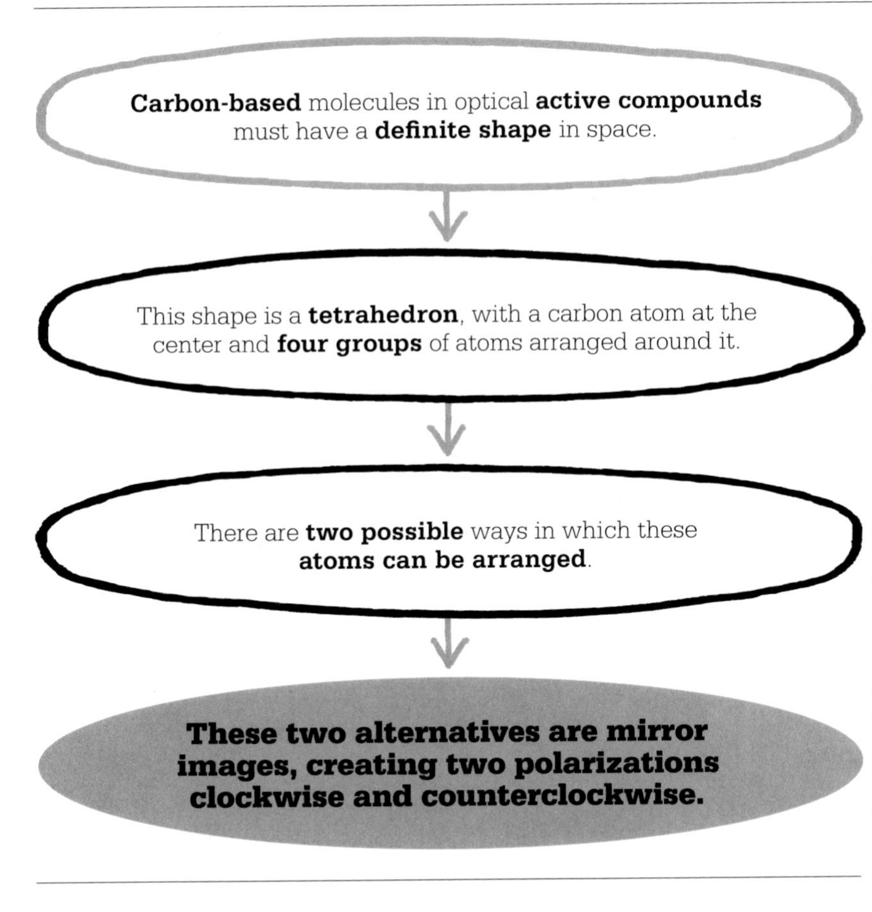

Carbon-based molecules in optical **active compounds** must have a **definite shape** in space.
This shape is a **tetrahedron**, with a carbon atom at the center and **four groups** of atoms arranged around it.
There are **two possible** ways in which these **atoms can be arranged**.
These two alternatives are mirror images, creating two polarizations clockwise and counterclockwise.

Gradually, however, the truth of what van 't Hoff and Le Bel were saying sank in. Chemists began to realize that the three-dimensional structure of molecules is key to their character, and the only way to understand their behavior is by visualizing how their atoms link to make a particular shape. Furthermore, these models were tactile and intuitive, and they allowed chemists to pick up and hold and move around molecules—albeit many times the size of the real thing—that had once been vague and imaginary.

Stereochemistry

During the next 50 years or so, scores of scientists set to work on establishing the science of stereo, or 3D, chemistry. This is the study of compounds that have the same molecular and constitutional formula—that is, the same combination of atoms arranged in the same order—but with their atoms arranged in different shapes. It became clear that this applies to carbon compounds, although silicon and a few other elements can occasionally form such molecules. Scientists also realized that these types of compound play a huge role in organic chemistry.

In 1904, Scottish physicist William Thomson, Lord Kelvin, introduced the term chirality, from the Greek for "hand," to describe stereoisomers. Then, in 1908, Scottish pharmacist Arthur Robertson Cushny noticed for the first time a difference in biological activity between the two chiral versions of a molecule, adrenaline, which acts as a vasoconstrictor (a narrower of blood vessels) in one version but not in its twin. As a direct consequence, by the 1920s, pharmaceutical companies had begun to develop new drugs based on stereoisomers. Today, 40 percent of all synthetic drugs are chiral, although most of those drugs include both versions of the molecule, so are said to be racemic.

In the 1960s, chirality became linked with the thalidomide disaster. Thalidomide is a drug that was introduced in Germany in 1957 and was prescribed to treat morning sickness in pregnant women. However, it caused damage to the unborn fetus, resulting in limb malformations. German clinician Widukind Lenz and Australian clinician William McBride realized that thalidomide's chiral nature was to blame, and the 10,000 infant victims became the driving force for stricter regulations on drug testing.

Despite this setback, chirality is fundamental to many of our most potent and useful drugs, and it has played a huge role in understanding how organic molecules behave. ■

> Jacobus van 't Hoff and Joseph Le Bel added a third dimension to our ideas about organic compounds.
>
> **John McMurry**
> ***Organic Chemistry* (2012)**

THE ENTROPY OF THE UNIVERSE TENDS TO A MAXIMUM

WHY REACTIONS HAPPEN

IN CONTEXT

KEY FIGURE
Josiah Gibbs (1839–1903)

BEFORE
1840s British physicist James Joule and Germans Hermann von Helmholtz and Julius von Mayer introduce the theory of conservation of energy.

1850, **1854** Rudolf Clausius publishes his first and second laws of thermodynamics.

1877 Austrian physicist Ludwig Boltzmann states the relationship between entropy and probability.

AFTER
1893 Swiss chemist Alfred Werner introduces the concept of coordination chemistry.

1923 American chemists Gilbert Lewis and Merle Randall publish a book placing free energy at the forefront of chemical reaction studies.

In the 1870s, chemistry and physics were viewed as completely separate sciences. However, in 1873, a remarkable paper by American mathematician and physicist Josiah Gibbs brought thermodynamics—the physics of heat and energy—into the heart of chemistry and initiated the new discipline of physical chemistry. This deals with the physics of chemical interactions.

The term "physical chemistry" dates back to 1752, when Russian polymath Mikhail Lermontov used it to explain what was happening in complex bodies through chemical operations—but it meant little until Gibbs's work.

See also: Conservation of mass 62–63 ▪ Catalysis 69 ▪ The ideal gas law 94–97 ▪ Coordination chemistry 152–153 ▪ Depicting reaction mechanisms 214–215

In a **chemical reaction**, there is a total amount of **energy** (enthalpy), plus free energy—the energy available to **make the reaction happen**.

During a **spontaneous chemical reaction**, enthalpy is reduced as **energy is lost**.

Energy is **also lost** as **entropy** or disorder **increases**.

The change in free energy available to continue the reaction is the difference between the changes in enthalpy and entropy.

As a young man, Gibbs traveled to study in Europe, becoming fluent in German and French. He learned first-hand from some of the leading scientists of the age, including German physicists Gustav Kirchhoff and Hermann von Helmholtz, and encountered cutting-edge ideas in mathematics, chemistry, and physics. The most exciting area was thermodynamics, where two profound laws had recently been established, as scientists such as German physicist Rudolf Clausius and Scottish engineer William Rankine tried to understand the relationship between heat, energy, and movement. The first law stated that although energy can move, it is always conserved and can never be created or destroyed. The second showed that energy naturally spreads out or dissipates, so that the entropy, or disorder, of a system always increases. As Clausius put it, "The entropy of the universe tends to a maximum."

Geometric model

Gibbs understood that the key to thermodynamics was mathematics, but while most physicists worked with algebra, he used geometry. In a key paper published in 1873, he stressed the importance of entropy, expanding on Scottish engineer James Watts's famous graph, which demonstrated the relationship »

Josiah Gibbs

Born in Connecticut in 1839, the fourth of five children, Josiah Gibbs had a keen interest in math when he was young. He studied at Yale University and at the Connecticut Academy of Arts and Sciences. At age 24, he received the first PhD in engineering awarded in the US. In 1866, he traveled to Europe, attending lectures in Paris, France, and in Berlin and Heidelberg, Germany.

In 1869, Gibbs returned to Yale, where he was appointed professor of mathematical physics and worked for the rest of his life, becoming America's first major theoretical scientist. His work in thermodynamics had an impact on chemistry, physics, and math, and Einstein called him "the greatest mind in American history." Gibbs died in 1903.

Key works

1873 "A Method of Geometrical Presentation of the Thermodynamic Properties of Substances"
1878 "On the Equilibrium of Heterogeneous Substances"

between pressure and volume, by adding a third coordinate for entropy. The result was that the relationship between volume, entropy, and energy could be represented graphically as three-dimensional geometric shapes.

Gibbs's approach was so radical that only a few understood it. One of those who did was James Clerk Maxwell, a Scottish mathematician. In 1874, he used Gibbs's 3D equations to build plaster models of the thermodynamic surface for an imaginary waterlike substance as a way of visualizing phase changes—when matter transitions between solid, liquid, or gas.

Making predictions

In 1878, Gibbs wrote another key paper that created a new field of science known as chemical thermodynamics. Gibbs realized that thermodynamics could be used to make general predictions about the behavior of substances. For chemists, his ideas were revolutionary because they helped show that an understanding of a chemical's behavior was possible without knowing details of its molecular structure. Gibbs showed how thermodynamic principles—and in particular the concept of entropy—works for everything from gases and mixtures to phase changes. His ideas revolutionized scientific understanding of every process involving heat and work, including chemical reactions.

Gibbs assumed that molecules exist evenly in all energy states and then worked out how properties of gases such as pressure and entropy average out. If there are, for example, three energy states, this can be imagined as the average position three marbles land in an egg carton when shaken trillions of times. Gibbs called such imaginary cartons "ensembles."

Key to his work was what he called "free energy" but is now known as Gibbs free energy—the thermodynamic energy available to do work. Just as any object lifted above the ground has gravitational potential energy because it can fall, so molecules have potential energy in their bonds. Weak bonds have high potential energy. Gibbs free energy is the energy available to make things happen, such as chemical reactions, so it indicates whether a reaction is likely to happen and, if so, how fast.

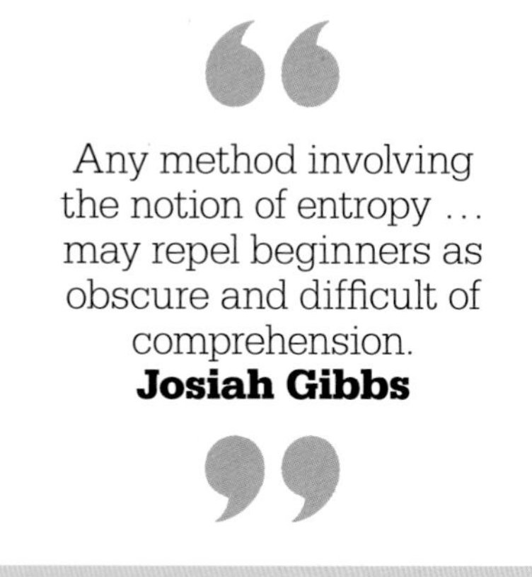

Any method involving the notion of entropy ... may repel beginners as obscure and difficult of comprehension.
Josiah Gibbs

A reaction can be spontaneous or nonspontaneous. The former happens by itself, using its own

A thermodynamic process: salt dissolving

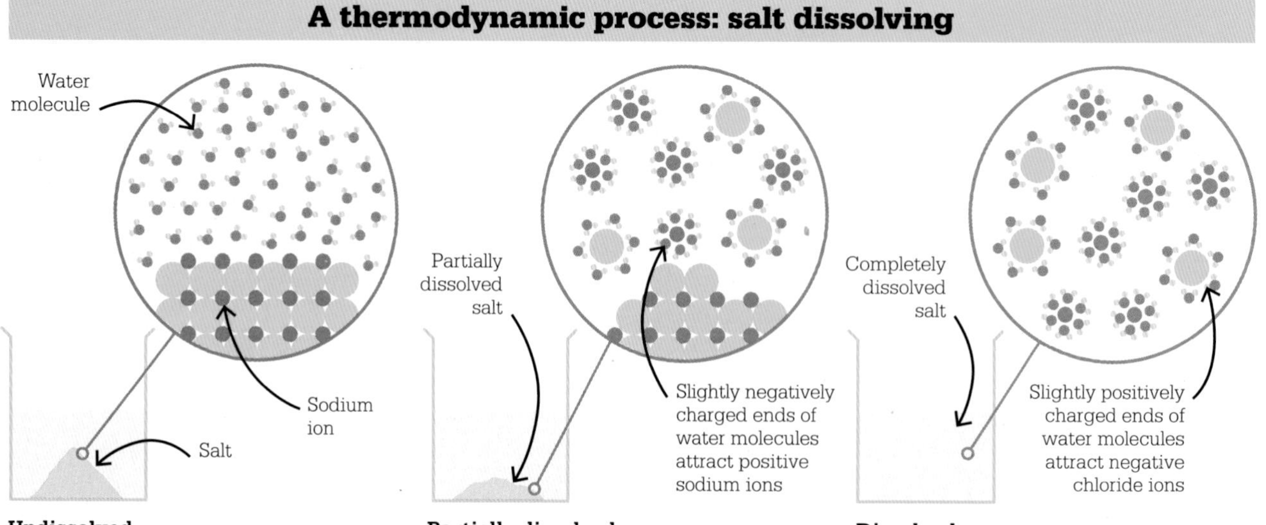

Undissolved
Salt (sodium chloride, NaCl) is an ionic compound—one atom essentially donates an electron to stabilize the other.

Partially dissolved
When sodium chloride dissolves in water, it separates into positive sodium ions (Na^+) and negative chloride ions (Cl^-).

Dissolved
The dissolution of the sodium chloride is an endothermic process that results in the temperature of the solution falling.

energy, although it may require a little "activation energy" to get it started—like the match you put to a firework. In contrast, a nonspontaneous reaction needs a constant input of energy.

Enthalpy and entropy

During a spontaneous chemical reaction, the total energy available, called the enthalpy (H), reduces over time—for example, coal loses energy as it burns. Similarly, the entropy (S), or disorder, increases, just as sugar spreads out more randomly as a sugar lump dissolves in tea. The more disordered a system becomes, the less energy is available. The energy left available, the Gibbs free energy, is the difference between the enthalpy and the entropy.

Gibbs's equation shows how these change in a reaction, where the symbol delta Δ shows change, G is Gibbs free energy, and T is temperature in Kelvin:

$$\Delta G = \Delta H - T\Delta S$$

This confirms that if enthalpy is reduced or entropy and temperature increase, then the Gibbs free energy must also be reduced. If the reaction is spontaneous, ΔG is less than zero, and the Gibbs free energy decreases. If the reaction is nonspontaneous, ΔG is greater than zero, and the Gibbs free energy decreases. Increasing or decreasing H or S creates four classes of reactions (see graph). If there is no change—ΔG is zero—then the system is in equilibrium.

Using this approach, chemists can calculate whether a reaction is very likely to be impossible or not. For instance, calculations showed that under extreme temperatures and pressures, graphene (an allotrope of carbon) might be converted to diamond, encouraging scientists to persist through many failures until they finally managed to accomplish the conversion.

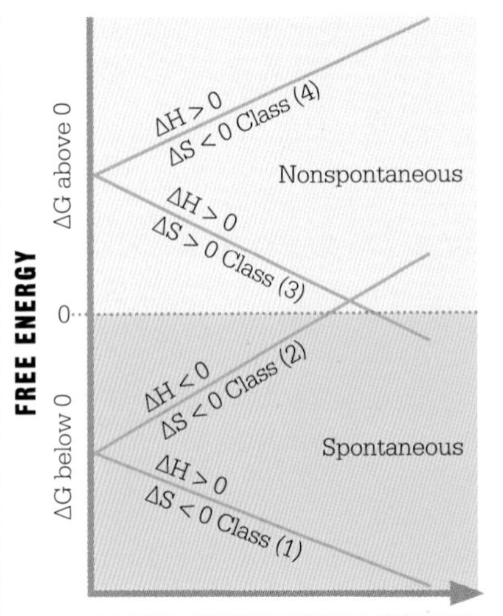

This graph shows the four scenarios for changes in enthalpy (H) and entropy (S) in spontaneous and nonspontaneous reactions, as demonstrated by Josiah Gibbs's equation.

Besides free energy, Gibbs also introduced another key concept: the phase rule. This shows the number of phases involved in a chemical system, taking into account the number of components (chemically independent constituents) and all the variables that affect the way they react—temperature, pressure, energy, and volume, described statistically as degrees of freedom. Gibbs showed how many phases can coexist and showed why water has a triple point, at which it can exist in three forms simultaneously as it melts—liquid, ice, and gas—known as three-phase equilibrium.

In 2020, a team of researchers from Eindhoven University of Technology and Paris-Saclay University showed that a five-phase equilibrium can also occur—a gas phase, two liquid-crystal phases, and two solid phases with "ordinary" crystals. Despite this, Gibbs's phase rule remains essentially true and has proved hugely valuable in many activities, such as predicting the melting points of alloys.

Gibbs's work provided a new set of tools for chemists, engineers, and theorists alike. Chemists could use them to predict whether chemical reactions would occur. Engineers could use his 3D diagrams to grasp thermodynamic rules in a simple practical way. And theorists could use them as a springboard for groundbreaking science. ■

Statistical mechanics

Gibbs realized that calculations in thermodynamics depend on the way atoms and molecules are considered. The Newtonian view assumes everything behaves in a precise way—enabling speed and trajectories to be calculated exactly. However, the interaction of billions of fast-moving atoms crashing into each other is so incalculable that it seems to be random.

Gibbs, along with Austrian physicist Ludwig Boltzmann, realized that atoms must be looked at as populations using a statistical approach. In 1884, he coined the term "statistical mechanics" for this. It is about probability; a scientist cannot predict the outcome for an individual atom but can use statistical mechanics to make a prediction for any whole reaction. This approach links macroscopic—large-scale, observable—and microscopic events. It helped drive the creation of quantum science in the early 20th century.

EVERY SALT, DISSOLVED IN WATER, IS PARTLY DISSOCIATED IN ACID AND BASE

ACIDS AND BASES

IN CONTEXT

KEY FIGURE
Svante Arrhenius
(1859–1927)

BEFORE
1766 Henry Cavendish discovers that reacting acids with metals releases hydrogen.

1809 German polymath Johann Wolfgang von Goethe explores ionic reactions in *Die Wahlverwandtschaften*.

1867 British surgeon Joseph Lister discovers that acids can kill germs, pioneering antiseptic surgery using carbolic acid.

AFTER
1912 Dutch chemist Willem Keesom describes dipole–dipole forces for the first time.

1963 American ecologist Gene Likens discovers acid rain.

1972 American astronomers G.T. Sill and A.T. Young show that the atmosphere of Venus is rich in sulfuric acid.

Acids and bases are key substances that have long attracted chemists' attention, partly because they cause extreme reactions. Pinning down what they are remains a problem, but the first modern definition, focusing on the role of ions, was introduced by Swedish chemist Svante Arrhenius in 1884.

In ancient times, people knew of things that tasted sour, such as vinegar, and the word acid comes from the Latin *acere*, meaning "to make sour." They also knew that some acids could dissolve metals. In Iran, around 800 CE, Jabir ibn Hayyan discovered hydrochloric and nitric acid, realizing that they combine to make aqua regia, which can dissolve even gold.

In 1776, Antoine Lavoisier asserted that it is the presence of oxygen that makes an acid, and he named oxygen after the Greek for acid-making. However, in 1810, Humphry Davy tested acids with metals and nonmetals and found that many did not involve oxygen at all. He hinted that hydrogen might be the key instead, and this was confirmed 20 years later by Justus von Liebig, who argued that an acid is a hydrogen-containing substance in which hydrogen can be replaced by a metal.

What are acids?

Liebig's definition worked but said little about what acids are. The answer came in 1884 in Arrhenius's doctoral dissertation, which focused on ions—atoms that become negatively or positively charged by gaining or losing electrons.

A simple experiment, in which vinegar (acetic acid) is mixed with baking soda (sodium bicarbonate), shows a powerful acid–base reaction.

See also: Inflammable air 56–57 ▪ Isolating elements with electricity 76–79 ▪ Sulfuric acid 90–91 ▪ Intermolecular forces 138–139 ▪ The electron 164–165 ▪ The pH scale 184–189 ▪ Depicting reaction mechanisms 214–215 ▪ The structure of DNA 258–261

The probability [is] that electrolytes can assume two different forms, one active, the other inactive.
Svante Arrhenius

It had previously been thought that ions only appear in a liquid when an electric current is passing through. Arrhenius argued that ions are always present in liquids. He also proposed that electrolytes—that is, chemicals that conduct electricity—can have an active state (that conducts) and an inactive state (that does not conduct).

This concept was so radical at the time that Arrhenius was vilified. But he persisted with it and in 1894 developed a new way of thinking about acids and bases. Acids, he said, are substances that add positively charged hydrogen ions (cations) to a solution. Bases are substances that add negatively charged hydroxyl ions (hydroxyl anions) to the solution. Arrhenius also suggested that acids and bases neutralize each other by forming water and a salt.

All modern definitions of acids have sprung from this idea, and it earned Arrhenius the 1903 Nobel Prize in Chemistry. In 1923, Danish chemist Johannes Brønsted and British chemist Thomas Lowry refined the idea. Focusing on protons, they defined an acid as a substance from which a proton can be removed and a base as a substance that can bind the proton from an acid. Chemists now talk of acids as "proton donors" and bases as "proton acceptors."

Also in 1923, American chemist Gilbert Lewis further developed the subject, focusing on electron pairs and covalent bonds. Both approaches were combined in the 1960s to give the modern picture of acids. ■

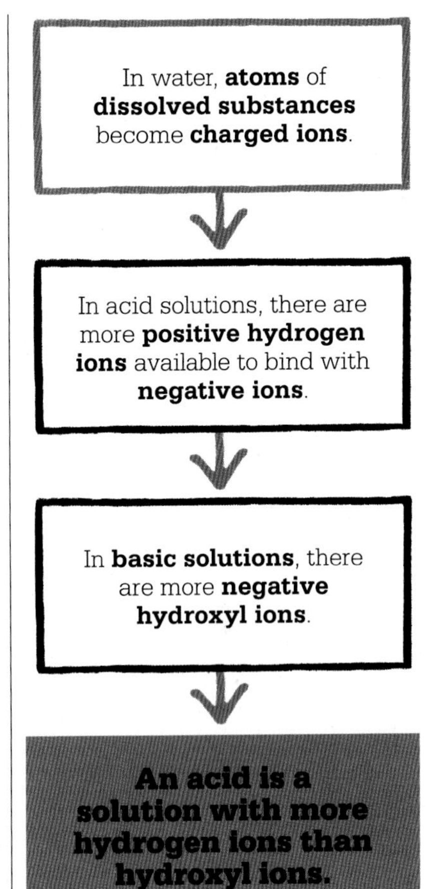

In water, atoms of dissolved substances become charged ions.
In acid solutions, there are more positive hydrogen ions available to bind with negative ions.
In basic solutions, there are more negative hydroxyl ions.
An acid is a solution with more hydrogen ions than hydroxyl ions.

Svante Arrhenius

Born in 1859 in Vik, near Uppsala in Sweden, Svante Arrhenius was a math prodigy as a young boy. After becoming dissatisfied with the teaching at Uppsala University, he moved to Stockholm and began undertaking work on his groundbreaking doctoral thesis exploring the role of ions in solutions. Despite the initial rejection of his ideas, he went on to pursue a successful scientific career.

Arrhenius's work spanned chemistry, physics, biology, and cosmology, and he made key contributions in three different areas of science. First, his theory of ions became the platform for our modern understanding of electrolytes and acids. Second, his study of geophysics gave the first scientific evidence of climate change. Third, he made key investigations into toxins and antitoxins. Arrhenius died in Stockholm in 1927.

Key work

1884 "Research on the Galvanic Conductivity of Electrolytes"

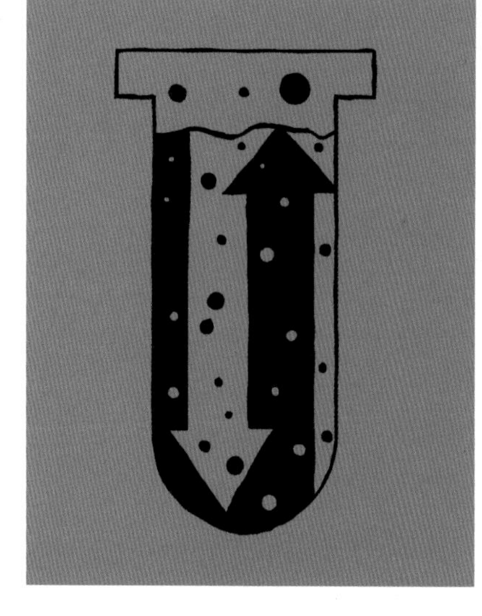

CHANGE PROMPTS AN OPPOSING REACTION

LE CHATELIER'S PRINCIPLE

IN CONTEXT

KEY FIGURE
Henry-Louis Le Chatelier (1850–1936)

BEFORE
1803 French chemist Claude Louis Berthollet discovers that some chemical reactions are reversible.

1864 Norwegian chemists Cato Guldberg and Peter Waage propose the law of mass action, which defines the rate of reactions and explains the behavior of solutions in dynamic equilibrium.

AFTER
1905 Fritz Haber proposes a way to mass-produce ammonia by applying Le Chatelier's principle.

1913 Industrial-scale production of ammonia begins, after German engineer Carl Bosch designs a practical method of applying Haber's ideas—the Haber-Bosch process.

Many chemical reactions are reversible, meaning that the products react to produce the original reactants. For example, ammonium chloride (NH_4Cl), a solid, forms gaseous ammonia (NH_3) and hydrogen chloride (HCl) when heated. When the two gases are cool enough, they react to form NH_4Cl again. This reversible reaction is expressed as $NH_4Cl \rightleftharpoons NH_3 + HCl$. A reaction is said to be moving to the right when the quantity of products increases and moving to the left when the quantity of reactants increases. It is in a state of dynamic equilibrium when there is no net change in the amounts of reactants and products.

French chemist Henry-Louis Le Chatelier proposed in 1884 that if a system at equilibrium is subjected to a change in conditions, the position of equilibrium will adjust to counteract the change. These conditions are the concentration of reactants and the temperature or pressure at which the reaction is occurring. For example, if a chemist increases the temperature of a reaction, the reaction will move to the right or to the left to reduce the temperature.

German chemist Fritz Haber applied Le Chatelier's principle when working out a method to maximize the production of ammonia from nitrogen and hydrogen, in the reaction $N_2 + 3H_2 \rightleftharpoons 2NH_3$. He calculated that the reaction should be conducted at a high pressure and low temperature. In practice, he succeeded only by using a catalyst, but Le Chatelier's breakthrough had been crucial to his work. ■

> I let the discovery of the ammonia synthesis slip through my hands.
> **Henry-Louis Le Chatelier**

See also: Catalysis 69 ▪ Why reactions happen 144–147 ▪ Fertilizers 190–191 ▪ Chemical warfare 196–199

HEAT-PROOF, SHATTER-PROOF, SCRATCH-PROOF

BOROSILICATE GLASS

IN CONTEXT

KEY FIGURE
Otto Schott (1851–1935)

BEFORE
1st century BCE People in the Roman province of Syria develop glassblowing techniques.

1830 French chemist Jean-Baptiste-André Dumas works out the best ratio of soda, lime, and silica for the optimum durability of glass.

AFTER
1915 Mass manufacture of heat-resistant borosilicate Pyrex bakeware begins.

1932 Norwegian-American physicist William Zachariasen explains the chemical structure of glass, distinguishing it from that of crystals.

Until the late 19th century, all glass was "soda-lime-silica" glass, made from silicon dioxide (SiO_2), sodium carbonate (NA_2CO_3), and calcium oxide (CaO), often with magnesium and aluminum oxides (MgO and Al_2O_3) added for durability. Known generally as "soda-lime," such glass works well for windows and bottles, but it has limitations: it distorts light and it expands and shatters at high temperatures. German chemist Otto Schott was fascinated by the relationship between the chemical composition of glass and its physical characteristics, and from 1887–93 he experimented with the ingredients, revolutionizing the material.

Laboratory use

Schott found that by adding lithium, he could produce glass with minimal optical aberrations, enabling a major advance in the quality of lenses in microscopes and telescopes. He also discovered that the addition of boron trioxide (B_2O_3) to the silicon dioxide created glass that was much more tolerant of heat and chemical exposure. These characteristics of the new borosilicate glass derive from its tightly bound chemical structure.

Although the manufacturing process was refined in the 20th century, borosilicate glass very similar to Schott's is used in laboratories and kitchens to this day. It can be heated to around 932°F (500°C) without shattering. ■

Borosilicate glass can withstand temperature differentials of up to 330°F (165°C), making it invaluable for use in science experiments.

See also: Making glass 26 ▪ X-ray crystallography 192–193

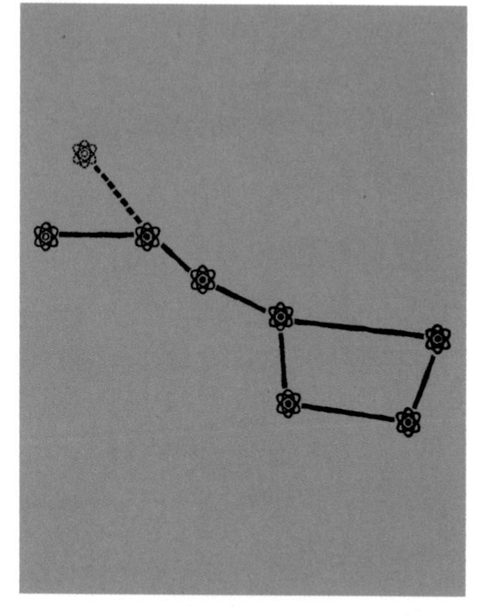

THE NEW ATOMIC CONSTELLATION

COORDINATION CHEMISTRY

IN CONTEXT

KEY FIGURE
Alfred Werner (1866–1919)

BEFORE
1832 Friedrich Wöhler and Justus von Liebig introduce functional groups of atoms within molecules.

1852 British chemist Edward Frankland introduces the idea of "combining power" to explain bonding in molecules.

1861 August Kekulé and Josef Loschmidt develop the concept of structural formulae.

1874 Jacobus van 't Hoff and Joseph Achille Le Bel argue that molecules have three-dimensional structures.

AFTER
1916 American chemist Gilbert Lewis proposes that chemical bonds form by the interaction of two shared electrons.

1931–1932 American chemist Linus Pauling publishes his valence bond theory.

In the 1880s, the concept of the "combining power" of atoms became known as valence (the number of hydrogen atoms that an atom will bond with). This was a major step in understanding how molecules are constructed, but it ran into problems with complex molecules that appeared to have multiple valences. Then, in 1893, Swiss chemist Alfred Werner proposed a radical explanation that opened up a new branch of chemistry: coordination chemistry.

Since the 1860s, most major advances in understanding molecular structure had been in organic chemistry, but Werner was looking at compounds of metals and, in particular, what we now call coordination compounds—complexes in which a metal atom is surrounded by nonmetal atoms or groups of atoms. Such compounds had been known for centuries (the pigment Prussian blue, introduced in 1706, is a coordination compound), but it was hard to pin down their chemical structure. Many were described as "double salts," since they seemed to be combinations of two salts, and so their formulae were written with a dot. For example, aluminum fluoride (AlF_3) and potassium fluoride (KF), which combine in a ratio of 1:3, was written as $AlF_3.3KF$.

> This concept [of coordination number] is destined to serve as a basis for the theory of the constitution of inorganic compounds.
> **Alfred Werner**

Secondary valence

Why did the elements combine in particular ratios and not others? Werner was obsessed with the problem, and one night he woke at 2 a.m. with a brainwave. By 5 p.m. that day, he had written a key paper explaining his theory, which was published in 1893. Werner saw metal complexes as a central metal atom combined with "ligands"—ions, atoms, or molecules that will bond with it. Ligands can be simple

See also: Isomerism 84–87 ▪ Functional groups 100–105 ▪ Structural formulae 126–127 ▪ Stereoisomerism 140–143 ▪ Chemical bonding 238–245

molecules, such as ammonia (NH_3) or water (H_2O), or much more complex ones.

Looking at his work on a complex of ammonia with platinum chloride ($PtCl_2.2NH_3$), Werner proposed two kinds of valence for a metal ion: a primary valence, given by the ion's positive charge; and a secondary valence, or "coordination number," which is the number of ligand-to-metal bonds that the metal can acquire. This concept enabled him to explore how coordination compounds were bonded. Facing skepticism, he set to work to create a new series of metal complexes that were predicted by the theory. In 1900, his British doctoral student Edith Humphrey succeeded in this task, preparing crystals of one of the predicted cobalt complexes.

Compound geometry

Like Dutch chemist Jacobus van 't Hoff, who had visualized carbon molecules in three dimensions as tetrahedrons, Werner thought of metal complexes in 3D. He worked out the configurations of certain complexes from the number and type of their isomers—compounds made of the same components but differently arranged. For example, he proposed that complexes of cobalt (III), which has a coordination number of six, were octahedral. Werner spent years analyzing these compounds to build evidence for his theory, but some crucial ones were missing. Then, in 1907, with the aid of American doctoral student Victor King, Werner managed to synthesize the highly unstable violeo tetraammines—$[Co(NH_3)_4Cl_2]X$—with their two isomers. His critics conceded defeat, and coordination chemistry began to have an impact.

Nearly all metals form complexes, and they are hugely important in many areas. Industry relies heavily on them, especially for catalysts, while complexes of transition metals are crucial in biological processes. Hemoglobin, which carries oxygen through the blood, is an iron complex, and other transition-metal complexes play a key role in enzymes, which are biological catalysts. Drugs containing metal complexes—metallodrugs—are used in the treatment of cancer. ■

A 6-coordinated metal complex might be in the form of a hexagon, a trigonal prism, or an octahedron. If the metal M has four ligands of type A and two of type B (MA_4B_2) and is a hexagon or trigonal prism, it will have three isomers; if it is an octahedron, it will have only two.

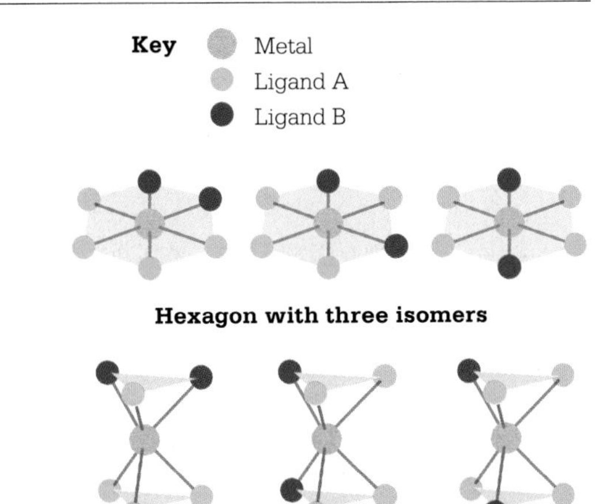

Hexagon with three isomers

Trigonal prism with three isomers

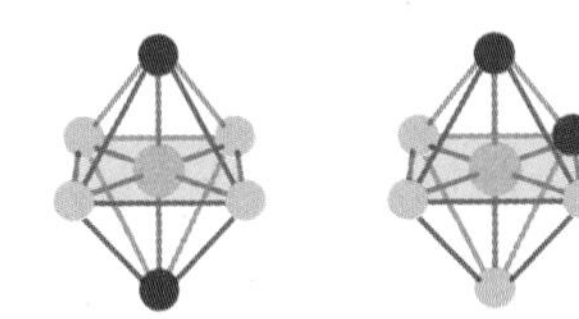

Octahedron with two isomers

Alfred Werner

The son of a factory worker, Alfred Werner was born in 1866 in Alsace, France, a town that was annexed by Germany four years later. Interested in chemistry at an early age, he conducted experiments in his bedroom. In 1889, he went to Switzerland to study chemistry, where he received his degree from the University of Zurich in 1890 and his doctorate two years later.

In 1895, at age 29, Werner was appointed professor of chemistry at Zurich, and in the same year, he became a Swiss citizen. He was a well-loved teacher who did a lot of his pioneering research with his students. These included, unusually for the time, a number of women, many of whom went on to become successful chemists—such as Edith Humphrey, thought to be the first British woman awarded a chemistry PhD. Werner was awarded the Nobel Prize in Chemistry in 1913 and died six years later.

Key work

1893 *Contribution to the Constitution of Inorganic Compounds*

A GLORIOUS YELLOW EFFULGENCE

THE NOBLE GASES

IN CONTEXT

KEY FIGURE
William Ramsay (1852–1916)

BEFORE
1860 Robert Bunsen and Gustav Kirchhoff suggest that new elements can be identified through spectral analysis.

1869 Dmitri Mendeleev arranges the 64 known elements into a periodic table and predicts that more will be discovered.

AFTER
1937 Russian physicist Pyotr Kapitsa and British physicists John Allen and Don Misener produce supercooled helium, the first superfluid with zero viscosity.

1962 British chemist Neil Bartlett synthesizes xenon hexafluoroplatinate, the first chemical compound to include a noble gas.

Dmitri Mendeleev's periodic table of 1869 organized the elements according to their atomic weights and showed how recurring characteristics of the elements occurred at regular intervals, or periods. Mendeleev's table appeared so solidly based that most scientists felt it would hold true no matter what new discoveries might be made. Mendeleev had even left gaps to be filled by these new elements when they did eventually come to light; in fact, when gallium and scandium were discovered in the 1870s, they slotted right in. However, some new elements would prove more of a challenge.

First hints

In 1783, British chemist Henry Cavendish published an account of his attempt to determine the composition of the atmosphere. His method consisted of adding nitric oxide (NO)—which combined with oxygen to form soluble nitrogen dioxide (NO_2)—to air and then measuring the reduction in volume. Cavendish was also able to remove the nitrogen from his sample, but to his surprise, he found that a small bubble of gas, about 0.8 percent of his original sample, remained. Cavendish was at a loss to explain what this was. The puzzle stayed unsolved for almost a century.

In August 1868, the year before Mendeleev published the first version of his table, French astronomer Pierre Janssen was examining the Sun's corona with his spectroscope during a total eclipse in India. He observed a bright yellow line that didn't match up with any known element. Two months later, and independently of Janssen, British astronomer Norman Lockyer also discovered this same yellow line. Lockyer did not hesitate to claim the discovery of a new element and named it helium, after Helios, the Greek god of the Sun. Further chemical analysis of helium was impossible at the time; it was the first instance of a chemical element to be discovered on an extraterrestrial body before it had been found on Earth.

Rayleigh and Ramsay

In 1892, Scottish chemist William Ramsay, a professor at University College London, heard of a puzzling finding by Lord Rayleigh (John William Strutt), previously professor

William Ramsay

Born in Glasgow, Scotland, in 1852, William Ramsay, the only child of a civil engineer father, studied at the University of Glasgow but left in 1870 without a degree. In 1871, he became a doctoral student under German chemist Rudolf Fittig at the University of Tübingen. After graduating in 1872, Ramsay returned to Glasgow. He was appointed professor of chemistry at University College, Bristol, in 1879 and, in 1881, became principal. He succeeded to the chair of chemistry at University College London (UCL) in 1887, and it was here that his most notable discoveries were made.

His work in isolating and identifying the noble gases required the addition of a new section in the periodic table. He was awarded the Nobel Prize in Chemistry in 1904 for this work. In 1912, Ramsay retired from his post at UCL. He died in 1916.

Key work

1896 *The Gases of the Atmosphere*

See also: Flame spectroscopy 122–125 ▪ The periodic table 130–137 ▪ Improved atomic models 216–221 ▪ Chemical bonding 238–245

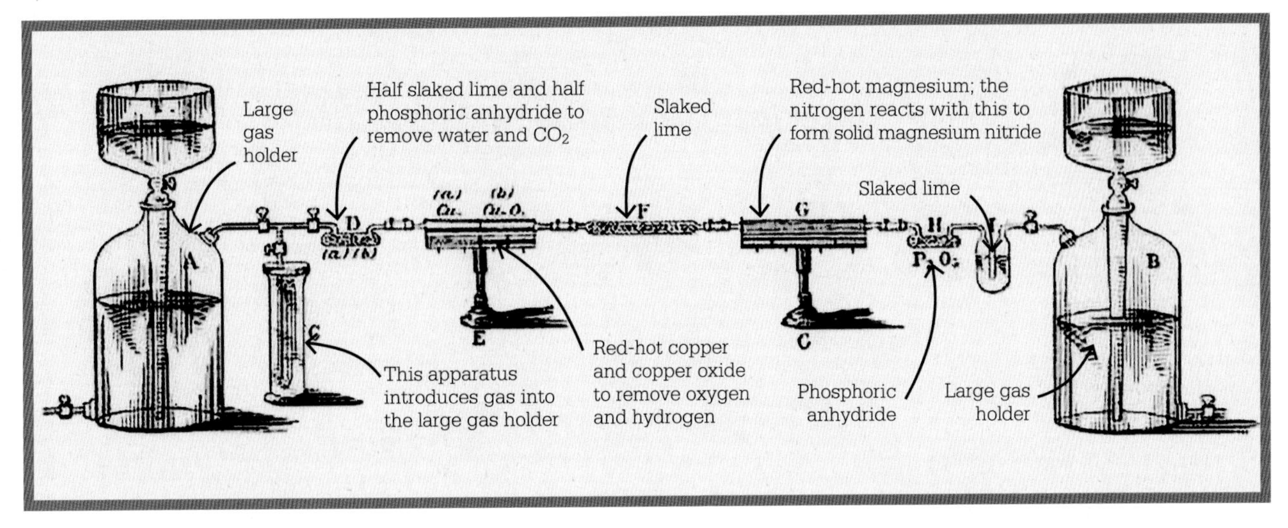

Ramsay used this apparatus to isolate argon. Nitrogen was pumped back and forth until it was absorbed. The tiny residue of argon left behind was then collected.

of physics at the Cavendish Laboratory in Cambridge but now an independent researcher. Rayleigh had discovered that atmospheric nitrogen was about 0.5 percent more dense than the nitrogen obtained from chemical compounds. Rayleigh later learned that Cavendish—after whom the Cambridge laboratory was named—had achieved a similar result many years before but had been unable to explain it.

Ramsay proposed that the sample might contain a previously unknown gas that had remained unaffected by the methods that were used to remove other gases. He isolated this gas by using red-hot magnesium to remove oxygen and nitrogen from the air. Experiments revealed that the mystery gas had to be, as Ramsay put it, "an astonishingly indifferent body" that could not be made to react with any other substance. Even highly reactive fluorine would not combine with it.

In 1894, Rayleigh and Ramsay announced the discovery of a new element at a meeting of the British Association for Science in London. They named it argon, from the Greek word for "idle." Although spectroscopic analysis by British chemist William Crookes confirmed that the new gas had a distinctive line pattern, some critics, among them Dmitri Mendeleev, disputed its status as an element and instead suggested that it must be a form of triatomic nitrogen, N_3. The main reason for this rejection was that argon did not fit easily into Mendeleev's periodic table, although Ramsay had written to Rayleigh on May 24, 1894, asking, "Has it occurred to you that there is room for gaseous elements at the end of the periodic table?"

I want to get back again from chemistry to physics as soon as I can. The second-rate men seem to know their place so much better.
Lord Rayleigh (1924)

Helium on Earth

Meanwhile, in 1882, while analyzing lava from Mount Vesuvius, Italian physicist Luigi Palmieri had spotted the same yellow spectral line that Janssen and Lockyer had observed in the spectrum of the Sun. It was the first indication of helium's presence on Earth—but Palmieri did not investigate further.

In 1895, British mineralogist Henry Miers informed Ramsay of the findings made by American chemist William Hillebrand in 1888. Hillebrand had discovered that heating cleveite, a uranium-containing mineral, with sulfuric acid generated an unreactive gas » that he thought was nitrogen but which Miers suspected was actually argon.

Ramsay ran the experiments again himself and collected the gas. Spectroscopic analysis confirmed that it was neither nitrogen nor argon. Instead, it was appearing in a spectroscope as a "glorious yellow effulgence," in the words of Lockyer, to whom Ramsay had sent a sample for verification. Its spectrum matched that of helium.

Independently, that same year, Swedish chemists Per Teodor Cleve and Nils Abraham Langlet in Uppsala, Sweden, also recovered helium from cleveite, collecting enough of the gas to accurately measure its atomic weight.

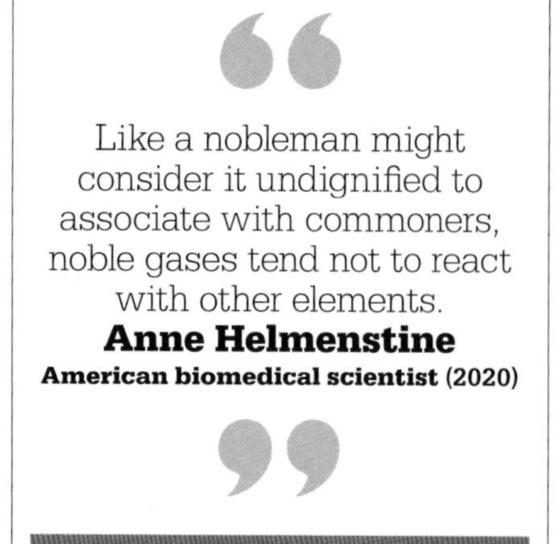

Like a nobleman might consider it undignified to associate with commoners, noble gases tend not to react with other elements.

Anne Helmenstine

American biomedical scientist (2020)

The search continues

The physical and chemical properties of helium and argon were so similar that it seemed they must belong to the same group of elements. Their differing atomic weights (helium 4 and argon 40) convinced Ramsay that there was likely to be at least one yet-to-be-discovered element to be found between them. After two years of fruitless searching in minerals, he decided to look to the air instead. If he was to isolate other atmospheric gases, Ramsay needed large-scale facilities for liquefying and fractionally distilling air. He approached British engineer William Hampson, who had patented an innovative process for liquefying gases, and in 1898, Hampson provided Ramsay with around 25 fl oz (0.75 liters) of liquid air.

In June 1898, Ramsay and his assistant, British chemist Morris Travers, carefully evaporated and distilled their liquid air sample. Once the nitrogen, oxygen, and argon were removed, they were left with a tiny residue of gas. Spectral analysis confirmed that they had found another new element but with an estimated atomic weight of 80; rather than being lighter than argon, it was actually heavier. It was named krypton, from the Greek for "the hidden."

Ten days later, Ramsay and Travers succeeded in isolating yet another gas, which they collected from a sample of argon. The atomic weight of this gas was 20, which did put it between helium and argon, as Ramsay had speculated. Spectral analysis of this gas, which was named neon, after the Greek for "the new," revealed a bright crimson light. By September 1898, Ramsay and Travers had separated a third gas from krypton, which they named xenon, from the Greek meaning "the stranger."

Identifying radon

In 1899, British physicist Ernest Rutherford and others reported a radioactive substance being emitted by thorium. That same year, French physicists Pierre and Marie Curie noted a radioactive gas emanating from radium, and in 1900, German physicist Friedrich Ernst Dorn saw the accumulation of a gas inside containers of radium. In each case, the substance turned out to be what was later named radon. In 1908, Ramsay collected enough radon to determine its properties and stated

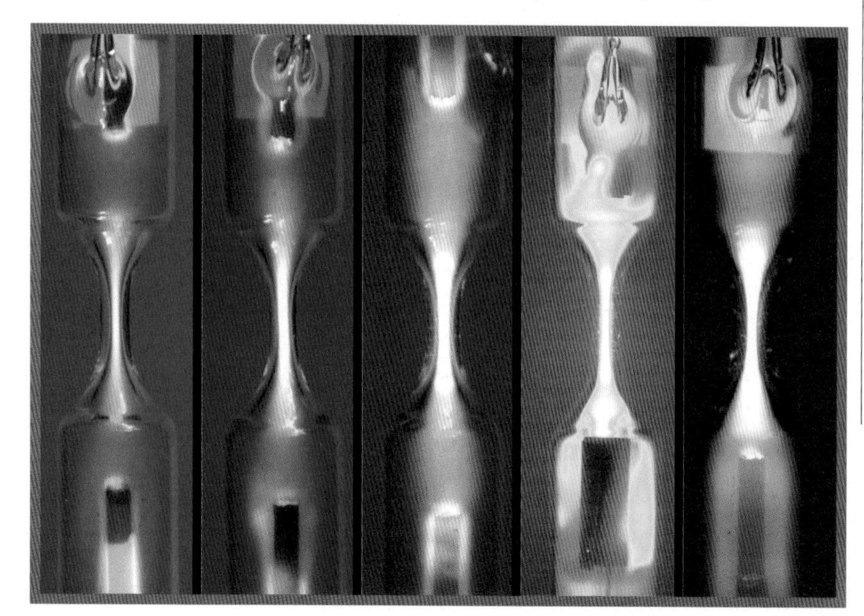

These gas discharge tubes (sealed glass tubes) show the bright colors produced by different noble gases. From left to right, they are xenon, krypton, argon, neon, and helium.

Noble properties

The noble gases are uniquely stable elements, apparently only taking part in chemical reactions under unusual conditions. This helps explain how chemical bonding works. In 1913, Danish physicist Niels Bohr suggested that electrons occupied energy "shells" that surround the nucleus of an atom. The electron capacities of the shells determined the numbers of elements in the rows of the periodic table. In the atoms of the noble gases, the outermost shell always contains eight electrons. (Helium, with just two electrons, is an exception.) American chemist Gilbert Lewis and German chemist Walther Kossel proposed that this octet of electrons was the most stable arrangement for an atom's outermost shell, and it was what atoms strived for when forming chemical bonds by donating, taking, or sharing electrons with other atoms. The noble gases already have a complete outer shell of eight electrons, so have no need to take part in chemical reactions to achieve stability.

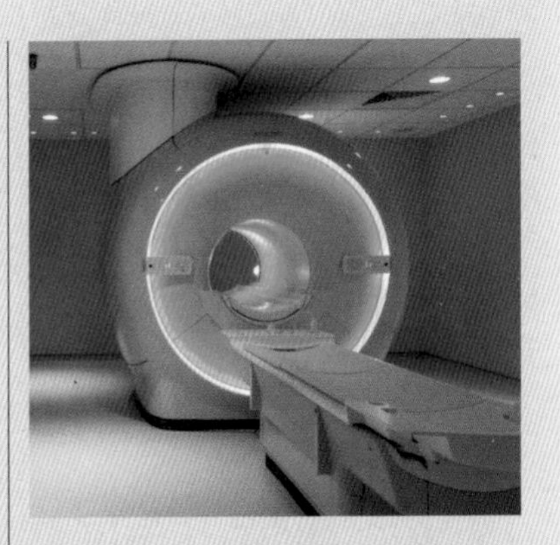

The superconducting magnets in MRI (magnetic resonance imaging) scanners are cooled through the use of large amounts of liquid helium.

that it was the heaviest gas then known. Assisted by British chemist Robert Whytlaw-Gray, Ramsay measured the density of radon with enough precision to establish that its atomic weight differed from that of its parent element, radium, by the weight of one helium atom.

Periodic placement

When it came to adding the gases to the periodic table, the sequence of atomic weights suggested they be placed between the halogens and the alkali metals, possibly in group 8. But the fact that these inert or noble gases, as they became known, were unreactive and did not form any compounds was a problem. William Crookes suggested, in 1898, that they be placed in a single column between the hydrogen group and the fluorine group. In 1900, Ramsay and Mendeleev met to discuss the new gases and their location in the table. Ramsay proposed a new group between the halogens and alkali metals. On the suggestion of Belgian botanist Léo Errera, in 1902, Mendeleev put the noble gases of helium, neon, argon, krypton, and xenon in a new group 0 (now group 18), which was positioned at the far right of the periodic table.

Today, the noble gases are used in our daily lives in industries such as welding, lighting, diving, and even medicine. ■

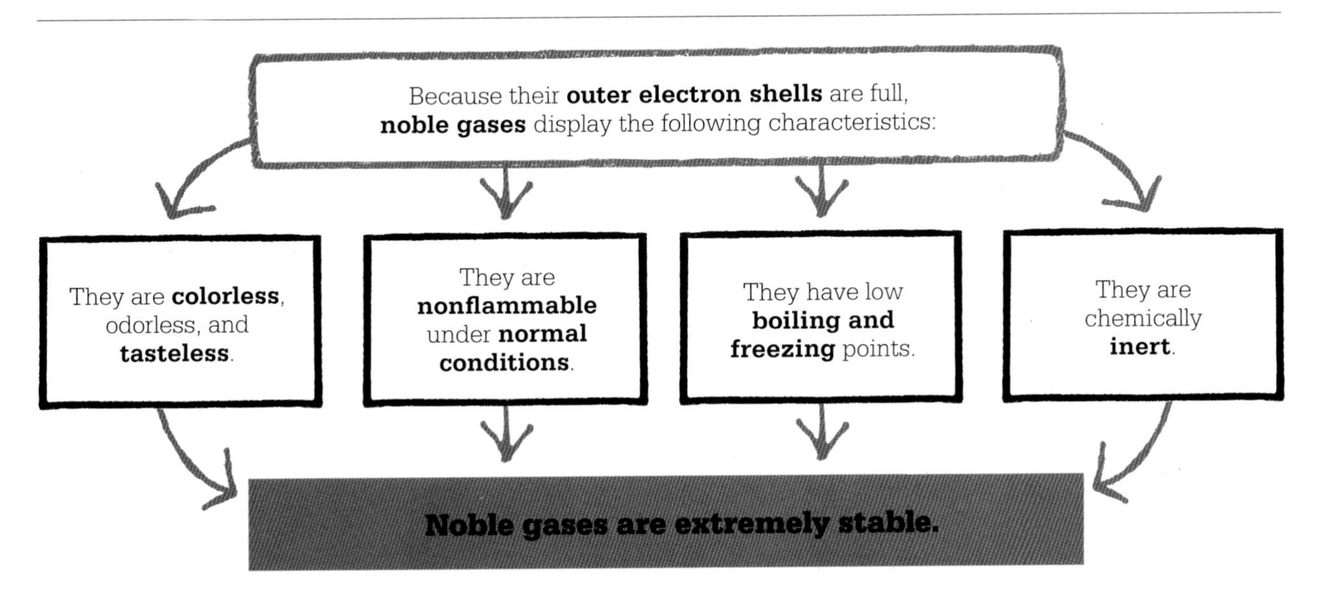

Because their outer electron shells are full, noble gases display the following characteristics:
They are colorless, odorless, and tasteless.
They are nonflammable under normal conditions.
They have low boiling and freezing points.
They are chemically inert.
Noble gases are extremely stable.

THE MOLECULAR WEIGHT SHALL HENCEFORTH BE CALLED MOLE

THE MOLE

IN CONTEXT

KEY FIGURE
Wilhelm Ostwald (1853–1932)

BEFORE
1809 French scientist Joseph Gay-Lussac publishes his law of combining volumes, which states that volumes of gases react with each other in ratios of small whole numbers.

1865 Josef Loschmidt calculates the number of particles in 1 cc of gas in standard conditions—this is the Loschmidt constant.

AFTER
1971 The International Committee for Weights and Measures defines 1 mole as the number of ^{12}C atoms in 12 grams of ^{12}C.

1991 Retired American chemistry teacher Maurice Oehler establishes the National Mole Day Foundation; chemistry students celebrate this annually on October 23.

The **numbers of molecules** involved in chemical calculations are **huge and awkward**.

To make **calculations**, scientists can use a **simple unit** for each substance: **the mole**.

A mole of any substance is its **molecular weight** expressed in **grams**.

Calculations are easier because you multiply or divide the moles rather than the number of atoms.

Atoms and molecules are at the heart of every chemical calculation, but they are tiny, and there are huge numbers of them in any volume of a substance. Dealing with such extremes has been made much more manageable by the concept of the mole, introduced by German chemist Wilhelm Ostwald in 1894.

The origins of the idea lie in Italian physicist Amedeo Avogadro's work with gases in the early 19th century. In 1811, he hypothesized that two equal volumes of gas at the same temperature and pressure always contain the same number of particles (which we now know to be atoms and molecules). This is called Avogadro's law, but scientists did not understand its full significance for more than half a century. It was not until the Karlsruhe Congress in Germany in 1860 that Italian chemist Stanislao Cannizzaro finally explained the implications of Avogadro's hypothesis by relating the weights of atoms to the weights of molecules. This led to an internationally agreed set of atomic weights for the elements.

Around the same time, James Clerk Maxwell's developing kinetic theory of gases highlighted the importance of numbers of

See also: Compound proportions 68 ▪ Catalysis 69 ▪ Isolating elements with electricity 76–79 ▪ Dalton's atomic theory 80–81 ▪ The ideal gas law 94–97 ▪ Atomic weights 121

molecules. In 1865, Austrian science teacher Josef Loschmidt estimated the number of particles in a cubic centimeter of gas in standard conditions: 2.6867773×10^{25} m^{-3}. In 1909, French physicist Jean Baptiste Perrin refined this figure to 6×10^{23} and called it Avogadro's number.

German chemist August von Hoffman had already introduced the word "molar" to describe changes at particle level, too small for the eye to see. In 1894, Ostwald realized he could unite atomic and molecular weights and numbers in a simple unit that he called the mole. He proposed that when the atomic or molecular weight of a substance is expressed in grams, its mass is 1 mole. Conversely, a mole is the amount of gas in 22 liters of air at ordinary temperatures and pressures. Perrin later linked this with Avogadro's number, suggesting the number of units in a mole is Avogadro's number.

The mole is a base unit that is used to count particles—atoms, molecules, ions, or electrons. A score of particles is 20 particles, a dozen is 12 particles, and so on, but chemists deal with much larger numbers: 1 mole is just over 6×10^{23}, or 600 billion trillion, particles.

To find the mass of 1 mole of molecules of any element or compound, add the masses of 1 mole of each atom involved. In water, for example, since the mass of 1 mole of hydrogen atoms is 1 g and that of 1 mole of oxygen atoms is 16 g, 1 mole of water molecules has a mass of 18 g—that is, 1 + 1 + 16.

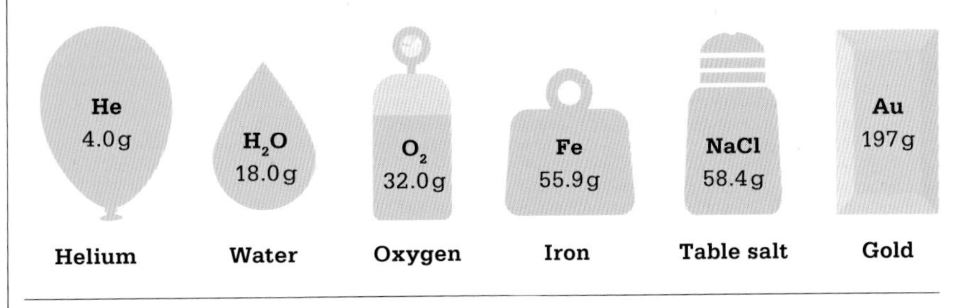

Calculations simplified

The use of the mole has simplified calculations. One mole of carbon atoms, for example, always weighs 12 g, while 12 g of carbon always contains 1 mole of carbon atoms. Similarly, since magnesium atoms are twice as heavy as carbon atoms, 1 mole of magnesium must weigh 24 g. It works in the same way with compounds. One mole of oxygen atoms weighs 16 g, so a mole of carbon dioxide (CO_2) must be 44 g (12 + 16 + 16).

The number 6×10^{23} is fine for approximations, but chemists have now pinned it down more precisely. In 1909, American scientists Robert Millikan and Harvey Fletcher experimentally calculated the charge on a single electron. By dividing the charge on 1 mole of electrons by this charge, it was possible to obtain an accurate figure for Avogadro's number, which is $6.02214154 \times 10^{23}$ particles per mole. ■

Wilhelm Ostwald

Born in Riga, Latvia, in 1853, Wilhelm Ostwald studied chemistry at the University of Dorpat (now Tartu, Estonia), before working there under eminent scientists Arthur von Oettingen and Carl Schmidt. In 1881, he became a chemistry professor at Riga before moving to Leipzig, Germany, where he was professor of physical chemistry. While there, he taught future Nobel Prize winners Jacobus van 't Hoff and Svante Arrhenius. Ostwald himself won the 1909 Nobel Prize in Chemistry for his work on catalysis. He was fascinated by chemical affinity and the reactions that formed chemical compounds. As well as being a pioneer of physical chemistry and establishing the concept of the mole, he contributed a law of dilution, stating how electrolytes become weaker, and conducted a groundbreaking analysis of color. Ostwald died in 1932.

Key works

1884 *Textbook of General Chemistry*
1893 *Handbook and Manual for Physico-Chemical Measurements*

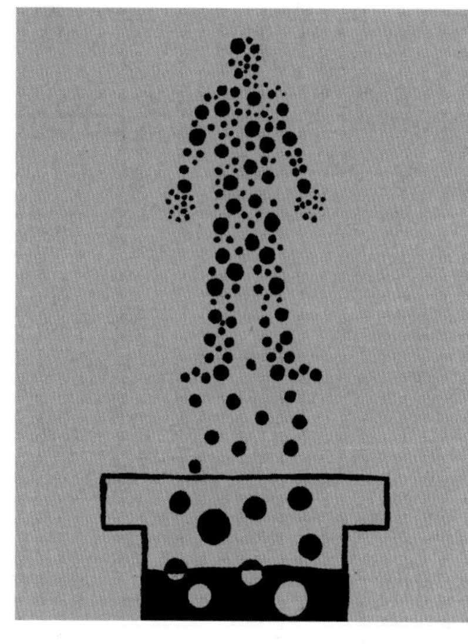

PROTEINS RESPONSIBLE FOR THE CHEMISTRY OF LIFE

ENZYMES

IN CONTEXT

KEY FIGURE
Eduard Buchner (1860–1917)

BEFORE
1833 Anselme Payen and Jean-François Persoz discover the first known enzyme, diastase.

1878 German physiologist Wilhelm Kühne coins the word enzyme, from the Greek for "leavened" (made with yeast).

AFTER
1926 James B. Sumner crystallizes urease, which leads to proof that enzymes are proteins.

1937 German-born British biologist Hans Krebs discovers the citric acid cycle—key to energy production in living things—and the role of enzymes in the cycle.

1968 Teams of researchers at Harvard and Johns Hopkins Universities identify "restriction" enzymes that recognize and cut short sections of DNA.

A chemical that speeds up chemical reactions is called a catalyst. This triggers reactions or speeds them up without being directly involved. In living organisms, catalysts are called enzymes. Countless processes in living things—from digestion to energy production—rely on these enzymes. In the 1890s, German chemists Eduard Buchner and Emil Fischer made breakthroughs in understanding how enzymes work.

Identifying enzymes

Enzymes had been used for thousands of years—such as rennet in cheese making—without a full understanding of their significance. However, in 1833, while working in a beet sugar factory, French chemists Anselme Payen and Jean-François Persoz identified a new enzyme involved in converting starch to malt. They called it diastase, from the Greek for separation. Soon after, German physician Theodor Schwann discovered another enzyme, pepsin, which is involved in digestion, and German chemist Eilhard Mitscherlich discovered invertase, which helps break fruit sugars into fructose and glucose.

In 1835, Jöns Jacob Berzelius identified catalysis in inorganic chemical reactions and then went on to suggest that enzymes might be the organic equivalent. However, it was unclear whether they were simply chemical catalysts or if they depended on living organisms.

Even during the 1850s, when French chemist Louis Pasteur was showing that the fermentation of

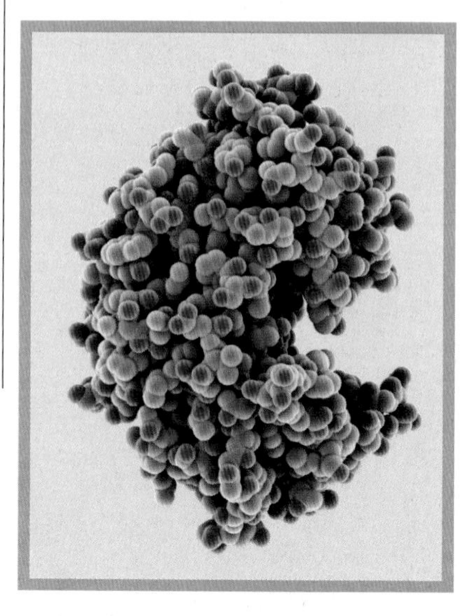

Pepsin is the digestive enzyme that breaks down proteins. Its molecule, modeled here with oxygen atoms in red, is built from 5,053 atoms. Proteins lock into the active site, the cleft on the right.

See also: Catalysis 69 ▪ The synthesis of urea 88–89 ▪ X-ray crystallography 192–193 ▪ Customizing enzymes 293

sugar into alcohol by yeasts in the act of beer making was catalyzed by "ferments," scientists were still convinced that processes such as fermentation, decomposition, and putrefaction depended on tiny living organisms.

The breakthrough came in 1897, when Buchner took liquid from crushed yeast cells and showed that this liquid could ferment sugar to make alcohol, with no live yeast at all. Fruit juice fermented in the same way. Buchner argued that the fermentation was caused by dissolved substances that he called zymases. Enzymes, then, while produced by living organisms, can operate without a living cell.

Lock-and-key theory

Enzymes seem only to have an effect on particular substances or "substrates," and in 1895, Fischer suggested why. An enzyme, he argued, has an active site. This site, he proposed, is like a lock, and the affected substrate is like a key. Only the correctly shaped substrate key fits the enzyme's lock. The theory has since been modified, but it provides a good starting point for understanding.

A host of new enzymes was discovered, each usually named after the substrate but with -ase at the end. For example, lactase is the enzyme that breaks down lactose. However, nobody really knew what enzymes were until 1926, when American chemist James B. Sumner succeeded in crystallizing an enzyme, which he called urease. Its analysis suggested that enzymes are proteins—something soon verified by American biochemists John Howard Northrop and Wendell Meredith Stanley with pepsin, trypsin, and chymotrypsin. Further examination using X-ray crystallography showed that most enzymes were balls of protein—with just a few turning out to be related to ribonucleic acid (RNA). Advances in biotechnology mean it is now possible to manipulate and enhance the power of enzymes in many different fields, including medical and industrial processes. ■

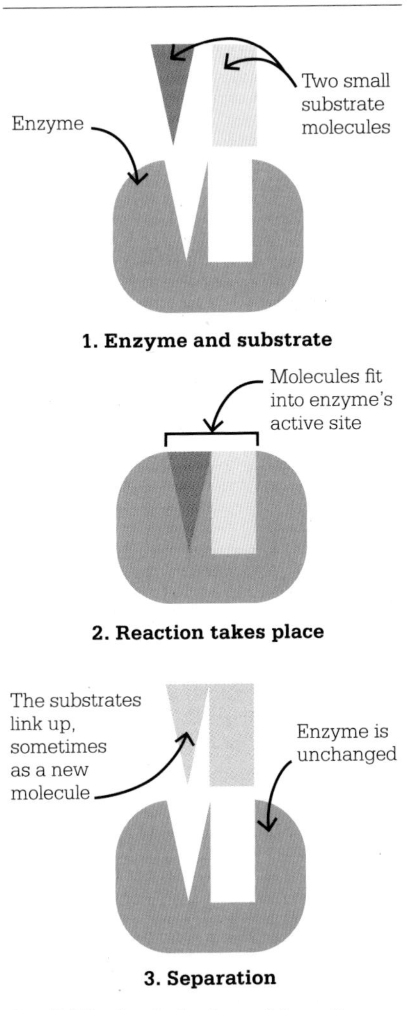
Two small substrate molecules
Enzyme
1. Enzyme and substrate
Molecules fit into enzyme's active site
2. Reaction takes place
The substrates link up, sometimes as a new molecule
Enzyme is unchanged
3. Separation

Emil Fischer's lock-and-key theory proposes that enzymes and substrates have complementary geometric shapes that fit one another.

Eduard Buchner

Born in Munich, Germany, in 1860, Eduard Buchner was the son of a physician. He was just 11 when his father died, and when he later went to Munich University, he had to support his studies by working at a canning factory. After being awarded his doctorate, Buchner became fascinated by fermentation chemistry, and his work has led to him being known as the "father of biochemistry in a test tube." Buchner completed his key study on enzymes at the University of Tübingen, where he had a chair in analytical and pharmaceutical chemistry. However, he spent much of his academic life between 1898 and 1909 at the Royal Academy for Agriculture in Berlin. An army major in World War I, Buchner was killed in action in 1917.

Key works

1885 "The Influence of Oxygen on Fermentations"
1888 "A New Synthesis of Trimethylene Derivatives"
1897 "On Alcoholic Fermentation in the Absence of Yeast Cells"

CARRIERS OF NEGATIVE ELECTRICITY

THE ELECTRON

IN CONTEXT

KEY FIGURE
J. J. Thomson (1856–1940)

BEFORE
1803 John Dalton sets out his atomic theory that elements are made of tiny particles called atoms.

1839 Michael Faraday believes that the structure of the atom must be related to electricity in some way.

AFTER
1909 American physicist Robert Millikan measures the charge on an electron in a famous experiment using oil drops.

1911 Danish physicist Niels Bohr proposes that electrons orbit the central nucleus of the atom.

1916 American physical chemist Gilbert Lewis realizes that chemical bonds are formed by pairs of electrons shared between atoms.

The 1820s saw a flurry of investigations into the then recently discovered phenomenon of electromagnetism. French physicist André-Marie Ampère proposed the existence of a new particle responsible for both electricity and magnetism, which he called the "electrodynamic molecule"—what we now know to be the electron, the atomic particle responsible for chemical reactivity.

Mystery rays

In 1858, German physicist Julius Plücker experimented with sending a high voltage between metal plates inside a glass tube from which most of the air had been removed. He found that this created a fluorescent glow in the tube. Plücker's pupil Johann Hittorf confirmed in 1869 that the cathode was the source of the mysterious green rays. A decade later, British physicist and chemist William Crookes discovered that the rays could be bent by a magnetic field and were apparently made up of negatively charged particles.

In 1883, German physicist Heinrich Hertz attempted to deflect cathode rays using an electric field, but with no success. He concluded (wrongly) that the rays were not charged particles but waves that could be bent by magnetic fields.

Thomson's discovery

British physicist J. J. Thomson began a series of experiments in 1894 that would settle the nature of the cathode rays once and for all. Using a cathode-ray tube with

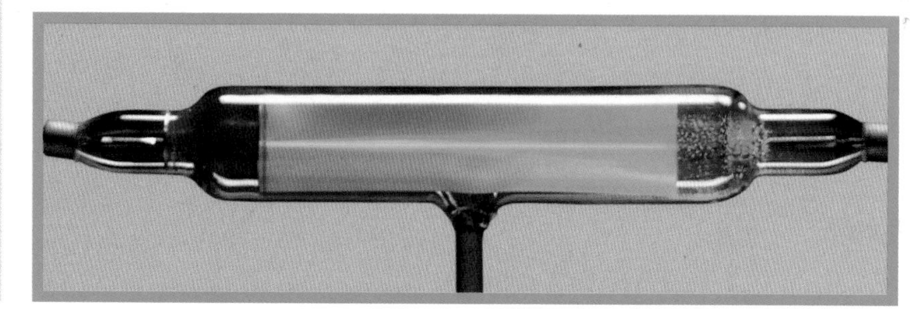

Experiments with a glass tube demonstrate that cathode rays travel in straight lines from the cathode (left) to the anode at the other end.

See also: Corpuscles 47 ▪ Dalton's atomic theory 80–81 ▪ Radioactivity 176–181 ▪ Improved atomic models 216–221

They are charges of negative electricity carried by particles of matter.
J. J. Thomson

the deflector plates inside rather than outside the glass tube, he discovered that cathode rays could indeed be deflected by an electric field.

Thomson's experimental setup also allowed him to determine the ratio of the charge of the mystery particle to its mass. He found that it remained the same regardless of the metal that was used for the electrodes or of the composition of the gas used to fill the tube. He deduced that the particles that made up the cathode ray must be something that was found in all forms of matter.

By 1897, Thomson had determined that the negatively charged particles of the cathode ray had a mass that was less than a thousandth of that of a hydrogen atom. This meant that the particles could not be charged atoms or indeed any other particle that was then known in the field of physics. Thomson referred to them as "corpuscles," but the name "electron," which had been proposed by Irish physicist George Stoney in 1891 for the fundamental unit of electrical charge, was soon adopted. In 1906, Thomson was awarded the Nobel Prize in Physics in recognition of his discovery.

The divisible atom

The next question that needed to be answered was precisely how Thomson's corpuscles fit into the structure of the atom. It was known that atoms were electrically neutral, so to balance the negative charge of the electrons, Thomson proposed that they were embedded in a positively charged cloud, like raisins in a cake—an image that led to his atom being dubbed the plum pudding model.

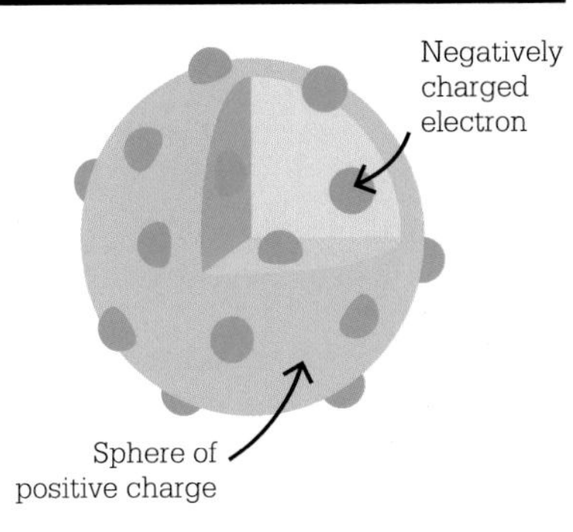

Thomson's plum pudding atomic model proposed that negatively charged electrons were embedded in a positively charged cloud.

Thomson's model was important in that it described, for the first time, the atom as something divisible and containing electromagnetic forces. This paved the way for a new model within a few years, when it emerged that the atom's positive charge was concentrated in a tiny volume in its center. ■

J. J. Thomson

Bookseller's son Joseph John (J. J.) Thomson was born in a suburb of Manchester, UK, in 1856. He began studying at Owens College (now the University of Manchester) at the unusually young age of 14 and went on to obtain a scholarship to the University of Cambridge in 1876.

Thomson worked at Cambridge for the rest of his life, carrying out experimental research at the Cavendish Laboratory, which was the scene of his groundbreaking discovery of the electron in 1897. Thomson was also an outstanding teacher: seven future Nobel Prize winners worked with him, including Ernest Rutherford. Outside physics, he had a great interest in plants and would search for rare specimens around Cambridge. In 1918, he was made master of Trinity College, a position he held until his death in 1940.

Key works

1893 "Notes on Recent Researches in Electricity and Magnetism"
1903 "Conduction of Electricity through Gases"

THE
MACHINE
1900–1940

AGE

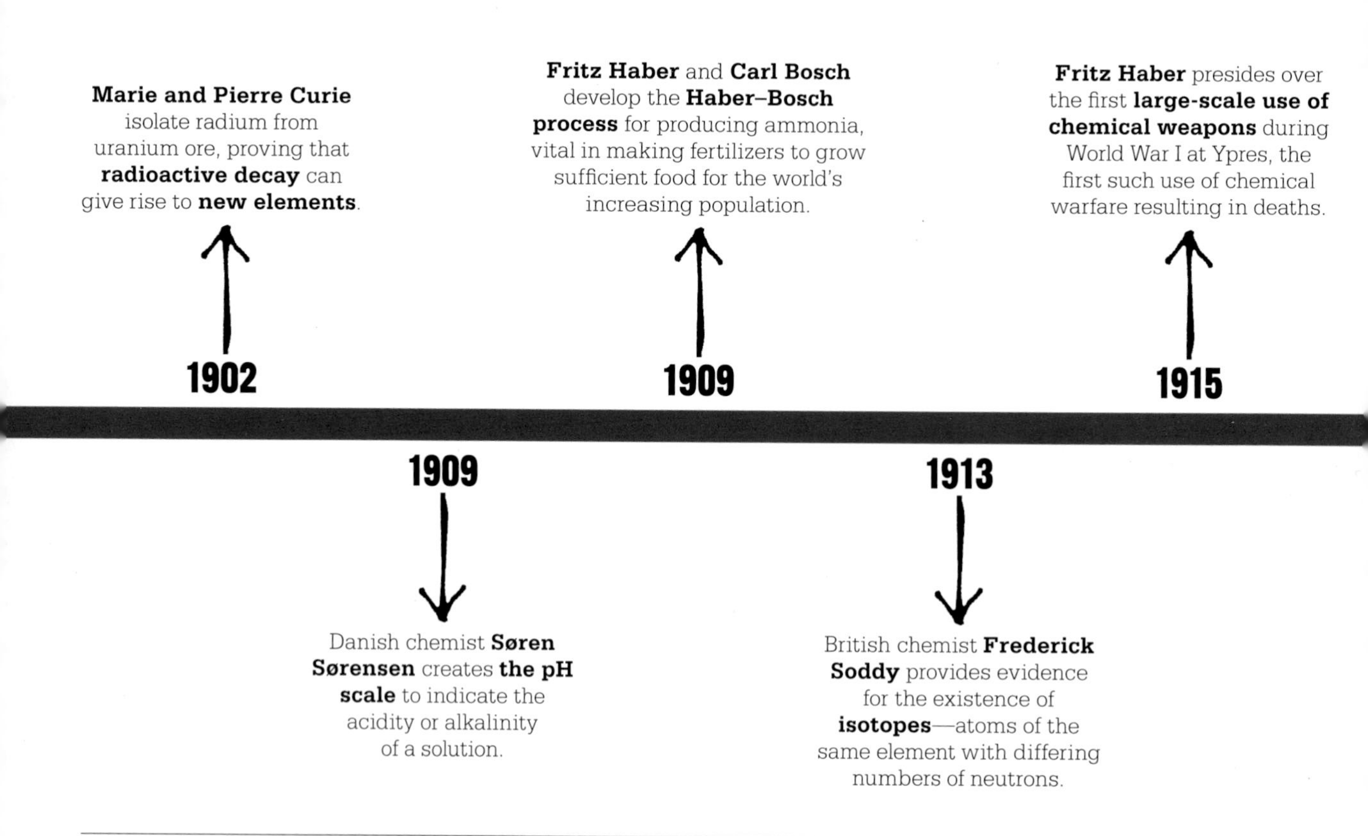

Marie and Pierre Curie isolate radium from uranium ore, proving that **radioactive decay** can give rise to **new elements**.
1902
Fritz Haber and **Carl Bosch** develop the **Haber–Bosch process** for producing ammonia, vital in making fertilizers to grow sufficient food for the world's increasing population.
1909
Fritz Haber presides over the first **large-scale use of chemical weapons** during World War I at Ypres, the first such use of chemical warfare resulting in deaths.
1915
1909
Danish chemist **Søren Sørensen** creates **the pH scale** to indicate the acidity or alkalinity of a solution.
1913
British chemist **Frederick Soddy** provides evidence for the existence of **isotopes**—atoms of the same element with differing numbers of neutrons.

The first half of the 20th century was a time of chaos and conflict. However, the backdrop of two World Wars, far from hampering scientific advancement, catalyzed it. Innovations in chemistry pushed the boundaries of what had been considered impossible just decades before. Some of these advances created entirely new fields of chemistry, while others would bring the ethics of certain chemical processes into sharper focus.

A quantum leap

The early 1900s saw a proliferation of atomic models. New insights into atomic composition required the development of increasingly complex models to explain atoms' structures accurately. The simple "plum pudding" model proposed by J. J. Thomson in 1904 was rendered redundant by Ernest Rutherford's discovery in 1911 that atoms have a nucleus where mass is concentrated. The Rutherford–Bohr model of the atom followed in 1913, only to be superseded in turn by Erwin Schrödinger's quantum model of the atom in 1926, in which the positions of electrons were defined as regions of probability rather than definite points.

These models gave a new clarity to atomic structure that aided the understanding of other areas of chemistry. A more detailed picture of the bonds holding substances together at the atomic level developed, culminating in 1939 with Linus Pauling's work on the nature of chemical bonds. And at the crossover between chemistry and physics, updated models of the atom assisted the theoretical understanding of radioactive decay.

Chemistry at war

The discovery of radiation at the turn of the 20th century revealed the possibility of one element transforming into another. This was not quite the sorcery of transmutation to which alchemists had aspired; in the 1970s, chemists discovered that lead could indeed be transformed into gold—but only fleetingly, in particle accelerators and at an atomic scale. However, the process of transforming elements hinted at the possibility of discovering further short-lived elements, and this would become the domain of the element hunters later in the century.

1920

Hermann Staudinger proposes that huge molecules can be formed by polymerization reactions; this established the field of **polymer science** and enabled the creation of plastics.

1928

Alexander Fleming discovers the naturally occurring antibiotic **penicillin**, whose use as a drug would transform how we treat bacterial infections and save millions of lives.

1934

Dorothy Hodgkin and **J. D. Bernal** record the **first X-ray diffraction image** of a crystallized protein, allowing its structure to be determined.

1938

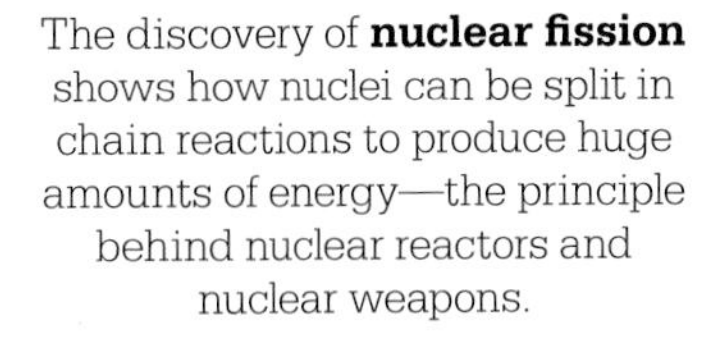

The discovery of **nuclear fission** shows how nuclei can be split in chain reactions to produce huge amounts of energy—the principle behind nuclear reactors and nuclear weapons.

Radiation's potential for creation could also be harnessed to cause destruction. The discovery of nuclear fission shortly before the outbreak of World War II showed that certain nuclei could be split by bombarding them with neutrons, triggering chain reactions that released a colossal amount of energy. These chain reactions would later be employed to provide nuclear power—but they were also harnessed by Allied powers to cause unprecedented death and destruction during the conflict in the form of nuclear weapons.

Chemistry had also played a part in warfare in World War I, and German chemist Fritz Haber embodied its divisive role better than anyone. Haber, along with his compatriot Carl Bosch, had spent the years before the war developing a process to fix nitrogen from the air in the form of ammonia. Ammonia was a vital precursor to fertilizers; using the Haber–Bosch process enabled industry to produce huge quantities of fertilizers at a time when demand was rocketing.

It is no understatement to say that conventional agriculture today could not be sustained without the Haber–Bosch process. However, while this achievement saw Haber win a Nobel Prize and should have seen him celebrated today as one of the greats of chemistry, his role in the development of chemical warfare during World War I would poison his legacy.

Plastic fantastic

In 1907, the first synthetic mass-produced plastic, Bakelite, was invented. At this point, the structure of plastics—materials composed of very large molecules called polymers—was a matter for heated debates, with chemists unable to agree on the structure of polymers. In 1920, Hermann Staudinger defined polymers as long chains of repeated molecular units; his work formed the basis of polymer science as a branch of chemistry in its own right.

This understanding of polymers accelerated the search for new plastic materials. The discovery of nylon in 1935 led to the creation of a range of artificial fabrics, while fluorinated polymers, discovered accidentally, found uses ranging from nonstick cookware to medical equipment. Super-strong plastics would also follow in the 1960s. As a result of these discoveries, plastics are inescapable in today's world. ■

LIKE LIGHT RAYS IN THE SPECTRUM THE DIFFERENT COMPONENTS ARE RESOLVED

CHROMATOGRAPHY

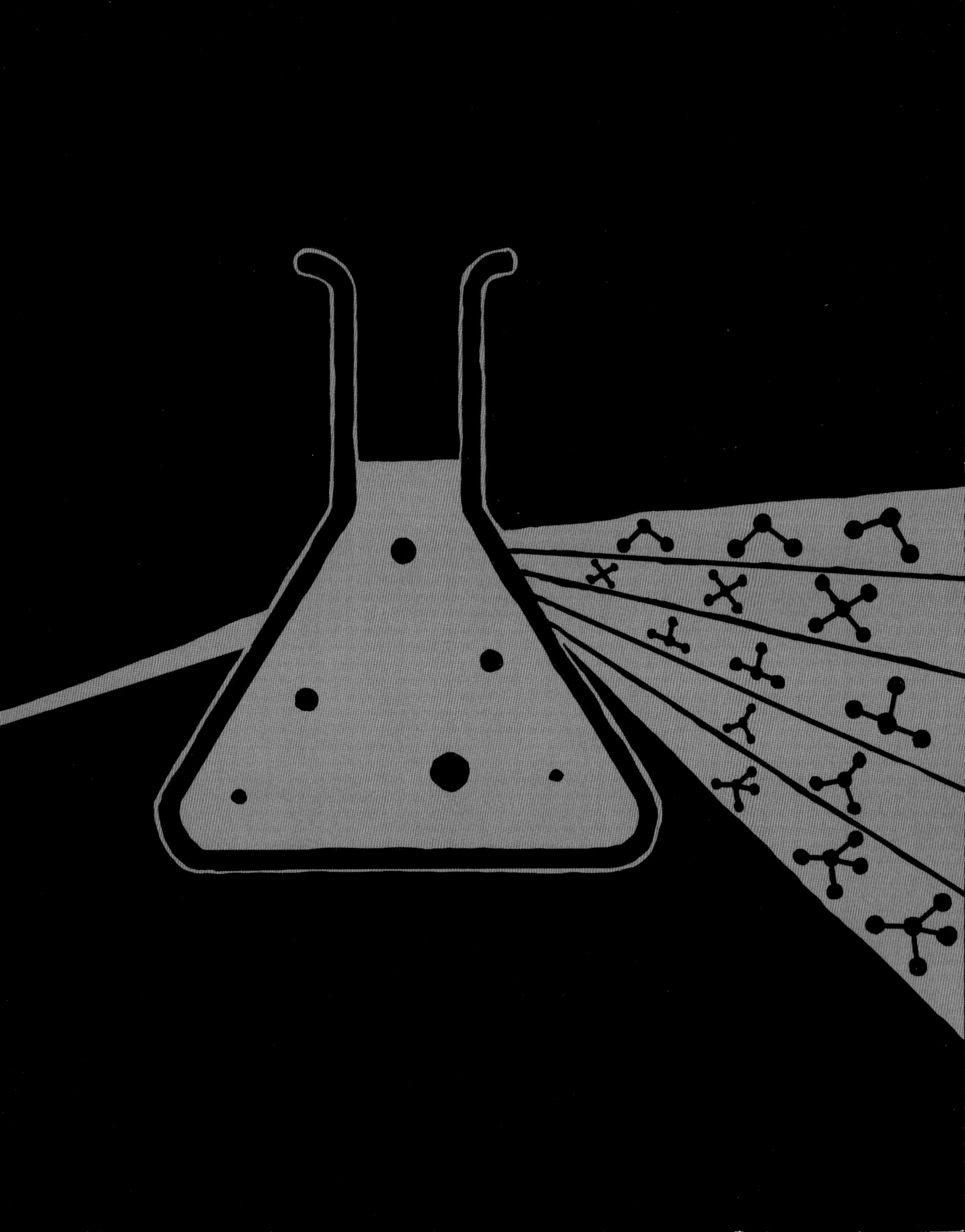

IN CONTEXT

KEY FIGURE
Mikhail Tsvet (1872–1919)

BEFORE
1556 German scholar Georg Bauer's *On the Nature of Metals* is published, in which he describes analyzing ores and methods for extracting and separating metals.

1794 Joseph Proust demonstrates the law of constant composition by showing that copper carbonate has the same ratio of elements whether it comes from natural or artificial sources.

1814 Spanish toxicologist Mathieu Orfila writes *Treatise on Poisons*, in which he urges the routine use of chemical analysis in cases of death from an unknown cause.

AFTER
1947 Erika Cremer, a German physical chemist, builds the first gas chromatograph.

1953 American flavorist Keene Dimick builds a gas chromatograph to analyze the essence of strawberries and improve the flavor of processed foods.

Chemical analysis is the science of separating, identifying, and quantifying chemical compounds. From the ancient metallurgist's use of a touchstone to assay precious metal alloys, to the impressive forensic laboratories of today, analytical chemistry is the cornerstone upon which all chemical disciplines build.

Paper chromatography

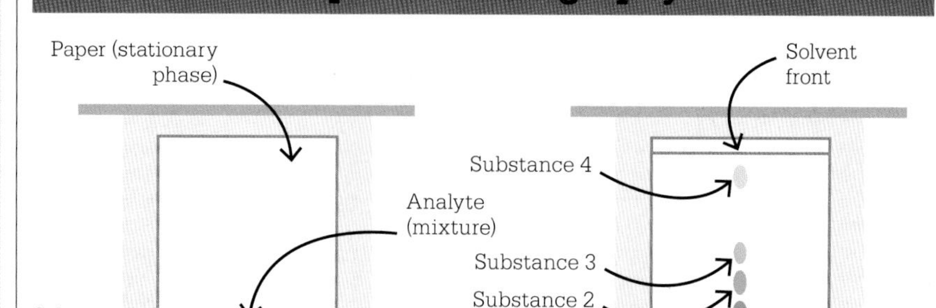

A sample spot of the mixture to be analyzed—the analyte—is placed near one end of a piece of paper, or the stationary phase. The paper is then suspended so that the bottom edge touches a solvent—the mobile phase.

The solvent gradually travels up the paper, carrying the mixture with it. The components in the mixture separate out based on their attractions to the paper, allowing each one to be individually analyzed and identified.

Typically, chemical analyses involve separation processes, followed by the identification of the purified chemicals. The challenge is to separate the components of a mixture without destroying their chemical identity. Medieval European and Islamic alchemists refined many methods of doing this, such as filtration, sublimation, and distillation. However, the workhorse of nondestructive separation techniques was not discovered until the turn of the 20th century, when Russian botanist and chemist Mikhail Tsvet developed chromatography—derived from the Greek words *chroma*, meaning "color," and *graphein*, meaning "to write."

Paper chromatography

The fundamental principle of chromatography can be explained in everyday terms. When a paper towel comes into contact with water, the water creeps up the paper towel. The same thing happens when a paper towel comes into contact with oil—but not as quickly. If a green-colored marker gets wet and comes into contact with a paper towel, the yellow component of the ink will travel farther than other components. Thus, the colored inks that are combined to make green will be separated. Materials with the highest intermolecular attractions for paper are drawn out of the mixture first and adhere to the paper towel, while those with

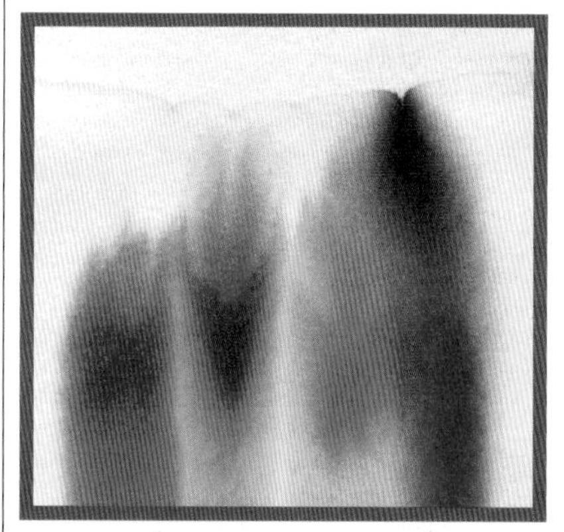

Qualitative paper chromatography can establish how many pigments are used in a color. The amounts of each component can be identified through modern quantitative chromatography.

See also: Isomerism 84–87 ▪ Flame spectroscopy 122–125 ▪ Intermolecular forces 138–139 ▪ The pH scale 184–189 ▪ Mass spectrometry 202–203

Separate thou the earth from the fire, the subtle from the gross sweetly with great industry.
Hermes Trismegistus
The Emerald Tablet
(c. 800 CE)

lower intermolecular attractions are carried farther. Observations such as these formed the basis of Mikhail Tsvet's chromatographic separation technique.

By the beginning of the 20th century, physicists and chemists, including Tsvet, were turning their attention to natural processes and finding fertile ground. They had discovered plants contained pigments that were believed to perform important biological functions. For example, the green pigment chlorophyll is a key part of photosynthesis, absorbing light and creating energy. At the time, the accepted view was that there were only two pigments in plants—green chlorophyll and yellow xanthophyll—but Tsvet believed there were more. Using standard separations techniques of the day, based on solubility and precipitation, Tsvet was able to separate chlorophyll into two parts, which he called alpha-chlorophyll and beta-chlorophyll. He managed to purify the alpha-chlorophyll, but the beta form stubbornly resisted purification.

Tsvet experiments

Possibly inspired by the methods of artisans to separate pigments to make paints and dyes, Tsvet decided to try adsorption to separate the components of beta-chlorophyll. In the adsorption process, the analyte (the mixture to be separated) is placed as a spot on a stationary phase (a solid support, such as a paper towel), and the stationary phase is put in contact with a mobile phase, such as water or oil. The mobile phase moves into the solid support and carries the analyte forward. The analyte compound with the greatest attraction to the support drops out of the mixture first, followed by the remaining compounds in order of adsorptivity.

After much trial and error, Tsvet settled on a column of chalk for his stationary phase and used light oils as his mobile phase. He placed plant materials onto the chalk, poured the oils on the top of the column, and observed the colored components of plant pigments separating as the oils crept down the solid support. He then cut the different regions from the support and extracted the separated pigments for analysis. He found both chlorophyll and xanthophyll separated into previously unknown compounds.

Tsvet presented his separation method to the Congress of Naturalists and Physicians in 1901 but did not name the process "chromatography" until 1906. »

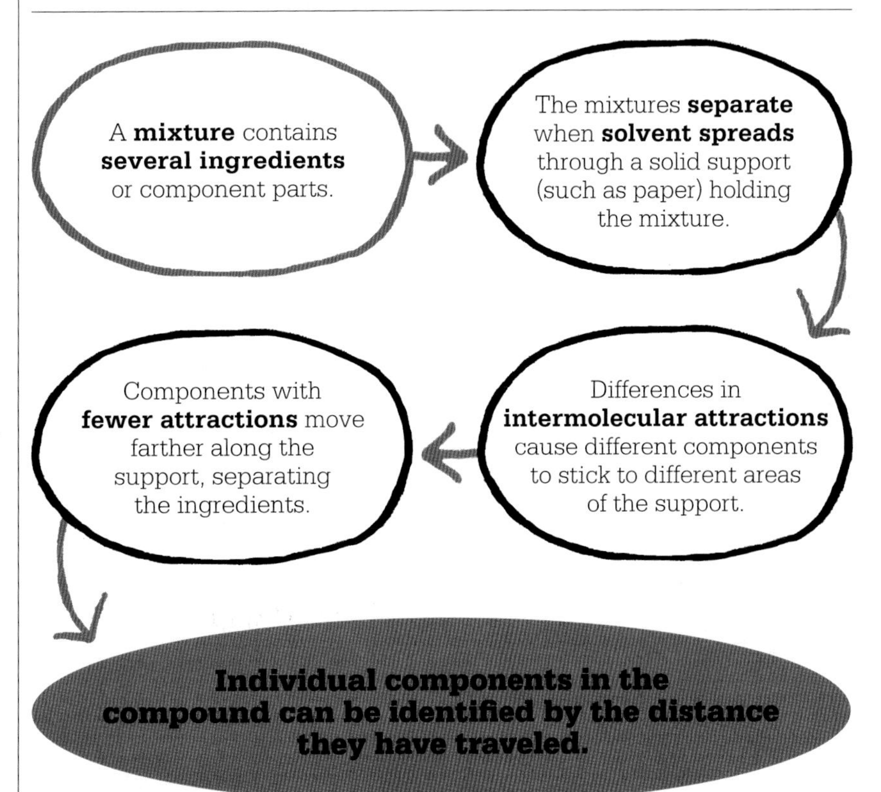

A **mixture** contains **several ingredients** or component parts.
The mixtures **separate** when **solvent spreads** through a solid support (such as paper) holding the mixture.
Differences in **intermolecular attractions** cause different components to stick to different areas of the support.
Components with **fewer attractions** move farther along the support, separating the ingredients.
Individual components in the compound can be identified by the distance they have traveled.

Mikhail Tsvet

Born a Russian citizen in 1872, Mikhail Tsvet was raised in Geneva, Switzerland. He studied chemistry, physics, and botany at the University of Geneva and completed a doctoral degree. He moved to Russia in 1896 but could not secure an academic position because his Swiss credentials were not recognized. He returned to his studies and received a degree from the University of Kazan, Russia, in 1901. At this time, chemistry and physics were just beginning to be applied to natural processes, and Tsvet engaged in this research at every opportunity. It was while working as a laboratory assistant in St. Petersburg that he developed chromatography. Tsvet's chromatography did not immediately receive the acclaim it deserved, and he died in 1919 at the age of 47, before the impact of his discovery had been fully realized.

Key works

1903 "On a new category of adsorption phenomena and on their application to biochemical analysis"

Chromatography evolves

Tsvet went on to use chromatography to separate a group of complex colored plant pigments that he named carotenoids. Compounds from this family have been found to contain vitamins and antioxidants. Unfortunately, Tsvet's work was interrupted several times by World War I, and when chromatography was explored in other laboratories, the outcome was not always favorable. However, in the 1930s, Austrian biochemist Richard Kuhn revived Tsvet's technique to study carotenoids and vitamins.

In the 1940s, chemist Archer Martin and biochemist Richard Synge, both British, also turned their attention to chromatography. Acknowledging Tsvet's work as their starting point, they developed a method in which the stationary phase and the mobile phase are both liquids. They demonstrated under tightly controlled conditions that the distance of the separated compound from the starting mixture could be used to identify the compound—for example, drugs such as heroin and cocaine could be identified by the distance they rose on a chromatographic column. This meant that chromatography could be employed as an analytical tool, as well as a separation method. Using the processes described by Martin and Synge, in 1955, British biochemist Frederick Sanger was able to unravel the structure of insulin, which has been critical to the fight against diabetes. From the efforts of these scientists, myriad chromatographic techniques evolved.

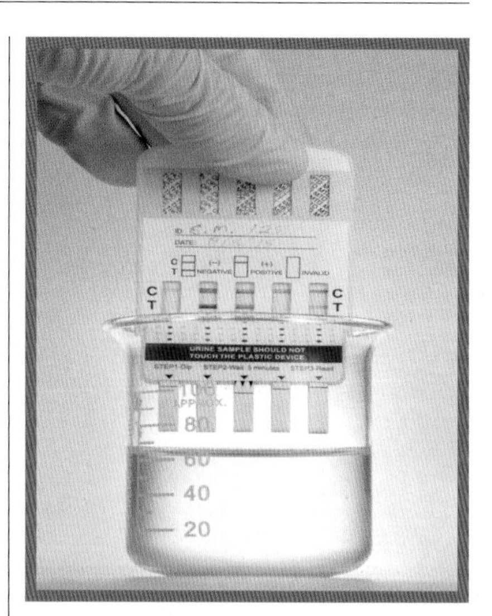

The compounds in urine separate as they move up strips testing for different drugs. Antibodies for a target drug at the top of each strip will trigger a color change, indicating if the drug is present.

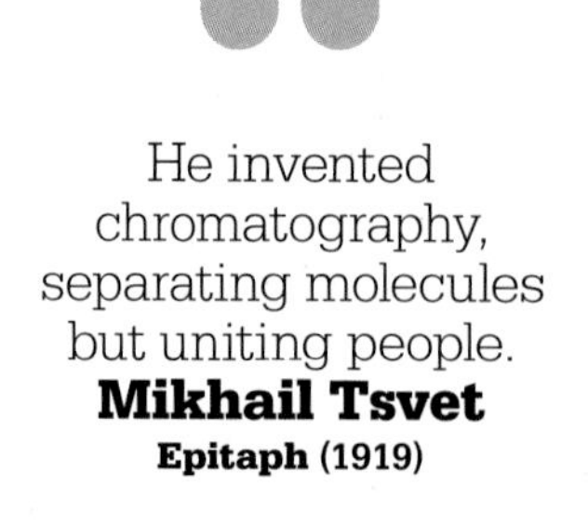

He invented chromatography, separating molecules but uniting people.
Mikhail Tsvet
Epitaph (1919)

Thin layer chromatography

In thin layer chromatography (TLC), a thin layer of a stationary absorbent, such as silica gel or cellulose, is placed on a support, such as glass, aluminum, or plastic. This overcomes the problem of lateral bleeds between liquids that can occur in paper chromatography and has the additional advantage of allowing multiple samples to be analyzed simultaneously, with better separation. This technique is particularly useful for drug testing and water purity testing.

Ion exchange chromatography

Used in the pharmacy and biotechnology industries, ion exchange chromatography (IEC) is a method by which charged particles can be separated. Many biologically active compounds—such as nucleotides, amino acids, and proteins—can carry an electrical charge at the normal pH of the body. Oppositely charged particles are put on ion exchange solid supports, and these particles attract and hold the separated compounds. The charged particles can be selectively removed from the solid support by changing the pH of the mobile phase.

High-performance liquid chromatography

In high-performance liquid chromatography (HPLC), the column is packed with adsorbent particles, which increases the surface area of the solid support and forces the mobile phase through the column. In this manner, separations can be carried out quickly and efficiently. For example, HPLC can be used to determine levels of hemoglobin A1C in the blood in the diagnosis of diabetes and prediabetes.

Gas chromatography–mass spectrometry

In analytical chemistry, once the compound has been separated, the chemicals need to be identified. Techniques such as spectrometry—the use of light to investigate chemical substances—and flame analysis can be used to assess the separate compounds leaving a chromatographic support, but mass spectrometry, in which the particles are separated by mass, is usually favored. When this method is combined with gas chromatography—in which a gas, generally helium, is used as the mobile phase—it is known as gas chromatography–mass spectrometry (GC-MS). This technique is used in virtually every research or analytical laboratory.

In GS-MS, a sample compound is injected into a chromatography column and carried through the column by a gas. The sample must be in the gas phase—or easily volatilized—but many mixtures meet this criterium. The gas chromatography column is packed with small grains of solid support, again creating a large surface area. The column can be several yards (meters) long, which increases the resolution (the separation of compounds). If there is poor resolution, the signal for the separate compounds might overlap, making it difficult to tell how much of each compound is present. The column is coiled inside a constant temperature chamber to ensure the materials stay in the gas phase. When the separated chemicals exit the column, they are immediately analyzed by a mass spectrometer, and the resulting spectrogram can be matched by computer to known compounds in a matter of seconds.

In addition to chemical research, GC-MS is employed by forensic specialists to detect accelerants in suspected cases of arson. It is also used to test food for additives and contaminants and to analyze for flavor and aroma compounds. Pharmaceutical companies use GC-MS for quality control and in the synthesis of new drugs. ■

It is surely no accident that … major advances … have coincided with the appearance of chromatography …

Lord Alexander Todd

Nobel lecture (1957)

Detecting mercury by thin layer chromatography

The presence of mercury in water and food supplies can lead to cancer, brain damage, and birth defects. Mercury can also disrupt and even destroy ecological systems, so it is imperative that mercury sources are identified and its presence in the environment detected at the lowest possible levels.

In 2007, Indian toxicologists Rakhi Agarwal and Jai Raj Behari used TLC to devise a method of detecting mercury at levels as low as 0.00002g per liter of analyte. They found they could detect mercury in samples from natural or complex systems—including body fluids, aquatic habitats, or waste streams—and in the presence of other heavy metals like lead and cadmium.

Agarwal and Behari's TLC tests are inexpensive and easily transportable, and—because they developed a color-change response to the presence of mercury—their test strips can be used by research technicians with minimal training.

THE NEW RADIOACTIVE SUBSTANCE CONTAINS A NEW ELEMENT

RADIOACTIVITY

IN CONTEXT

KEY FIGURES
Marie and Pierre Curie
(1867–1934, 1859–1906)

BEFORE
1858 German physicist Julius Plücker discovers cathode rays, which show as a green glow when he passes a high voltage across metal plates embedded in an evacuated glass tube.

1895 Wilhelm Röntgen discovers X-rays while investigating cathode rays.

AFTER
1909 British physicist Ernest Marsden and Hans Geiger, working with Ernest Rutherford, carry out an experiment that hints for the first time at the existence of the atomic nucleus.

1919 Ernest Rutherford "splits the atom," changing nitrogen into oxygen by bombarding it with alpha particles.

I developed the photographic plates … expecting to find the images very weak. Instead the silhouettes appeared with great intensity.
Henri Becquerel

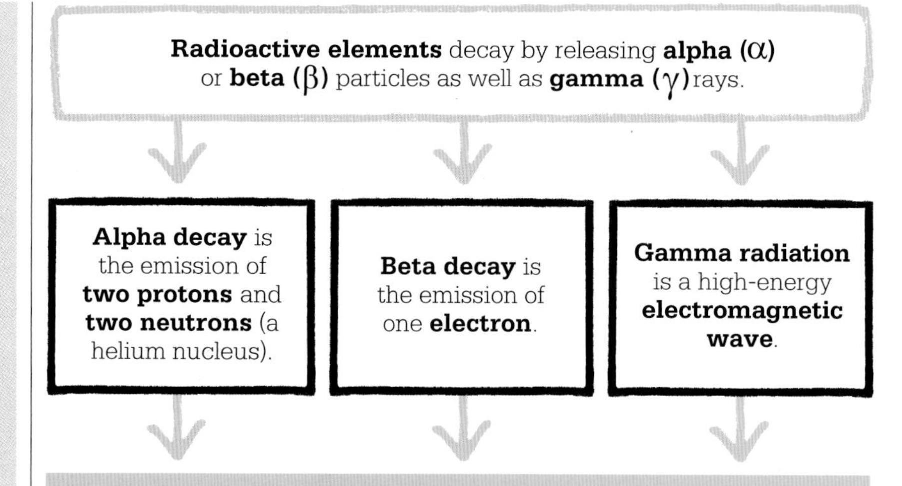

Radioactive elements decay by releasing alpha (α) or beta (β) particles as well as gamma (γ) rays.
Alpha decay is the emission of two protons and two neutrons (a helium nucleus).
Beta decay is the emission of one electron.
Gamma radiation is a high-energy electromagnetic wave.
Radioactive decay leads to the formation of a new element.

Around the same time that J.J. Thomson was discovering the electron, French physicist Henri Becquerel was making discoveries of his own. In 1896, he was studying the properties of X-rays, which had been discovered the previous year by German engineer and physicist Wilhelm Röntgen. Working with a cathode-ray tube in his laboratory, Röntgen noticed that a nearby fluorescent screen began to glow. He concluded that an unknown ray was being emitted from the tube. Further experiments established that this "X-radiation," as he called it, could pass through many substances, including the soft tissue of humans, but not denser materials such as bone or metal. This discovery would win Röntgen the first Nobel Prize in Physics, which was awarded in 1901.

Uranium radiation

Becquerel believed that uranium absorbed the sun's energy and then emitted it as X-rays. He planned to expose a compound containing uranium to sunlight and then place it on photographic plates wrapped in black paper. When developed, he predicted that the plates would reveal an image of the uranium compound. His experiment was frustrated by overcast weather, but Becquerel decided to develop his photographic plates anyway, expecting to find only a weak image, if any at all. To his surprise, the outlines left by the compound

The cross pattern at the bottom of this photographic plate is the result of a metal Maltese cross being placed between the uranium and the plate during Becquerel's 1896 experiment.

See also: Atomic weights 121 ▪ The electron 164–165 ▪ Isotopes 200–201 ▪ Improved atomic models 216–221 ▪ Synthetic elements 230–231 ▪ Nuclear fission 234–237

were strong and clear, proving that the uranium emitted radiation without the need for an external source of energy, such as the sun. The energy emitted by the uranium compound seemed not to diminish over time, even over a period of several months, and pure metallic uranium worked even better.

The Curies' discoveries

Although Becquerel had discovered radioactivity, he had no idea what the nature of his discovery was. The word was actually coined by Polish-French physicist Marie Curie in 1898. Further investigations by Becquerel, Marie Curie, her husband, Pierre, and others established that the amount of radioactivity emitted by a uranium mineral was proportional to the amount of uranium present.

Other substances were found to have radioactive properties as well. For example, the Curies discovered that samples of pitchblende, a mineral that contains uranium, appeared to produce more radioactivity than pure uranium itself. They surmised that there had to be another radioactive substance present in the sample. Eventually, the Curies isolated a new chemical element that was over 300 times more radioactive than uranium; they named it polonium. They also discovered that the waste that was left behind after the polonium had been extracted was itself still highly radioactive.

After some years of arduous work grinding, filtering, and dissolving 44-lb (20-kg) samples of pitchblende from which the uranium had already been extracted, in 1902, Marie Curie isolated a small amount of the element she called radium, which was also present in the mineral.

Curie calculated that 1 oz (28 g) of radioactive radium would produce 4,000 calories of heat per hour, apparently indefinitely, and she wondered where this energy was coming from. The answer would have to wait a few more years until Albert Einstein published his special theory of relativity, in 1905. According to Einstein, mass and energy are equivalent, as summed up in the iconic equation $E = mc^2$. As the radium radiated heat, it should also be losing mass. Unfortunately, the equipment available at the time was not accurate enough to measure the tiny amount of mass that was being converted into energy, so there was no way to verify Einstein's explanation experimentally.

In 1903, Marie Curie and her husband were jointly awarded the Nobel Prize in Physics with »

[Marie Curie had] a devotion and tenacity in execution under the most extreme hardships imaginable.
Albert Einstein

Marie Curie

Born Maria Skłodowska in Warsaw, Poland, in 1867, Marie received early training in science from her father, a secondary school teacher. Her involvement in a students' revolutionary organization saw her leave Warsaw for Krakow, which was then under Austrian rule. In 1891, she left to continue her studies at the Sorbonne in Paris, France. It was there, in 1894, that she met Pierre Curie, a professor at the School of Physics, and they married the following year. Their research, often performed together and under difficult conditions, led to the isolation of the elements polonium (1898) and radium (1902).

After Pierre's death in 1906, Marie took his place as professor of general physics, the first time a woman had held this position. She promoted the therapeutic use of radium in World War I and developed mobile X-ray units for use at the front. She died in 1934 from leukemia, possibly the result of exposure to radiation. In 1935, her daughter Irène was awarded the Nobel Prize in Chemistry for her work with radioactive elements.

Henri Becquerel for their work on radioactivity. And in 1910, Marie Curie was also awarded the Nobel Prize in Chemistry for discovering radium and polonium, so becoming the first person to receive two Nobel Prizes.

Different kinds of radiation

In 1898, using a simple experimental setup at Cambridge, UK, New Zealand–born physicist Ernest Rutherford discovered that there were different types of radioactivity. He used an electroscope (a device that detects the presence of electric charge on a body) and a sample of uranium as his radioactive source, then placed increasing thicknesses of aluminum foil between them. He measured the intensity of the radiation by making a note of the time required to discharge the electroscope. Rutherford discovered that there were, in fact, at least two distinct types of radiation, which he called alpha (α) and beta (β). He established that the beta rays were about 100 times more penetrating than the alpha rays. Further investigation showed that beta rays were deflected by a magnetic field, thereby indicating that they were negatively charged particles, similar to cathode rays.

In 1903, Rutherford discovered that alpha rays were deflected slightly in the opposite direction, which suggested that they were massive, positively charged particles. He later proved, in 1908, that alpha rays are in fact the nuclei of helium atoms. He did this by detecting the buildup of helium in an evacuated tube in which alpha rays were collected over a period of days.

Some years earlier, in 1900, French chemist Paul Villard had identified a third type of radiation. Called gamma radiation, this was even more penetrating than alpha radiation. Gamma rays were later shown to be a high-energy form of electromagnetic radiation, similar to X-rays, although with a much shorter wavelength.

Decay chains

While studying the radioactivity of the metallic element thorium in 1902, Rutherford and British chemist Frederick Soddy found that one radioactive element can decay into another. This discovery would earn Rutherford the 1908 Nobel Prize in Chemistry.

Interest in radioactivity grew further with the Curies' discovery of the previously unknown polonium and radium. Many other

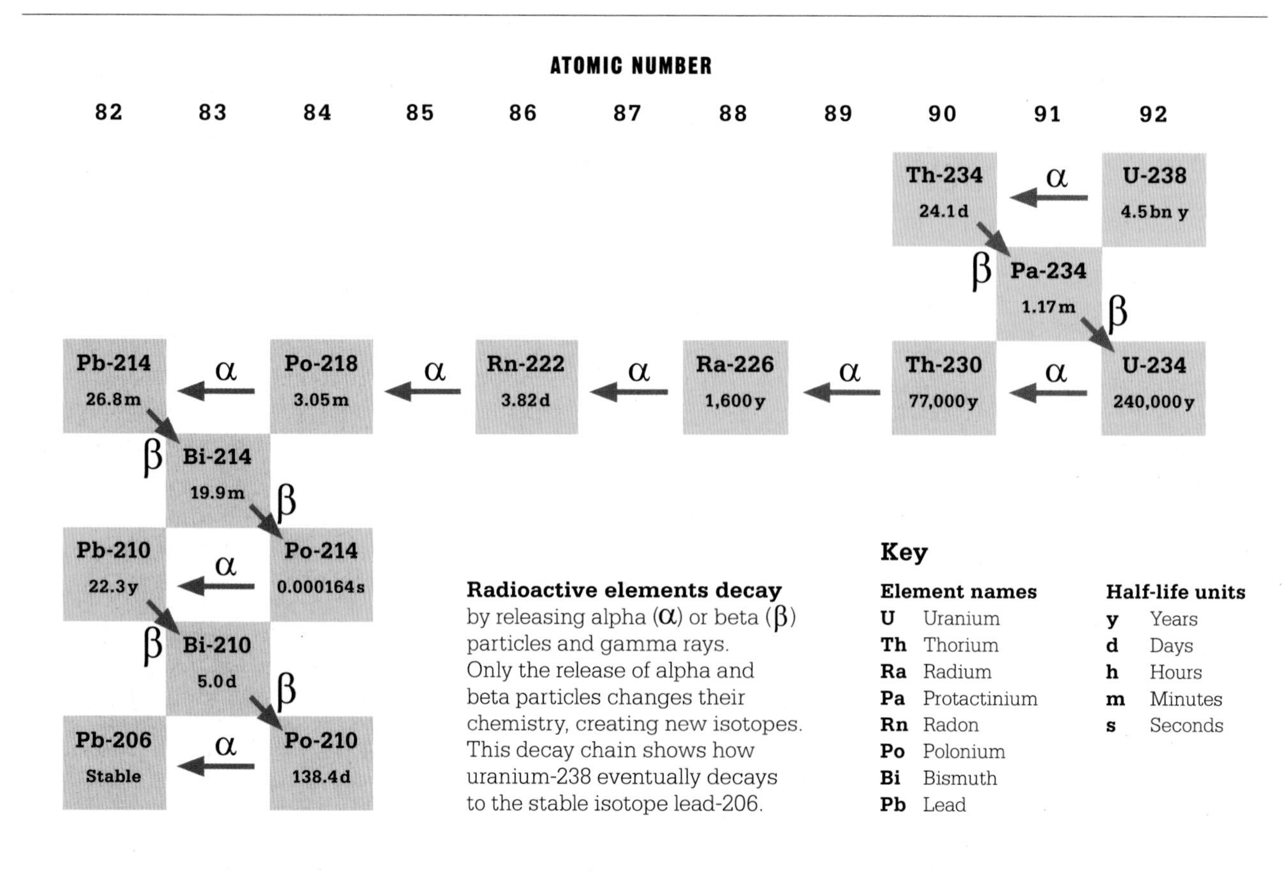

Radioactive elements decay by releasing alpha (α) or beta (β) particles and gamma rays. Only the release of alpha and beta particles changes their chemistry, creating new isotopes. This decay chain shows how uranium-238 eventually decays to the stable isotope lead-206.

Radioactive decay

The elements of the periodic table can come in more than one form, some of which are more stable than others. Not surprisingly, the most stable form of an element is generally the most common in nature. All elements have an unstable form, which is radioactive and emits ionizing radiation. Some elements, such as uranium, have no stable form and are always radioactive. These unstable elements decay, transforming into different elements called decay products and emitting radiation as alpha particles, beta particles, and gamma rays in the process. If the decay product is itself unstable, the process continues until a stable, nonradioactive form is reached.

Only 28 of the 38 radioactive elements occur naturally. The others have been created in laboratories. One of these—oganesson—was synthesized for the first time in 2002 and is believed to have a half-life of less than 1 millisecond.

substances isolated from uranium and thorium seemed to be new elements. Several researchers had noticed that the radioactivity of thorium salts seemed to vary randomly and that, rather oddly, this variation appeared to be related to how drafty the laboratory was. Evidently, there was some "thorium emanation" that a breeze could dislodge from the surface of the thorium. Rutherford speculated that it was a thorium vapor and repeatedly measured its ability to ionize air in order to assess its radioactivity. He was surprised to discover that there was an exponential decrease in radioactivity with time.

Rutherford also noticed that the walls of the vessels he was using became radioactive. This "excited activity" diminished regularly over time, and after 11 hours had fallen by half. Unknown to Rutherford, he was witnessing a decay chain and was measuring the decay of a radioactive form of lead, the result of the decay of the radon-220 isotope.

Rutherford and Soddy suggested that elements were undergoing "spontaneous transformation," a phrase that they hesitated to use, feeling it hinted at alchemy. It seemed evident that radioactivity resulted from changes at the subatomic level, although a full explanation of these changes would have to await the coming of quantum mechanics. According to Rutherford and Soddy, a radioactive element transforms into another element in the sense that it changes from a "parent element" into a different "daughter element." The actual atoms in the substance change randomly but at a rate that depends on the element involved. This process of change for a radioactive element came to be termed its half-life—the time it takes for one half of the radioactive sample to decay. Rutherford had assessed the half-life of the "thorium emanation" he had discovered to be 60 seconds, and that proved to be remarkably close to the 55.6-second half-life of radon-220 that scientists later discovered it to be. ■

Ernest Rutherford (right), after his radioactive decay discoveries, later worked with German physicist Hans Geiger (left) to develop an electrical counter for detecting ionized particles.

“The best sprinters in this road of investigation are Becquerel and the Curies.”

Ernest Rutherford
Letter to his mother (1902)

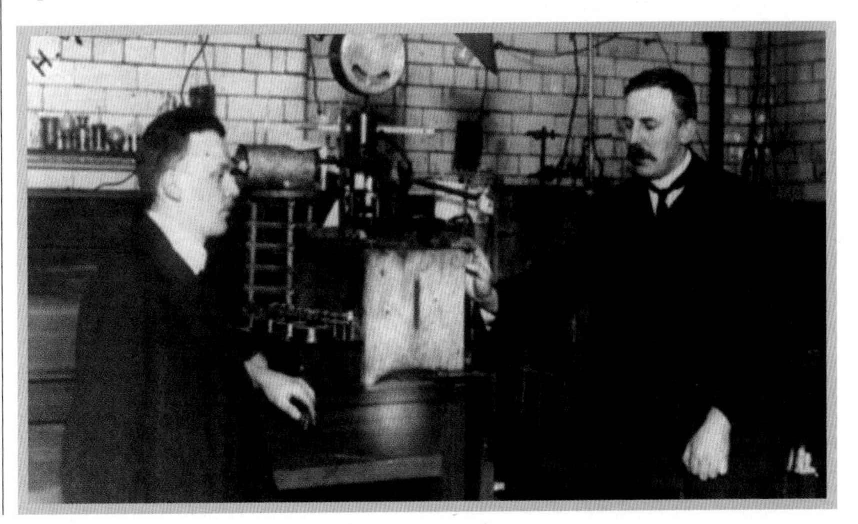

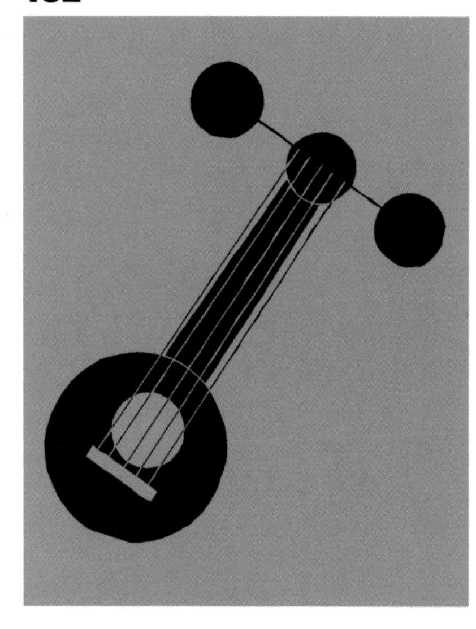

MOLECULES, LIKE GUITAR STRINGS, VIBRATE AT SPECIFIC FREQUENCIES

INFRARED SPECTROSCOPY

IN CONTEXT

KEY FIGURE
William Coblentz
(1873–1962)

BEFORE
1800 German-born astronomer William Herschel detects light outside the visible spectrum and names it infrared light.

***c.*1814** Joseph von Fraunhofer, a German physicist, builds a spectrometer and finds dark lines in the visible spectrum.

1822 French physicist Joseph Fourier devises a mathematical tool used to obtain information from spectra.

AFTER
1969 A team of engineers at Digilab in the UK builds the first Fourier Transform Infrared (FTIR) spectrometer.

1995 The European Space Agency creates the Infrared Space Observatory, which finds water in the atmospheres of planets in the solar system and Orion Nebula.

Infrared spectroscopy—which analyzes the absorption, emission, or reflection of infrared light by primarily organic molecules—was still in its infancy when US research student William Coblentz began his exploration at Cornell University in 1903. To investigate the absorption of infrared light by organic compounds, he built an infrared spectrometer with a light source on a movable arm, which shone light through a transparent cell holding the compound. The light then traveled down the arm to a prism, which separated the light into different wavelengths. Depending on the position of the arm, specific wavelengths were focused onto a light meter, which gave information on the amount of light absorbed versus wavelength.

Coblentz's data, published in 1905, revealed that certain functional groups consistently absorb infrared light at characteristic wavelengths. This provided researchers with a powerful method for identifying known compounds and for discerning the structure of new ones.

With the advent of computers, Fourier Transform Infrared (FTIR) spectroscopy became possible. In FTIR instruments, a complete range of infrared wavelengths is shown on the sample at once and sorted into absorption versus wavelength by computer. Spectra can, therefore, be collected in seconds. Today, FTIR is used in forensic laboratories to quickly identify counterfeit money and documents. ■

Whether the same chemical elements as those of our earth are present throughout the universe, was most satisfactorily settled in the affirmative.

Sir William Huggins
British astronomer (1824–1910)

See also: Functional groups 100–105 ■ Flame spectroscopy 122–125

THIS MATERIAL OF A THOUSAND PURPOSES

SYNTHETIC PLASTIC

IN CONTEXT

KEY FIGURE
Leo Baekeland (1863–1944)

BEFORE
1872 German chemist Adolf von Baeyer experiments with mixing phenol and formaldehyde, forming a tarlike and insoluble black solid.

1899 Having discovered that formaldehyde transforms milk into a solid material, Bavarian chemist Adolf Spitteler patents the milk-based plastic to make buttons and buckles.

AFTER
1926 Another phenol formaldehyde resin, Catalin, is patented. Unlike Bakelite, it is colorless and transparent, which means it can be produced in bright colors.

1940s Other synthetic plastics, such as polythene and polyvinyl chloride (PVC), replace Bakelite in the majority of its applications.

At the turn of the 20th century, moldable plastics were little more than an ambition. Chemists investigated various combinations of phenol and formaldehyde, but they invariably produced hard, insoluble black masses. One of those involved was Belgian Leo Baekeland. Initially, he focused on finding a cheaper and more durable replacement for shellac, used at the time to insulate electrical cables. He produced a soluble shellac known as Novolak, but it was not a commercial success.

Undeterred, Baekeland turned his mind to strengthening wood with a synthetic resin. By heating phenol and formaldehyde in the presence of a catalyst and controlling the pressure and temperature, in 1907 he produced a hard but moldable thermosetting plastic. He named it "Bakelite" after himself.

Material of 1,000 uses

Although plastics produced from existing materials already existed—such as celluloid—Bakelite was the first completely synthetic plastic.

Bakelite was used to make a vast array of household items and consumer goods—from clocks, telephones, and radios to lamps and kitchenware.

Importantly, it was heat-resistant and could be molded into useful shapes. Baekeland took out more than 400 patents relating to his invention, which soon became ubiquitous. But Bakelite was costly and complex to produce, and it was also quite brittle. After two very successful decades, it began to be superseded by newer plastics with more favorable properties, such as polythene or polyvinyl chloride (PVC). Today, Bakelite retains some automotive and electrical uses. ■

See also: Polymerization 204–211 ▪ Nonstick polymers 232–233 ▪ Super-strong polymers 267 ▪ Renewable plastics 296–297

THE MOST MEASURED CHEMICAL PARAMETER

THE pH SCALE

IN CONTEXT

KEY FIGURE
Søren Sørensen (1868–1939)

BEFORE
c.1300 Arnaldus de Villa Nova uses litmus to study acids and alkalis.

1852 British chemist Robert Angus Smith uses the term "acid rain" for the first time in a report on the chemistry of rain around the city of Manchester.

1883 Svante Arrhenius proposes that acids produce hydrogen ions and alkalis produce hydroxide ions in solution.

AFTER
1923 American chemist G. N. Lewis introduces his theory that an acid is any compound that will attach itself to an unshared pair of electrons in a chemical reaction.

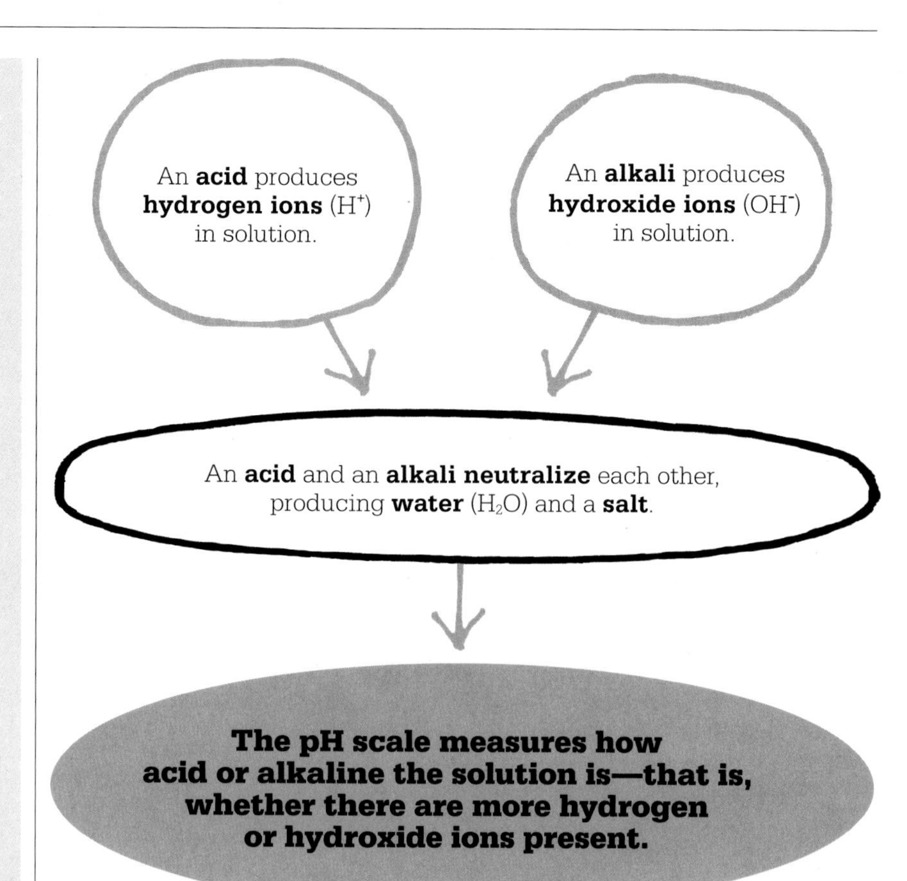
An **acid** produces **hydrogen ions** (H^+) in solution.
An **alkali** produces **hydroxide ions** (OH^-) in solution.
An **acid** and an **alkali neutralize** each other, producing **water** (H_2O) and a **salt**.
The pH scale measures how acid or alkaline the solution is—that is, whether there are more hydrogen or hydroxide ions present.

"pH measurement is often deceptively easy … pH measurement can also be exasperatingly difficult."
G. Mattock
"pH Measurement and Titration" (1963)

Acids and alkalis are familiar and well-understood chemicals both in the laboratory and in the home. The pH scale—known to chemists, gardeners, and brewers and widely used in the food industry and in the manufacture of fertilizers—is the means of measuring acidity or alkalinity. The pH scale was first developed by a chemist experimenting with beer production in 1909.

The acid test

Alchemists had tests for acids and alkalis centuries ago. Around 1300, Spanish alchemist Arnaldus de Villa Nova discovered that a purple dye extracted from lichens would turn red when combined with an acid—the stronger the acid, the darker it became. Litmus, as it became known, would turn blue when it came into contact with an alkali, making it the first acid–alkali indicator. In the 17th century, Robert Boyle discovered that acids and alkalis also caused other plant-derived substances to change color. These compounds opened up a way for chemists to determine the relative strengths of acids and alkalis by comparing the proportions of each that would neutralize the other.

Isolating the essence

In the late 18th century, chemists generally accepted the definition of alkalis as substances that could neutralize acids. In 1776, Antoine

See also: Brewing 18–19 ▪ Inflammable air 56–57 ▪ Electrochemistry 92–93 ▪ Acids and bases 148–149 ▪ Depicting reaction mechanisms 214–215

Litmus paper was the first-ever indicator to distinguish whether a solution was acid or alkaline. It is made from dye extracted from lichens.

Lavoisier attempted to isolate the "essence" in acids that gave them their unique properties, wrongly concluding that it was oxygen. Around 1838, Justus von Liebig discovered that acids reacted with metals to produce hydrogen and reasoned that hydrogen was common to all acids. Then, 45 years later, in 1883, Svante Arrhenius proposed that the properties of acids and alkalis were the result of the action of ions in solution. Acids, he declared, are simply substances that release hydrogen ions, H^+, in solution. Conversely, alkalis release hydroxyl ions, OH^-. Acids and alkalis neutralize each other because the H^+ and OH^- ions combine to form water.

In 1923, British chemist Thomas Lowry and Danish chemist Johannes Brønsted independently proposed a modified form of Arrhenius's definitions, agreeing that acids are releasers of protons (hydrogen ions) but simply defining alkalis as substances capable of binding protons. This cemented the idea that the strength of an acid could be determined by the quantity of hydrogen ions it released in solution.

The pursuit of better beer

In 1893, German chemist Hermann Walther Nernst developed a theory to explain how ionic compounds break down in water. He suggested that the positive and negative ions in the compound lose contact with each other, allowing them to move freely through the water and to conduct an electric current. That same decade, Latvian chemist Wilhelm Ostwald invented electrical conductivity equipment that could determine the quantity of hydrogen ions in a solution by measuring the current generated by those ions migrating to oppositely charged electrodes. There was, however, no universally accepted way of expressing hydrogen ion concentrations.

In 1909, Danish chemist Søren Sørensen, director of the chemical department at the Carlsberg Laboratory in Copenhagen—created by the brewer of the same name—was studying fermentation and the effect of ion concentration. He discovered that the hydrogen ion content played a key role in the crucial enzyme reactions involved in the production of beer, one of the world's oldest chemical industries.

Sørensen needed a method that would allow him to measure extremely low H^+ concentrations without chemically affecting the enzymes he was studying. Chemists had been aware of the technique »

Søren Sørensen

The son of a farmer, Søren Sørensen was born in Havrebjerg, Denmark, in 1868. At age 18, he entered the University of Copenhagen, initially planning to study medicine but then opting for chemistry. Most of his training was in inorganic chemistry. While working on his doctorate, he assisted in a geological survey of Denmark, acted as assistant in chemistry at the laboratory of the Danish Polytechnic Institute, and served as a consultant at the naval dockyard. In 1901, Sørensen became director of the chemical department of the Carlsberg Laboratory in Copenhagen, where he remained for the rest of his life. Here, he began addressing biochemical problems, and it was at Carlsberg in 1909 that he devised the pH scale. His wife Margrethe Høyrup Sørensen assisted him in much of this work. After a period of poor health, Sørensen retired in 1938 and died the following year.

for determining the acidity of a solution by titration—gradually adding a solution of an alkali of known concentration until the acid is neutralized—since the 18th century, but this was unsuitable for Sørensen's purposes. He therefore decided to make his measurements using electrodes rather than chemically, employing the methods developed by Nernst and Ostwald as the basis for his approach.

Sørensen's original setup for measuring pH values involved a lengthy and somewhat tedious process. It required a source of hydrogen gas and several pieces of equipment, including a resistance potentiometer and a highly sensitive galvanometer. Calculating the hydrogen ion content from the results was a time-consuming process using a rather complex equation developed by Nernst.

Sørensen first introduced the term pH in his paper of 1909, in which he discussed the effect of H^+ ions on enzymes. He defined pH as the -log[H^+], or the negative logarithm of the hydrogen ion content. (The log of a number is the power to which 10 must be raised to give that number.) The genius

> Brewing and science, especially chemistry, have been intertwined throughout history.
>
> **Maria Filomena Carões**
> **"A Century of pH Measurements" (2010)**

of taking this approach was that the logarithmic scale took the cumbersome need of expressing very small quantities of hydrogen ions that vary over a wide range and turned it into something that was quick and easy to understand. Rather than say, for example, that for pure water the concentration of hydrogen ions equals 1.0×10^{-7} M (moles per liter), Sørensen simply took the negative of the log (in this case, -7) to give a pH of 7. A solution with a hydrogen ion content of 1.0×10^{-4} M, a concentration 1,000 times higher than that of pure water, has a pH of 4. Every step up (or down) the pH scale means an increase (or decrease) of the hydrogen ion concentration by a factor of 10.

pH calculations

In 1921, a specialized calculator—looking something like a circular slide rule—was produced by the company Leeds & Northrup of Philadelphia for the purpose of calculating pH. From the late 1920s, H^+ sensitive glass electrodes gradually replaced Sørensen's hydrogen electrodes, cutting out the need for a hydrogen gas source. By the late 1930s, it was possible to pack all the necessary electronic components into a compact meter that could give an immediate pH reading. Today, the most common way of taking measurements in the laboratory is to use a pH meter.

The scale of pH values is generally given as between 0 and 14, with 0 being the value for concentrated hydrochloric acid, 7 the value for pure water (neutral pH), and 14 the value for concentrated sodium hydroxide. With very high concentrations, it

Efficient breathing helps runners maintain their carbonic acid levels by expiring the waste CO_2 produced by muscle activity.

Biological buffers

Most of the enzymes that the cells in the human body depend on to function operate within a narrow pH range, typically from 7.2 to 7.6. (An exception is stomach protease, a digestive enzyme that operates in the 1.5–2.0 pH range.) Straying from this narrow band would be extremely harmful, even fatal. The body maintains a safe pH through a system of buffers, the main one being the bicarbonate buffer. When sodium bicarbonate reacts with a strong acid, it forms carbonic acid (a weak acid) and salt. When carbonic acid reacts with a strong alkali, bicarbonate and water are formed. In normal conditions, bicarbonate ions and carbonic acid are present in the blood in a 20:1 ratio, meaning that the buffering system is most efficient at dealing with excess acid. This makes sense, as most of the body's metabolic wastes are acids, such as lactic acid and ketones. Carbonic acid levels in the blood are controlled by expiring CO_2 through the lungs. The level of bicarbonate in the blood is controlled through the renal system.

The pH scale shows that a variety of substances that we use on a daily basis have a wide range of acid and alkaline levels. The pH is a property of a solution, so differences in the concentration of a substance lead to different pH values.

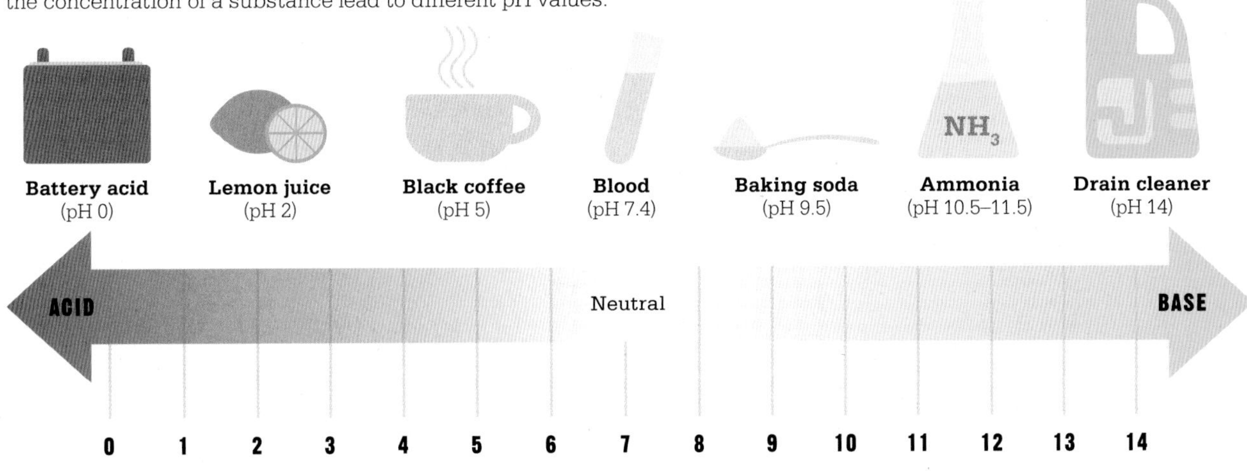

is possible to get a pH value of -1, which appears to be the limit of acidity, and a pH of 15, the limit of alkalinity. In pure water, the concentration of H^+ ions, 10^{-7} M, is balanced by the concentration of OH^- ions, also 10^{-7} M.

Gardeners and others who do not need quite such accurate results as a meter provides may use color-changing litmus paper or indicator dyes. The indicators are themselves actually weak acids that change color according to the quantity of hydrogen ions they produce. Indicators are generally specific to a certain range of pH values. For example, phenolphthalein reacts over a range of about 8 to 10, while methyl red ranges from 4.5 to 6.

Power, potenz, potential

No one has ever been exactly certain what the "p" in pH stands for. Sørensen himself was never clear about it. Some sources, such as the Carlsberg Foundation, say it stands for the "power" of hydrogen. German sources say it is *potenz* (which also means power), French chemists have suggested it stands for *puissance* (also "power"), and some would have it that it is from the Latin *potentia hydrogenii* (capacity of hydrogen). In his lab notes, Sørensen used the subscripts p and q to distinguish the two electrodes in his system; q related to a reference electrode, and p referred to the positive hydrogen electrode, so he may have simply been referring to the concentration of hydrogen ions at the p electrode.

The pH scale has advanced our world in many ways. In agriculture, it is crucial for indicating whether crops will grow in certain soils; in medicine, doctors can use it to diagnose conditions, such as kidney problems; and in the food industry, it aids all processing stages. ■

A pH meter is useful for testing soil acidity if plants or crops need particular conditions to thrive. Gardeners can adjust soil acidity with additives such as compost, mulch, or fertilizer.

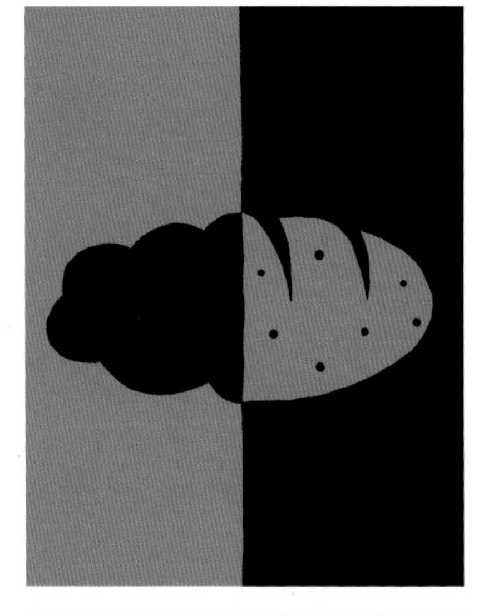

BREAD FROM AIR

FERTILIZERS

IN CONTEXT

KEY FIGURES
Fritz Haber (1868–1934),
Carl Bosch (1874–1940)

BEFORE
1861 Potassium fertilizers are manufactured in Germany.

1902 German physical chemist Wilhelm Ostwald develops a process for making nitric acid from a series of reactions that begins with oxidizing ammonia at a high temperature.

AFTER
1913 Large-scale production of ammonium nitrate fertilizer begins in Oppau, Germany.

1920s French, British, and US manufacturers start to produce ammonia.

1968 William Gaud of the US Agency for International Development (USAID) coins the term "Green Revolution" to describe the increase in food production partly brought about by synthetic fertilizers.

In addition to water, carbon dioxide from air, and sunlight, plants need mineral nutrients from the soil, principally nitrogen (N), phosphorus (P), and potassium (K). Fertilizers provide these nutrients to help crops grow bigger and faster, and increase yields. Since plant roots absorb nutrients from water in soil, chemical compounds in fertilizers must be water-soluble. Natural fertilizers have a history as old as civilization: Neolithic people probably used animal dung to stimulate plant growth.

Until the early 19th century, farmers could rely only on manure (contains N, P, K, and traces of magnesium/Mg), bone meal (N, P, and calcium/Ca), and ash (P, K, and Mg) to fertilize their crops. Then, in the 1830s, companies began to mine the huge deposits of nitrate-rich guano—bird droppings—on Peru's coastal islands to supply farmers in North America. European farmers later used guano from southwest Africa's Namibian coast.

The first synthetic fertilizers were made in the early 19th century by treating bones with sulfuric acid, which increases the level of water-soluble phosphorus. Then, in 1861, German chemist Adolph Frank was granted a patent for creating potassium-based fertilizers from potash, a type of rock salt.

With growing populations in North America and Europe, demand for fertilizers was incessant. In 1908, German chemist Fritz Haber, examining how high temperatures and pressures affected chemical reactions, succeeded in fixing nitrogen from the air to ammonia—a nitrogen compound that plants can absorb. In 1909, Haber's compatriot Carl Bosch succeeded in converting the bench-top process to an industrial scale.

> Civilization as it is known today could not have evolved, nor can it survive, without an adequate food supply.
>
> **Norman Borlaug**
> **Nobel lecture (1970)**

See also: Sulfuric acid 90–91 ▪ Explosive chemistry 120 ▪ Le Chatelier's principle 150 ▪ Chemical warfare 196–199 ▪ Pesticides and herbicides 275–275

The Haber-Bosch process quickly produces large quantities of ammonia, using nitrogen from air and hydrogen from natural gas. The reaction is conducted at 752–1022°F (400–550°C) and a pressure of 150–300 atm with an iron catalyst. Then, using the Ostwald process, the ammonia (NH_3) as well as oxygen (O_2) are passed over a platinum catalyst to yield nitric oxide (NO) and nitrogen dioxide (NO_2). The NO_2 is then dissolved in water (H_2O) to produce nitric acid (HNO_3). Nitric acid is the key component of the fertilizer ammonium nitrate (NH_4NO_3), which stores and transports easily as a white solid in temperate conditions.

Manufacturing takes off

German chemical company BASF began manufacturing fertilizers using the Haber-Bosch process in 1913. However, after World War I broke out, production temporarily switched in 1915 to making nitric acid, the main starting reagent for nitro-based explosives. Synthetic fertilizer production increased from the late 1940s and was a core component of the "Green Revolution," which aimed to abolish famine. Nitrogen fertilizers are still the most widely used, and demand continues to rise, but phosphate and potassium fertilizer production has also increased. The minerals fluorapatite and hydroxyapatite are treated with sulfuric or phosphoric acids to create soluble phosphates; mined potassium-bearing minerals, such as sylvite, form the raw material for potassium fertilizers.

Nitrogen fertilizer production pollutes groundwater, contributes to greenhouse gases, and is very energy-hungry, using 5 percent of global natural gas production. Thanks to fertilizers, millions of people were fed in the late twentieth century. However, in 2015, UN scientists warned that these farming methods are unsustainable; they projected that within decades topsoil would be eroded and the land infertile. ■

Production of synthetic nitrogen fertilizers grew from 14.3 million tons (13 million metric tons) in 1961 to 124.5 million tons (113 million metric tons) in 2014. In the same period, the global population increased from 3 billion to 7.2 billion. By 2020, fertilizers supported about 48 percent of the human population.

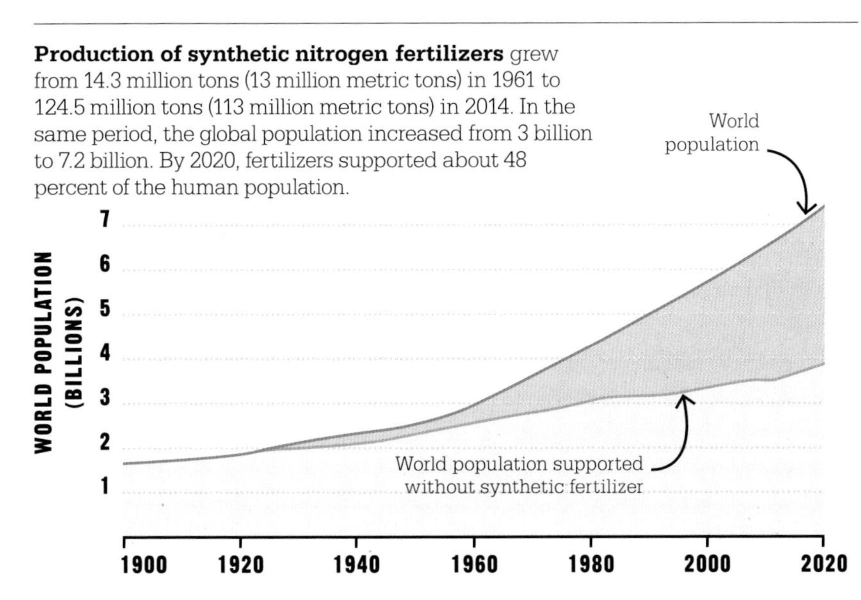

Carl Bosch

Born in Cologne, Germany, in 1874, Bosch studied chemistry at Leipzig University. He developed the Haber-Bosch process while at the chemical company BASF. During World War I, he helped develop a technique to mass-produce nitric acid, which was mostly used to make munitions. In 1923, Bosch invented a process for converting carbon monoxide and hydrogen into methanol for use in the production of formaldehyde. Two years later, he co-founded I.G. Farben, which became one of the world's largest chemical corporations.

Bosch shared the 1931 Nobel Prize in Chemistry with German chemist Friedrich Bergius, for their work on high-pressure chemistry. Although his company helped finance the Nazis, Bosch criticized their anti-Semitic policies and fell out of favor with the government. He suffered with depression and died in 1940.

Key work

1932 "The development of the chemical high-pressure method during the establishment of the new ammonia industry."

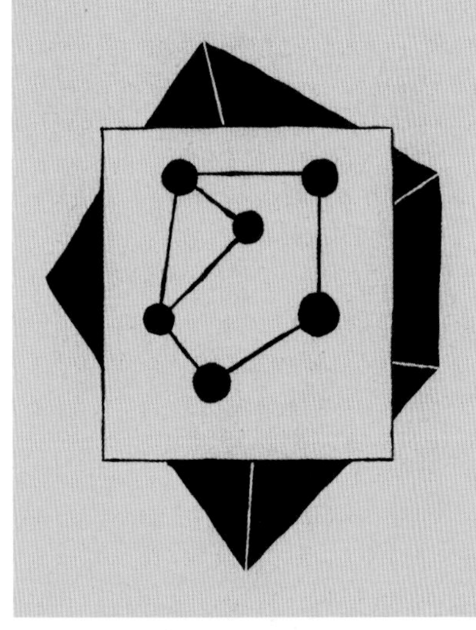

THE POWER TO SHOW UNEXPECTED AND SURPRISING STRUCTURES

X-RAY CRYSTALLOGRAPHY

IN CONTEXT

KEY FIGURES
William Bragg (1862–1942),
Lawrence Bragg (1890–1971)

BEFORE
1669 Danish scientist Nicolas Steno shows that the angles between crystal faces are constant for every different crystal type.

1895 Wilhelm Röntgen discovers X-rays.

AFTER
1934 J.D. Bernal produces the first X-ray diffraction image of a crystallized protein.

1945 Dorothy Hodgkin maps the molecular structure of penicillin, making it easier to manufacture the drug.

1952 Rosalind Franklin takes "Photo 51," the XRC image later used to determine the 3D structure of DNA.

2012 NASA's Curiosity rover carries out X-ray diffraction analysis of Martian soil.

In 1912, at the suggestion of Max von Laue, a German physicist who was investigating the structure of crystals, German researchers Walter Friedrich and Paul Knipping aimed a fine beam of X-rays at a crystal of zinc sulfate and achieved a regular pattern of spots on a photographic plate. This demonstrated that X-rays were diffracted when passed through a crystal but also had more far-reaching implications.

Later in 1912, British chemist William Bragg discussed this so-called von Laue effect with his son Lawrence. They believed that the patterns reflected the underlying crystal structure, but wondered why—when there were so many directions in which the X-ray beam could be diffracted—only a limited number of spots appeared on the photographic plate. Lawrence thought this was a product of the properties of the crystal and carried out experiments on crystals of various kinds of rock salt, calcite, fluorspar, iron pyrites, and zinc blende. He found different diffraction patterns and proposed they were produced by different arrangements of atoms.

X-rays fired at a crystal are diffracted, and the resulting diffraction pattern is an interpretation of the crystal's atomic structure.

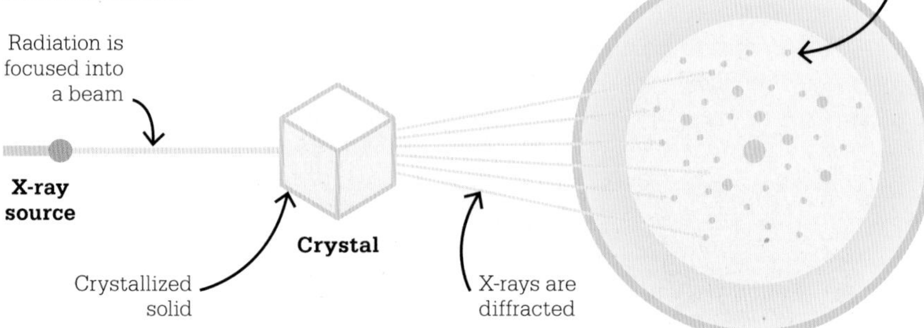

See also: The electron 164–165 ▪ Mass spectrometry 202–203 ▪ Improved atomic models 216–221 ▪ Chemical bonding 238–245 ▪ The structure of DNA 258–261 ▪ Protein crystallography 268–269 ▪ Atomic force microscopy 300–301

I should not like to leave an impression that all structural problems can be settled by X-ray analysis or that all crystal structures are easy to solve.

Dorothy Hodgkin
Nobel lecture (1964)

Bragg's equation

In 1913, Lawrence formulated an equation (later known as Bragg's law) to predict at what angles X-rays will be diffracted by a crystal when the X-ray's wavelength and the distance between the crystal atoms are known. In other words, if the wavelength and the angle of diffraction are known, the distance between the atoms can be calculated. This was the foundation of the new discipline of X-ray crystallography (XRC): the use of X-rays to determine the atomic and molecular structure of crystals.

Many materials, including salts, metals, and minerals, can form crystals. The atoms in each are arranged regularly, and each has a unique geometry. When X-rays are fired into a crystal, they scatter as they interact with the atoms' electrons. An X-ray striking an electron produces a spherical secondary wave, which spreads out in all directions from the electron. As a crystal has a regular array of atoms and electrons, the passage of X-rays produces a regular array of secondary waves. They create complicated arrangements of destructive and constructive interference patterns. In the destructive patterns, the waves cancel each other out, and in the constructive patterns, they reinforce each other. An image of the latter is recorded on photographic film. Modern computers can convert a series of 2D images into a 3D model, showing the density of electrons within the crystal. From this, a crystallographer can determine the position of the atoms and the nature of their chemical bonds.

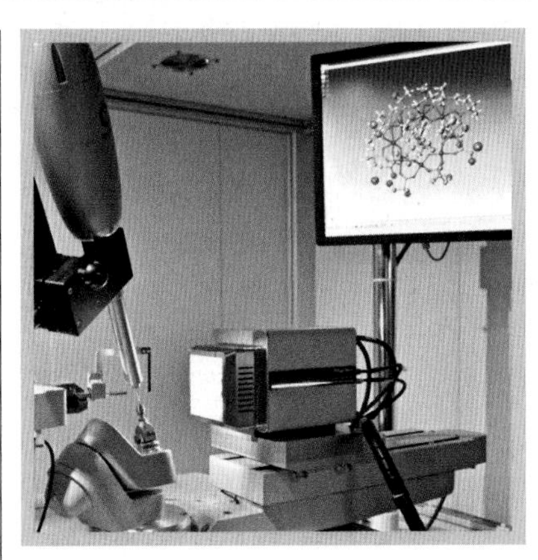

This modern X-ray crystallography setup shows a diffractometer, comprising a radiation source, a monochromator to select wavelength, a sample, a detector, and a monitor screen.

Greater complexity

Irish chemist J.D. Bernal had suggested in 1929 that since biological macromolecules such as proteins have regular structural organization, XRC should be able to interpret this. In 1934, working with Dorothy Hodgkin, one of his students at Cambridge University, he recorded the first X-ray diffraction image of a crystallized protein. Hodgkin later made a number of ground-breaking discoveries using XRC. For example, she used the technique to tackle the complexity of vitamin B_{12}. Never before had XRC been applied successfully to such a complex substance, but Hodgkin prevailed and revealed the secrets of vitamin B_{12} in 1954.

The discovery in 1953 of the double helical structure of DNA (deoxyribonucleic acid) relied on British chemist Rosalind Franklin's XRC image of DNA. This breakthrough gave a vital clue that DNA could produce exact copies of itself and carry genetic information. ■

Diamond and graphite

The chemical composition of diamond and graphite is identical—pure carbon—yet they look very different. Whereas diamond is the hardest naturally occurring substance, graphite readily cleaves along parallel planes. The explanation for this was revealed by Bernal in 1924, using X-ray crystallography. He showed how the atoms in diamond form a tetrahedral structure held together by covalent bonds (where atoms share electrons), making it very strong. In graphite, the atoms are arrayed in stacked sheets with covalent bonds between atoms on the same sheet, but not between the sheets. There are, therefore, lines of weakness between the sheets, which make graphite easy to split.

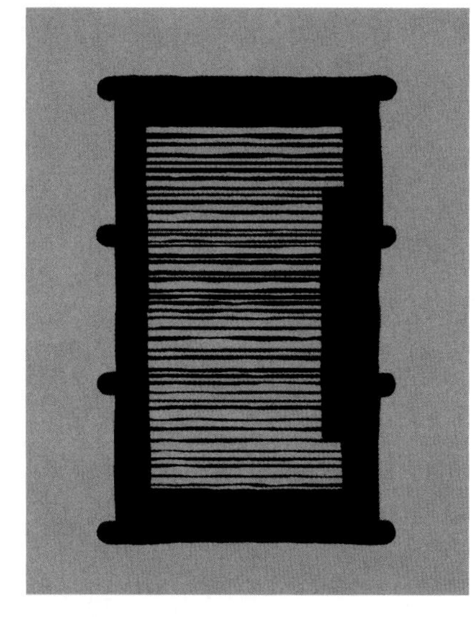

GAS FOR SALE

CRACKING CRUDE OIL

IN CONTEXT

KEY FIGURE
Eugene Houdry (1892–1962)

BEFORE
1856 Ignacy Łukasiewicz, a Polish pharmacist and engineer, builds the world's first modern oil refinery in Ulaszowice, Poland.

1891 Vladimir Shukhov patents the world's first thermal cracking process.

1908 American industrialist Henry Ford invents the Model T car, popularly known as Tin Lizzie. Its success accelerates the search for fuel.

AFTER
1915 American pioneer Almer M. McAfee develops the first catalytic cracking process, but the cost of the catalyst prevents widespread use.

1942 The first commercial fluid catalytic cracking plant begins operation at the Standard Oil Company refinery in Baton Rouge, Louisiana.

In 1913, a chemical process was patented that would change the course of aerial battles—not in World War I, but almost a quarter of a century later, in World War II. The process, known as cracking, allowed Allied countries to produce sufficient quantities of superior aviation fuel to give their planes a distinct advantage over those of the Axis forces.

Fractional distillation

In the mid-19th century, early oil refineries used fractional distillation to increase the quantities of useful products obtained from crude oil. By heating the oil to evaporation point, then condensing the resulting vapors at varying temperatures to separate products with different boiling points, fractions (groups) of hydrocarbons with specific uses could be obtained.

Initially, kerosene was in highest demand, for lamp oil. However, the invention of the car in the late 19th century drastically increased the need for different fractions of crude oil, such as gasoline and diesel, previously considered waste. Demand quickly outstripped

Fractional distillation is the process used to separate the molecules in crude oil. Larger molecules have higher boiling points. The fractionating column is hottest at the bottom, where the largest molecules condense, and coolest at the top, where the lightest molecules condense.

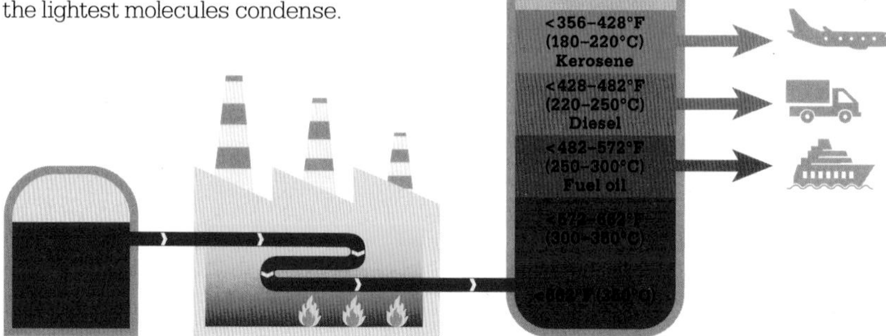

See also: Catalysis 69 ▪ The greenhouse effect 112–115 ▪ Polymerization 204–211 ▪ Leaded gasoline 212–213 ▪ Carbon capture 294–295

supply, and methods to produce more of these fractions from crude oil supplies were required.

Thermal cracking

"Cracking" refers to the process of breaking down larger, long chain hydrocarbons into more useful smaller, shorter chain hydrocarbons. Originally, this was done using heat and was termed "thermal cracking." The first thermal cracking process was patented in 1891 by Vladimir Shukhov, a Russian engineer, but it was not until American chemists William Burton and Robert E. Humphreys patented a similar process in the US in 1913 that wider commercial use took off.

Problems with thermal cracking limited its application. The huge amounts of energy needed to reach the high temperatures required and the fact that many less useful long chain hydrocarbons were left over meant that chemists continued to search for a better procedure. In response, French engineer Eugene Houdry introduced catalysts to the cracking process in the early 1920s.

Three catalytic oil cracking units on the site of the first commercial fluid catalytic cracking plant at Baton Rouge refinery, Louisiana, in 1944.

Catalytic cracking

Catalysts increase the rate of chemical reactions without being consumed themselves. Houdry initially worked on a process for turning brown coal into high-quality gasoline but turned to crude oil when this failed. He identified an aluminosilicate catalyst, which was effective even with hard-to-crack fractions. His collaboration with oil companies in the 1930s meant that by 1937 a commercial plant that could produce 15,000 barrels of gasoline per day began operation in Marcus Hook, Pennsylvania.

The timing proved prescient. World War II commenced two years later, and the aviation fuel produced by Houdry's process was advantageous because it had superior antiknocking properties. Knocking is an engine issue that happens when fuel combustion is not in sync with the engine cycle. Octane ratings indicate how well fuel avoids this problem, with the compounds n-heptane (0, readily causes engine knocking) and isooctane (100, resists engine knocking) used as a reference. Houdry's aviation fuel was 100 octane, compared to 87 to 90 octane for the Axis forces' aviation fuel.

Today, Houdry's process has been replaced by more economical fluid catalytic cracking. However, this method is built on the principles Houdry laid down and still uses aluminosilicate catalysts. ■

Environmental impact

The more reliable supply of fuels enabled by Houdry's process is not without issues. The additional carbon dioxide (CO_2) emissions generated lead to anthropogenic global warming and climate change. In 2016, transport accounted for around one-fifth of global CO_2 emissions, behind only those produced by power generation and industry. Houdry also recognized air pollution as a problem. In 1950, he formed a company—Oxy-Catalyst—to try and produce catalytic preventative solutions to tackle declining air quality and health problems caused by car emissions. He invented the first catalytic converter, for which he was granted a patent in 1956. While his invention was not widely used due to the tetraethyl lead that was added to gasoline poisoning the catalyst, his work preceded the three-way catalytic converters that made it to the market in 1973. These devices are now fitted in all gasoline-fueled cars to reduce the emissions of nitrogen oxides and carbon monoxide.

Few of us have the vision to anticipate industrial needs and single-mindedly pursue their satisfaction as Eugene Houdry did.

Heinz Heinemann
Houdry Award address (1975)

THE THROAT SEIZED AS BY A STRANGLER

CHEMICAL WARFARE

IN CONTEXT

KEY FIGURE
Fritz Haber (1868–1934)

BEFORE
1812 John Davy discovers phosgene, used to make dyes.

1854 Scottish scientist Lyon Playfair's proposal of using cacodyl cyanide artillery shells in the Crimean War is rejected.

1907 The Hague Convention forbids the use of "poison or poisoned weapons" in war.

AFTER
1925 Sixteen countries pledge in the Geneva Protocol not to use chemical agents in war.

1993 The UN Chemical Weapons Convention, signed by 130 nations, bans the making of many chemical weapons.

2003 A US-led invasion of Iraq to destroy "weapons of mass destruction" fails to find them.

In spring 1915 in World War I, there was stalemate on the Western Front. Near Ypres in Belgium, Canadian, Belgian, and French-Algerian soldiers dug into trenches, with German troops in facing trenches. On April 22, the Germans opened valves on more than 5,000 pressure tanks to release about 165 tons (150 metric tons) of toxic chlorine gas. A huge, yellow-green cloud rapidly formed. Heavier than air, the chlorine spread close to the ground, wafted by the breeze across "no-man's land" to Allied lines.

When inhaled, chlorine gas (Cl_2) reacts with water in the lungs to produce hydrochloric acid (HCl) and constricts the chest, tightens the

See also: Gunpowder 42–43 ▪ Gases 46 ▪ Why reactions happen 144–147 ▪ Le Chatelier's principle 150 ▪ Fertilizers 190–191

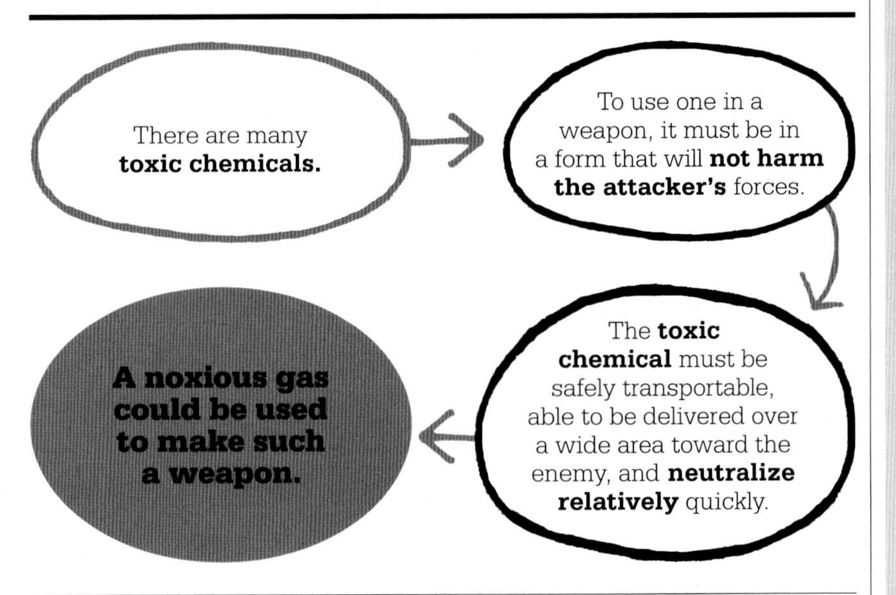

There are many **toxic chemicals.**
To use one in a weapon, it must be in a form that will **not harm the attacker's** forces.
The **toxic chemical** must be safely transportable, able to be delivered over a wide area toward the enemy, and **neutralize relatively** quickly.
A noxious gas could be used to make such a weapon.

throat, and causes suffocation. One part per 1,000 in air can cause death in minutes; many soldiers breathed in greater concentrations. Some died on the spot, and many more fled in panic. Few escaped completely: there were about 15,000 casualties, including 1,100 deaths. A new era of warfare had dawned.

Deadly chemicals

In 1914, the French had fired shells that were filled with a tear gas (ethyl bromoacetate, $C_4H_7BrO_2$) at German lines, but they had made little impact. The Ypres attack was the first large-scale deployment of poison gas in conflict resulting in deaths and was followed by many more attacks. Combatants were later equipped with protective gas masks, but surprise attacks meant they could not be deployed in time.

By 1916, all the major combatant nations were using poison gas and chemists had devised even deadlier chemical weapons. An estimated 132,000 tons (120,000 metric tons) of poison gas were produced during the war, causing at least 91,000 deaths and 1.3 million casualties. Delivery generally relied on the wind, so it was impossible to control the path of the gas; it often drifted over settlements, resulting in up to 260,000 civilian casualties.

The mastermind of the Ypres chlorine attack was Fritz Haber, who in 1908 had developed a process that enabled the fixation of atmospheric nitrogen into ammonia-based fertilizer. His process for accelerating fertilizer »

> We heard the cows bawling and the horses screaming.
>
> **Willi Siebert**
> **German soldier, Ypres (1915)**

Fritz Haber

Born in Breslau, Prussia (now Wroclaw, Poland), in 1868, Fritz Haber was the son of a Jewish dye importer. After gaining a chemistry doctorate from the University in Berlin in 1891, he applied high temperatures and pressures to chemical reactions, enabling mass production of fertilizers. In World War I, Haber worked at the Kaiser Wilhelm Institute for Physical Chemistry and Electrochemistry (now the Fritz Haber Institute) in Berlin. The German army asked him to develop chemical weapons, as head of the Chemistry Section of the Ministry of War.

Haber controversially won the 1918 Nobel Prize in Chemistry for synthesizing ammonia from its elements. Scientists under his direction later developed Zyklon-B. The Nazis attacked Haber, despite his conversion to Christianity, as well as the Institute for harboring Jewish scientists. Haber fled Germany in 1933 and died a year later.

Key works

1913 "The production of synthetic ammonia"
1922 "Chemical warfare"

German chemist Fritz Haber (second left) directs the release of chlorine gas by German forces on the Western Front, in Ypres in Belgium, on April 22, 1915.

production also could be applied to explosives. During the war, Haber urged politicians, industrial leaders, generals, and scientists to join forces in developing new processes for mass production of traditional armaments and chemical weapons to gain advantage over the enemy.

After the chlorine attack, Haber worked on even more deadly chemical weapons. He developed Haber's Law: that the severity of a toxic effect depends on the total exposure, which is the exposure concentration (c) multiplied by duration time (t), so $(c \xi t)$. Longer exposure of a weak concentration can have the same effect as a short exposure to a high concentration.

Other poison gases

In December 1915, the Germans weaponized phosgene gas at Ypres. This difficult-to-spot, colorless gas with an odor of hay is produced by carbon monoxide (CO) and chlorine (Cl) reacting to form $COCl_2$. The army was able to deliver phosgene in smaller, concentrated amounts in artillery shells rather than relying on the wind. Soldiers found the gas hard to detect at low concentrations, and severe symptoms often took hours to reveal themselves. The gas reacts with proteins in the lung alveoli (air sacs) to disrupt blood–oxygen exchange so that fluid builds up in the lungs and results in suffocation. Some estimate that phosgene was responsible for up to 85 percent of the 91,000 deaths from poison gas in World War I.

Mustard gas ($C_4H_8Cl_2S$) was first synthesized in the early 19th century and was the most deployed chemical weapon in the conflict. Like phosgene, it was generally delivered in artillery shells. Mustard gas differs from chlorine and phosgene in that it is an aerosol, not a gas, but toxic aerosols are often called noxious gases. Smelling like garlic and mustard-hued, mustard gas caused chemical burns. Mortality rates, at 2–3 percent, were much lower than from chlorine or phosgene, but casualties were hospitalized for a long time.

Despite the horrors of World War I, some Geneva Protocol signatories continued to develop chemical weapons in secret, while other nations simply ignored it. The Imperial Japanese Army deployed chemical weapons, including phosgene, mustard gas, and lewisite, against Chinese soldiers and civilians in the Second Sino-Japanese War (1937–1945). Lewisite had been developed in 1918, but not in time for use in World War I. This liquid—often dispersed as a heavy, colorless gas—is produced by reacting arsenic trichloride ($AsCl_3$) with acetylene (C_2H_2). Lewisite ($C_2H_2AsCl_3$) damages the skin, eyes, and respiratory tract.

In 1916, the French had utilized hydrogen cyanide (HCN) gas—previously used by American farmers to fumigate citrus trees against insect pests. After World War I, it was marketed as Zyklon-B for fumigating clothing and freight trains. From early 1942, after using Zyklon-B to murder thousands of Russian prisoners of war, the Nazis deployed the gas in concentration camps on an industrial scale as a key weapon in the Holocaust, the genocide of 6 million Jews as well as millions from other racial groups. More than 1 million people were executed by Zyklon-B gas alone.

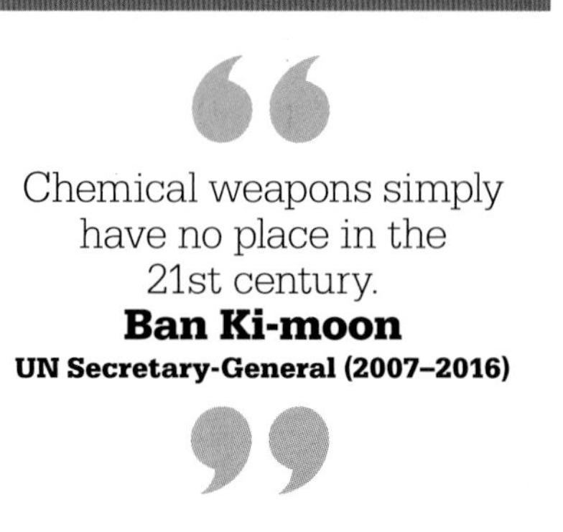

Chemical weapons simply have no place in the 21st century.

Ban Ki-moon

UN Secretary-General (2007–2016)

Nerve agents

In 1938, German chemists, while attempting to develop stronger pesticides, manufactured an extremely toxic, clear, colorless, tasteless liquid compound. Naming it sarin, they found that it readily evaporates to form an equally toxic, heavier-than-air gas. Its military implications were recognized at once but, for reasons unknown, the Nazis never deployed it.

Sarin ($C_4H_{10}FO_2P$) is a nerve agent—a toxin that disables the enzyme acetylcholinesterase (AChE), which activates the body's "off" switch for muscles and glands, so they are stimulated constantly. A small drop of liquid sarin on human skin causes sweating and muscle twitching. Exposure to the aerosol or vapor leads to loss of consciousness, convulsions, paralysis, and respiratory failure. In its pure form, sarin is 500 times more deadly than chlorine gas and much more potent than phosgene or Zyklon-B.

In 1988, the Iraqi Air Force attacked the ethnic-Kurdish city of Halabja with chemical bombs, including sarin, killing up to 5,000 civilians. Terrorists left leaking packages containing sarin on the Tokyo underground in 1995, killing 12 commuters and injuring more than 5,000. The Syrian Air Force used it in several attacks between 2013 and 2018—again with deadly consequences for civilians—during the Syrian civil war.

Several novichok nerve gases were developed in the Soviet Union and Russia from the 1970s. One was used in an assassination attempt on a former Russian spy and his daughter in the UK in 2018.

Tear gas

Halogens are a group of highly reactive elements not found in their pure forms in nature. Some are used to make two forms of tear gas—the synthetic organic halogens chloroacetophenone (C_8H_7ClO) and chlorobenzylidene malononitrile ($C_{10}H_5ClN_2$), also known as mace. Neither are gases, but fine liquids or solids, dispensed through sprays or grenades by law-enforcement agencies to control demonstrations and riots. Tear gas causes a temporary eye irritation and a burning sensation in the respiratory tract and is used almost daily somewhere in the world. ■

Classes of chemical weapons

Chemical weapons are usually deployed in the form of gas (like air, being neither solid nor liquid) or an aerosol (very fine, liquid particles that spread through air), but some can be dispersed as liquids or dusts. They are classified into several groups.

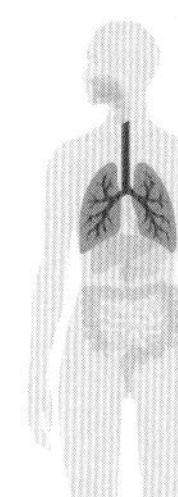

Choking agent
Absorbed by: lungs
Primarily attacks: lung tissue
Effects: pulmonary edema—excess fluid floods lungs and then "drowns" victim
Examples: chlorine, phosgene
Toxicity: high mortality

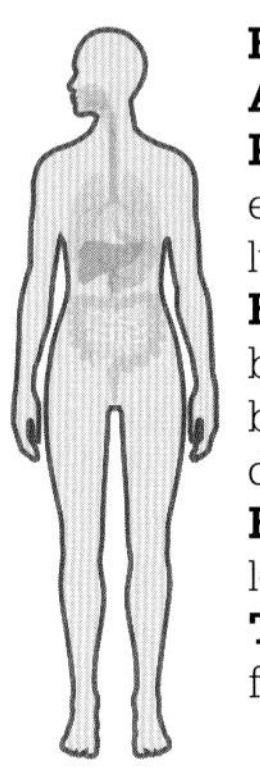

Blister agent
Absorbed by: lungs, skin
Primarily attacks: skin, eyes, mucous membranes, lungs
Effects: burns and blisters that may result in blindness or respiratory damage
Examples: mustard gas, lewisite
Toxicity: unlikely to be fatal without high exposure

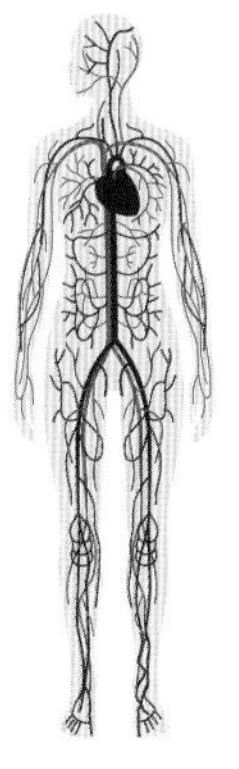

Asphyxiant/blood agent
Absorbed by: lungs
Primarily attacks: all vital organs
Effects: interferes with the ability of cells, often blood cells, to absorb oxygen so that the body suffocates, damaging vital organs
Examples: hydrogen cyanide, as in Zyklon-B
Toxicity: rapidly fatal

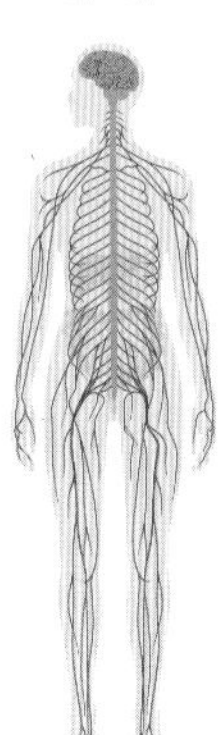

Nerve agent
Absorbed by: lungs, skin
Primarily attacks: nervous system
Effects: hyperstimulation of muscles, glands, and nerves, causing seizures, paralysis, respiratory failure
Examples: sarin, novichok
Toxicity: high chance of death

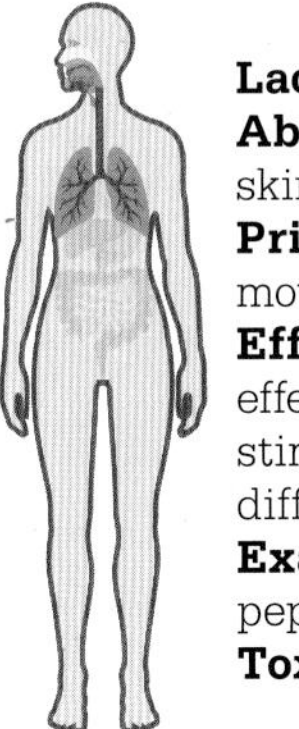

Lachrymatory agent
Absorbed by: lungs, skin, eyes
Primarily attacks: eyes, mouth, throat, lungs, skin
Effects: temporary effects of blindness, stinging eyes, breathing difficulties
Examples: tear gas, pepper spray
Toxicity: very rarely fatal

THEIR ATOMS HAVE IDENTICAL OUTSIDES BUT DIFFERENT INSIDES

ISOTOPES

IN CONTEXT

KEY FIGURE
Frederick Soddy (1877–1956)

BEFORE
1896 French physicist Henri Becquerel makes the first observations of naturally occurring radioactivity.

1899 Ernest Rutherford discovers that radioactivity takes at least two forms: alpha rays and beta rays.

AFTER
1919 Rutherford bombards nitrogen gas with alpha particles and obtains protons and atoms of an isotope of oxygen. It is the first artificially induced nuclear reaction.

1931 American chemist Harold Urey discovers deuterium, an isotope of hydrogen with an atomic mass of 2, which is later shown to be one proton and one neutron.

All elements have **isotopes**.

Isotopes of an element **cannot be distinguished** chemically.

Isotopes of an element have the same number of **protons** but different numbers of **neutrons**.

Isotopes can be stable or unstable (radioactive).

The discovery, at the beginning of the 20th century, that radioactivity involved the decay of one radioactive element into another was a huge step forward in our understanding. However, this breakthrough from British physicist Ernest Rutherford and British chemist Frederick Soddy raised new problems, too. Rutherford, Soddy, and others, such as German chemist Otto Hahn and Austrian physicist Lise Meitner, recorded nearly 40 new elements in the first two decades of the century, forming the links in three decay chains: the radium series that began with uranium, the thorium series, and the actinium series. These new elements often existed in tiny traces too minute to measure and could be identified only by their differing half-lives (the time it takes for half of the sample to decay).

Where in the table?

Deciding where these new elements fit into the periodic table was a challenge. There were only 11 places on the periodic table between uranium and lead into which nearly 40 new elements had to be squeezed. These new radioelements were given names such as radiothorium; radium A, B, C, D, E, and F; and uranium X, each with a different half-life. Chemists trying to separate radiothorium from thorium failed to find a chemical technique to accomplish

See also: The periodic table 130–137 ▪ The electron 164–165 ▪ Radioactivity 176–181 ▪ Improved atomic models 216–221

the task. Similarly, mesothorium was not shown to be chemically distinguishable from radium.

In 1910, Soddy observed that it was impossible to separate these new elements chemically because many of them, despite their slightly different atomic masses, were, in fact, the same element. Radium D and thorium C, for instance, were actually two different forms of lead and behaved the same as lead chemically; therefore, they belonged in the same position in the periodic table as lead. British doctor Margaret Todd suggested that these similar elements be denoted by the Greek for "same place"—*iso-topos*—and Soddy agreed. What had once been a defining characteristic of an element—its atomic mass—would now be seen as a variable quantity. British physicist James Chadwick later discovered the neutron, in 1932, which was responsible for these varying atomic masses.

Fajans's and Soddy's law

In 1913, Soddy set out the rules of transmutation—at the same time that Polish-American chemist Kazimierz Fajans and British chemist Alexander Russell also discovered them, all three working independently. When an atom emits an alpha particle, it moves back two places in the periodic table (so uranium-238 becomes thorium); when an atom emits a beta particle, it goes forward one place (carbon-14 becomes nitrogen). These rules, also known as the radioactive displacement law, determined the progression of the decay chains to their end points as stable lead.

One of the predictions of this law was that lead resulting from the decay of uranium would have a different atomic weight than naturally occurring lead. Czech-Austrian chemist Otto Hönigschmid was asked by Fajans and Soddy to carry out the work to prove this. He then recruited Polish chemist Stefanie Horovitz for the task of obtaining an uncontaminated sample of lead chloride ($PbCl_2$) from uranium-rich mineral pitchblende. Analysis of her sample proved that lead resulting from radioactive decay had a lower atomic weight than that of typical lead. It was the first physical proof that isotopes exist. ▪

Stefanie Horovitz carried out painstaking work that helped confirm the existence of isotopes, separating, purifying, and accurately measuring lead ready for analysis.

The discovery [of the neutron] is of the greatest interest and importance.
Ernest Rutherford

The neutron

The atomic number of an element—that is, the number of protons within its nucleus—defines it. An atom with six protons, for example, is always carbon. The discovery of isotopes, however, showed that an element's atomic mass could vary. It seemed that there must be something else besides protons in the nucleus.

Ernest Rutherford hinted that there could be a particle consisting of a paired proton and electron, which he called a neutron, that had a mass similar to that of the proton but had no charge. Meanwhile, French chemists Frédéric and Irène Joliot-Curie had been studying the particle radiation emitted by beryllium and believed the radiation took the form of high-energy photons. The experiments of James Chadwick, in 1932, established that the radiation emitted by beryllium was, in fact, a neutral particle similar in mass to the proton. Chadwick was awarded the 1935 Nobel Prize in Physics for proving the existence of neutrons.

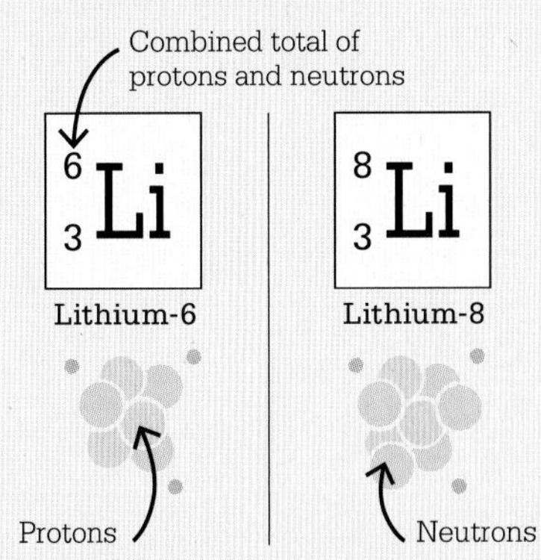

Lithium always has the atomic number 3, referring to how many protons it has. The number of neutrons can vary: the isotope lithium-6 has three neutrons, and there are five neutrons in lithium-8.

EACH LINE CORRESPONDS TO A CERTAIN ATOMIC WEIGHT

MASS SPECTROMETRY

IN CONTEXT

KEY FIGURE
Francis Aston (1877–1945)

BEFORE
1820 Danish chemist Hans Christian Ørsted discovers that magnetic and electric fields interact.

1898 German physicist Wilhelm Wien discovers that beams of positively charged particles can be deflected by magnetic and electric fields.

1913 Frederick Soddy discovers radioactive forms of the same element (isotopes), where the atoms are identical chemically but have a different atomic mass.

AFTER
1930s American physicist Ernest Lawrence uses mass spectrometry to identify new isotopes and elements.

1940 American physicist Alfred Nier separates uranium-238 from uranium-235 using mass spectroscopy.

At the beginning of the 19th century, the proposal by John Dalton that all atoms of an individual element have equal mass was a momentous step forward in the understanding of matter. In 1897, J. J. Thomson made a breakthrough discovery by finding evidence for electrons, negatively charged particles that appeared to be part of atoms. This opened the possibility of there being other parts within an atom. Thomson focused on developing his model for the atom and other investigations, but when Francis Aston joined Thomson's research group in 1910, the investigations into developing mass spectrometry—a technique that is now used in qualitative analysis to determine what compounds are present in a mixture—took off.

> Make more, more, and yet more measurements.
> **Francis Aston**

Spectrometers and isotopes

In 1912, Thomson and Aston built their first mass spectrometer, an instrument that created ions in a gas discharge tube and projected the ions through parallel magnetic and electric fields. The fields created parabolic beams, the shape of which depended on mass, charge, and velocity. At first, Aston expected the mass spectrometer to support Dalton's theory that all atoms of an element were the same mass, but when they used it to measure the mass of neon, they found two different parabolas: 22 mass units and 20 mass units. The mass of neon reported on the periodic table was 20.2 mass units.

Thomson hypothesized a new type of neon and began to examine this prospect. Aston was assigned the less likely options of there being a new compound of neon, or neon being composed of two particles of different masses, isotopes.

Aston built a balance that could weigh very small masses and found that every naturally occurring

See also: Dalton's atomic theory 80–81 ▪ Atomic weights 121 ▪ The periodic table 130–137 ▪ The electron 164–165 ▪ Radioactivity 176–181 ▪ Isotopes 200–201

sample of neon he measured gave the same mass: 20.2. World War I then called Aston away to serve on the Admiralty Board of Invention and Research, but he continued to think about his investigations. He returned to Thomson's lab after the war, and in 1919, built a new mass spectrometer that used a magnetic field to disperse ionized particles the way a prism disperses light. The position of the ions on a photographic plate depends on their mass: the smaller the mass of an ion or the more electric charge it has, the more it is deflected, so it can be focused by changing the strength of the magnetic field. This design removed the dependence on velocity and also allowed for the measurement of intensities.

Aston measured the intensities of the two beams of ions—20 mass units and 22 mass units. He found the ratio of their intensities, and so their ratio of abundances, was 10:1. This meant the average mass of neon, corrected for abundance, would be 20.2 mass units—as recorded on the periodic table.

Aston had found evidence for the existence of isotopes in stable elements. He went on to identify 212 other isotopes, essentially instigating the Atomic Age.

Fragmentation patterns

By 1935, chemists had identified the principal isotopes and their relative abundances for most of the elements. They speculated on other applications of mass spectrometry to analytical chemistry, but the mass spectrometer seemed like such a delicate instrument that most thought there was little possibility for analyzing large molecules in solvent. Such molecules would have to be put into the gas phase and would fragment into many pieces once hit by an electron beam.

In the 1950s, chemists such as William Stahl in the US managed to volatilize fruit flavor molecules and identify them by matching them to the known fragmentation patterns of individual molecules. This work opened the door to the application of mass spectrometry to analytical chemistry. ■

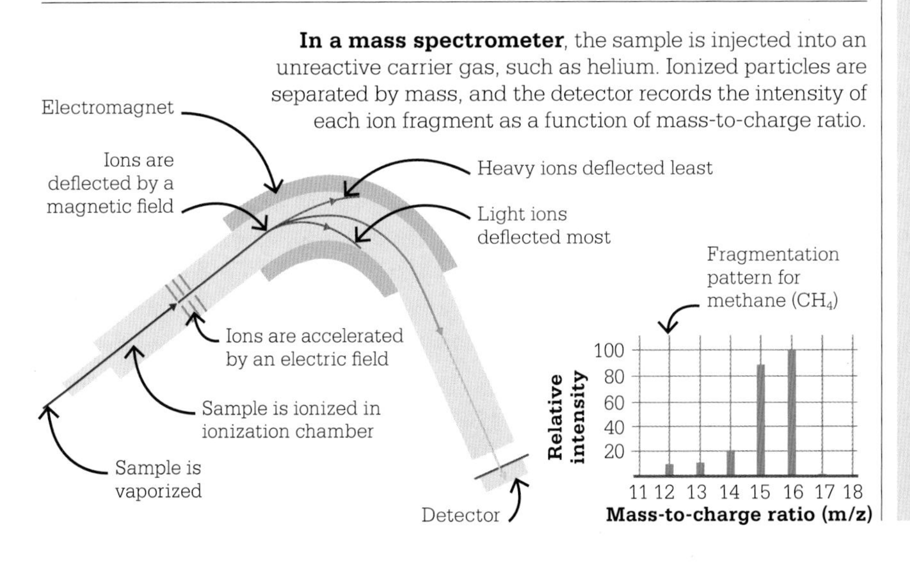

In a mass spectrometer, the sample is injected into an unreactive carrier gas, such as helium. Ionized particles are separated by mass, and the detector records the intensity of each ion fragment as a function of mass-to-charge ratio.

Francis William Aston

Born in Birmingham, UK, in 1877, Francis William Aston had an early interest in chemistry. He researched organic compounds in his family home and carried out mentored studies on tartaric acid. He also studied fermentation and worked for three years in a brewery. In 1903, Aston left the brewery for a research position at the University of Birmingham, and his interests turned to physics. Working with British physicist John Poynting, he constructed equipment for investigating the dark space between the cathode and anode in discharge tubes. In 1910, he moved to Cambridge to work with J. J. Thomson at the Cavendish Laboratory. His research was interrupted by World War I, but in 1922, he received the Nobel Prize in Chemistry in part for his discovery of isotopes in a large number of nonradioactive elements. He died in 1945.

Key works

1919 "A positive ray spectrograph"
1922 *Isotopes*
1933 *Mass-Spectra and Isotopes*

THE BIGGEST THING CHEMISTRY HAS DONE

POLYMERIZATION

IN CONTEXT

KEY FIGURE
Hermann Staudinger
(1881–1965)

BEFORE
1832 Jöns Jacob Berzelius uses the term "polymer," although not in the modern sense.

1861 Thomas Graham suggests that starches and cellulose are formed from small molecules clumping together in aggregates.

1862 Alexander Parkes creates Parkesine, later known as celluloid, from cellulose.

AFTER
1934 Wallace Carothers invents nylon.

1935 Canadian chemist Michael Perrin develops the first industrial-scale synthesis of polyethylene.

1938 American chemist Roy Plunkett invents Teflon, the first fluoropolymer.

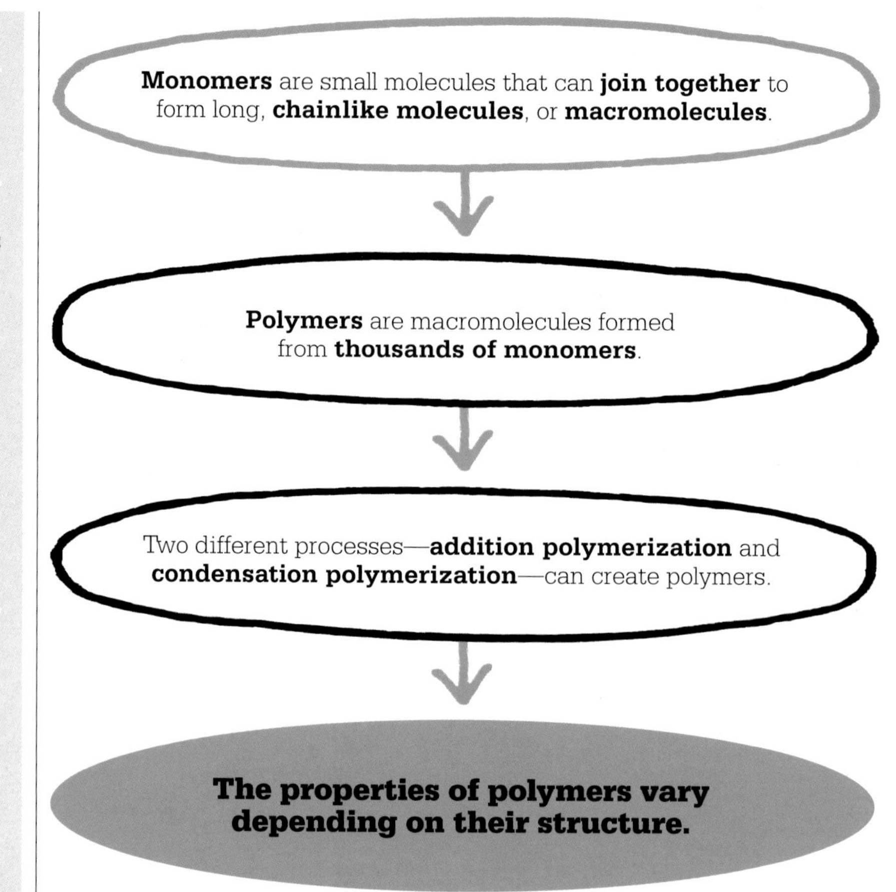

Monomers are small molecules that can **join together** to form long, **chainlike molecules**, or **macromolecules**.
Polymers are macromolecules formed from **thousands of monomers**.
Two different processes—**addition polymerization** and **condensation polymerization**—can create polymers.
The properties of polymers vary depending on their structure.

Plastics are a fact of modern life. We wear clothes made from plastic fibers, we buy food and other goods packaged in plastic, we pay for our groceries with plastic cards, and we carry it home in plastic bags. Today, due to a key discovery in 1920, we know that all plastics are polymers: long, chainlike molecules made of smaller repeating units, or monomers.

Polymers are often explained with the analogy of a chain of paperclips: the individual paperclips represent the monomers, while the connected chain of a large number of paperclips represents the polymer. Plastics are synthetic or semisynthetic polymers, but polymers are also common in the natural world. Examples include rubber, or latex, from plants such as the rubber tree (*Hevea brasiliensis*); cellulose, the main structural fiber of plants; DNA, in which the genetic material of all living things is encoded; and the proteins that DNA contains the instructions to build.

By the start of the 20th century, the word "polymer" had been in use for almost 70 years, but not in the sense in which we understand it now. When Jöns Jacob Berzelius introduced the term "polymer" in 1832, he used it to refer to organic compounds that had the same atoms in the same proportions but different molecular formulae. Today, we would describe these as being part of the same homologous series

> The macromolecular compounds include the most important substances occurring in nature.
> **Hermann Staudinger**
> **Nobel Prize lecture (1953)**

See also: Functional groups 100–105 ▪ Intermolecular forces 138–139 ▪ Synthetic plastic 183 ▪ Cracking crude oil 194–195 ▪ Nonstick polymers 232–233 ▪ Super-strong polymers 267

of compounds. For example, ethane (C_2H_6) and butane (C_4H_{10}) are both members of the alkane homologous series: they have the same general formula—the number of their hydrogen atoms is double that of their carbon atoms, plus two—but different molecular formulae.

Polymer production

By the mid-19th century, certain polymers had been discovered, synthesized, and marketed. In 1839, American engineer Charles Goodyear found that natural rubber could be hardened using sulfur, which meant it could be used in applications from machinery to bicycle tires. A year earlier, French chemist Anselme Payen had identified and isolated cellulose from wood, and in 1862, British chemist Alexander Parkes used cellulose nitrate to make a plastic material that he called Parkesine.

Parkes's creation is considered by many to mark the birth of the modern plastics industry. Although Parkes did not find commercial success with his invention—his material was expensive and not very resilient—it was improved on by others, renamed celluloid, and used for products ranging from camera film to billiard balls. Decades later, in 1907, Leo Baekeland invented Bakelite, which was the first mass-produced, fully synthetic plastic. However, despite the widespread use of polymer products, chemists still disagreed about the exact structure of these materials.

Bakelite, a hard resin created from phenol and formaldehyde, was the first completely synthetic plastic to be commercially produced. Bakelite products still have "retro" appeal today.

Most leading chemists at the time subscribed to the association theory proposed by Scottish chemist Thomas Graham back in 1861. He stated that substances such as rubber and cellulose were made up of clusters of small molecules held together by intermolecular forces. The idea that these substances might be made up of larger molecules was dismissed out of hand since most chemists thought such a thing was impossible: extremely large molecules could not possibly be stable.

Macromolecules

One person, German organic chemist Hermann Staudinger, sought to challenge the polymer orthodoxy. Highly regarded in the field of small-molecule chemistry, Staudinger had his curiosity piqued by isoprene, the main polymer component of rubber. In 1920, he published a paper proposing that natural substances such as »

Hermann Staudinger

Born in 1881 in Worms, Germany, Hermann Staudinger originally studied botany at the University of Halle. He took courses in chemistry to improve his understanding, and it became his primary interest. One of his earliest research discoveries was ketenes—highly reactive compounds that have a range of applications in organic synthesis.

Staudinger's groundbreaking work in macromolecular chemistry led to him establishing the first European research institute solely focusing on polymer research in 1940. He also set up the first journal exclusively focused on polymer chemistry. In 1953, he was awarded the Nobel Prize in Chemistry. He died in 1965.

Staudinger was an advocate for peace; in the 1930s, he publicly questioned the authority of the Nazis, which resulted in all his requests to travel abroad being rejected. In 1999, his work on polymer chemistry was designated an International Historic Chemistry Landmark.

Key works

1920 "On Polymerization"
1922 "On Isoprene and Rubber"

rubber have molecules far larger than was believed at the time, with molecular weights measuring in the millions. Staudinger outlined how these huge molecules are formed from polymerization reactions that link together a large number of small molecules. He would later refer to these enormous molecules as "macromolecules." His 1920 paper "On Polymerization" is considered the starting point for the field of macromolecular science.

Proponents of the association theory remained unconvinced by Staudinger's proposals. They argued that the properties of rubber that Staudinger attributed to his macromolecules could be explained in terms of weak intermolecular interactions. To disprove this, Staudinger needed experimental evidence that would directly contradict such claims.

Two other chemists, German Carl Harries and Austrian Rudolf Pummerer, had previously claimed that rubber was composed of aggregates of many small molecules of isoprene and that this gave the material its colloidal properties. These properties included forming a suspension rather than a solution

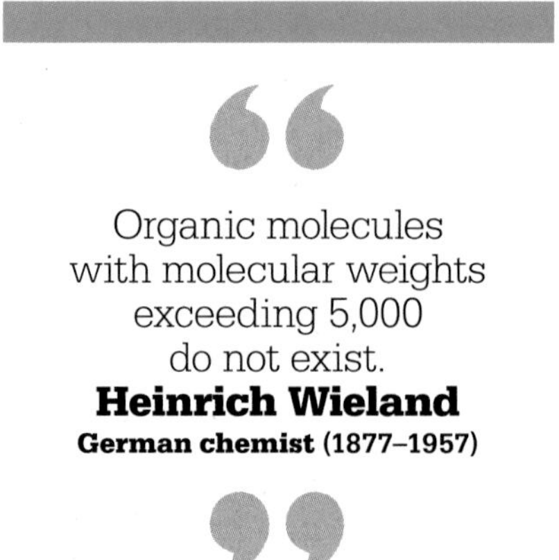

"Organic molecules with molecular weights exceeding 5,000 do not exist."

Heinrich Wieland
German chemist (1877–1957)

in solvents, since the particles in a colloid are large and do not dissolve. Harries and Pummerer thought that these properties were explained by the double carbon–carbon bonds having a "partial valence"—that is, a weak force holding the molecules together in aggregates.

To challenge this theory, Staudinger decided to hydrogenate (add hydrogen to) the double carbon–carbon bonds in rubber. This caused a hydrogen atom to attach to each carbon atom, leaving only a single bond between the carbon atoms. If the theory of Harries and Pummerer was correct, this should break up the aggregates in rubber and change its properties. Staudinger observed no such thing: the hydrogenated rubber behaved no differently from natural rubber.

Triumph of theory

Although Staudinger's experiments seemed to clarify that his theory of macromolecules was correct, his peers were not convinced. So Staudinger turned to trying to create macromolecules directly. Using small molecules such as styrene as a starting point, he and his colleagues made a range of different polymers. They identified a relationship between the molecular weights of these polymers and their viscosity (resistance to changing shape), which was further evidence for the macromolecule model.

For a time, there was still resistance to Staudinger's ideas. But as the evidence mounted and other scientists were able to more accurately determine the large masses of polymer molecules, the macromolecular theory slowly became accepted. Staudinger was vindicated years later, when he was awarded the 1953 Nobel

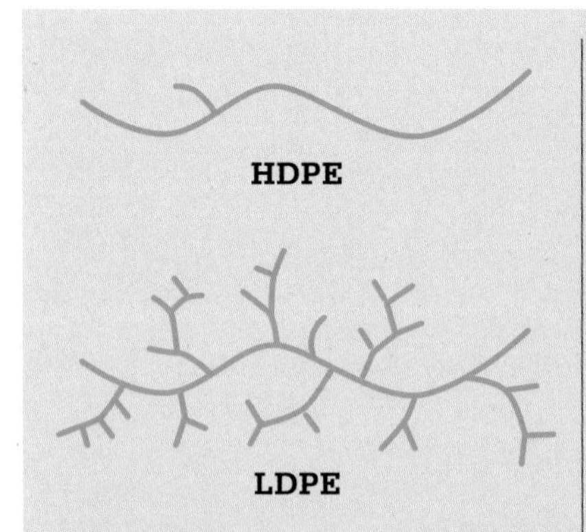

HDPE and LDPE are the same polymer but are differently structured. The branching structure of LDPE makes it less dense than HDPE.

Polymer properties

The considerable size of polymer molecules means that the same polymer can have different properties depending on its structure. An example of this is polyethylene, which occurs as high-density (HDPE) or low-density polyethylene (LDPE).

The structure of HDPE is like one long chain with very few branches. This enables the molecules to pack closely together, with strong intermolecular forces. HDPE is a hard, rigid plastic used for plastic bottles and piping. LDPE has a more branched structure, with more loosely packed molecules. The weaker intermolecular forces lead to a softer plastic, most commonly used for plastic bags.

Polymer behavior also varies with temperature. When heated, polymers eventually reach their "glass-transition temperature" (T_g), becoming soft and flexible. Polymers with a T_g above room temperature are hard and brittle, while those with a T_g below room temperature are flexible.

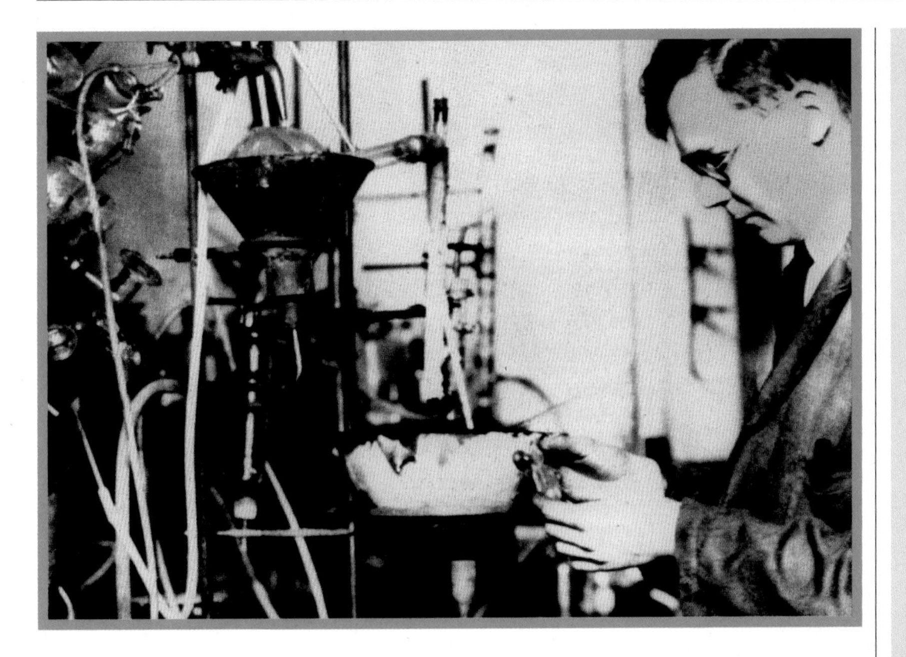

Prize in Chemistry in recognition of "his discoveries in the field of macromolecular chemistry."

Staudinger's work was largely a triumph of theory, in that he played little part in the development of any industrial polymerization processes. But the work inspired other chemists to investigate the practicalities of polymerization and opened up a new world of chemical possibilities. The large sizes of macromolecules meant that the potential number and variety of different molecules was huge. Even for the same molecule, properties could vary significantly with differences in structure and the conditions in which it was used.

Building superpolymers

One person inspired by Staudinger's theories was American organic chemist Wallace Carothers. From 1928, he was group leader at the research facility of the DuPont chemical company in Delaware. While Staudinger had focused on analyzing natural polymers, Carothers took a more practical approach to the investigation into macromolecules—pioneering new ways to produce polymers.

Wallace Carothers is best known as the inventor of nylon. He discovered how polymers are formed and synthesized new ones of huge molecular weight.

Carothers was the first chemist to define the two main categories of polymerization that we still refer to today: addition and condensation. Addition polymers are the more straightforward of the two. As the name suggests, they are formed by a simple linking together of many smaller molecules (monomers). The monomers must contain a double carbon–carbon bond, which becomes a single bond as the monomers join in a chain reaction.

Condensation polymers, in contrast, are formed from monomers that have two different functional groups. If the monomers are identical, they have one group at each end. These functional groups react with each other to form the polymer chain, and a small molecule is lost as a result »

Methods of polymerization

In addition polymerization, an initiation step triggers the reaction. Then, in propagation steps, monomers add on to the chain one after another before an eventual termination step. Termination occurs randomly and at different points for different chains, which means that the polymer chains vary in length.

The monomers used to make addition polymers can be identical, as in the case of polyethylene, which creates a homopolymer. But more than one type of monomer can also be used, which produces a copolymer. Regardless of the type of monomers, the polymer is the only product of addition polymerization.

During condensation polymerization, monomers join together in a reaction in which a small molecule is eliminated. This side product is often water—hence "condensation." In this type of polymerization, two different monomers are usually used.

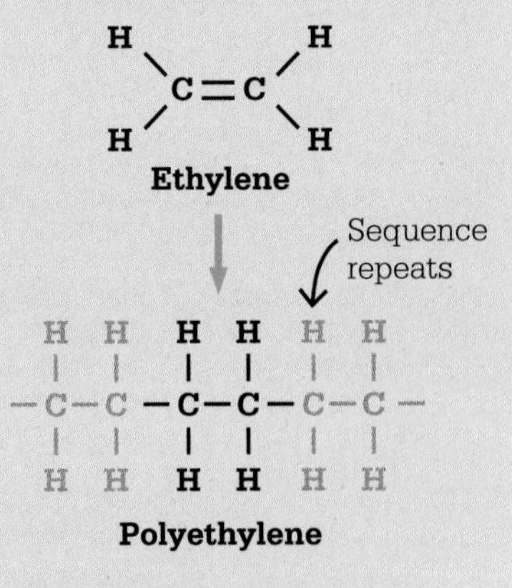

Polyethylene is an addition polymer built up from ethylene monomers, where double carbon bonds are replaced by single ones.

of the reaction. It was this type of polymerization that Carothers wanted to pursue.

Starting with compounds with a low molecular weight, Carothers aimed to use established organic chemistry reactions to join them together one by one to eventually produce macromolecules. He surmised that, with knowledge of the structure of the original molecules and the nature of the reactions being carried out, it would be possible to predict the structure of the macromolecule that would be obtained. His other goal was to make a molecule as large as possible, surpassing what was considered at the time to be the limit of molecular weight.

In March 1930, Carothers's team produced a polymer of chloroprene, which behaved like rubber and was later named neoprene. Then in April of the same year, the group succeeded in producing polyesters with molecular weights as high as 25,000. They also noticed that these "superpolymers" could be drawn out into threadlike fibers and that, when cooled, these fibers could be pulled to extend them even further. This increased their strength and elasticity.

Synthetic textiles

It seemed obvious that the new polyesters could have applications for textile fabrics. However, this was hard to achieve in practice. Various candidate polyester polymers were produced, but they all had faults that made them unsuitable for commercialization: they melted at too low a temperature or dissolved too easily.

After a lull in polymer research, Carothers's team changed tack and pursued polyamide fibers. In 1934, using a dicarboxylic acid and a diamine, which each contained six carbon atoms, they produced a strong elastic fiber that did not dissolve in most solvents and had a high melting point. In 1935, this "fiber 66" was chosen by DuPont for full-scale production. It would later be named nylon.

It took another three years for DuPont to devise ways to make the two reactants. Nylon finally went on sale in 1940 and was an instant success. In the first year, 64 million pairs of nylon stockings were sold. Nylon was also used in a wide range of applications during World War II, including for tents and parachutes.

We have not only a synthetic rubber but something theoretically more original—a synthetic silk. ... that will be enough for one lifetime.

Wallace Carothers (1931)

Sadly, Carothers did not live to see the success of the polymer that he and his team had produced. For years, he had struggled with periods of depression, and in April 1937, he took his own life. Had he still been alive at the time, it seems likely he would have shared

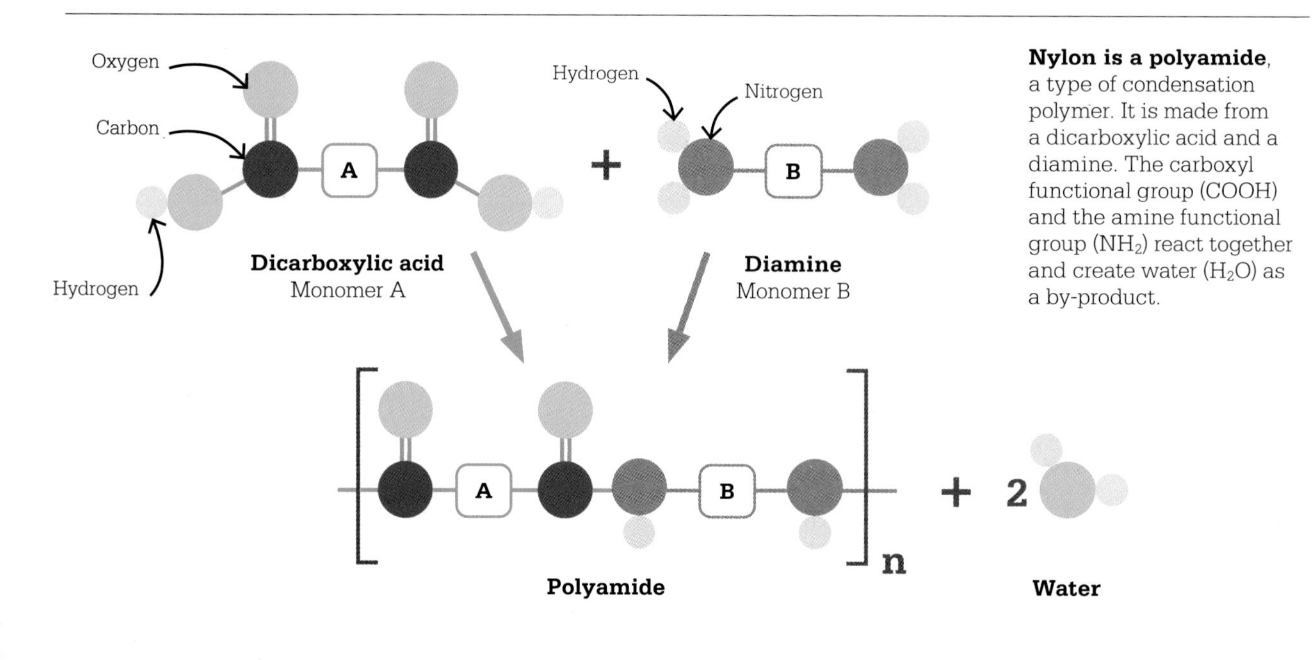

Nylon is a polyamide, a type of condensation polymer. It is made from a dicarboxylic acid and a diamine. The carboxyl functional group (COOH) and the amine functional group (NH_2) react together and create water (H_2O) as a by-product.

Nylon stockings went on sale to the public in May 1940. In the first four days, 4 million pairs were sold. Silk from Asia was unavailable during wartime, and nylon became symbolic of the era.

in Staudinger's 1953 Nobel Prize for his own contribution to our understanding of macromolecules.

Plastic pollution

In the decades following the work of pioneers such as Staudinger and Carothers, plastics became part of everyday life. In 2020, the world produced 404 million tons (367 million metric tons) of plastic. In many ways, these convenient materials have made the previously impossible possible—but since the 1960s, we have also become increasingly concerned about the impact that plastic is having on our planet.

The first problem is that the molecules required to make many plastics are derived from fossil fuels, and the process of extracting them generates various pollutants. If the world's governments are serious about eventually phasing out fossil fuels, we need to develop ways of making plastics that are not tied to them as a source material.

Plastics also have harmful effects when they are disposed of. Poor plastic recycling provision and the culture of single-use plastics mean that an ever-increasing mass of plastic waste is accumulating worldwide, estimated to reach a total of 13 billion tons (12 billion metric tons) by 2050. Plastic has not been around long enough for us to know whether it will ever decompose fully. Plastic waste causes the deaths of millions of marine animals every year as a result of them becoming entangled in or eating the plastic.

Additionally, as plastics break down, they disintegrate into tiny particles known as microplastics. Scientists have found microplastics almost everywhere they have looked for them, in the farthest reaches of our planet. The impact of microplastics is an emerging field of study, but there are concerns about adverse effects on human health, as well as the environment.

Plastic could take up to 1,000 years to decompose, and less than 20 percent of plastic is recycled. This usually involves melting it down for reuse, which can only be done a limited number of times.

Sustainable plastics

The challenge for chemists is to find ways to mitigate the harm caused by plastics. Some progress is already being made. When plastic is recycled mechanically, it is melted down to be reused as lower-grade plastic. But methods for "unpicking" plastics may make it possible to break polymers down into monomers that can be used to make different types of plastic. Ongoing research aims to create polymers that are easier to deconstruct in this way.

If we are to have any chance of dealing with the problem of plastic pollution, both producers and consumers must be committed to implementing solutions. Bioplastics made from plant materials and fully biodegradable plastics already exist but represent just a tiny fraction of total plastic production. ■

DEVELOPMENT OF MOTOR FUELS IS ESSENTIAL

LEADED GASOLINE

IN CONTEXT

KEY FIGURES
Thomas Midgley (1889–1944), **Clair Cameron Patterson** (1922–1995)

BEFORE
1st century BCE The Roman civil engineer Vitruvius warns of the dangers of poisoning from lead water pipes.

1853 German chemist Carl Jacob Löwig synthesizes tetraethyl lead.

1885 German engineers Karl Benz and Gottlieb Daimler each invent motorized, gasoline-fueled vehicles.

AFTER
1979 American pediatrician Herbert Needleman reports a link between high lead levels in children and their poor academic performance and behavioral problems.

2000 The sale of leaded gasoline is banned in the UK, having been gradually phased out since the 1980s.

After World War I, demand for cars soared: around 15 million were registered in the US by 1924. Their internal combustion engines worked by igniting gasoline mixed with air to produce energy, as well as carbon dioxide and water exhaust by-products. However, a problem known as "knocking"—when some of the fuel mix ignites prematurely—created noise, damaged the engine, and reduced its efficiency.

In 1921, working for the General Motors Corporation (GM), Thomas Midgley found a solution by adding tetraethyl lead (TEL) to the gasoline. When TEL combusts, it produces carbon dioxide, water, and lead. The lead particles prevent knocking

Knocking in a gasoline engine

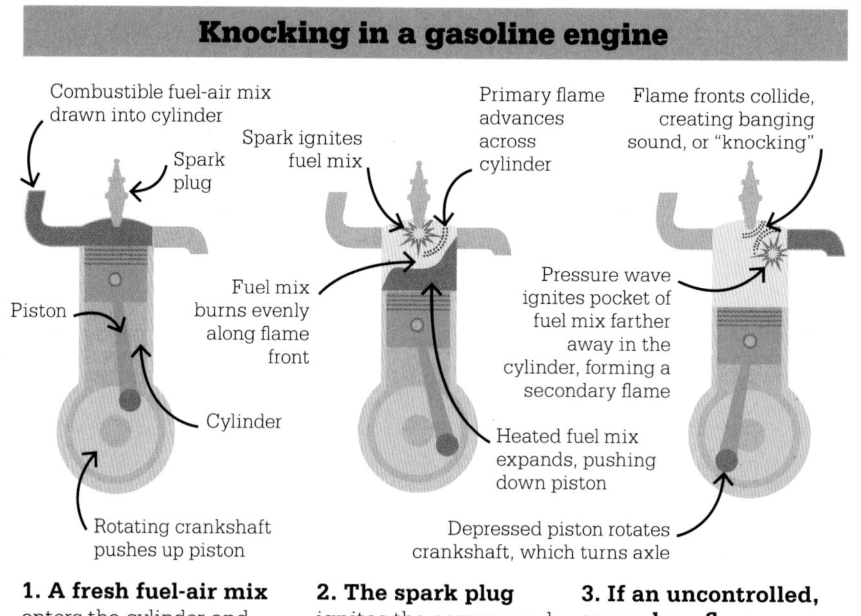
Combustible fuel-air mix drawn into cylinder
Spark plug
Piston
Cylinder
Rotating crankshaft pushes up piston
Spark ignites fuel mix
Primary flame advances across cylinder
Fuel mix burns evenly along flame front
Heated fuel mix expands, pushing down piston
Flame fronts collide, creating banging sound, or "knocking"
Pressure wave ignites pocket of fuel mix farther away in the cylinder, forming a secondary flame
Depressed piston rotates crankshaft, which turns axle

1. A fresh fuel-air mix enters the cylinder and is compressed by the rising piston.

2. The spark plug ignites the compressed fuel mix to create a primary flame.

3. If an uncontrolled, secondary flame occurs, the primary and secondary flame fronts collide.

See also: The new chemical medicine 44–45 ▪ Gases 46 ▪ Cracking crude oil 194–195 ▪ The hole in the ozone layer 272–273

Where there is lead, some case of lead poisoning sooner or later develops …
Alice Hamilton
American expert in industrial medicine (1869–1970)

by raising the temperature and pressure at which premature ignition occurs. They also react with oxygen to form lead oxide.

A known poison

Although the toxicity of TEL was well known, GM and the Standard Oil Company of New Jersey created the Ethyl Gasoline Corporation in 1921 to manufacture and market leaded gasoline. Five workers died and 35 others suffered severe lead poisoning at one production plant, while at another site, workers suffered hallucinations. Despite this, and US public health officials warning of the dangers of lead in the atmosphere, the first tankful was sold in 1923. Better working practices made leaded gasoline safer to produce, but campaigners remained convinced of the dangers of lead emitted from car exhausts.

In 1964, American geochemist Clair Cameron Patterson analyzed ice cores from Greenland and found there had been a 200-fold increase in lead deposits since the 18th century; most of the increase had occurred in the previous three decades. Antarctic ice cores yielded similar results in 1965. Patterson was sure that most of the lead came from leaded gasoline—and that such high levels were dangerous. In 1966, he submitted evidence to a US Congress subcommittee on Air and Water Pollution. The 1970 Clean Air Act directed the newly formed US Environmental Protection Agency to regulate TEL in gasoline. It was capped at 0.1 g/gallon in 1986 and banned completely in 1996.

In 1997, the US Center for Disease Control and Prevention found that average lead levels in the blood of children and adults had fallen by more than 80 percent in the prior 20 years. By 2002, only 82 countries allowed the sale of leaded gasoline. A UN Environment Programme (UNEP) report in 2011 estimated the global phase-out had avoided more than 1.2 million premature deaths. In 2021, Algeria became the last nation in the world to ban the toxic fuel. ■

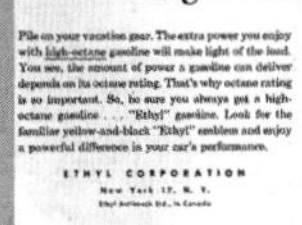

Despite being aware of its toxicity, the Ethyl Gasoline Corporation promoted the benefits of leaded gasoline to American families in the 1950s.

Thomas Midgley

The son of an inventor, Midgley was born in Pennsylvania in 1889. He graduated from Cornell University with a degree in mechanical engineering and began working for the General Motors Corporation in 1919. Having pioneered leaded gasoline by discovering the additive TEL in 1921, he later became seriously ill with lead poisoning. However, this did not stop him from promoting it to US regulators and the public.

In 1928, Midgley led a research team that developed dichlorofluoromethane—a chlorofluorocarbon (CFC) known as Freon 12—as a nonflammable alternative to the flammable refrigerants in use at the time. The damaging effects of CFCs on the ozone layer would not be discovered until the 1980s. Midgley, who received several prestigious rewards for his discovery, contracted polio in 1940 and was left seriously disabled. He died in 1944.

Key works

1926 "Prevention of fuel knock"
1930 "Organic fluorides as refrigerants"

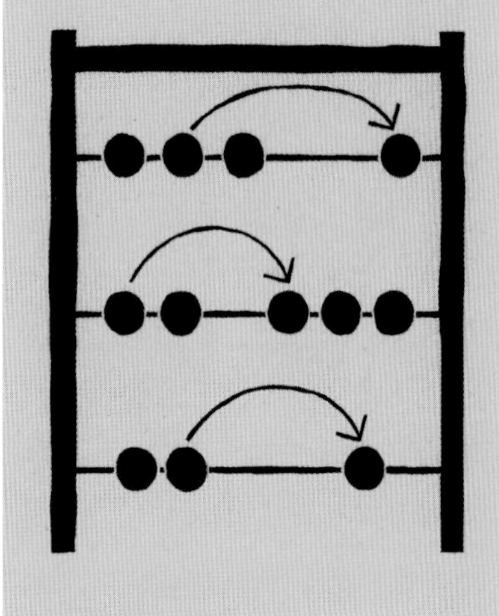

CURLY ARROWS ARE A CONVENIENT ELECTRON ACCOUNTING TOOL

DEPICTING REACTION MECHANISMS

IN CONTEXT

KEY FIGURE
Robert Robinson
(1886–1975)

BEFORE
1857–1858 August Kekulé's theory of chemical structure helps determine the bond order in molecules.

1885 German chemist Adolf von Baeyer contributes the strain theory of triple bonds and small carbon rings to theoretical organic chemistry.

AFTER
1928 Linus Pauling proposes resonance as a model for the seeming oscillation of bonds from single to multiple bonds.

1934 British chemist Christopher Ingold presents reactions as a sequence of steps, or reaction mechanisms.

1940 American physical chemist Louis Hammett identifies a new field of study: physical organic chemistry.

Ever since German chemist Friedrich Wöhler synthesized urea in 1828, disproving the notion that organic materials can only be made in a living body, the goal of many organic chemists has been to find methods to synthesize materials for pharmaceuticals, plastics, fuels, and research. Chemical bonds are formed or broken by the movement of electrons, so knowing where the electrons are or where they are needed is important for understanding—and predicting—chemical reactions.

> Hardly a day goes by when a modern organic chemist does not use curly arrows to explain a reaction mechanism or to plan a synthetic route.
>
> **Thomas M. Zydowsky**
> ***Chemistry Explained* (2021)**

Reaction mechanisms—theoretical, schematic diagrams of how a reaction might proceed—are critical to successful synthesis. However, these mechanisms lacked clarity until 1922, when British chemist Robert Robinson devised the curly arrow symbol.

Curly arrows

In 1897, British physicist Joseph John Thomson established the electron as a part of the atom, and chemists realized rearrangements of these electrons were most likely integral to the mechanisms for producing a new compound from the reaction of other chemicals. However, the description of the movements of electrons was difficult to understand without visualization. In response, in a 1922 paper co-authored with British chemist William Ogilvy Kermack, Robinson introduced curly arrows to depict how electrons—represented by dots—might move in a reaction of organic compounds and the resulting molecular structures.

See also: The synthesis of urea 88–89 ▪ Structural formulae 126–127 ▪ Benzene 128–129 ▪ Intermolecular forces 138–139 ▪ Why reactions happen 144–147 ▪ Improved atomic models 216–221 ▪ Chemical bonding 238–245

How curly arrows work

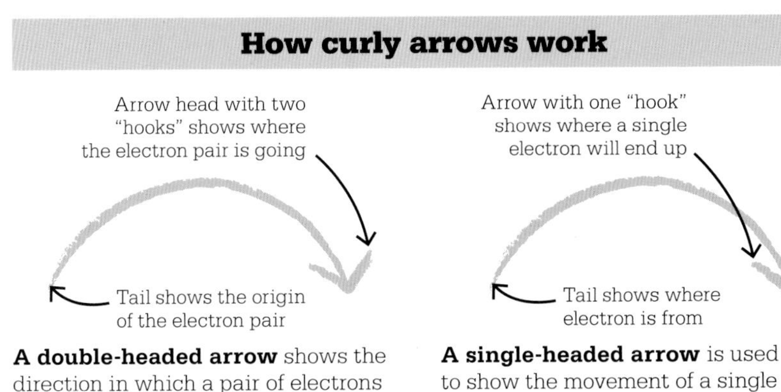

Arrow head with two "hooks" shows where the electron pair is going
Tail shows the origin of the electron pair
Arrow with one "hook" shows where a single electron will end up
Tail shows where electron is from

A double-headed arrow shows the direction in which a pair of electrons move during a chemical reaction.

A single-headed arrow is used to show the movement of a single electron during a reaction.

This method of visualizing the movement of electrons helped chemists understand the mechanisms of reactions and design potentially useful new reaction pathways based on the previous reactivity of compounds.

Initially, however, the curly arrows confused rather than enlightened chemists. Part of the problem was that the electronic theory of organic chemistry—the role of electrons in organic bonding—was very new and still developing. In addition, Robinson's early papers did not always make it clear whether the arrow showed the movement of one or two electrons.

Pioneers of the new field of physical organic chemistry, such as American chemists Linus Pauling and Gilbert N. Lewis, resolved many of the conceptual problems in the use of curly arrows by showing that all chemical bonds consist of two electrons. From 1924, Robinson refined his method to take this into account, using one arrow to describe the movement of two electrons, as shown below.

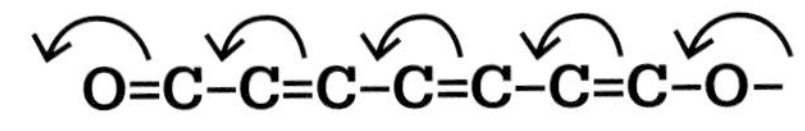

While it is now understood that chemical reactions need not occur as separate steps, curly arrows are still used and taught to chemistry students as a visual device to help them understand the driving forces in chemical synthesis. The validity of curly arrows was disputed by the field of quantum mechanics, which depicts molecular structures in terms of wave theory. However, in 2018, Australian chemists Timothy Schmidt and Terry Frankcombe at the University of New South Wales connected the two by means of a series of inspired quantum chemical calculations. ■

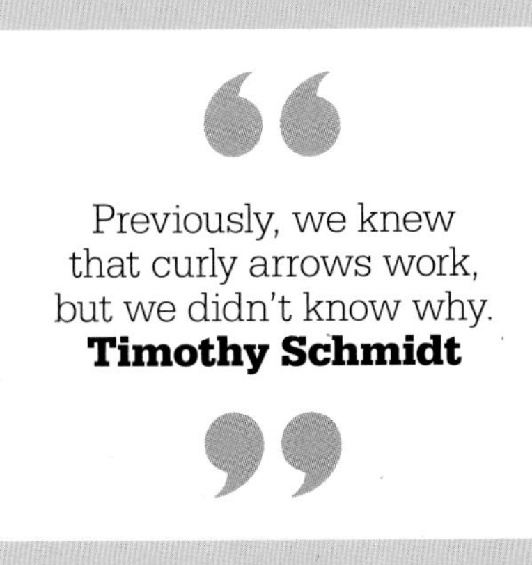

Previously, we knew that curly arrows work, but we didn't know why.
Timothy Schmidt

Curly arrows represent pairs of electrons that move from one bond to another in a reaction mechanism using ethylene (C_2H_4) and hydrobromic acid (HBr) to make bromoethane (C_2H_5Br).

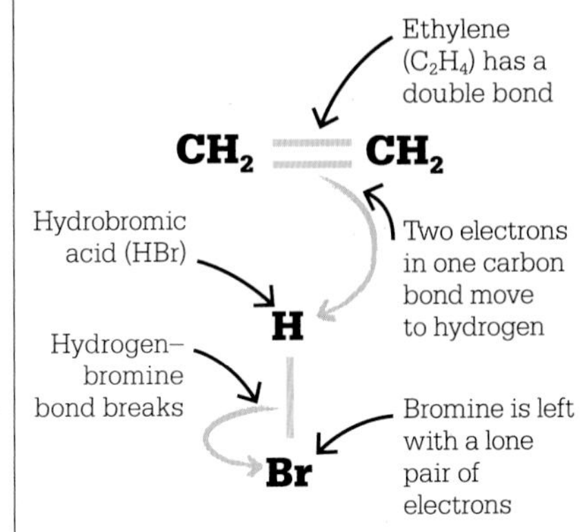

Hydrogen–bromine bond breaks

1. A pair of electrons in the double carbon bond shift to hydrogen. This breaks the bond between hydrogen and bromine and creates a positive charge on the nonreacting carbon.

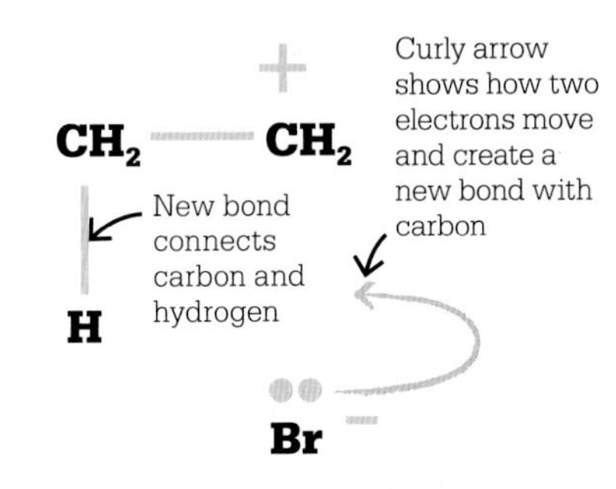

New bond connects carbon and hydrogen

2. The two remaining electrons on the bromide ion—now represented by dots—are attracted to the positively charged carbon. This creates a new bond between carbon and bromine.

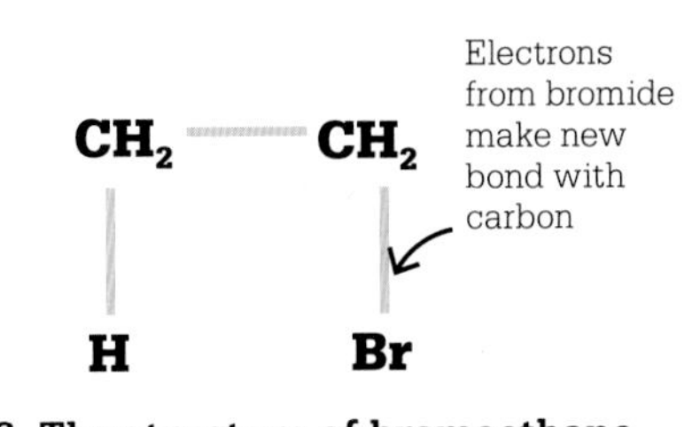

Electrons from bromide make new bond with carbon

3. The structure of bromoethane (C_2H_5Br)—also known as ethyl bromide—the final product of the reaction mechanism.

SHAPES AND VARIATIONS IN THE STRUCTURE OF SPACE

IMPROVED ATOMIC MODELS

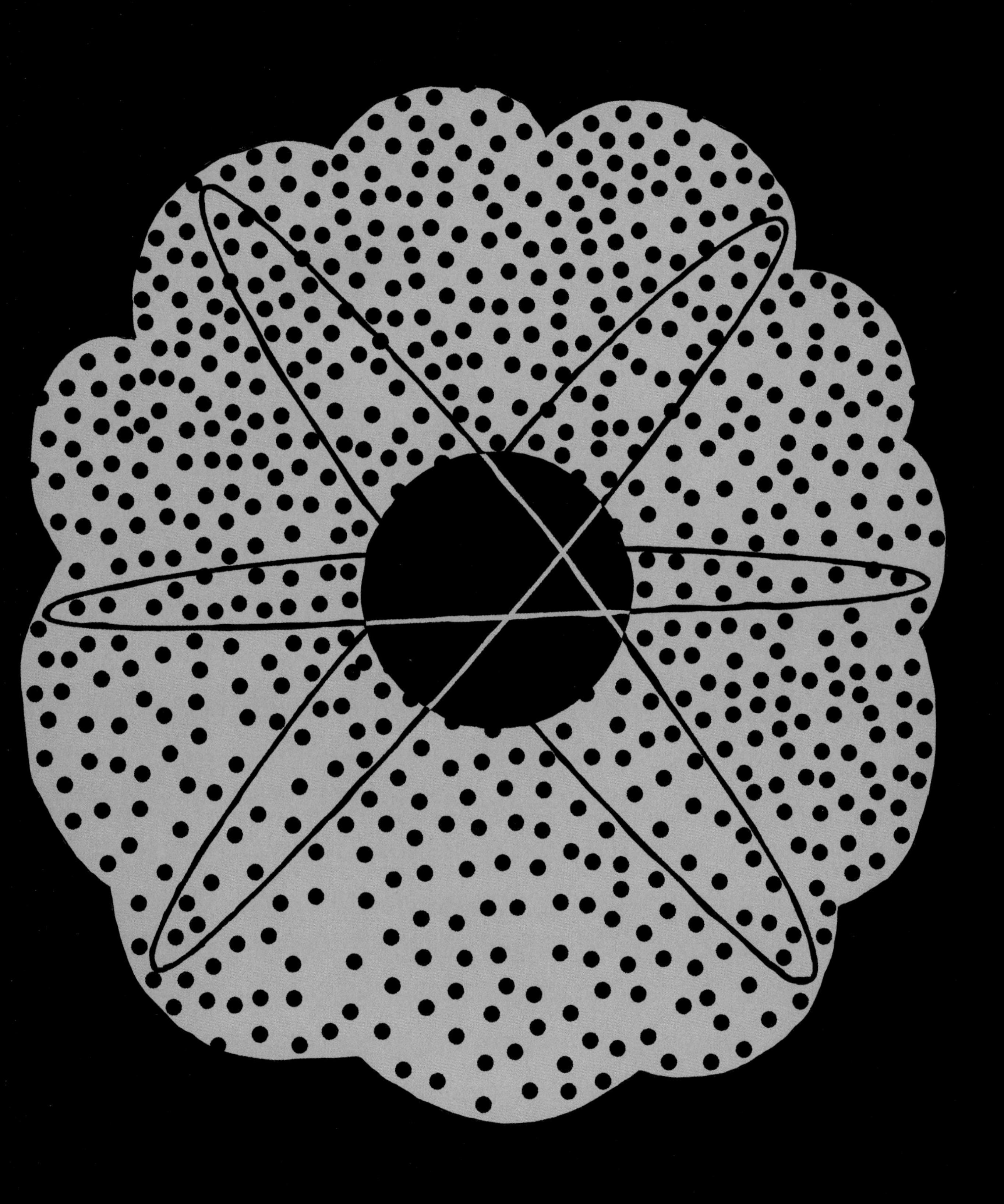

IN CONTEXT

KEY FIGURE
Erwin Schrödinger
(1887–1961)

BEFORE
1805 John Dalton's atomic theory of matter unites chemical processes with physical reality, enabling him to calculate atomic weights.

1896 Nuclear radiation is discovered by French physicist Henri Becquerel, forcing scientists to change their ideas about atomic structure.

1900 Max Planck accurately models the distribution of radiation from a black body by assuming that light has particulate properties.

AFTER
1927 Niels Bohr, Werner Heisenberg, and others develop the Copenhagen interpretation of wave mechanics, which models Schrödinger's waves as probability waves.

1932 British physicist James Chadwick discovers the neutron, a particle in the nucleus that has mass but no charge.

1938 German chemists Otto Hahn, Fritz Strassman, and Lise Meitner split the atomic nucleus.

The discovery at the turn of the 20th century that the atom could be split into smaller fragments revolutionized thinking about atomic structure. The ensuing succession of atomic models would result in Erwin Schrödinger creating his quantum model in 1926; this model is still accepted today.

In 1897, when electrons were first identified by Joseph John Thomson, the question became where to locate them within the structure of the atom. In 1904, Thomson proposed that the electrons—which have a negative charge—might be embedded in the nucleus, which has a positive charge, an idea called the plum pudding model. However, that idea was soon set aside.

In 1911, New Zealand–born British physicist and chemist Ernest Rutherford suggested that electrons existed outside a densely packed nucleus and that most of the volume of the atom was empty space. Rutherford based this idea on the results of experiments carried out by German physicist Hans Geiger and British physicist Ernest Marsden in 1909 at the University of Manchester. In these experiments, they projected radioactive alpha particles at an ultra-thin gold foil and found that most of these heavy particles went straight through or were deflected at low angles—but some bounced back toward the source. On the basis of this rebound phenomenon, Rutherford proposed that the atom contained a small, dense, positively charged nucleus around which the negative electrons orbited.

Rutherford's interpretation was not well received until 1913, when a young Danish student, Niels Bohr, applied an obscure mathematical

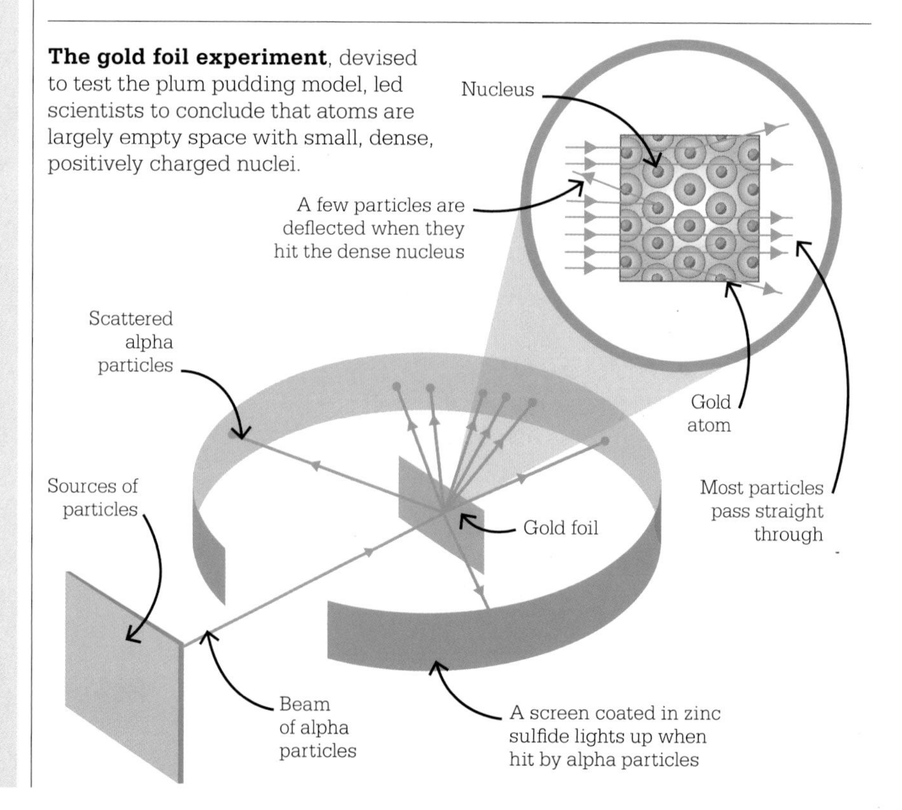
Nucleus
A few particles are deflected when they hit the dense nucleus
Scattered alpha particles
Gold atom
Sources of particles
Gold foil
Most particles pass straight through
Beam of alpha particles
A screen coated in zinc sulfide lights up when hit by alpha particles

The gold foil experiment, devised to test the plum pudding model, led scientists to conclude that atoms are largely empty space with small, dense, positively charged nuclei.

See also: The atomic universe 28–29 ▪ Dalton's atomic theory 80–81 ▪ Coordination chemistry 152–153 ▪ The electron 164–165 ▪ Chemical bonding 238–245

formula to the quantum concept of German physicist Max Planck, and the subatomic pieces fell into place.

Fixed orbits

The obscure formula was the work of Johann Balmer, a German lecturer at the University of Basel in Switzerland. In 1885, Balmer had devised a formula that predicted the positions of four visible lines in the emission spectrum of the hydrogen atom. The lines appeared at specific intervals rather than continuously, but the reason why the mathematical model worked remained unexplained.

In 1900, Planck came up with a model to explain the distribution of light from a heated black body. To do so, however, he had to assume that light energy came in packets, now called quanta or photons. Up to this point, light had generally been thought of as continuous; now, he was proposing that light could be modeled as a wave under some circumstances, but in others it was best modeled as a particle. Sound also behaves as a particle or a wave. When sound is of the right frequency and intensity, it can act like a bullet shattering glass. Even so, sound is behaving as a wave when it bends around corners and travels from room to room.

We even believe that we have an intimate knowledge of the constituents of the individual atoms.
Niels Bohr
Nobel Lecture (1922)

In this example of wave-particle duality, electrons are fired through a barrier with two slits in it. Over time, interference patterns of light and dark bands are produced, just as would happen with light waves. This shows that particles have wavelike properties and exhibit wavelike behavior.

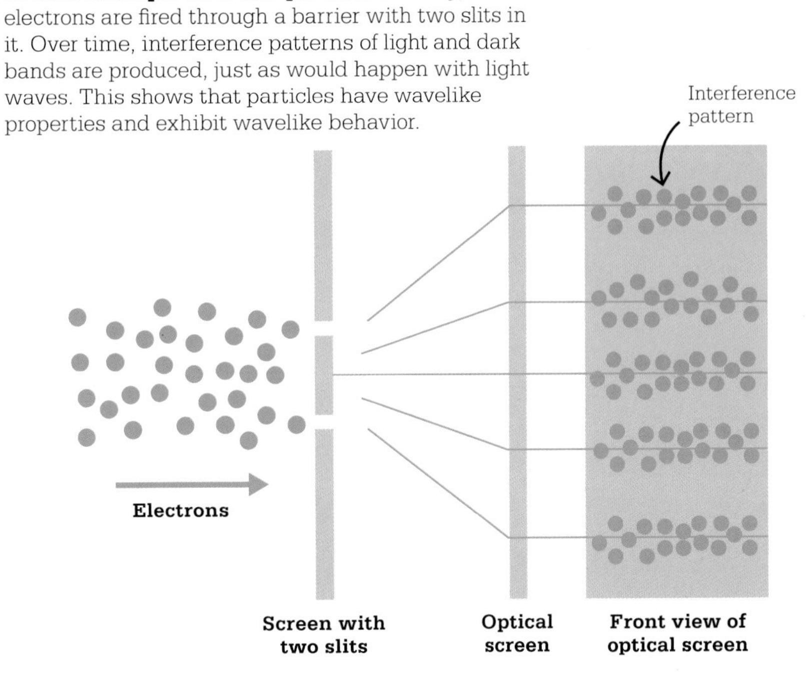

In the hands of Niels Bohr, the quantum idea showed how an atomic model where electrons could only orbit at certain fixed distances from the nucleus could agree with Balmer's equation for predicting the position of hydrogen's spectral emission lines. By absorbing a quantum of light of just the right energy, an electron could "jump" from a lower to a higher energy orbit; or, when falling from a higher to a lower orbit, an electron would release the same amount of light energy. These changes in energy fit the pattern predicted by Balmer's formula.

The Bohr model explained many measurements, but there were several problems, perhaps the most troublesome being that the electrons in the Bohr model were moving, and a moving electron should lose energy and spiral into the positive nucleus. Moreover, Bohr's model could not predict spectral lines for any neutral atom (those with an equal number of electrons and protons) other than hydrogen, nor the intensities for hydrogen lines, nor could it be used to make predictions for any molecule—including the simplest, the hydrogen molecule, H_2.

Wave-particle duality

In 1923, French physicist Louis de Broglie proposed that matter behaves as particles—and also as waves. Wave-particle duality for light may have been difficult to accept, but to apply the same situation to matter was a step too far for many scientists. But not for »

Schrödinger. In a rapid series of four papers in 1926, collectively called "Quantization as an Eigenvalue Problem," he laid out a wave theory of quantum mechanics. This is a system of mechanics that describes the physical behavior of atomic-scale particles like electrons, atoms, and molecules, just as classical mechanics describes the behavior of macroscopic objects like soccer balls, automobiles, and planets. The difference is that the properties of quantum-scale particles can only be inferred and not directly measured.

In the first paper, Schrödinger presented what has become known as the Schrödinger equation to describe the behavior of a quantum mechanical system:

$$i\hbar \frac{\partial}{\partial t}\Psi = \hat{H}\Psi$$

Essentially, the Schrödinger equation describes the behavior of wave functions (Ψ). When applied to the form of the wave function that best describes a system, it will yield measurable energies for that system. Schrödinger used his equation to analyze a hydrogen-like system and reproduced the energy levels for hydrogen. The year before, German physicists Werner Heisenberg, Max Born, and Pascual Jordan had developed a system for describing the electronic structure of an atom that was based on fairly complicated matrix mathematics, but Schrödinger's wave theory was more intuitive and easier to present visually.

Probability waves

Hardly had Schrödinger's concept been digested when Heisenberg, in 1927, propounded the Heisenberg uncertainty principle. This, loosely stated, is that the position and momentum of an electron cannot both be known at the same time.

This conclusion has to do with the size of the measuring instrument compared to the object being measured. For instance, laser guns or radar guns can measure the velocity of a car by sending light beams that bounce off the car and are reflected back to the instrument. However, when light is used to measure the velocity of an electron, the light knocks the electron off course. It would be like trying to measure the velocity of a car with a cannonball. To resolve this problem, Heisenberg essentially rejected the possibility of locating electrons in space and time as though they were macroscopic objects.

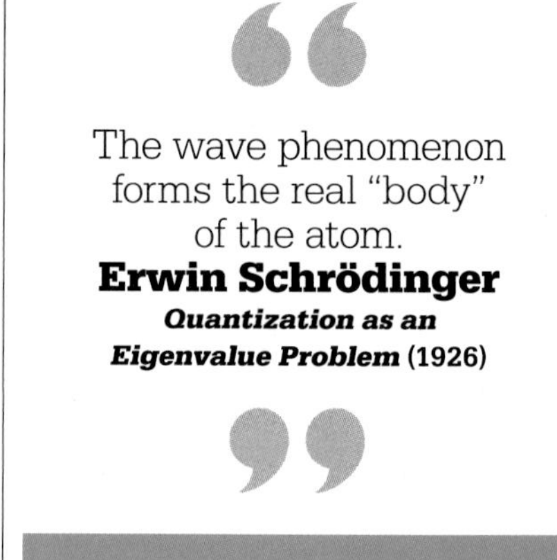

The wave phenomenon forms the real "body" of the atom.
Erwin Schrödinger
Quantization as an Eigenvalue Problem (1926)

Therefore, the question became: if the electron could not be located physically, what were the waves in Schrödinger's wave mechanics? In 1926, Max Born had offered an explanation: they were probability waves. The waves showed where the probability of finding an electron in a certain position was large, small, or nonexistent.

Erwin Schrödinger

Of Austrian and British heritage, Schrödinger was born in Vienna in 1887. He studied theoretical physics at the city's university, and he was also passionate about poetry and philosophy. After serving in World War I, he moved to Germany and then to the University of Zurich, Switzerland.

In 1927, he moved to Berlin, then a center for physics. He left in 1933, in protest of the Nazi regime, and took a post at the University of Oxford, UK. That year, he was jointly awarded the Nobel Prize in Physics with British theoretical physicist Paul Dirac. He returned to Austria, but had to flee the Nazis again in 1938. Friends provided safe passage to Dublin, Ireland, where he spent 17 years as Director of the School for Theoretical Physics at the Dublin Institute for Advanced Studies. He retired to Austria in 1956 and died in 1961.

Key works

1926 "Quantization as an Eigenvalue Problem"
1926 "An undulatory theory of the mechanics of atoms and molecules"

As it turned out, the Heisenberg uncertainty principle became an essential tool to explain and predict many quantum phenomena. The orbits of the electrons were now referred to as "orbitals" in order to reflect their nebulous nature. In contrast to the well-defined orbits of the Bohr model, the orbitals were visualized as electron "clouds," and it was posited that the probability for the electron existing in that area was highest where the cloud was at its most dense.

Yet Born's concept of probability waves did not find favor with Schrödinger. It was to ridicule the concept that he came up with the famous Schrödinger's cat thought experiment, which he explained to his close friend Albert Einstein. He imagined a cat sealed in a box that contains a vial of poison linked to a radioactive source. If the source decays and emits a particle of radiation, a mechanism will release a hammer that breaks the vial and releases the poison, killing the cat. There is an equal probability of the atom decaying or not decaying. The only way to know if the cat is alive or dead is to look inside the box. Schrödinger concluded that as long as the system is unobserved, the cat is alive and dead at the same time. Ironically, this analogy is now used to explain Born's probability waves, not to deride them.

The quantum mechanical model of the atom quickly became a powerful tool for explaining atomic phenomena. In 1926, German physicist Lucy Mensing was able to model diatomic molecules, such as hydrogen, using quantum mechanics, a feat that had been impossible with Bohr's atomic model. And in 1927, chemistry was brought into the process when the German physicist Walter Heitler was able to show how a covalent bond, formed when two atoms share a pair of electrons, would be an outcome of Schrödinger's wave equations. Today, virtually all chemistry students are taught quantum mechanics in terms of Schrödinger's wave equation. ■

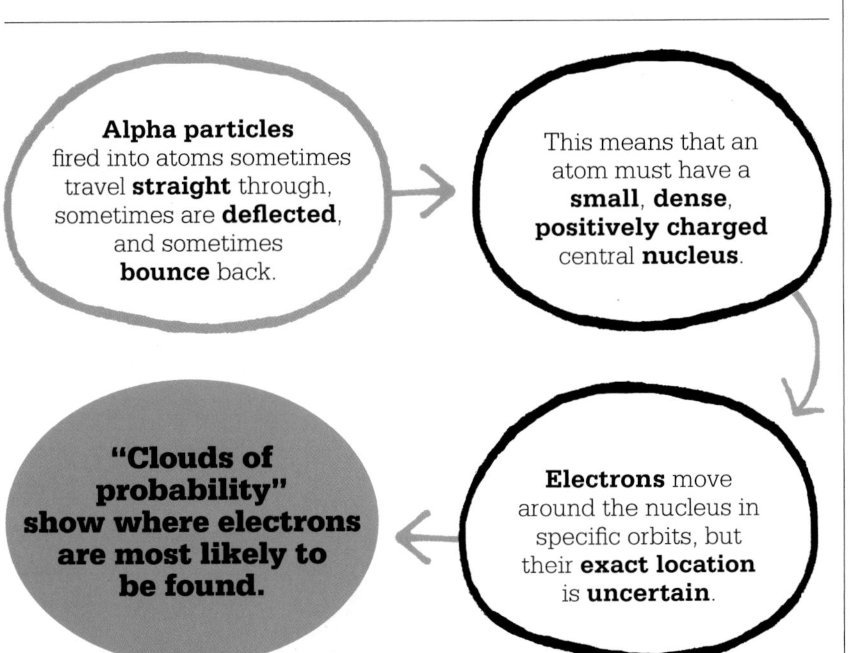

Alpha particles fired into atoms sometimes travel straight through, sometimes are deflected, and sometimes bounce back.
This means that an atom must have a small, dense, positively charged central nucleus.
Electrons move around the nucleus in specific orbits, but their exact location is uncertain.
"Clouds of probability" show where electrons are most likely to be found.

Evolution of Atomic Models

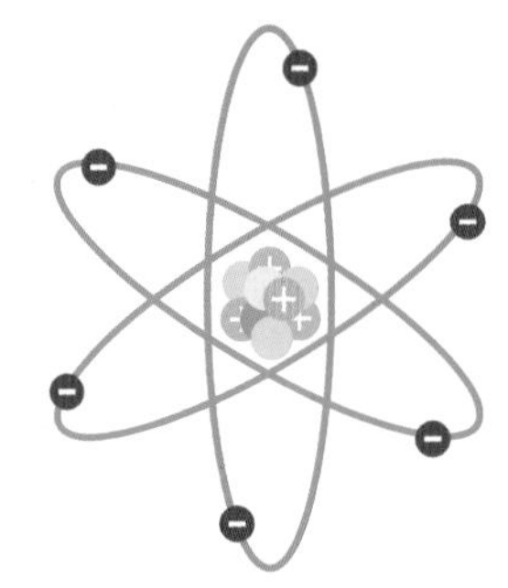

Rutherford's Nuclear Model
In his 1911 model, Rutherford placed electrons outside a dense, positively charged nucleus in the center of the atom but not in specific orbits.

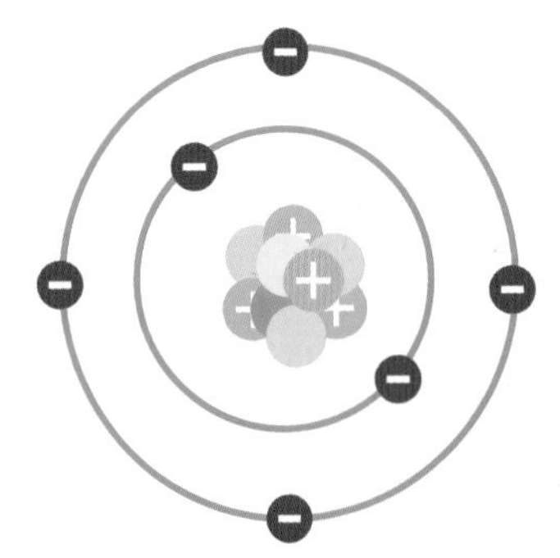

Bohr's Planetary Model
In 1913, Bohr modified Rutherford's model, placing the electrons in fixed orbits around the positively charged nucleus.

Schrödinger's Quantum Model
In 1926, Schrödinger described the electron orbits as 3D waves rather than moving around the nucleus in fixed paths.

PENICILLIN STARTED AS A CHANCE OBSERVATION

ANTIBIOTICS

IN CONTEXT

KEY FIGURE
Alexander Fleming
(1881–1955)

BEFORE
1670s Dutch microbiologist Antonie van Leeuwenhoek sees bacteria, or "animalcules," through a microscope.

1877 French chemist Louis Pasteur finds that airborne bacteria can render harmless the soil bacteria anthrax.

1909 Paul Ehrlich develops the first synthetic antibacterial treatment, Salvarsan.

AFTER
1942 The mass production of penicillin begins.

1960 In the UK, pharmaceutical company Beecham launches methicillin, a new antibiotic for penicillin-resistant pathogens.

2020 The WHO warns that the misuse of antibiotics will cause deaths from common infections.

Scottish bacteriologist Alexander Fleming was studying *Staphylococcus* bacteria at St. Mary's Hospital in London in 1928 when he discovered the first naturally occurring antibiotic, which would later be produced for therapeutic use and transform the treatment of infections. *Staphylococcus* commonly—and harmlessly—live on human skin, but they are pathogenic if they enter a person's bloodstream, lungs, heart, or bones.

The bacteria are responsible for a range of illnesses, some serious and even fatal. Ailments include relatively minor boils, skin rashes, and sore throats; more serious cellulitis and food poisoning; and potentially fatal septicemia (blood poisoning) and infections of the internal organs. In the early 20th century, varieties of bacterium from the genera *Staphylococcus* and *Streptococcus* were responsible for millions of deaths every year. Even minor scratches could prove to be fatal if they became infected, and pneumonia and diarrhea—now considered relatively straightforward to treat—were the top causes of death in the developed world.

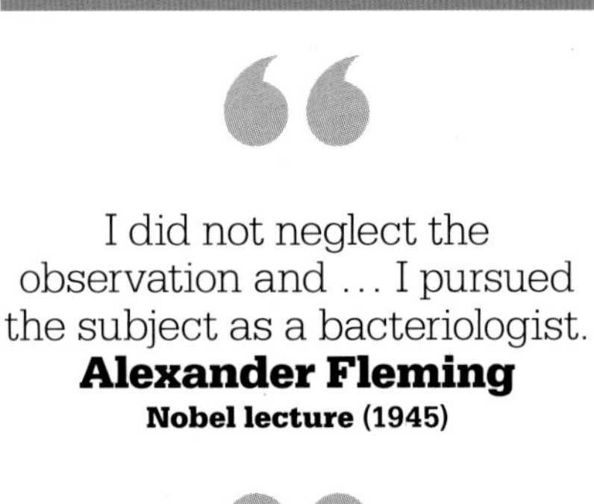

I did not neglect the observation and ... I pursued the subject as a bacteriologist.
Alexander Fleming
Nobel lecture (1945)

A chance discovery

On his return from a vacation in 1928, Fleming noticed that the *Staphylococcus* he had cultivated in one of his petri dishes had been invaded by a fungus—with no bacteria growing in the zone around the invading fungal mold. He isolated the mold and found it was *Penicillium notatum* (which is now called *Penicillium chrysogenum*). Fleming was on the lookout for a "perfect antiseptic"; he had reasons for supposing that penicillin would not work as an antibiotic, and used it instead for another purpose.

Alexander Fleming

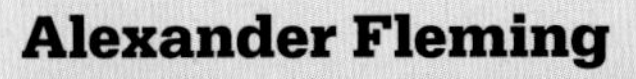

Born in rural Ayrshire, Scotland, in 1881, Alexander Fleming moved to London at age 14 to live with his brother. There, he later studied medicine at St. Mary's Hospital Medical School and developed an interest in immunology. While serving in the Army Medical Corps during World War I (1914–1918), he realized that the antiseptics used to fight infection often did more harm than good because they damaged the immune system.

Returning to St. Mary's, Fleming became professor of bacteriology in 1928, the same year that he discovered penicillin; for this work, he received the 1945 Nobel Prize in Physiology or Medicine, alongside Howard Florey and Ernst Chain. In 1946, he became head of St. Mary's inoculation department. Fleming was awarded honorary degrees from multiple universities. He died as a result of a heart attack in 1955.

Key work

1929 "On the Antibacterial Action of Cultures of a Penicillium"

See also: Anesthetics 106–107 ▪ Synthetic dyes and pigments 116–119 ▪ Enzymes 162–163 ▪ Polymerization 204–211 ▪ Rational drug design 270–271 ▪ New vaccine technologies 312–315

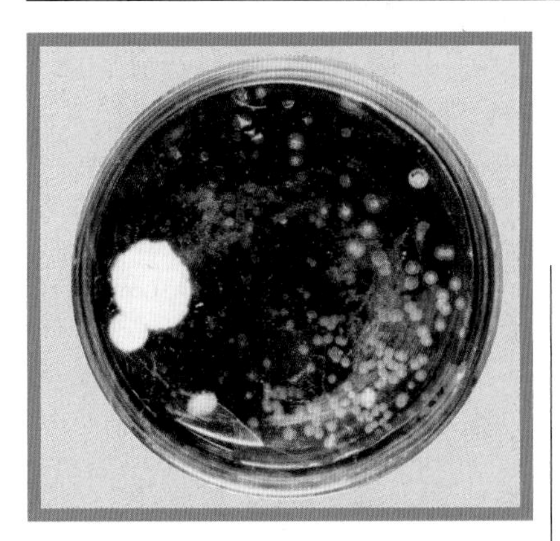

The original culture plate on which Alexander Fleming first observed the growth of the fungus *Penicillium notatum* in 1928 has the penicillin colony visible on the left.

Fleming published his findings the following year, but his attempts to isolate and purify the compound in "mold juice" responsible for the antiseptic effect failed. He sent the mold to other bacteriologists in the hope that they might have more success, but 10 years passed before there was another breakthrough, when penicillin began to be mass produced.

Gram-positive bacteria

The life-threatening illnesses of pneumonia, meningitis, and diphtheria are all caused by Gram-positive bacteria, which Danish bacteriologist Hans Christian Gram discovered in 1884. These bacteria have a peptidoglycan membrane outside the cell wall. Peptidoglycan is a polymer (a complex molecule) of amino acids and sugars that creates a meshlike structure around a bacterial cell's plasma membrane, strengthening the cell walls and preventing external fluids and particles from entering.

In Gram-negative bacteria responsible for diseases such as typhoid and paratyphoid, the peptidoglycan layer is beneath the protective outer membrane. In early 1929, Fleming demonstrated that penicillin could kill bacteria that were Gram-positive but not the Gram-negative species. The structural difference between Gram-positive and Gram-negative bacteria was crucial in whether penicillin was effective or not.

Ancient and synthetic

Natural antibiotics had been known for millennia, since the time of the ancient Egyptians, who applied moldy bread to wounds to curb infection. This treatment often worked, although they could not explain how. Early physicians elsewhere used a range of other natural remedies, but their effectiveness was a hit-or-miss affair. It was not until microscopy advanced in the 1830s that major progress could be made.

In the early 1880s, Gram discovered that certain chemical dyes colored some bacterial cells but not others. German medical scientist Paul Ehrlich concluded from this that it should be possible to selectively target bacterial cells. In 1909, he developed a synthetic »

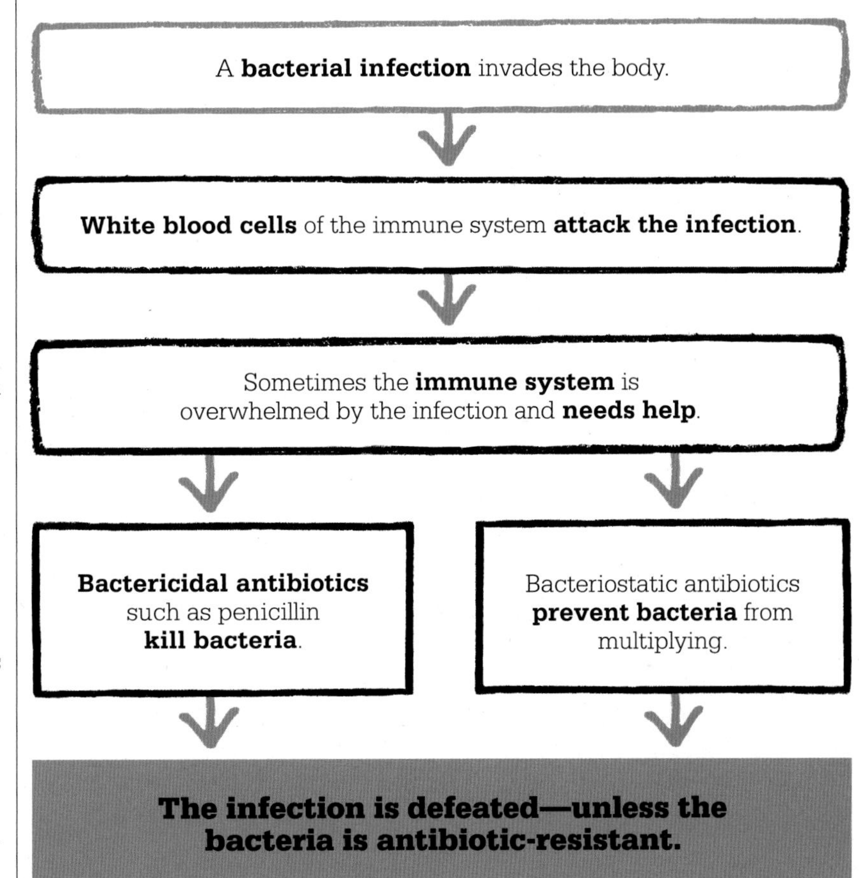

A bacterial infection invades the body.
White blood cells of the immune system attack the infection.
Sometimes the immune system is overwhelmed by the infection and needs help.
Bactericidal antibiotics such as penicillin kill bacteria.
Bacteriostatic antibiotics prevent bacteria from multiplying.
The infection is defeated—unless the bacteria is antibiotic-resistant.

arsenic-based drug, Salvarsan, to kill the bacterium *Treponema pallidum*, which caused syphilis.

In the early 1930s, German chemist Gerhard Domagk and his team were conducting research into the infection-fighting potential of compounds related to synthetic dyes. They very quickly made a landmark discovery in the control of bacterial infections. Domagk tested hundreds of compounds and, in 1931, found that one—KL 730—showed strong antibacterial effects on diseased laboratory mice. The compound was a sulfa drug, or sulfonamide ($C_6H_8N_2O_2S$), and a German pharmaceutical company patented this synthetic compound as Prontosil the following year. Physicians used it to effectively treat patients with *Staphylococcus* and *Streptococcus* infections. Domagk successfully treated his own daughter, who had a serious infection in her arm that could have resulted in amputation. He was awarded the Nobel Prize in Medicine and Physiology in 1939.

From Oxford to the US

In 1939, a team of biochemists at the University of Oxford, UK, set about trying to transform penicillin into a life-saving drug, something that no one else had previously been able to do. Led by Australian pathologist Howard Florey and British biochemist Ernst Chain, the team needed to isolate and purify the substance and process up to 880 pints (500 liters) of mold filtrate every week. Storage space was at a premium, and they were forced to use food tins, milk churns, baths, and even bedpans. The team processed the filtrate in a chemical compound derived from an acid—the ester amyl acetate—and water, then purified it prior to clinical

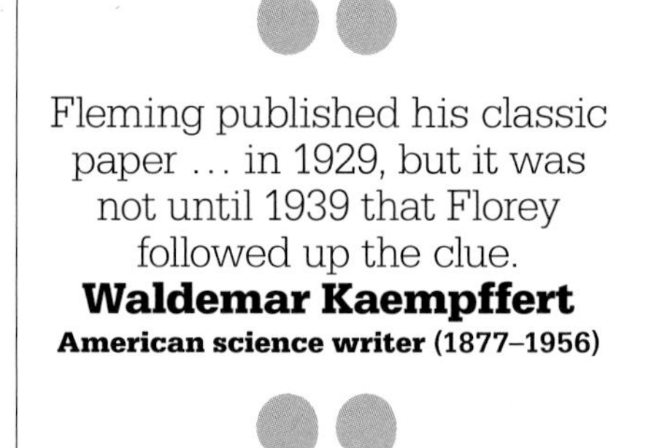

trials. Florey then showed that penicillin protected mice against infection. The next challenge was to trial it on people.

Fleming published his classic paper … in 1929, but it was not until 1939 that Florey followed up the clue.
Waldemar Kaempffert
American science writer (1877–1956)

The opportunity for a test case came with Albert Alexander, a 43-year-old British policeman who had developed large, life-threatening abscesses on his face and in his lungs after scratching his mouth while pruning roses. In 1941, he became the first person to be treated with penicillin. Alexander recovered well after being injected, but the drug was in short supply; as there was not enough to continue the treatment, he eventually died.

With the UK's chemicals sector now fully occupied with war production, Florey sought help in the United States to mass-produce penicillin. The US Department of Agriculture's Northern Regional Research Laboratory at Peoria, Illinois, agreed to take up the challenge. Chemists there found that substituting the sugar compound lactose for the sucrose

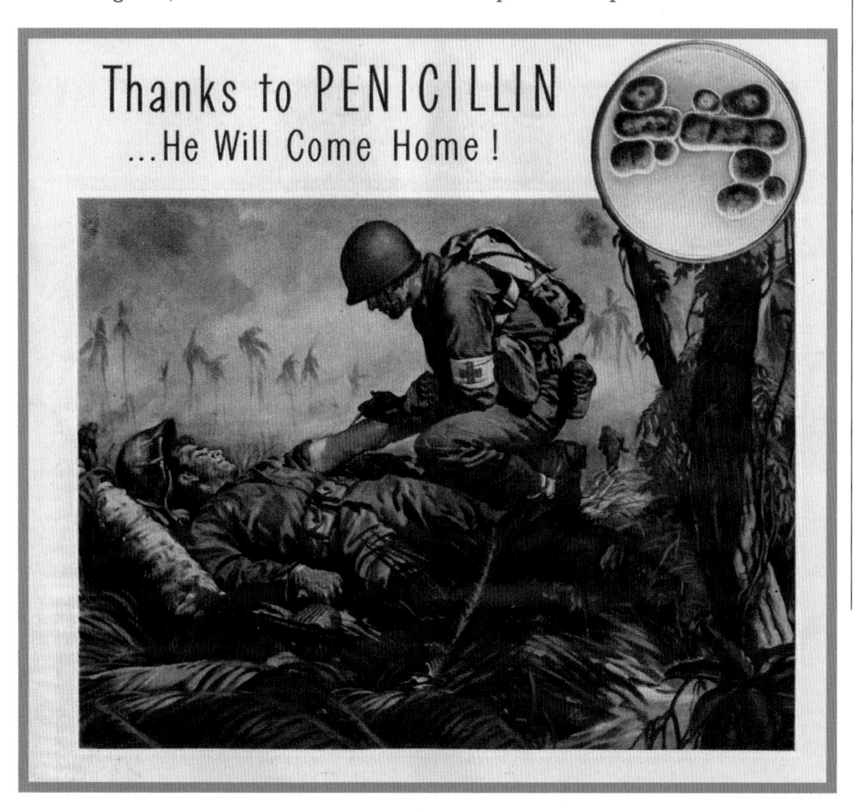

During World War II, injured soldiers were injected with penicillin. Troops also carried powdered Prontosil in their medical kits to treat bacterial infections such as sepsis.

Penicillin attacks the peptidoglycan layer in bacteria. In Gram-positive bacteria, this layer forms part of the cell wall, so penicillin can easily attack it. In Gram-negative bacteria, it is internal, so it is difficult to access.

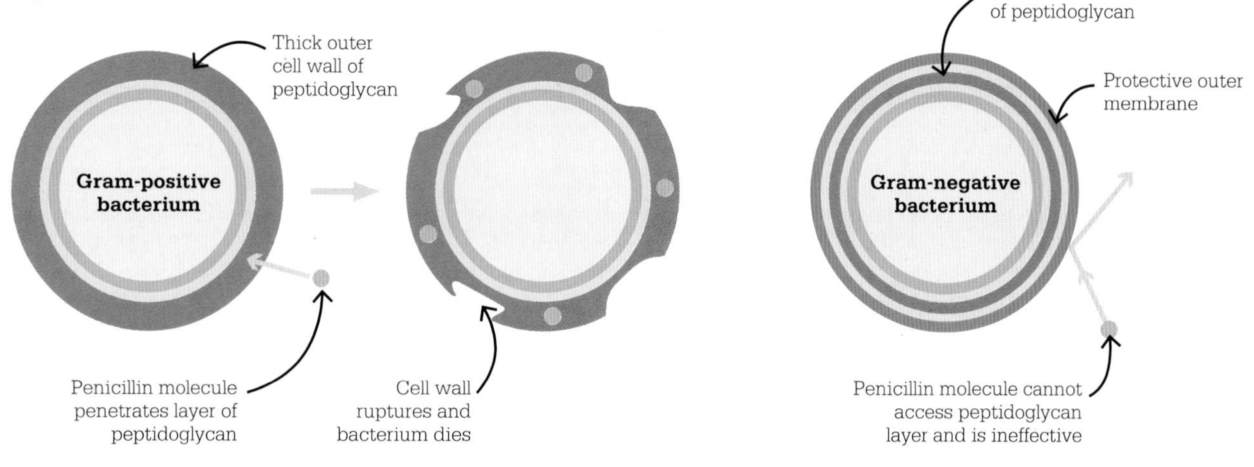

Cell wall ruptures and bacterium dies

used by the Oxford team in its culture medium increased its rate of production. Then American microbiologist Andrew Moyer discovered that by adding corn-steep liquor (a viscous mix of amino acids, vitamins, and minerals) to the mold culture, the yield could be accelerated tenfold.

After a series of meetings, Florey persuaded the US pharmaceuticals industry to back the penicillin project. In March 1942, an American woman, Anne Miller, contracted a severe infection after a miscarriage and was close to death. Doctors injected her with the drug, and she became the first person to recover fully as a result of the treatment.

Mass production of penicillin began in 1943 and accelerated exponentially. By this time, the drug had been shown to be effective against syphilis, which was common among soldiers. The immediate aim was to increase production dramatically before the D-Day invasion of France in June 1944, which would inevitably result in an enormous number of casualties. Some 21 billion units were manufactured in 1943, climbing to 6.8 trillion in 1945 and 133 trillion in 1949. Over the same period, costs per 100,000 units fell from $20 to 10¢. In 1946, penicillin became available as a prescription-only drug for the first time in the UK.

How penicillin works

It is now understood that penicillin drugs work by causing the walls of bacterial cells to burst. Penicillin

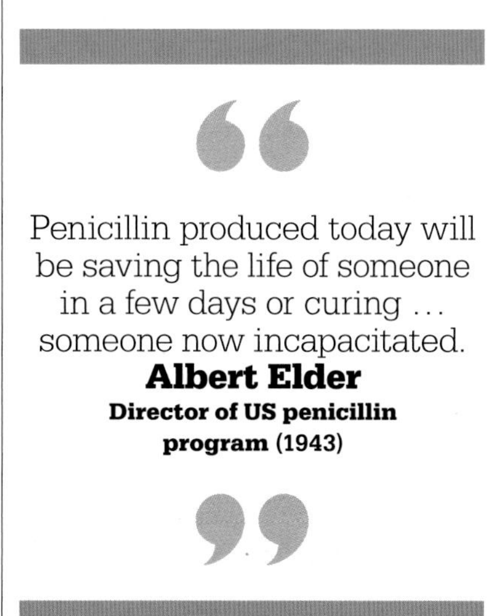

molecules act directly on the peptidoglycan outer cell walls of Gram-positive bacteria. Bacterial cells are like salty bubbles in a less salty environment, so if fluid could pass through the cell wall, osmosis—the movement of fluid through a membrane—would cause it to flow into the cell to even out the saltiness between it and its environment. If this happened, the inflow would cause the cell to burst and die.

Penicillin produced today will be saving the life of someone in a few days or curing … someone now incapacitated.
Albert Elder
Director of US penicillin program (1943)

Peptidoglycan prevents this because it strengthens the cell walls and does not allow external fluids to enter. However, when a bacterial cell divides, small holes open in its wall. Newly produced peptidoglycan fills these holes to rebuild the wall, but if penicillin molecules are present, they will block the protein struts that link the peptidoglycan. This prevents the holes from closing, allowing water to flood in and burst the cell.

In Gram-negative cells, the peptidoglycan layer is protected by an outer membrane, so it is harder for the penicillin to act on it. And penicillin does not burst » healthy human cells, because they do not have an outer sheath of peptidoglycan. Because it kills the target pathogen, penicillin is known as a bactericidal antibiotic.

A golden age

In 1945, British chemist Dorothy Hodgkin revealed the chemical structure of penicillin and published her findings four years later. Contrary to the view of many contemporary scientists, she showed that it contains a beta-lactam ring in its molecular structure. This discovery enabled scientists to modify the compound's molecular structure and create a whole family of derivative bactericidal antibiotics. This marked the dawn of a golden age in the development of antibiotics during the 1950s and 1960s.

Biochemists developed a range of new fungal-based antibiotic compounds, some derived from penicillin, as well as others such as fusidic acid and cephalosporine. Russian-American biochemist Selman Waksman, who defined an antibiotic as "a compound made by a microbe to destroy other microbes," pioneered research into the antibiotic potential of anaerobic actinobacteria, especially those from the genus *Actinomyces.* Tetracyclines, glycopeptides, and streptogramins are all antibiotics derived from these bacteria.

Tetracyclines are used to treat pneumonia, some forms of food poisoning, acne, and some eye infections. They work in a different way from penicillin, inhibiting bacterial growth by preventing the synthesis of protein within the pathogenic bacterial cell. This process occurs on structures within the cell called ribosomes. Tetracyclines can pass through the cell wall, then accumulate within the cell's cytoplasm and bind to a site on a ribosome to prevent the lengthening of protein chains. Because they prevent the pathogens from multiplying instead of killing them, these antibiotics are known as bacteriostatic.

Chemists also produced many new synthetic antibiotics, including more sulfonamides, quinolones, and thioamides. The first group works by inhibiting the enzyme dihydropteroate synthase. Unlike human cells, bacterial cells need this enzyme to produce folic acid, which all cells require to grow and divide.

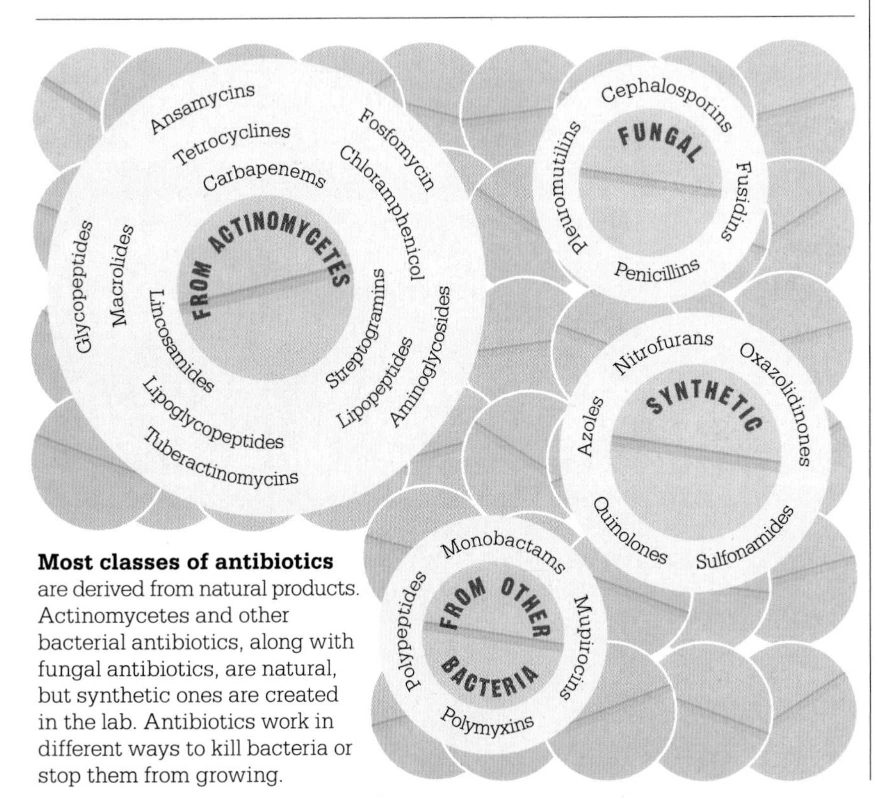

Most classes of antibiotics are derived from natural products. Actinomycetes and other bacterial antibiotics, along with fungal antibiotics, are natural, but synthetic ones are created in the lab. Antibiotics work in different ways to kill bacteria or stop them from growing.

Superbugs

Antibiotics cannot kill viruses—such as those responsible for colds, flu, chickenpox, and COVID-19—because they have a different structure from bacterial infections and replicate differently. Even so, experts estimate that antibiotics have saved more than 200 million lives worldwide by defeating a vast array of bacterial pathogens and curing countless infections.

The widespread adoption of antibiotics has created its own problems. Scientists first noticed resistance to penicillin in *Staphylococcus aureus* as early as 1942. This Gram-positive bacterium is responsible for some skin infections, sinusitis, and some food poisoning. Resistance meant that the antibiotic was not always effective. Biochemists noted a growing trend of resistance as more antibiotics were introduced. Vancomycin, an actinomycete-derived antibiotic, was introduced in 1958, but a vancomycin-resistant form of *Enterococcus faecium*, which causes neonatal meningitis, and a vancomycin-resistant *Staphylococcus aureus* were later discovered.

A kind of "arms race" developed. When, in 1960, a penicillin derivative (methicillin) was introduced to get around the problem of penicillin resistance, it was not very long before methicillin-resistant *Staphylococcus aureus* (MRSA) began to appear. Methicillin was rendered clinically useless, and the MRSA "superbug" became a serious problem, quickly spreading in hospitals. By 2004, 60 percent of *Staphylococcus* infections in the US were caused by MRSA, thousands proving fatal. Epidemiologists suspect that tens of millions of people worldwide may carry MRSA.

Causes of antibiotic resistance

Overprescribing of antibiotics: The more antibiotics are used, the more bacteria adapt to survive.

Patients not finishing their treatment: This may allow some of the bacteria causing the infection to survive.

Overuse of antibiotics in livestock and fish farming: This increases the risk of transmitting drug-resistant bacteria to humans.

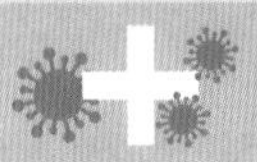

Poor infection control in hospitals: Bacteria can travel if infected patients and staff do not keep spaces clean.

Lack of hygiene and poor sanitation: Poor hygiene leads to the spread of infections and use of more antibiotics.

Lack of new antibiotics being developed: It is expensive to develop new antibiotics, so it is not always deemed cost-efficient.

Resistance to antibiotics

Bacteria evolve to fight antibiotic attacks in several ways. Once they become resistant, they multiply, making antibiotics less effective. Bacteria can restrict an antibiotic's access by making the peptidoglycan sheath more effective. Some change so they can pump the antibiotic from their cells; this has happened with beta-lactam compounds such as penicillin. Others alter the chemistry of the antibiotic. For example, *Klebsiella pneumoniae*, one of the bacteria responsible for pneumonia, can produce beta-lactamase enzymes to break down beta-lactams. To overcome this resistance, biochemists have manufactured beta-lactam antibiotics with beta-lactamase inhibitors, such as clavulanic acid. Still more bacteria reproduce by invading other cells and also develop new cell processes so they are not affected by the sites that are targeted by antibiotics. *Escheria coli* (*E. coli*), which causes chronic food poisoning, can add a compound to the outside of its cell wall that prevents the antibiotic colistin from latching onto it.

Future research

The World Health Organization (WHO) reported in 2019 that 750,000 people die each year from drug-resistant infections, with this figure projected to rise to 10 million annually by 2050. Therefore, it is not enough to rely on antibiotics that have already been developed. The search for new ones has to continue, despite the huge costs involved. Chemists are constantly researching new synthetic drugs, and because most antibiotics in current use are derived from living organisms, biochemists continue to screen bacteria, fungi, plants, and animals for 21st-century "mold juices" that may well be used to develop the next generation of pathogen-fighting drugs. ■

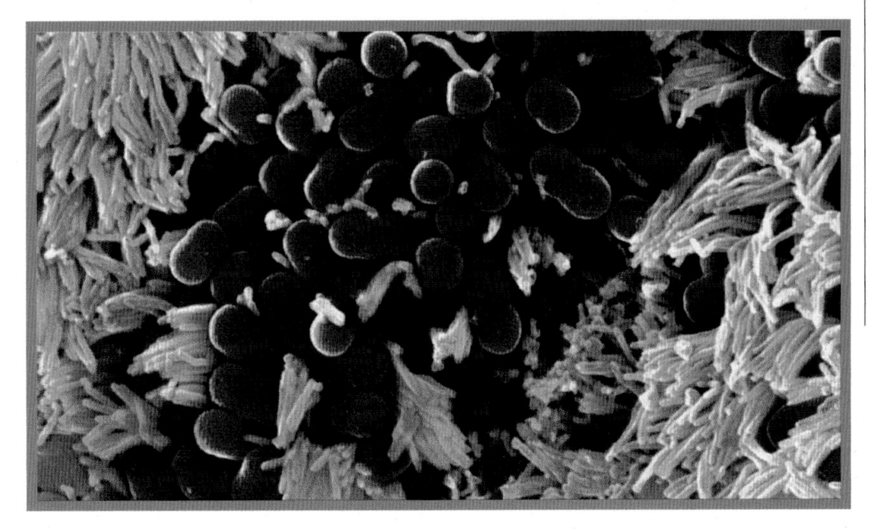

***E. coli* bacteria**, seen here in red, are commonly found in the gut of animals and humans. Most strains are harmless, but some can cause illness. Antibiotics are not recommended to treat it.

OUT OF THE ATOM SMASHER

SYNTHETIC ELEMENTS

IN CONTEXT

KEY FIGURE
Emilio Segrè (1905–1989)

BEFORE
1869 Dmitri Mendeleev's periodic table predicts several undiscovered elements.

1875 French chemist Paul Lecoq de Boisbaudran isolates predicted element 68 (later named gallium) in zinc ore.

1909 Masataka Ogawa claims to have discovered element 43, although his findings cannot be replicated.

1925 Walter Noddack, Otto Berg, and Ida Tacke claim the discovery of element 43 and name it masurium.

AFTER
1940 Emilio Segrè and Carlo Perrier create the second synthetic element, astatine.

2009 A Russian–US collaboration creates synthetic element 117, tennessine.

Mendeleev's periodic table of 1869 is known for the gaps left for predicted but not yet discovered elements. Some—including germanium, gallium, and scandium—were found over the next two decades, proving him correct. But at the time of Mendeleev's death in 1907, one placeholder, called eka-manganese, had still not been isolated.

The road to its discovery was pitted with false starts. In 1909, Japanese chemist Masataka Ogawa discovered an unknown element in a rare thorium oxide mineral. Believing it to be the missing element 43, he named it nipponium, but nobody else could replicate his discovery. Later research suggests Ogawa had actually identified another missing element—number 75 (rhenium)—but in not realizing this, he missed the opportunity to name it.

In 1925, it seemed as though German chemists Walter Noddack, Otto Berg, and Ida Tacke had made the breakthrough. When analyzing platinum and columbite ores, they claimed to have produced X-ray spectroscopy evidence of two undiscovered elements, 43 and 75. They confirmed their discovery of element 75—which Ogawa had unwittingly found previously—by isolating greater quantities of it from a molybdenite ore. They named it rhenium. Although they tried to isolate element 43, which they named masurium, they were unsuccessful. It remained elusive.

> Experimental complication to me is more an unavoidable evil to be tolerated in order to obtain the results than a stimulating challenge.
> **Emilio Segrè**

Collaborative effort

In 1936, Italian professor of physics Emilio Segrè visited the US, where he spent time at physicist Ernest Lawrence's laboratory in Berkeley, California. There, he saw firsthand the cyclotron, a particle accelerator

See also: Isomerism 84–87 ▪ The periodic table 130–137 ▪ Isotopes 200–201 ▪ The transuranic elements 250–253 ▪ Completing the periodic table? 304–311

that was used to bombard atoms of various elements with high-speed particles, creating different isotopes of lighter elements.

Disputed claims

Intrigued by the possible range of radioactive products generated, Segrè persuaded Lawrence to send discarded parts of the cyclotron to his lab in Palermo, Italy. In 1937, Segrè and Italian mineralogist Carlo Perrier analyzed some radioactive molybdenum foil from the cyclotron and isolated two isotopes. After excluding niobium and tantalum as possible sources of the radiation, they concluded that some of the radiation was being produced by element 43—but still they were unable to isolate it.

Shortly after, Segrè returned to Berkeley, where he worked with American chemist Glenn Seaborg. Segrè discovered another isotope of what he believed was element 43 and one of its nuclear isomers (an atom with the same number of protons and neutrons but differing energy and radioactive decay). This was the final corroboration needed to announce the discovery of element 43. Since Noddack, Berg, and Tacke failed to renounce their claim to have found element 43, Segrè and Perrier delayed suggesting a name for it. Finally, in 1947, they proposed that it be called technetium.

In 1961, a single nanogram of technetium was isolated from pitchblende, a uranium ore, found in what is now the Democratic Republic of the Congo. This minuscule sample was produced by the fission of uranium-238 in the ore. The discovery showed that technetium is not a completely artificial element, although it was still the first undiscovered element to be produced in a laboratory.

Today, technetium is not just a curiosity. The nuclear isomer that was found by Segrè and Seaborg, technetium-99m, is commonly used as a radioactive tracer in nuclear medicine to image parts of the body. The discovery of technetium heralded the beginning of the age of synthetic element discoveries. In the following years, many more new elements would be created and identified in a lab. ■

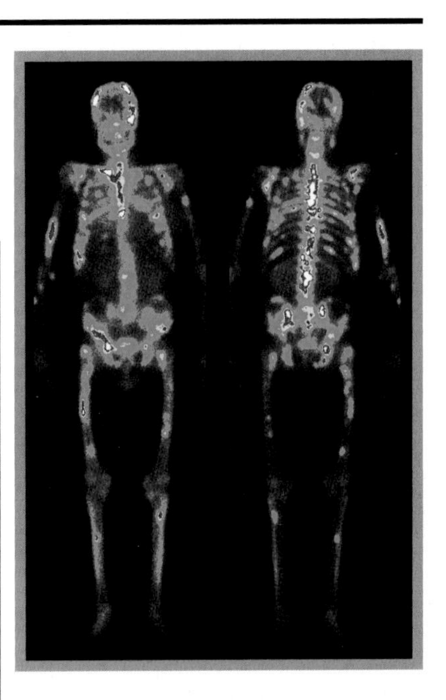

Bone cancer—shown in red on these scans—can be identified by injecting the radioactive isotope technetium-99m. This tracer material becomes concentrated in the cancerous tissues.

Emilio Segrè

Born into a Jewish family in Tivoli, Italy, in 1905, Emilio Segrè enrolled at the University of Rome as an engineering student but later transferred to physics. As director of the physics laboratory at the University of Palermo, he discovered technetium with his colleague Carlo Perrier. In 1938, while visiting Berkeley, California, to carry out further work on technetium, anti-Semitic laws passed in Italy forced him to remain in the US permanently. Segrè later discovered another "missing" element in the periodic table, astatine. He also worked on the Manhattan Project and established conclusive evidence of the existence of antiprotons, for which he and American physicist Owen Chamberlain won the 1959 Nobel Prize in Physics. Segrè died in 1989.

Key works

1937 "Some Chemical Properties of Element 43"
1947 "Astatine: The Element of Atomic Number 85"
1955 "Observation of Antiprotons"

TEFLON TOUCHES EVERY ONE OF US ALMOST EVERY DAY

NONSTICK POLYMERS

IN CONTEXT

KEY FIGURE
Roy Plunkett (1910–1994)

BEFORE
1920 German chemist Hermann Staudinger proposes that substances like rubber are made up of huge molecules formed by polymerization reactions, inspiring efforts to make more human-made polymers.

1930s American chemist Wallace Carothers invents the polymers nylon and neoprene.

AFTER
1967 After the deadly Apollo 1 launchpad fire, NASA incorporates a PTFE-coated fabric into its spacesuits to make them more durable and noncombustible.

2015 The use of perfluorooctanoic acid (PFOA) is phased out in the US due to concerns about the environmental persistence of long-chain fluorochemicals.

Polymer science and the discovery of polytetrafluoroethylene (PTFE) owe a lot to serendipity. In 1938, American chemist Roy Plunkett was working on making new chlorofluorocarbon refrigerants at the chemicals company DuPont by reacting gaseous tetrafluoroethylene (TFE) with hydrochloric acid. Plunkett and his research assistant, Jack Rebok, stored the TFE in small cylinders with valves that released it when needed, but when Rebok opened the valve of one of the cylinders, no gas came out. After confirming, by weighing, that the cylinder was not empty, Plunkett and Rebok shook the cylinder and small flakes of a white, waxy substance fell out. Perplexed, they cut open the cylinder to find its insides coated by a white solid. Plunkett realized that the tetrafluoroethylene had polymerized—that is, the individual molecules had reacted together to form long chains—and that the white material was the polymer it had formed.

Plunkett carried out a series of tests on the white solid to determine its properties. He established that the polymer had a high melting point (620°F/327°C), was resistant to reaction with almost anything, and was incredibly slippery. Plunkett was granted a patent for TFE polymers in 1941, but had no further involvement in their development. He was not a polymer chemist and was moved on to other DuPont projects.

Initially, the expense of making PTFE prohibited any potential applications, but this changed during World War II. Launched in

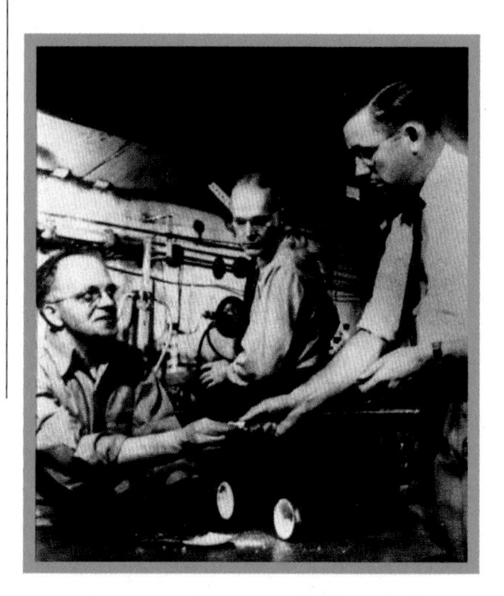

The accidental discovery of Teflon in 1938 is reenacted by (left to right) Jack Rebok, Robert McHarness, and Roy Plunkett.

See also: Intermolecular forces 138–139 ▪ Polymerization 204–211 ▪ Super-strong polymers 267 ▪ The hole in the ozone layer 272–273

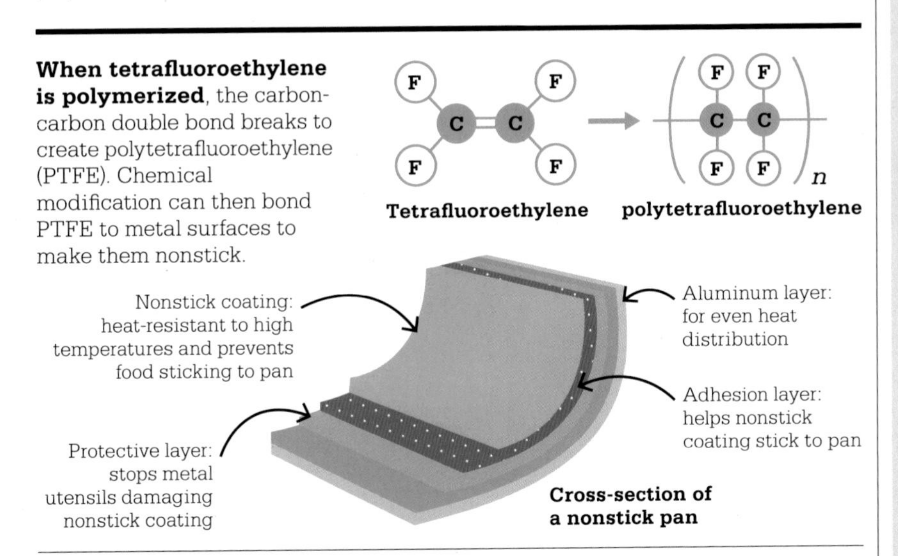

When tetrafluoroethylene is polymerized, the carbon-carbon double bond breaks to create polytetrafluoroethylene (PTFE). Chemical modification can then bond PTFE to metal surfaces to make them nonstick.

Cross-section of a nonstick pan

the US in 1942, the Manhattan Project recruited thousands of scientists in the race against Nazi Germany to produce the first functional nuclear weapon. Uranium enrichment was key, but this process used uranium hexafluoride, which corroded seals and gaskets made of virtually any material. PTFE could be used to resist this chemical attack.

In the kitchen and beyond

After World War II, PTFE started its transition from warfare to cookware, patented by DuPont with the more familiar name "Teflon" in 1946. The challenge, however, was getting the nonstick substance to stick to anything. Several approaches were tested, including using high temperatures, applying resins, and sandblasting or etching surfaces to make them rougher. Today, Teflon is chemically modified to break away some fluorine atoms from its structure, allowing it to bond easily with metal surfaces.

DuPont used perfluorooctanoic acid (PFOA) to polymerize TFE; studies submitted in evidence to a class action law suit concluded that there was an association between PFOA exposure and health effects including cancers. DuPont hesitated to use Teflon on cookware, but in the meantime, a French couple, Marc and Colette Grégoire, took the initiative. In 1956, the couple started their own business—Tefal—and millions of their pans have been sold worldwide.

Today, PTFE is used in waterproof fabrics, lubricants, cosmetics, food packaging, wiring insulation, and more. Its discovery opened the door to the creation of other fluoropolymers with similarly useful properties, employed in a variety of applications to make materials water, heat, and stain proof. Unfortunately, the chemical properties that make fluoropolymers useful also pose a problem. They are so inert that they do not break down in the environment for thousands of years. There are growing concerns about how they may accumulate in the environment and our bodies. In recognition of this, long chain fluorochemicals are being phased out from nonessential use. ■

What makes Teflon nonstick?

Teflon's nonstick properties are in part due to the strong bond between the carbon and fluorine atoms, which is, in fact, the strongest bond to a carbon atom that can be made. This makes fluoropolymers such as Teflon incredibly unreactive. It is just not possible for the molecules in food to form bonds with the carbon atoms in Teflon chains. Even fluorine gas, which is very reactive, does not react with the polymer. Fluorine's high electronegativity also makes it difficult for molecules to stick to Teflon because they are easily repelled. Even geckos, lizards whose sticky toepads enable them to cling to any surface, cannot grip onto Teflon; the van der Waals intermolecular forces they rely on to do this are not strong enough with Teflon. Although Teflon itself is inert, at temperatures above those commonly used for cooking it can degrade, breaking down and releasing toxic fluorine-containing compounds.

> It was obvious immediately to me that the tetrafluoroethylene had polymerized and the white powder was a polymer of tetrafluoroethylene.
>
> **Roy Plunkett**

I WILL HAVE NOTHING TO DO WITH A BOMB!

NUCLEAR FISSION

IN CONTEXT

KEY FIGURE
Lise Meitner (1878–1968)

BEFORE
1919 New Zealand–born British physicist Ernest Rutherford pioneers a technique for bombarding atom nuclei with smaller particles.

AFTER
1945 On July 16, the first atomic explosion is conducted in the New Mexico desert. Three weeks later, on August 6, an atomic bomb is dropped on Hiroshima in Japan.

1954 Obninsk, in the USSR, becomes the first nuclear power station to generate electricity for a power grid.

1997 Meitnerium, the heaviest known element, with atomic number 109, is named in Lise Meitner's honor.

When British physicist James Chadwick discovered the existence of neutrons in 1932, he could not have envisaged the enormous impact they would have on society.

Italian physicist Enrico Fermi understood that neutrons were powerful new tools to further his own research into atomic structure. He deduced that since neutrons carried no charge, they should pass into atomic nuclei without resistance (unlike positively charged protons). He and his team bombarded 63 stable elements with neutrons and produced 37 radioactive elements—elements

See also: Atomic weights 121 ▪ Radioactivity 176–181 ▪ Isotopes 200–201 ▪ Improved atomic models 216–221 ▪ The transuranic elements 250–253

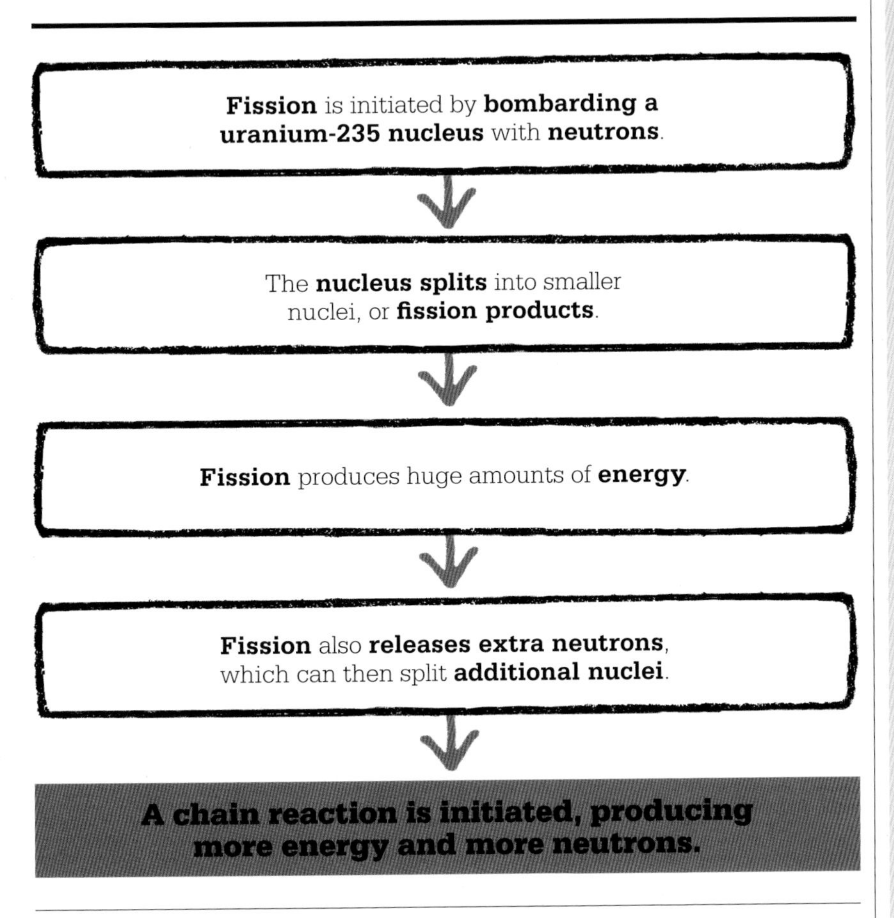

whose nuclei were unstable and dissipated excess energy in the form of radiation. Fermi had discovered nuclear fission without realizing it. In fact, he believed that his neutron bombardment of the metal uranium (then the heaviest known element, atomic number 92) may have produced the first transuranic elements—those with atomic numbers greater than 92. However, German chemist Ida Noddack suggested an alternative explanation that we now know to be correct: the uranium had actually split apart into lighter elements.

Fission is initiated by **bombarding a uranium-235 nucleus** with **neutrons**.
The **nucleus splits** into smaller nuclei, or **fission products**.
Fission produces huge amounts of **energy**.
Fission also **releases extra neutrons**, which can then split **additional nuclei**.
A chain reaction is initiated, producing more energy and more neutrons.

In Berlin, German radiochemists Otto Hahn and Fritz Strassmann conducted similar experiments, firing neutrons at the nuclei of a variety of elements. In late 1938, the pair discovered traces of the lighter element barium (atomic number 56) when they bombarded uranium. The uranium nuclei had split into two roughly equal pieces. Significantly, these each had less than half the mass of the original nucleus.

Team effort

Hahn decided to seek the advice of former colleague Lise Meitner. Over Christmas 1938, Meitner's nephew Otto Frisch, also a nuclear physicist, visited her, and the two pondered Hahn and Strassmann's findings. Frisch suggested they should consider the nucleus as »

Lise Meitner

Born in Vienna, Austria, in 1878, Lise Meitner became interested in science from an early age. She studied at the city's university, where, in 1905, she became one of the first women in the world to receive a doctorate in physics.

Moving to Berlin, Meitner researched radioactivity with physicist Max Planck and Otto Hahn at the Kaiser Wilhelm Institute for Chemistry. In 1938, because she was Jewish, Meitner was forced to flee Germany and continue her work in Stockholm, Sweden, but she secretly remained in touch with Hahn to plan the experiments that would demonstrate nuclear fission. However, she was overlooked when the 1944 Nobel Prize in Chemistry was awarded to Hahn and Strassmann for their research. Meitner retired to Cambridge, UK, where she died in 1968.

Key works

1939 "Disintegration of Uranium by Neutrons: A New Type of Nuclear Reaction"
1939 "Physical Evidence for the Division of Heavy Nuclei under Neutron Bombardment"

being like a drop of liquid—an idea previously advanced by Ukrainian physicist George Gamow and Danish physicist Niels Bohr.

After a neutron bombardment, the target nucleus would become stretched, pinched in the middle, and split into two drops, which would be driven apart by the force of electrical repulsion. Since the two "daughter" nuclei were known to have less mass than the original uranium nucleus, Frisch and Meitner did some calculations.

According to Albert Einstein's famous equation $E = mc^2$ (where E is energy, m is mass, and c is the speed of light), the loss of mass resulting from the splitting process must have been converted into kinetic energy that could, in turn, be converted into heat. This, Meitner and Frisch believed, was the process Hahn and Strassmann had enacted—Frisch coined the term fission to describe it—and they realized that it had enormous implications for producing energy. Hahn and Strassmann had discovered it, but Meitner and Frisch had provided the theoretical explanation.

Explosive potential

Nuclear fission had the potential to release large amounts of energy—and to release more neutrons as the two main fragments of the uranium atom split. Scientists began to investigate how these secondary neutrons might be able to create a chain reaction that, if contained, could produce a supply of energy for power and heat. But with the world on the brink of World War II, the discovery took on a greater relevance: the chain reaction had the potential to generate the most powerful explosions ever created.

News spread rapidly of Hahn and Strassmann's experiments and Meitner and Frisch's calculations. Scientists seeking to exploit nuclear fission needed to know more about the atomic structure of uranium. It is a very heavy metal, 18.7 times denser than water, and it occurs in three isotopes: U-238 (with 92 protons and 146 neutrons in the nucleus), U-235 (92 protons and 143 neutrons), and U-234 (92 protons and 142 neutrons).

American physicist John Dunning and his team at Columbia University in New York discovered that only U-235 was capable of fission. When the nucleus of a U-235 atom captures a moving neutron, it splits in two—or fissions—and releases heat energy. Either two or three additional neutrons are thrown off. The challenge for the scientists was that uranium consists of 99.3 percent U-238, only 0.7 percent U-235, and just a trace of U-234. Somehow, enough U-235 needed to be separated from the U-238 to achieve the necessary critical mass and produce a chain reaction. Chemical separation would be impossible since the

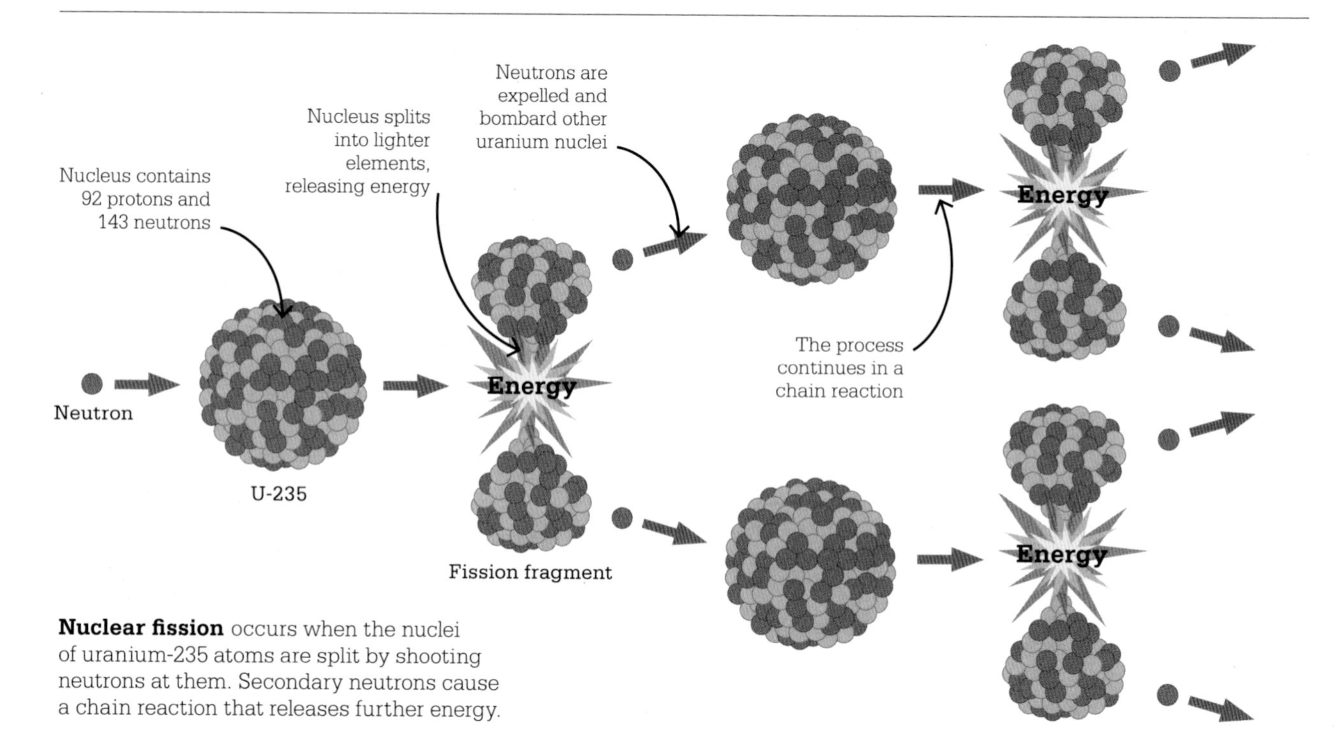
Nucleus contains 92 protons and 143 neutrons
Nucleus splits into lighter elements, releasing energy
Neutrons are expelled and bombard other uranium nuclei
Neutron
U-235
Energy
Fission fragment
The process continues in a chain reaction
Energy
Energy

Nuclear fission occurs when the nuclei of uranium-235 atoms are split by shooting neutrons at them. Secondary neutrons cause a chain reaction that releases further energy.

isotopes are chemically identical, and physical separation would be extremely difficult because their mass differs by less than 1 percent.

The Manhattan Project

As early as August 1939, Einstein wrote to President Franklin D. Roosevelt to warn him that Nazi Germany was planning to develop an atomic bomb. Einstein played no direct part in developing the American response, but his warning was heeded. Roosevelt established the Manhattan Project in 1942, with the aim of creating a viable fission bomb.

Thousands of scientists were employed to find ways to enrich uranium to increase the percentage composition of U-235. Different teams invented three processes: gaseous diffusion, liquid thermal diffusion, and electromagnetic separation. All three were employed to enrich uranium for the atomic bomb that, after several years of development, was dropped on Hiroshima in 1945.

Peaceful application

Although several nations built atomic bombs after World War II, they were never used in anger, and research turned to developing nuclear fission to produce energy for wider use. To enhance uranium for use in a nuclear power plant, engineers usually use the gaseous enrichment process, very similar to that used to produce U-235 for the first atomic bombs.

The process involves converting uranium oxide (U_3O_8) to uranium hexafluoride gas (UF_6). This is then fed into centrifuges with thousands of rapidly spinning tubes that separate the U-235 and U-238 isotopes. Two streams are thereby produced—one of enriched uranium, the other of depleted uranium—and this process increases the proportion of U-235 from its natural level of 0.7 percent to 4–5 percent of the total. In the core of a nuclear reactor, the U-238 is "fertile," which means it can capture neutrons thrown off by U-235 nuclei. In the process, it becomes plutonium-239, which (like U-235) is fissile and can produce energy.

Nuclear power remains a controversial subject because of the dangers of fission and of the difficulties of disposing of radioactive waste, but in 2020, it provided about 10 percent of the world's electricity. Since nuclear energy is one of several alternatives to fossil fuels, its contribution may grow as part of the drive to decarbonize energy production in the face of climate change. ■

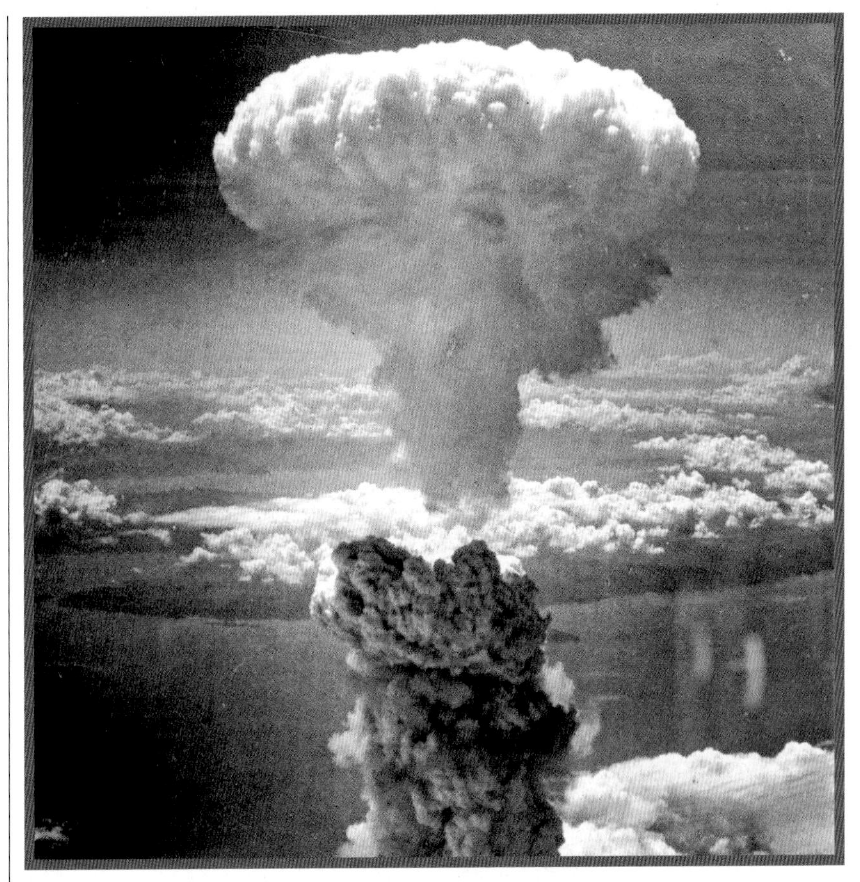

The atomic bombs dropped on the Japanese cities of Hiroshima and Nagasaki (above) in 1945 are thought to have killed between 129,000 and 226,000 people.

> Now whenever mass disappears energy is created. . . . So here was the source for that energy; it all fitted!
>
> **Otto Frisch (1979)**

CHEMISTRY DEPENDS UPON QUANTUM PRINCIPLES

CHEMICAL BONDING

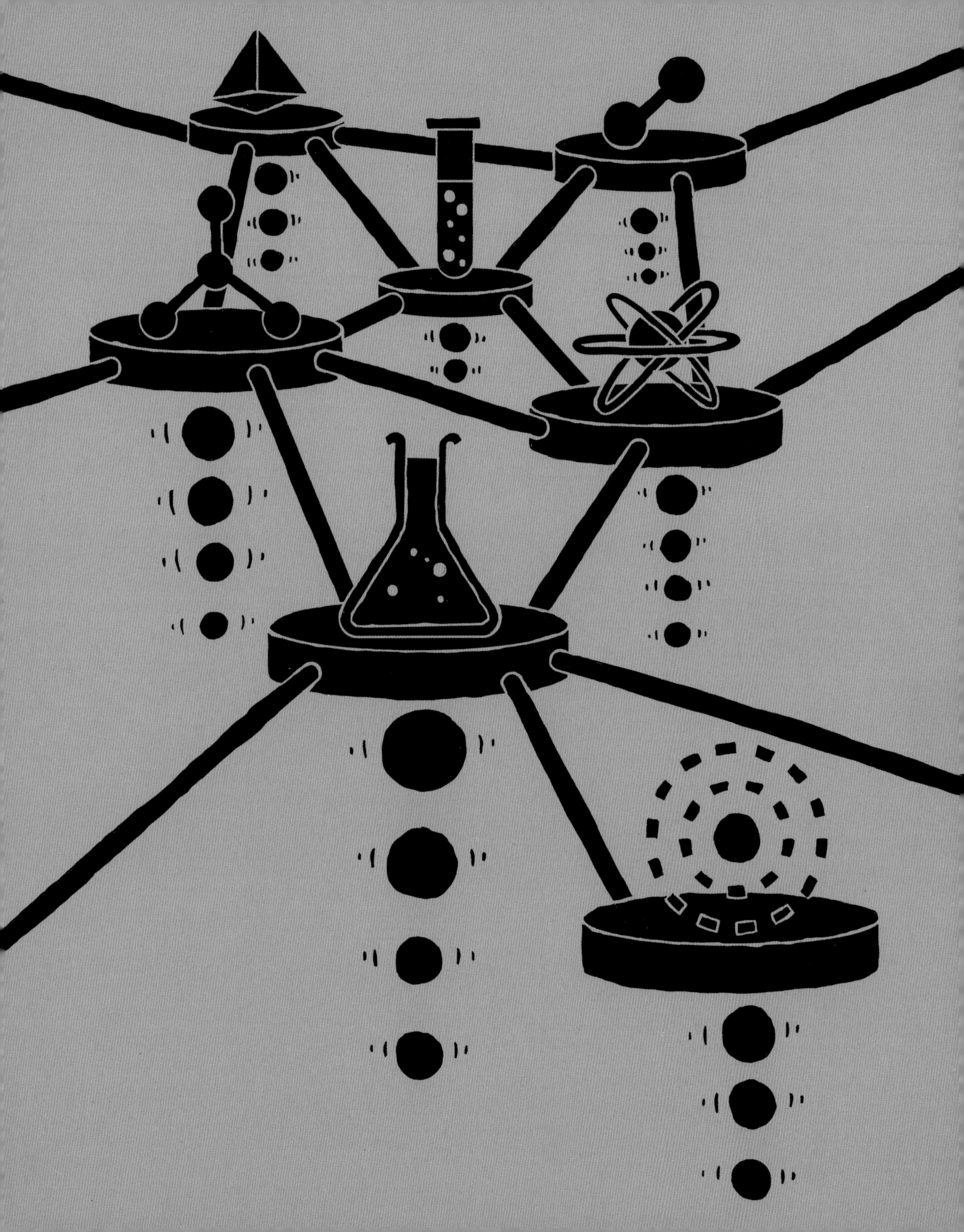

IN CONTEXT

KEY FIGURE
Linus Pauling (1901–1994)

BEFORE
1794 Joseph Proust shows that in compounds elements always combine in fixed ratios, hinting that the process must follow rules.

1808 John Dalton starts to lay out the properties of atoms that determine how they combine in compounds.

AFTER
1964 In the US, theoretical physicists Walter Kohn and Pierre Hohenberg develop density functional theory to solve quantum mechanical equations that show bonding in complex substances.

2021 Russian chemist Artem Oganov and Italian chemist Christian Tantardini modify Pauling's formula for electronegativity so that it performs better.

My work on the chemical bond probably has been most important in changing the activities of chemists all over the world.
Linus Pauling

Atoms share electrons to connect via **covalent bonds**.

Electrons move between covalently bonded atoms, so molecules **can resonate** between different structures, which **makes them more stable**.

Atoms can share electrons unevenly, causing a magnetlike attraction and **covalent bonding**, which holds together more strongly.

Resonance allows electrons to mix together and to form hybrid orbitals that determine the shapes of molecules.

One of the main goals in chemistry is to establish what substances are made of. At the start of the 19th century, it had become clear that atoms were key building blocks. After that, the question became, "How do these atoms connect to give chemical structures?" A breakthrough came in 1939 with American chemist Linus Pauling's work on the nature of the chemical bond, after 100 years of slow progress. An early advance in 1852 came when British chemist Edward Frankland proposed that an atom of an element can only connect with a certain number of atoms of other elements. He called the number of possible connections an element's valence.

Understanding chemical bonding required the 1897 discovery by J. J. Thomson of tiny electrically charged particles, which he called electrons. Then, in 1900, German theoretical physicist Max Planck suggested treating energy as if it came only in small packets of a consistent size called quanta, with each packet called a quantum. Quanta explained the amount of energy produced when objects shine ultraviolet light, which physicists had previously been unable to do.

Dipole moments

In 1911, Dutch-American physicist and physical chemist Peter Debye started work at the University of Zurich, and he soon made the discovery that would occupy him for 40 years and earn him the 1936 Nobel Prize in Chemistry. Debye built on prior discoveries that molecules could be made to behave

See also: Compound proportions 68 ▪ Dalton's atomic theory 80–81 ▪ Functional groups 100–105 ▪ Structural formulae 126–127 ▪ Coordination chemistry 152–153 ▪ The electron 164–165 ▪ Improved atomic models 216–221

like magnets when put into an electromagnetic field that was formed when electrons flow through wires. In 1905, French physicist Paul Langevin had proposed that this was because the electromagnetic field had temporarily shifted, or polarized, the molecules' electrons. With that imbalance of electric charge, the molecules behaved like magnets, called electrical dipoles. In 1912, Debye suggested that there could be permanent polarizations in how electrons are distributed around molecules called dipole moments.

Sharing ideas

Meanwhile, at the University of California, American chemist Gilbert Lewis proposed that chemical bonds arise from atoms sharing pairs of electrons. Take, for example, the single bond between two carbon atoms or a carbon atom and a hydrogen atom, which are represented by single lines in structural formulae. Lewis proposed that this bond consists of a pair of electrons held jointly by the two atoms that are bonded together.

Bonding in the Lewis cubical atom model

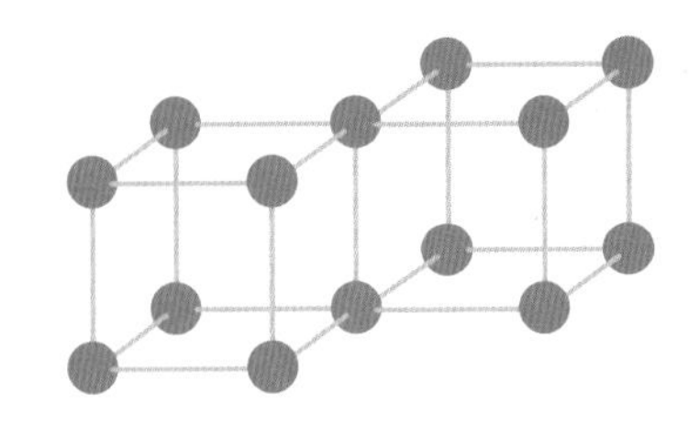

Single bonds form when two atoms share an edge. This results in the sharing of two electrons in a covalent bond.

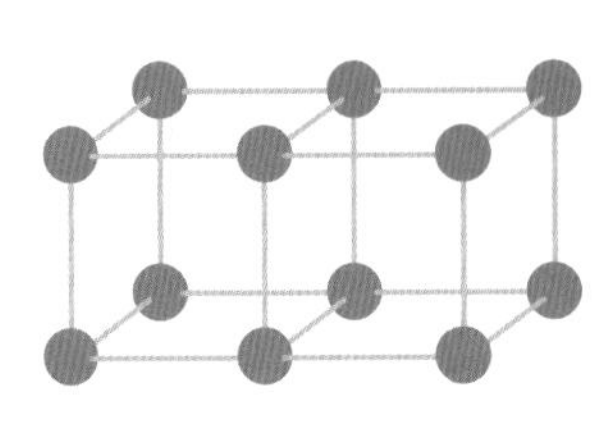

Double bonds form when two cubic atoms share a face. This results in the sharing of four electrons in a covalent bond.

Atoms that are bonded by sharing electrons are more stable than the atoms would be individually. In a 1916 paper, he portrayed atoms as cubes with electrons at their corners. Lewis argued that they accumulate an electron at every corner by sharing edges with other atoms. American chemist Irving Langmuir helped popularize the idea, calling this kind of bonding "covalent."

What Lewis proposed was a dramatic shift from the prevailing idea that chemical bonding arose from electromagnetic attractions between ions with opposite electrical charges. The word ion refers to a charged atom, which has either gained or lost one or more electrons. Not everyone liked the idea of covalent rather than ionic bonding, but it fascinated American chemistry student Linus Pauling.

Pauling's contributions

Over the next two decades, Pauling showed how electron sharing could be described by quantum theory, making Lewis's picture central to modern chemical bonding theory. »

Linus Pauling

Born in Portland, Oregon, in 1901, Linus Pauling had an impoverished upbringing due to his father dying in 1910. He got his first degree, in chemical engineering, from Oregon State College in 1922 and became a tutor and graduate student at the California Institute of Technology (Caltech). He worked with American chemist Roscoe G. Dickinson until 1925 to determine crystal structures and develop theories about the nature of the chemical bond. He continued his research in the 1930s and summarized his discoveries in his famous publication of 1939. He was awarded the Nobel Prize in Chemistry in 1954 for his work on chemical bonding.

Later in life, Pauling became a peace activist, and in 1963, he received the Nobel Peace Prize. He died in 1994.

Key works

1928 "The Shared-Electron Chemical Bond"
1939 *The Nature of the Chemical Bond and the Structure of Molecules and Crystals*
1947 *General Chemistry*

Pauling began his quest soon after he started work as a postgraduate student at the California Institute of Technology in 1922. From 1929, this role enabled him to spend several weeks each year for five years in Berkeley, California, as a visiting lecturer in physics and chemistry, talking in depth with Lewis.

By then, scientists had taken Planck's quantum theory further. In 1913, Danish physicist Niels Bohr proposed that electrons orbited the nucleus at the center of an atom with energies that were set at specific quantum levels. Only a few electrons could sit at each level, but nobody yet knew why they could not all be at the same level.

Pairs of electrons

In 1924, Austrian theoretical physicist Wolfgang Pauli proposed a previously unknown quantum property of electrons that explained this separation into different levels. The new property had some things in common with the angular momentum everyday objects have when they rotate, so scientists called it spin. Electrons could adopt one of only two opposite quantum values of spin. They could exist in pairs of electrons with opposite spin values, but once paired up, no other electrons could join that pair. This was one of the important discoveries that in 1925 gave rise to quantum mechanics—in particular, the wave equation devised by Austrian physicist Erwin Schrödinger. Schrödinger's equation included a wave function that mathematically describes the quantum properties of particles.

It quickly became clear that Schrödinger's equation could be applied to atoms, and therefore quantum mechanics could be a reliable basis for the theory of molecular structure. In 1927, German physicist Walter Heitler showed how two hydrogen atom wave functions joined together to form a covalent bond. However, it also soon became obvious that the Schrödinger wave equation was too complicated to easily describe more complex molecules.

As a result, chemists such as Pauling had to devise their theories of molecular structure and chemical bonds based upon their own experimental observations and referring to quantum mechanical principles. These principles showed how a covalent bond required more than just two atoms with an electron each to share. The electrons must

Linus Pauling's handwritten notes indicate how he derived a tetrahedral set of hybrid orbitals using s, p, and d functions. These letters are linked to the behavior of the electrons in each orbital.

Pauling realized that if a molecule can be drawn with different arrangements of its bonds, such as benzene, then that molecule could continually change, or resonate, between the different arrangements.

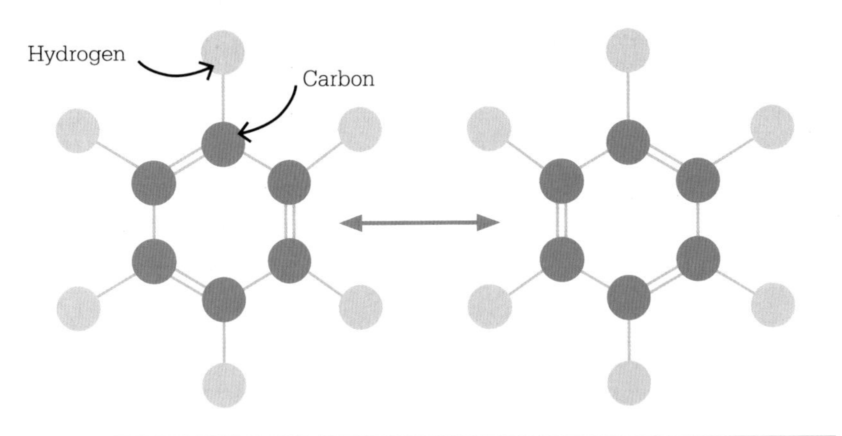

have opposite spins, and the atoms must also each have a stable energy level, called an electronic orbital, for the electrons to occupy.

Valence bonding theory

Pauling used Lewis's pair bonding idea and quantum mechanics to develop three other key concepts in what became known as valence bonding theory. For the first, in 1928, he introduced the idea that electrons in bonds could move between the various atoms in a molecule. This was significant when there are two possible ways of drawing the bonds in a structure, and it enabled chemists to make calculations that predicted how molecules would behave.

A key example is the chemical benzene, which has six carbon atoms arranged in a hexagonal ring. It is constructed from three single covalent bonds and three double covalent bonds. In each double bond, two carbon atoms share four electrons. The single and double bonds alternate, and there are two ways to draw the arrangement. If electrons can move, benzene can be thought of as existing in both structures, resonating from one to the other and back. As such, Pauling called the effect resonance. The extended sharing of electrons means that bonds in which resonance is possible are more stable than the same bonds without resonance.

Hybrid orbitals

For the second concept, one evening in December 1930, Pauling worked out how to explain some of the period's most puzzling problems in chemistry. Bohr's ideas of electrons orbiting atoms at specific levels did not allow atoms to share as many electrons as they shared in real molecules. Perhaps the most important puzzle was why carbon atoms could often form four single covalent bonds, equally spaced in a tetrahedral shape. If a central carbon atom is bonded to four atoms that are all the same, the bonds are all equivalent. That did not fit with the ideas of the time.

The problem arose from how scientists had further developed the idea of electron orbitals, which they distinguished using three quantum properties: charge, spin, and orbital angular momentum. Orbital energy levels started at the lowest value of one and increased. Differing values »

The chemical bonding spectrum

A bond between two atoms of the same element is primarily covalent. Each atom shares one or more of its electrons with its neighbor evenly. The two shared electrons come together in an orbital that helps hold the atoms together.

When atoms of two elements bond, their relative electronegativities determine the extent to which they share electrons. An atom of an element with a very low electronegativity tends to lose one or more electrons to atoms of elements whose electronegativity is very high. The two atoms end up with opposite electrical charges, which attract each other and hold the atoms together. This happens with sodium, which has low electronegativity, and chlorine, which has high electronegativity.

When atoms of different elements have similar electronegativities, they can form bonds that are partly covalent and partly ionic. This can happen in bonds between hydrogen and chlorine, for example. Sometimes, having both forms of attractive interaction means that these bonds can be even stronger.

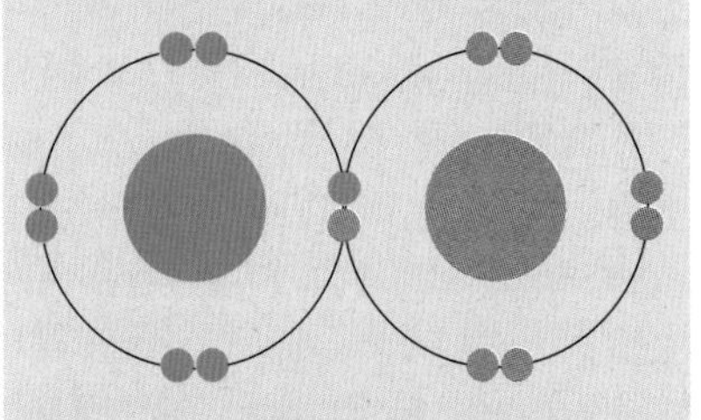

This covalent bond illustrates that bonding electrons are shared equally between two atoms, and there are no charges on the atoms.

of angular momentum could be described by the letters s, p, d, and f. The letters relate to lines seen when the light produced by some metals as they are heated is spread out through a prism; they stand for sharp, principal, diffuse, and fundamental. The lines link back to how the electrons in each orbital behave. Atoms of carbon should have two electrons in their 2s orbital, already paired up and seemingly unavailable for bonding. It also has two lone electrons in 2p orbitals. These lone electrons suggest that carbon should be able to form two bonds. Chemists wondered where carbon's other two bonds came from.

In December 1930, Pauling realized that resonance meant that carbon's two 2s electrons could be shared with its 2p orbitals. He worked out a mathematical way to treat the orbitals as if they had all mixed together and formed four new orbitals that were equivalent and tetrahedrally arranged. Pauling called the new orbitals hybrids, and therefore the mixing process was hybridization. Other hybrid orbitals could explain other equally puzzling shapes, such as flat squares and octahedra, seen with other atoms.

Electronegativity scale

In 1932, Pauling revealed the third concept, which was a connection between ionic and covalent bonding, helping explain Debye's dipole moments. He noticed that, in chemical compounds, bonds between similar elements were not as strong as bonds between dissimilar elements. Pauling suggested that this was because these bonds were partly covalent and partly ionic. The strength of the bonds depended on how much the atoms of an element attract electrons from their surrounding environment—how greedy they are for electrons, essentially. In 1811, as part of his electrochemical theory, Jöns Jacob Berzelius had called this property electronegativity.

Although chemists such as Berzelius had previously studied electronegativity, their ideas were limited. Part of the problem is that there is no measurable constant value for electronegativity, making it difficult to compare the different elements. In a paper published in 1932, Pauling based relative electronegativity values on the energy released when molecules formed and burned, which in turn relates to the strength of their bonds. He determined that the bond between lithium and fluorine was almost 100 percent ionic, so he put lithium at the least electronegative end of his scale and fluorine at the most electronegative. Later refining his methods, Pauling estimated each electronegativity value as the covalent contribution to an element's bond subtracted from the measured bond energy. Many have tried to improve on this, but Pauling's 1932 scale is still the most widely used.

Pauling's electronegativity scale was much less well supported by theory than resonance or hybrid orbitals, but it was one of his most influential ideas. With electronegativity, chemists can make interesting predictions about bonds and molecules without referring to the bond's complicated wave equation. For example, Pauling predicted that fluorine was so electronegative that it could form compounds with the gas xenon, which is otherwise unreactive. In

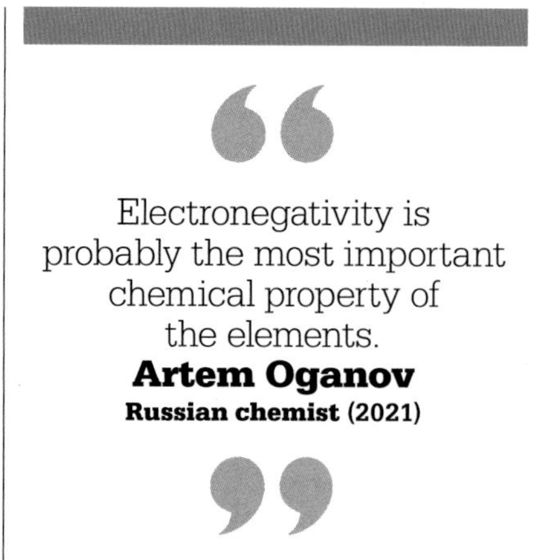

Electronegativity is probably the most important chemical property of the elements.
Artem Oganov
Russian chemist (2021)

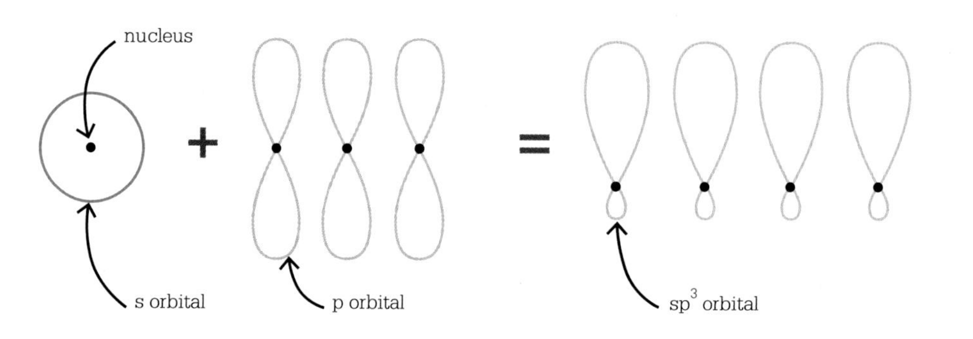

Pauling discovered that mixing an s orbital and three p orbitals gave four hybrid orbitals, called sp^3 orbitals. They are all at the same energy level and need to be distributed as symmetrically around an atom as possible, which means that they form tetrahedral shapes. This process is hybridization.

Linus Pauling pioneered the use of physical models to reveal chemical structures, such as using paper sheets to show that protein molecules could form this coiled alpha helix structure.

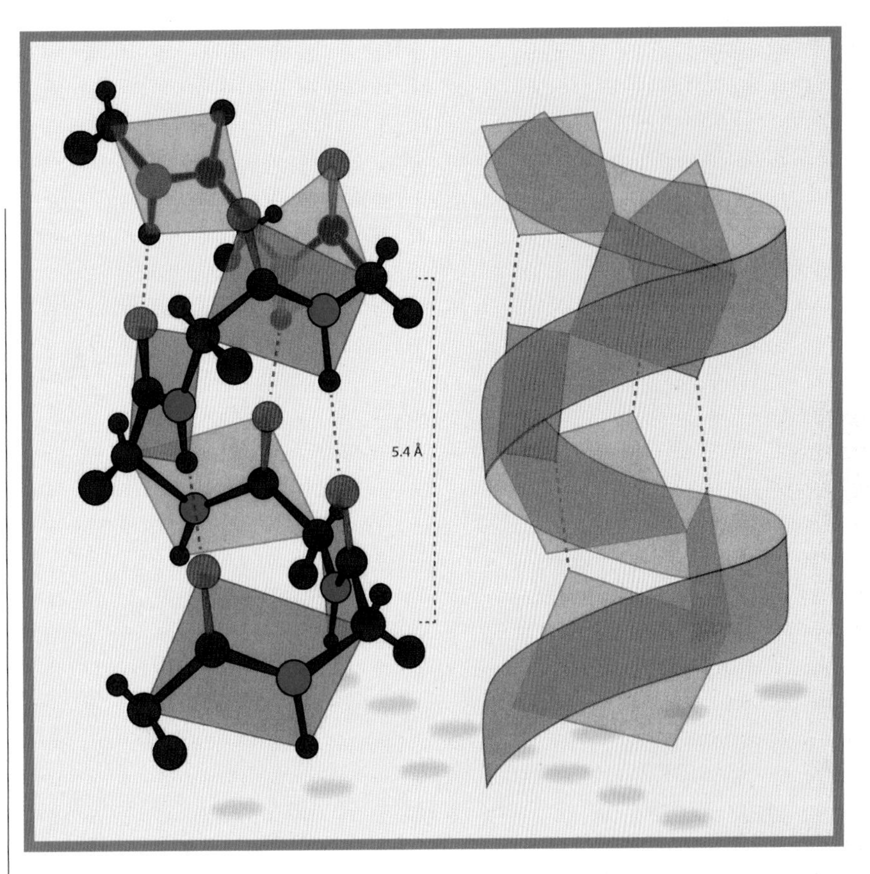

1933, one of Pauling's colleagues set out to test the prediction but was unsuccessful. Pauling was not proven correct until 1962, when separate American and German teams of scientists succeeded within months of each other in making xenon difluoride.

Illustrating atoms

Pauling's three concepts of resonance, hybrid orbitals, and an electronegativity scale—along with his many other ideas—transformed how chemists understood molecular structure. Pauling continued bringing these ideas together, and his 1947 textbook *General Chemistry* became a bestseller. It was a new way to teach chemistry to degree students, combining quantum physics, atomic theory, and real-world examples in explaining basic chemical principles.

This landmark textbook also included the clearest pictures yet of atoms and molecular bonding, representing these invisible objects directly. A large number of the drawings were made by American artist, inventor, and architect Roger Hayward.

Creating models

Using illustrations and models to visualize molecules was another transformative step, for Pauling and more broadly. For example, when Pauling was a visiting lecturer at the University of Oxford in 1948, he had a sudden insight into a problem whose solution had eluded him for more than a decade. Proteins are the long-chain molecules that give cells their structure and form the machinery inside them that drives life. Chemists knew that proteins were formed from molecules called amino acids linked together, but nobody had yet worked out the details of the structure of proteins. One night, Pauling coiled up a protein chain from a sheet of paper, folding it according to experimental evidence and to the bonding principles he had helped establish. He called the coil that emerged from his paper model an alpha helix, and additional experiments soon showed that proteins did in fact coil up in precisely that way.

Pauling further developed his modeling methods with his Caltech colleague, American biochemist Robert Corey. In 1952, the pair published details of how to make wooden kits for constructing three-dimensional models. These methods would quickly prove influential, and in 1953, they helped American chemist James Watson and British molecular biologist Francis Crick make their groundbreaking discovery of the iconic twin coil, known as the double helix, of the molecule DNA, which carries genetic information.

Pauling had not only dived deep into the most complex physics of how atoms connected, he also presented the information more clearly and accessibly than ever before, using striking images and models. Later scientists have since improved on his findings and techniques, but Pauling was the one who opened up the vast range of possibilities for them to explore. ■

THE
NUCLEAR
1940–1990

AGE

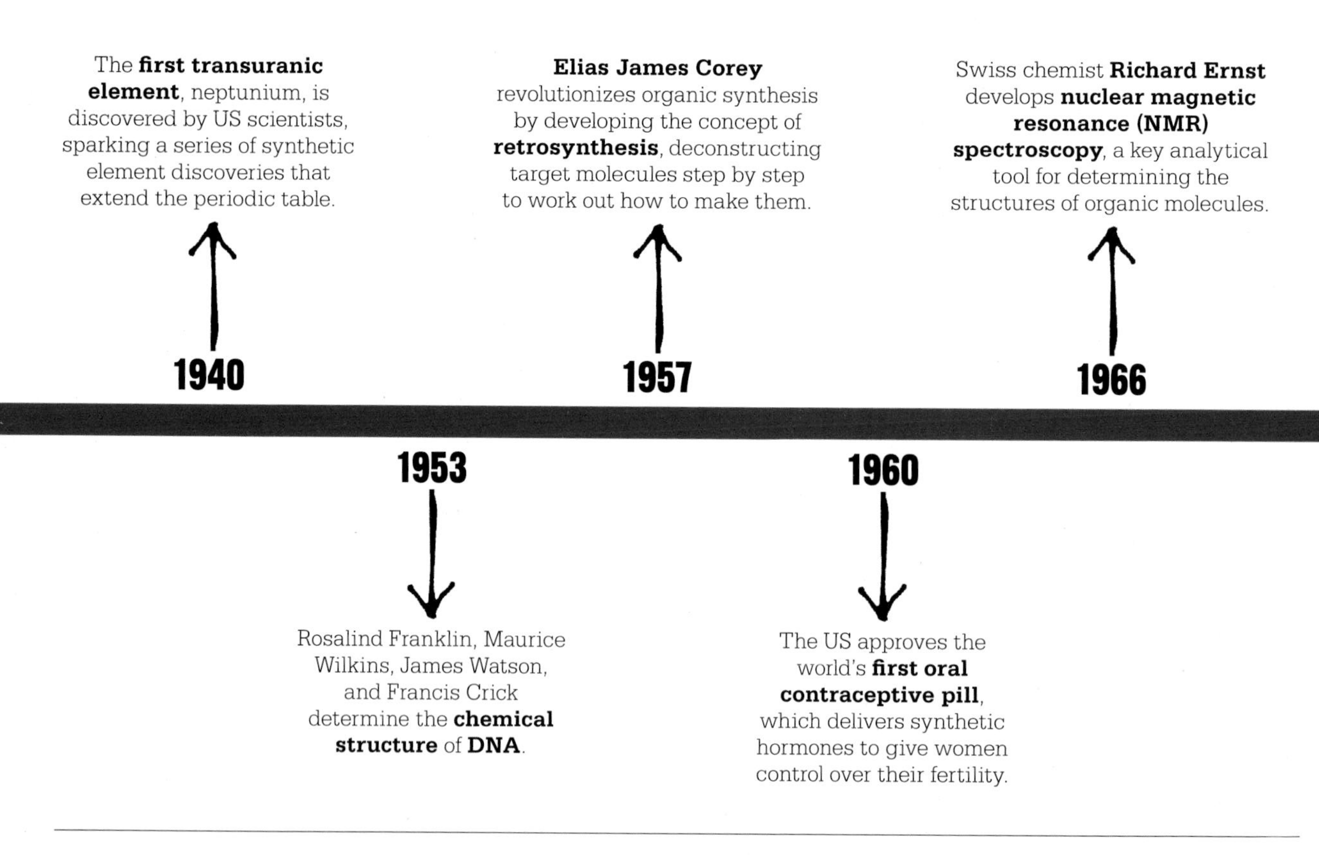

1940
The **first transuranic element**, neptunium, is discovered by US scientists, sparking a series of synthetic element discoveries that extend the periodic table.
1953
Rosalind Franklin, Maurice Wilkins, James Watson, and Francis Crick determine the **chemical structure** of **DNA**.
1957
Elias James Corey revolutionizes organic synthesis by developing the concept of **retrosynthesis**, deconstructing target molecules step by step to work out how to make them.
1960
The US approves the world's **first oral contraceptive pill**, which delivers synthetic hormones to give women control over their fertility.
1966
Swiss chemist **Richard Ernst** develops **nuclear magnetic resonance (NMR) spectroscopy**, a key analytical tool for determining the structures of organic molecules.

In 1900, the average worldwide life expectancy was just 32 years. By 1990, it had doubled to 64 years. While not the sole factor, advancements in medicine, including the development of new drugs to treat previously incurable diseases, played a significant part. But as medicines provided longer living through the application of new chemical techniques, these decades would also see an increase in scrutiny of the health and environmental consequences of other chemical advances.

Drugs by design

The discovery of antibiotics in the 1920s may be considered by many people to mark the starting point of modern chemical medicine. However, it was in the second half of the 20th century that progress in organic chemistry would really revolutionize the way we treat a range of diseases.

In 1957, the development of retrosynthesis—the process of working backward from a target molecule step by step in order to identify potential routes to make it—changed how chemists approached molecular synthesis. This process made it possible for chemists to make synthetic versions of complex natural molecules from commonly available reagents.

The creation of new synthetic analogs of natural hormones in the human body would produce even greater change. In 1960, the US Food and Drug Administration (FDA) approved the first oral contraceptive pill; this marked the point at which women were given reproductive autonomy, a liberating advance that transformed society. Today, an estimated 151 million women worldwide use the pill.

Another advance was in the analysis of natural substances so that they could be manufactured and utilized more easily. Dorothy Crowfoot Hodgkin was single-handedly responsible for several of these developments, using X-ray diffraction to map the molecular structures of penicillin, vitamin B12, and finally insulin.

With the organic chemists' armory greatly enhanced, rational drug design provided the next step toward better medicines. While early drug development often involved little more than trial and error, by the 1960s, chemists were starting to consider how they could selectively target specific

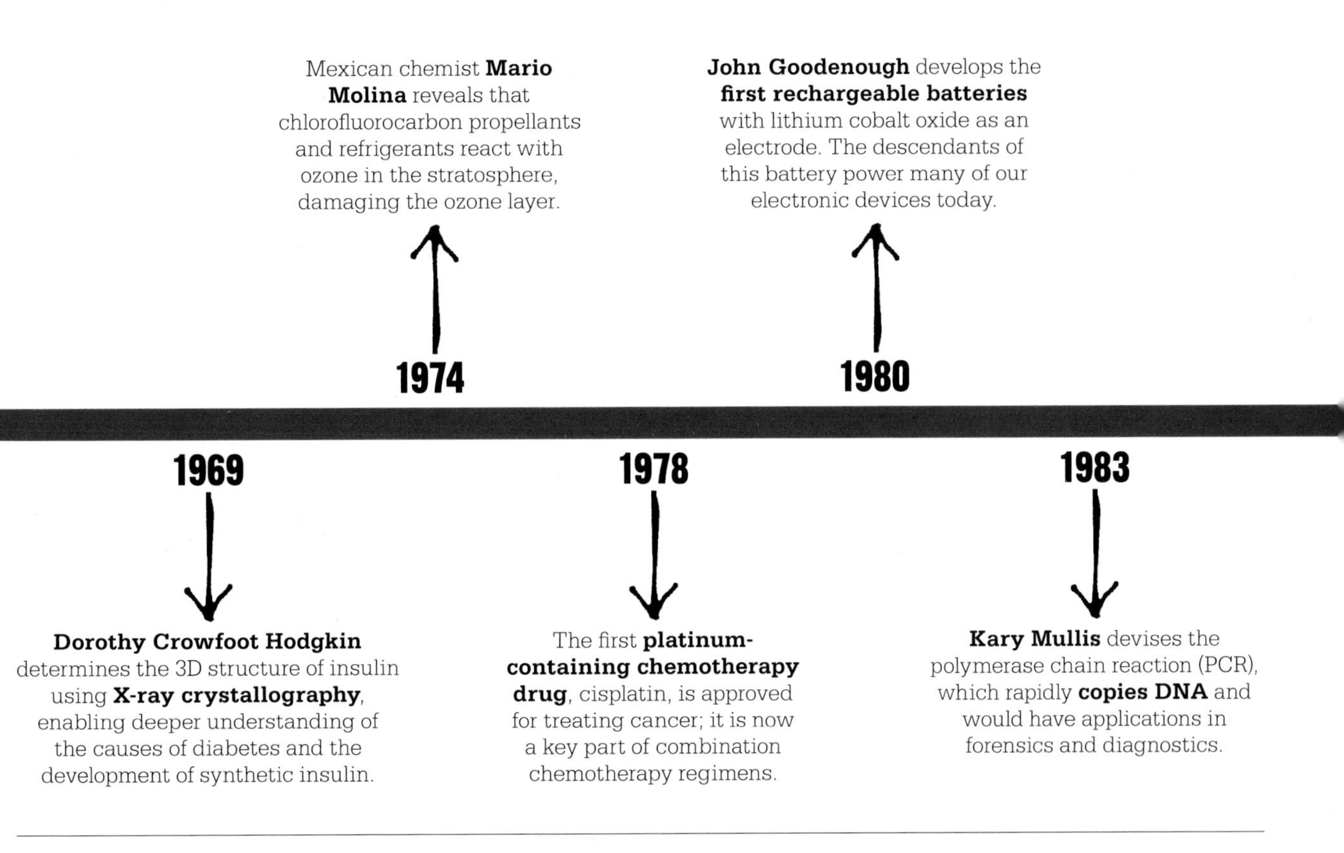

1969
Dorothy Crowfoot Hodgkin determines the 3D structure of insulin using X-ray crystallography, enabling deeper understanding of the causes of diabetes and the development of synthetic insulin.
1974
Mexican chemist Mario Molina reveals that chlorofluorocarbon propellants and refrigerants react with ozone in the stratosphere, damaging the ozone layer.
1978
The first platinum-containing chemotherapy drug, cisplatin, is approved for treating cancer; it is now a key part of combination chemotherapy regimens.
1980
John Goodenough develops the first rechargeable batteries with lithium cobalt oxide as an electrode. The descendants of this battery power many of our electronic devices today.
1983
Kary Mullis devises the polymerase chain reaction (PCR), which rapidly copies DNA and would have applications in forensics and diagnostics.

biochemical mechanisms within cells. This led to more efficient drug design for a wide range of diseases.

Chemical consequences

As the applications of chemistry proliferated, it was difficult to look past the benefits that they brought. However, as scrutiny fell on certain commonly used substances, a variety of environmental and health problems began to be revealed.

In some cases, these problems were known but had been covered up. The potential adverse effects of tetraethyl lead, a gasoline additive that increased the efficiency of car engines, had been postulated shortly after its introduction in the 1920s, but it was not until the 1960s that research definitively showing its toxicity in humans led to it being phased out from the 1970s. Even then, it was only in 2021 that the final stocks of leaded fuel were exhausted in Algeria. Atmospheric lead levels in towns and cities across the world still remain higher than expected background levels.

Other problems triggered more urgent action. In 1974, researchers showed that chlorofluorocarbons (CFCs), used in propellants and refrigerators, were the cause of depleting ozone in the stratosphere. The concern that this could lead to a permanent hole in the ozone layer at Earth's poles spurred an agreement in 1987 to phase out CFC use worldwide, and today the ozone layer is on the mend.

After the development of fertilizers early in the 20th century, synthetic pesticides and herbicides were the next advances to improve crop yields. These compounds, too, are controversial. The biggest-selling herbicide—glyphosate, approved in 1974—has been mired in a court battle since 2019 over its alleged potential to cause cancer, while research in 2017 showed that neonicotinoid pesticides can be toxic to honeybees.

Better batteries

Some advances of the late 20th century attracted better publicity. The development of rechargeable batteries, culminating in John Goodenough's lithium ion battery precursor, defined the course of technological development into this century. Almost all of the electronic devices we use today, from electric cars to smartphones, rely on the descendants of Goodenough's batteries, and it is hard to imagine the modern world without them. ■

WE CREATED ISOTOPES THAT DID NOT EXIST THE DAY BEFORE

THE TRANSURANIC ELEMENTS

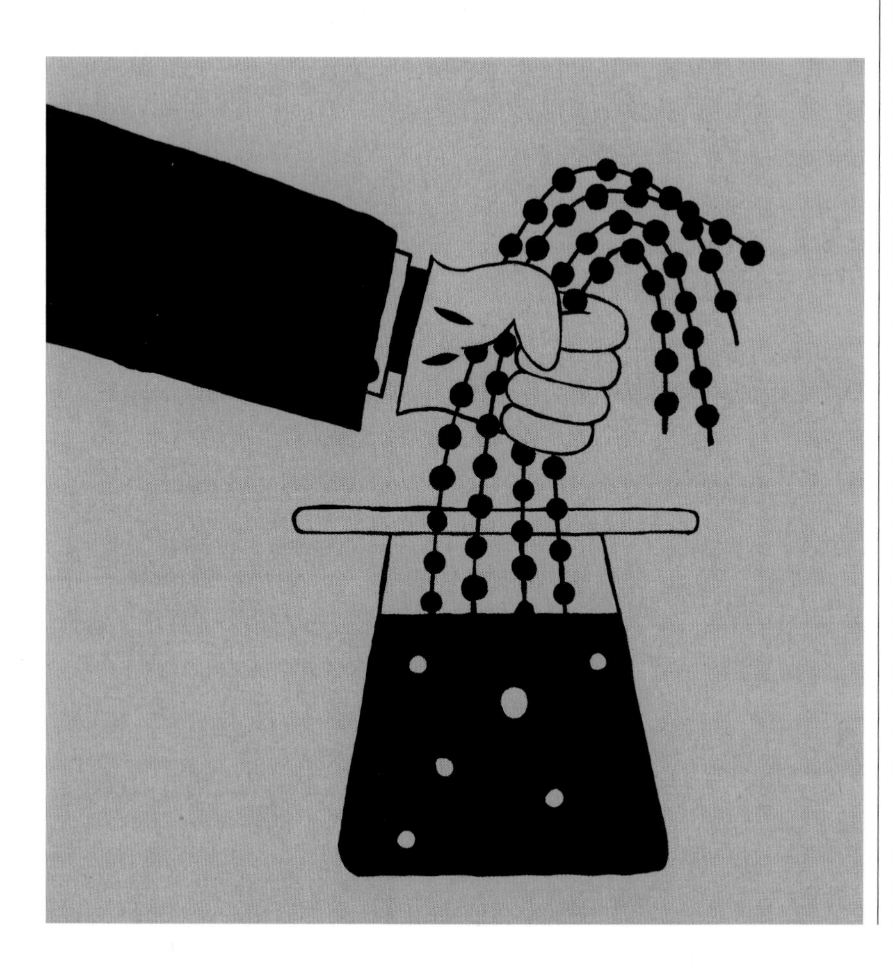

IN CONTEXT

KEY FIGURE
Glenn Seaborg (1912–1999)

BEFORE
1913 Frederick Soddy and others realize that elements can have different forms, known as isotopes. Previously, some isotopes had been mistaken for new elements.

1937 Emilio Segrè discovers element 43, technetium, in molybdenum discarded from a cyclotron. This is the first element created in this way.

AFTER
1977 *Voyager 1* and *Voyager 2* spacecraft start their journey to Jupiter, Saturn, Uranus, and Neptune, powered by radioactive decay of plutonium.

1981–2015 Teams in the US, Russia, Germany, and Japan make elements 107 to 118 using methods pioneered by Seaborg and Albert Ghiorso.

On August 9, 1945, a US bomber plane dropped a "Fat Man" atomic bomb on Nagasaki, Japan. It killed 22,000 people instantly, and the radiation killed four times that number by the end of the year. This terrible event was also a scientific milestone because the bomb contained about 13 lb (6 kg) of enriched plutonium, an element whose existence had been kept secret until then. The bomb was intended to hasten Japan's surrender and to end World War II.

Plutonium made its mark on history, and it would shape politics for decades to come—mainly thanks to chemists working at the Metallurgical Laboratory of the

See also: Rare-earth elements 64–67 ▪ The periodic table 130–137 ▪ Radioactivity 176–181 ▪ Isotopes 200–201 ▪ Synthetic elements 230–231 ▪ Nuclear fission 234–237

Elements always have the same number of protons but can exist as **isotopes**, with differing numbers of neutrons.

The **invention of the cyclotron** enables scientists to make new isotopes by **smashing atoms** together.

Smashing atoms together in cyclotrons can also create **new elements**.

The new synthetic elements fit into the periodic table after uranium.

Manhattan Project in Chicago, Illinois, led by Glenn Seaborg. Seaborg had first created plutonium in 1940, and he went on to discover nine more transuranic elements with atomic numbers higher than that of uranium. In 1951, his discoveries led to him receiving the Nobel Prize in Chemistry, awarded jointly to Seaborg and Berkeley physicist Edwin McMillan.

Isotope separation

Seaborg's work with radioactive isotopes began at the University of California, Berkeley. In 1937, Italian-American physicist Emilio Segrè isolated the new element technetium from molybdenum that had been exposed to high-energy radiation in a cyclotron, a type of particle accelerator. The previous year, Berkeley physicist Jack Livingood had asked Seaborg to help separate and identify the isotopes of different elements that the cyclotron was producing. They were simply trying, at that time, to discover new isotopes of existing elements. Many of these would go on to be useful in medical diagnoses and treatments. As Seaborg later recalled, "I demonstrated the usefulness of a chemist in this area dominated by physicists." The collaboration, he said, "steered me into my life's work."

Nuclear fission

Research underway in Berlin, Germany, in 1938 would also help steer Seaborg into his life's work. German chemists Otto Hahn and Fritz Strassmann were, for the first time, measuring radioactivity arising from the nuclei of uranium atoms splitting into smaller ones—the process of nuclear fission. »

Glenn Seaborg

Born in 1912 in Michigan, Glenn Seaborg completed his chemistry degree at the University of California, Los Angeles (UCLA), in 1933, then his PhD at UC Berkeley in 1937. Berkeley's renowned dean of chemistry Gilbert Lewis invited Seaborg to be his personal laboratory assistant, and they published many papers together.

Best known for his discoveries of transuranic elements and his crucial role on the Manhattan Project, Seaborg also contributed to the discovery of more than 100 isotopes of elements, some of which became essential to medicine. In 1980, he transmuted a tiny quantity of bismuth-209 into gold—the ancient alchemists' dream.

Seaborg authored or coauthored more than 500 scientific papers and was also scientific adviser to 10 US presidents. He died in 1999.

Key works

1949 "A New Element: Radioactive Element 94 from Deuterons on Uranium"
1949 "Nuclear Properties of $^{238}94$ and $^{238}93$"

However, they had ruled out fission as an explanation. Early in 1939, Hahn's former colleague, Austrian-Swiss physicist Lise Meitner, together with her nephew Otto Frisch, pieced together the evidence showing that this unbelievable outcome had really occurred. Gradually, the world's scientists became aware of the potential of fission. It releases vast amounts of energy, which could be harnessed for peaceful purposes—or for weaponry.

Plutonium for the atomic bomb was produced during World War II at the Hanford Engineer Works (above) in Washington.

The nuclear arms race

At first, scientists were mainly curious to understand how fission worked. Seaborg's colleague Edwin McMillan started experimenting with uranium in the Berkeley cyclotron. In 1940, by bombarding uranium-238—its most common isotope—with neutrons, McMillan discovered the first transuranic element, neptunium. Soon after, he departed to begin military research on radar, leaving Seaborg to lead the research team at the age of just 28.

With fellow chemists Arthur Wahl and Joseph Kennedy, Seaborg switched to a different approach, bombarding uranium-238 with deuteron particles, which contain one proton and one neutron. In 1940, they discovered tiny amounts of another new element, which they called plutonium. Despite the small quantities, they quickly established that, if used in a bomb, plutonium would explode with inconceivable force. With World War II already underway, they decided to keep their discovery a secret.

In 1941, in response to Japan's attack on Pearl Harbor, the US entered the war, and in spring 1942, Seaborg was ordered to the University of Chicago to develop an atomic bomb for the Manhattan Project. Seaborg invited nuclear scientist Albert Ghiorso, with whom he had worked at Berkeley, to join him in Chicago. Ghiorso would prove essential to research on transuranic elements, inventing ways to isolate and identify heavy elements atom by atom—a scale Seaborg described as "ultra-microchemistry." To develop weapons, however, tiny quantities were far from enough.

The Berkeley cyclotron used two huge magnets to create an electromagnetic force that spun particles around in a circular path.

The cyclotron

When Ernest Rutherford first split the nuclei of nitrogen in 1919, a door opened into an exciting new field. But to be able to split other atoms, physicists needed to give them more energy. Many machines were built to accelerate charged particles, the most famous of which was the cyclotron, the brainchild of Ernest Lawrence from the University of California, Berkeley.

Lawrence realized that with a circular design, particles could be accelerated more than once. In his cyclotron, particles were injected into a space formed by two hollow semicircular pieces of metal, separated by a gap. Above and below this space were two powerful magnets.

The pieces of metal were connected to a high-frequency alternating electric current, which energized the particles every time they crossed the gap. With each energy boost, the particle's path spiraled outward. Eventually, the particle would leave the cyclotron and slam into a target, combining with an atom in the target and forming a new isotope or element.

> We created isotopes that did not exist the day before, with uses yet to be discovered.
> **Glenn Seaborg**

To source enough plutonium for a weapon, Manhattan Project scientists used a neutron-producing fission chain reaction of the uranium-235 isotope. As well as triggering fission in other uranium-235 atoms, the neutrons could also turn uranium-238 into plutonium-239 on a bigger scale.

Seaborg's team became expert at isolating elements in their pure form by oxidizing them into salts, then using further techniques to separate them from the highly radioactive uranium and other fission products. For plutonium, for example, they added bismuth phosphate ($BiPO_4$). This made a precipitate with one form of the plutonium salt. After isolating that precipitate, the chemists could then oxidize plutonium further and separate it from the precipitate. In 1943, the US built a factory to make plutonium in this way.

Adding to the periodic table

After their success with plutonium, Seaborg, Ghiorso, and their colleagues began to look for more transuranic elements. They returned to their cyclotron ultra-microchemistry approach but found that the new elements could not be oxidized as easily as plutonium. It took nearly a year to separate the next two elements: americium, made by bombarding plutonium-239 with neutrons, and curium, made by bombarding plutonium with helium ions. From there, Seaborg and his colleagues would go on to make berkelium, californium, einsteinium, fermium, mendelevium, and nobelium.

Seaborg proposed that uranium and the transuranic elements form a new row in the periodic table, along with actinium, thorium, and protactinium. This family became known as the actinides (from *aktis*, the Greek word for beam), named after the mode in which cyclotrons smashed beams of atoms together.

Superheavy elements

In 1961, using the cyclotron methods independently from Seaborg, Ghiorso and other Berkeley researchers discovered lawrencium, the last and heaviest of the 15 actinides. In 1969, Ghiorso's group—now including James Harris, the first African American credited with finding a new element—discovered rutherfordium, the first of the transactinides, or "superheavy elements." This was followed by dubnium in 1970. Seaborg later rejoined the team, which in 1974 discovered element 106 and named it seaborgium in his honor. That find was confirmed by another Berkeley chemist, Darleane Hoffman.

Hoffman had also made another remarkable discovery. In 1971, she extracted a tiny amount of plutonium-244 from rock samples several billion years old. This isotope has a half-life of 80 million years, so the plutonium must have been primordial, made before Earth was formed, by nuclear fusion reactions during supernova explosions. It seemed, then, that the heaviest naturally occurring element was not in fact uranium but plutonium. ■

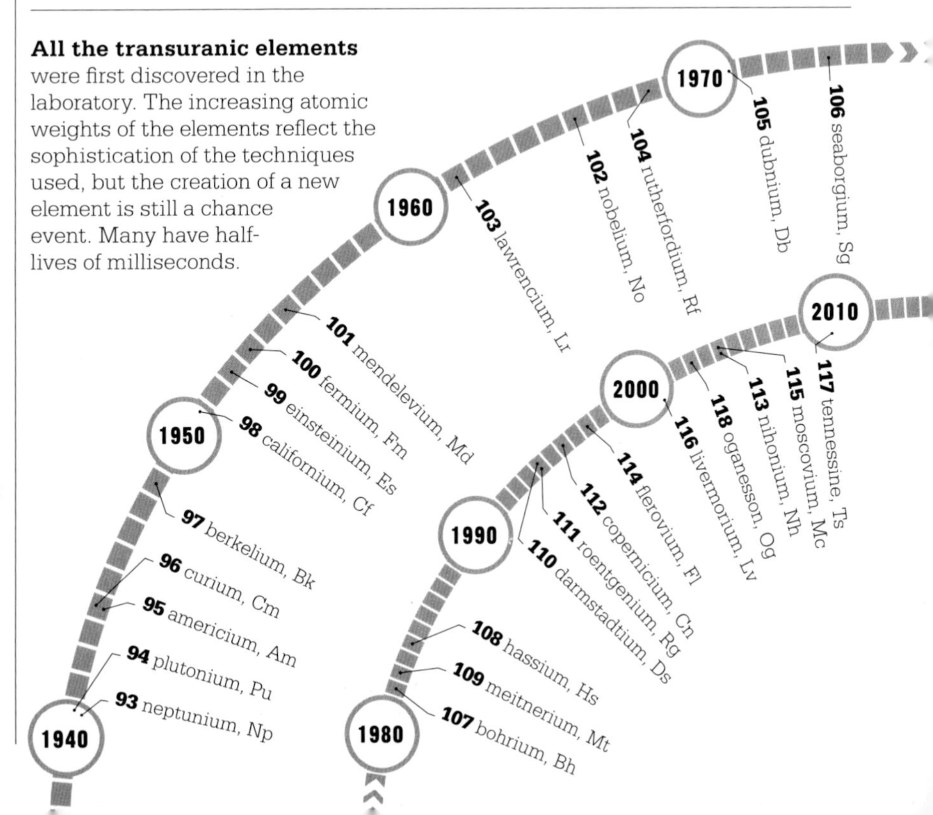

All the transuranic elements were first discovered in the laboratory. The increasing atomic weights of the elements reflect the sophistication of the techniques used, but the creation of a new element is still a chance event. Many have half-lives of milliseconds.

DELICATE MOTION THAT RESIDES IN ORDINARY THINGS

NUCLEAR MAGNETIC RESONANCE SPECTROSCOPY

IN CONTEXT

KEY FIGURE
Richard Ernst (1933–2021)

BEFORE
1919 Francis Aston, a British chemist and physicist, publishes his mass spectrograph for identifying chemical elements.

1924 Austrian quantum physics pioneer Wolfgang Pauli proposes that atomic nuclei behave like spinning magnets.

AFTER
1971 An American chemist, Paul Lauterbur, invents 3D nuclear magnetic resonance imaging (MRI), which is widely used to see inside human patients.

1989 Swiss scientist Kurt Wüthrich devises an NMR method to study highly complicated dissolved protein structures.

Working out the structural formula of an organic compound has always been vital to chemistry because it provides information about its properties and how it reacts with other compounds. However, it was an extremely difficult task until Swiss chemist Richard Ernst developed a technique for high-resolution nuclear magnetic resonance (NMR) spectroscopy in 1966.

Determining structures

Typically, to find out the structure of a molecule, chemists used reactions to break it down into smaller parts. They then isolated and identified those parts by looking at how they reacted and by establishing their melting points, for example. Once the smaller molecules had been identified, chemists needed to work out how they had been linked originally. Although one method for doing this—mass spectrometry—was invented in 1913, it did not become widespread until the 1960s.

A new technique emerged in 1945 from two groups of physicists. Swiss-American Felix Bloch led a team at Stanford University, while American Edward Purcell headed a group at Harvard University. Independently, they carried out the first successful NMR experiments. In NMR, "nuclear" refers to the properties of the nuclei at each atom's center, and "magnetic" refers to how scientists readied the nuclei for analysis by placing chemical samples in strong magnetic fields. The term "resonance" arose because they put the samples in weak electromagnetic radio waves and gradually changed their frequencies. Many objects have natural frequencies and resonate when exposed to an external force at that frequency. For example, you can make a wine glass "sing" by rubbing your

> We are dealing not merely with a new tool but with a new subject, a subject I have called simply nuclear magnetism.
> **Edward Purcell**

See also: Structural formulae 126–127 ▪ Infrared spectroscopy 182 ▪ X-ray crystallography 192–193 ▪ Mass spectrometry 202–203 ▪ Chemical bonding 238–245

wet finger around the rim. Scientists used this principle to detect when the radio wave's frequency matched the characteristic resonance frequency of the nuclei. Tracking the strength of the electromagnetic signal across different frequencies gives an NMR spectrum.

Game changer

In 1966, Ernst made a major breakthrough that brought NMR into everyday usage for chemists. While working at a leading manufacturer of NMR instrumentation, he swapped a radio frequency sweep for short, intense radio frequency pulses. He then measured the signal for some time after each pulse, with gaps of a few seconds between pulses. Ernst analyzed the resonance frequencies in the signals and converted them to NMR spectra, using computers to do a mathematical operation known as a Fourier transform. This boosted NMR sensitivity 10- to 100-fold. Such a boost made it possible to study small amounts of material and chemically interesting isotopes that do not occur in large amounts naturally, such as carbon-13. Today, all routine NMR chemical spectroscopy is based on Fourier transforms. Ernst and his colleagues also used Fourier transforms to develop the technique so that it can detect hydrogen nuclei in living bodies for medical NMR imaging. We know this as magnetic resonance imaging (MRI).

NMR instruments are now even more sensitive because they use superconducting wires that have no electrical resistance and conduct very high electrical currents. These wires can produce magnetic fields almost 600,000 times stronger than the one that forms naturally between Earth's poles. This allows chemists to study large, complex, moving molecules. For example, highly sensitive NMR has shown how an enzyme involved in leukemia and other cancers switches between active and inactive states. ■

Making atoms resonate

1. Atomic nuclei behave like spinning magnets, usually all jumbled up.

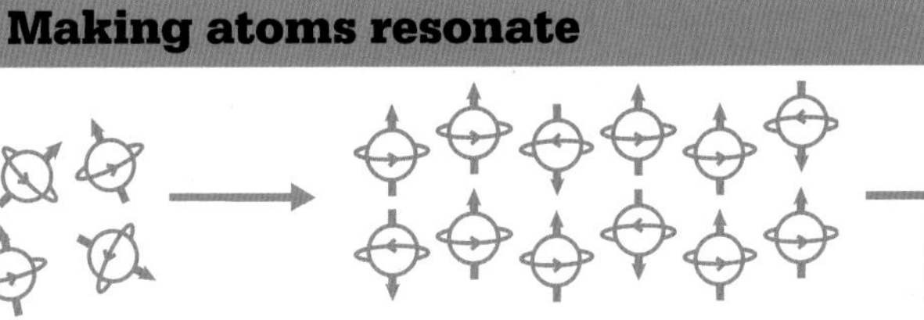

2. In a magnetic field, nuclei spins align either with or against the magnetic field.

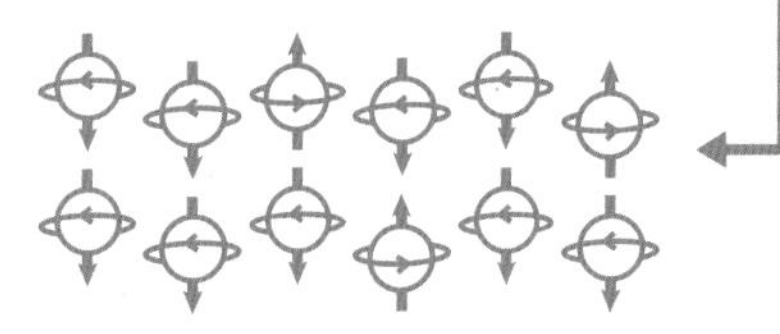

3. When radio waves of the right frequency hit nuclei aligned with the field, some of them flip their alignment.

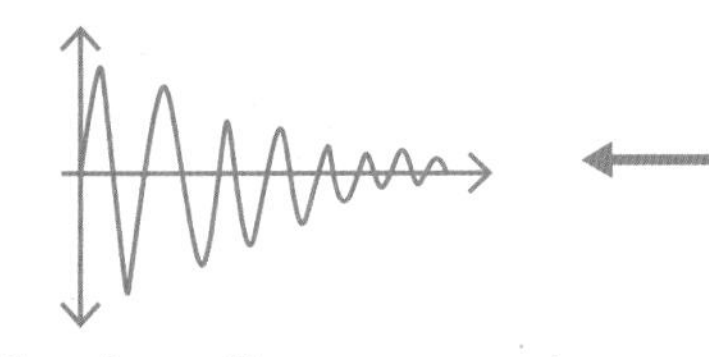

4. After the radio waves stop, the nuclei spins realign with the magnetic field, emitting radio signals as they do so. The signals provide structural information.

Richard Ernst

Born in 1933 in Winterthur, Switzerland, Richard Ernst became interested in chemistry at the age of 13, when he found a box of chemicals left by a late uncle in his attic. He studied chemistry at ETH Zurich, where NMR pioneer Felix Bloch had also studied, and graduated in 1957. Ernst remained at the institution to study for a PhD, and during this time, he built important components for two NMR spectrometers. After completing his PhD in 1962, he moved to California to work for NMR instrument maker Varian, where he developed the Fourier transform technique. Ernst then returned to Switzerland to take over the NMR research group at ETH Zurich and continued to advance NMR research for the rest of his career. In 1991, Ernst won the Nobel Prize in Chemistry. He died in 2021.

Key works

1966 "Application of Fourier transform spectroscopy to magnetic resonance"
1976 "Two-dimensional spectroscopy: Application to nuclear magnetic resonance"

THE ORIGIN OF LIFE IS A RELATIVELY EASY THING

THE CHEMICALS OF LIFE

IN CONTEXT

KEY FIGURES
Stanley Miller (1930–2007),
Harold Urey (1893–1981)

BEFORE
1828 Friedrich Wöhler makes urea from ammonium cyanate.

1871 Charles Darwin proposes that life began in a small body of water rich in chemicals.

AFTER
1962 American biologist Alexander Rich proposes the concept of the "RNA world," in which RNA (ribonucleic acid) molecules copied themselves before DNA (deoxyribonucleic acid) or proteins evolved.

1971 Scientists find amino acids in the Murchison meteorite, proving that they form before life exists.

2009 British chemist John Sutherland and his team show how building blocks of RNA could form under conditions that might have existed on early Earth.

Inorganic chemicals can react to form **organic ones**, usually made by living things.

Early Earth had water in oceans and **ammonia**, **hydrogen**, **and methane** in the atmosphere.

Chemicals present on early Earth **created** the conditions for the **building blocks of life** to appear.

These building blocks could have naturally led to the emergence of living organisms.

In 1924, Russian biochemist Alexander Oparin proposed the theory of abiogenesis—that life on Earth originated in a primordial soup, where simple chemicals reacted to form the carbon-based compounds needed for life. Oparin's compatriot, Dmitri Mendeleev, suggested that as early Earth's atmosphere cooled, metals solidified first, followed by other elements, including carbon, leaving an atmosphere that contained light gases such as hydrogen. Oparin proposed that carbon could react with superheated steam and form hydrocarbons. Could amino acids—the building blocks of life—then be created from them?

Making amino acids

In 1952, American chemist Harold Urey and his PhD student Stanley Miller came up with an experiment that would test this theory. Miller tried to mimic early Earth in a sealed glass apparatus, with two hollow spheres connected by tubing. One

See also: Isolating elements with electricity 76–79 ▪ The synthesis of urea 88–89 ▪ Enzymes 162–163 ▪ Retrosynthesis 262–263

The real question is whether or not there are very chance elements in the formation of life.
Stanley Miller

sphere held water, like that in Earth's oceans. The second contained hydrogen, ammonia, and methane—the gases they believed to be in the atmosphere. To start a chemical reaction, Miller simulated lightning with electric sparks, repeatedly putting energy into the mixture. After a day, the water in the flask had turned pink, and after a few days, a yellow slick appeared on the sides. Analyzing the substances he had made, Miller found five amino acids, including 3 of the 20 that are essential for life on Earth.

Doubters converted

When Miller and Urey published their findings in 1953, many scientists did not believe them. However, the experiment was so simple that their doubters tried it for themselves and were silenced.

The results supported the theory that the conditions on early Earth could provide the chemicals needed for the creation of organic life. But the experiment was not a complete success. It did not produce other important biological molecules that enabled living organisms to evolve, such as the nucleic acids RNA and DNA that carry genetic information.

Throughout his career, Miller continued to lead and innovate in the field, but admitted that fully explaining life's origins was far more complex than his 1952 experiment had made it seem. One critical issue is that scientists will never know for sure the exact conditions on Earth when life appeared. And even if researchers do create simple life by simulating the assumed conditions, they will not be able prove that it is by the same process that led to the first life forms on Earth. ■

Stanley Miller

Born in Oakland, California, in 1930, Miller studied chemistry at the University of California, Berkeley, then moved to the University of Chicago in 1951, where he studied for his PhD supervised by Urey. Miller initially preferred theory over "messy," "time-consuming" experiments, but he would ultimately be best known for his innovative origin of life experiments. Miller completed a one-year postdoctoral fellowship at the California Institute of Technology in 1954 before moving to Columbia University in 1955.

In 1960, Urey recruited him to the new San Diego campus of the University of California. He spent the remainder of his career there until his death in 2007. Although researching the origin of life remained Miller's main focus, he also contributed to other areas, including anesthesia.

Key works

1953 "A production of amino acids under possible primitive earth conditions."
1972 "Prebiotic synthesis of hydrophobic and protein amino acids."

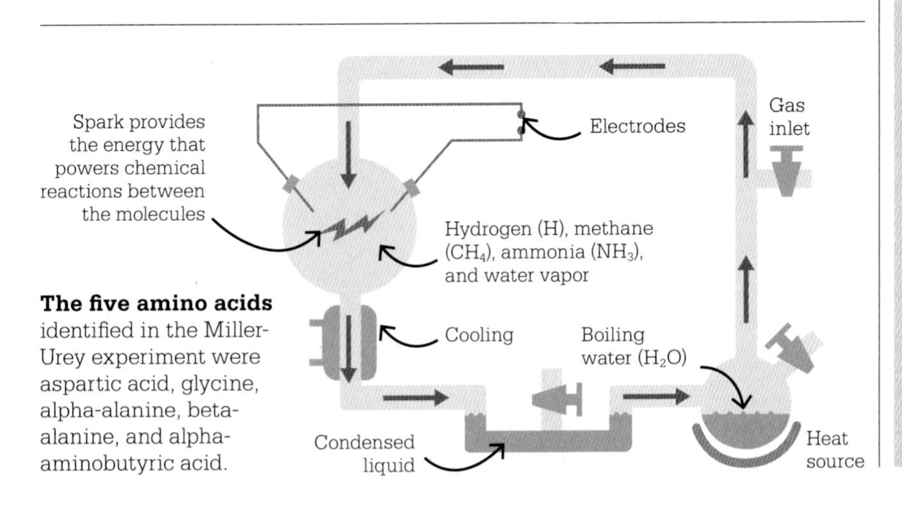

The five amino acids identified in the Miller-Urey experiment were aspartic acid, glycine, alpha-alanine, beta-alanine, and alpha-aminobutyric acid.

THE LANGUAGE OF THE GENES HAS A SIMPLE ALPHABET

THE STRUCTURE OF DNA

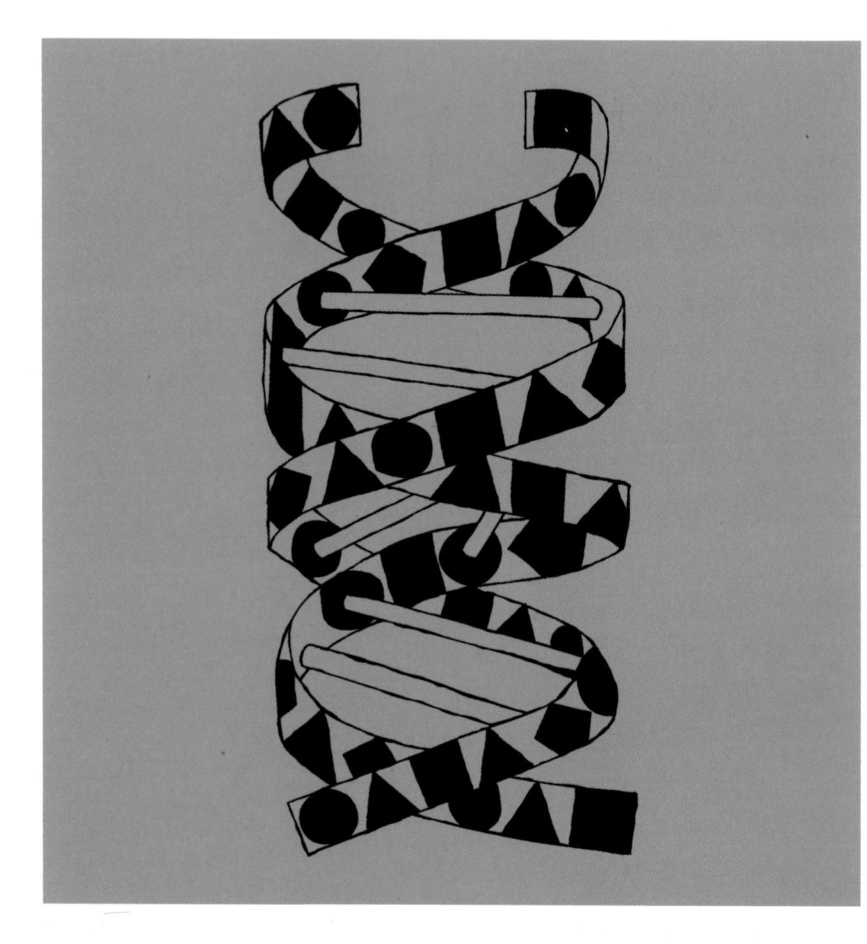

IN CONTEXT

KEY FIGURES
Francis Crick (1916–2004), **Rosalind Franklin** (1920–1958), **James Watson** (1928–), **Maurice Wilkins** (1916–2004)

BEFORE
1869 Friedrich Miescher extracts acidic phosphorus-rich substances from blood cells, including nucleic acids.

1885–1901 Albrecht Kossel isolates five essential building blocks of nucleic acids: adenine (A), guanine (G), cytosine (C), thymine (T), and uracil (U).

AFTER
2000 The Human Genome Project and Celera Genomics separately announce the first working draft of all information held in human DNA.

2012 Jennifer Doudna and Emmanuelle Charpentier develop the CRISPR-Cas9 system for easy DNA editing.

The way organisms pass on the instructions to enable them to develop, survive, and reproduce is truly one of nature's greatest wonders. After various clues emerged over nearly a century, the answer to how this happens literally took shape in 1953, in the form of a coil of molecules of deoxyribonucleic acid (DNA), called a double helix.

Long quest

Although DNA was discovered by Swiss physician Friedrich Miescher and formally named by German biochemist Albrecht Kossel in the 19th century, scientists still thought that protein molecules carried the

See also: Intermolecular forces 138–139 ▪ Stereoisomerism 140–143 ▪ X-ray crystallography 192–193 ▪ Polymerization 204–211 ▪ Chemical bonding 238–245 ▪ The polymerase chain reaction 284–285

instructions for life. In 1928, Fred Griffith, a bacteriologist working for Britain's Ministry of Health, mixed harmless live bacteria with dead disease-causing ones. He injected the mixture into mice, which then became infected with live disease-causing bacteria, now called pathogens. Griffith suggested that a "transforming principle" was responsible for the change in the bacteria, such as an unknown chemical substance passing from the dead bacteria to the live ones.

Other scientists thought Griffith must have made a mistake. In 1944, however, Canadian-American medical researchers Oswald Avery and Colin MacLeod and their American colleague Maclyn McCarty successfully repeated the experiment at the Rockefeller Institute of Medical Research in New York. They also biochemically tested the substance passing the instructions that caused the change. After eliminating all other possible substances, the team concluded that DNA was the transforming principle.

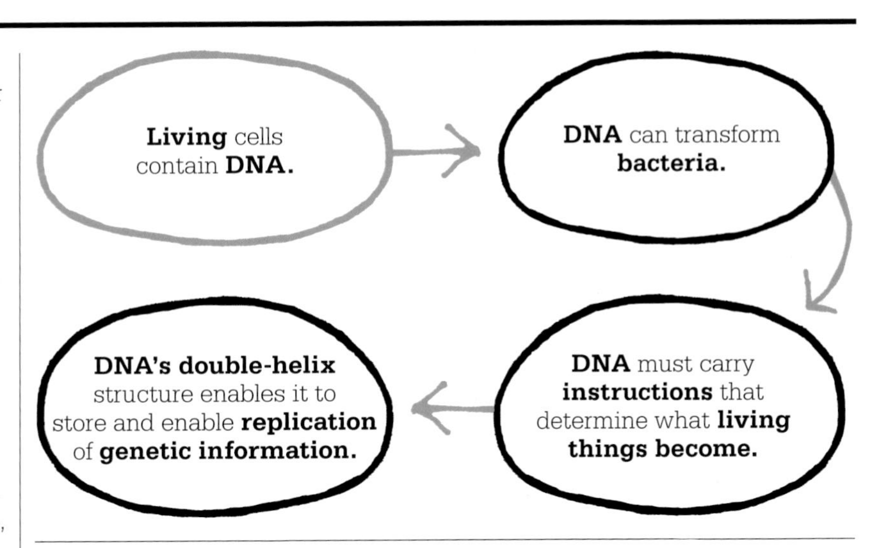

Living cells contain **DNA.**
DNA can transform **bacteria.**
DNA must carry **instructions** that determine what **living things become.**
DNA's double-helix structure enables it to store and enable **replication** of **genetic information.**

Also at the Rockefeller Institute, Russian-American biochemist Phoebus Levene, who had worked with Kossel, studied DNA in detail from 1905 to 1939. He revealed that DNA molecules are long chains made up primarily of a sugar building block, called deoxyribose. Every deoxyribose was joined to the next by phosphorus and oxygen atoms in a phosphate group. He also found that every deoxyribose is linked to one of four nucleic-acid building blocks, adenine (A), guanine (G), cytosine (C), and thymine (T), now called nucleobases, or bases. However, no one knew how DNA molecules organized themselves in living things.

An ideal information store

The powerful X-ray diffraction technique, which enabled scientists to determine atoms' positions from how crystals of a substance scatter X-ray beams, »

Rosalind Franklin

Born on July 25, 1920, in London, Rosalind Franklin obtained her PhD in physical chemistry from the University of Cambridge in 1945, with funding from a research fellowship. She worked at the Centre National de la Recherche Scientifique in Paris in 1947–1951, studying X-ray diffraction. She then moved to King's College London. Due partly to her strained relationship with Maurice Wilkins and partly to how King's treated women, she moved to Birkbeck College, London, in 1953. She also left the field of DNA research, instead publishing 17 papers on the structure of helical and spherical viruses, whose impact later helped in fighting the COVID-19 pandemic. Franklin was diagnosed with ovarian cancer in 1956 and died two years later, at age 37.

Key works

1953 "Molecular Configuration in Sodium Thymonucleate"
1953 "Evidence for 2-chain Helix in Crystalline Structure of Sodium Deoxyribonucleate"

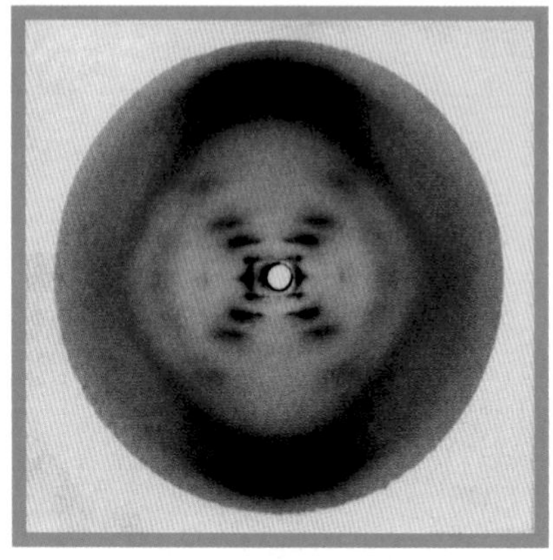

Photograph 51, an X-ray diffraction image taken by Rosalind Franklin in 1952, provided crucial proof that DNA was structured as a double helix.

would reveal DNA's structure in the early 1950s. American biochemist Linus Pauling from the California Institute of Technology (CalTech) seemed most likely to solve the DNA puzzle. He led the world in physical chemistry, deciphering the structure of many biological molecules by using X-ray diffraction data to build physical models. Instead, the answer began to emerge at King's College London in May 1950, when British biophysicist Maurice Wilkins received high-quality DNA crystals from Swiss chemist Rudolf Signer. Studying the crystals, Wilkins and PhD student Raymond Gosling showed that the DNA molecules were well organized, making them ideal for storing and transferring information.

British chemist Rosalind Franklin, an expert at collecting and analyzing X-ray diffraction patterns, joined King's College in January 1951. She took over from Wilkins the X-ray work on Signer's DNA sample and supervision of Gosling's PhD. Wilkins worked on DNA using a different sample and in May 1951 presented his results at a conference in Naples.

Enter the outsiders

In Naples, young American researcher James Watson was one of the few to see the significance of Wilkins's results. Later that same year, Watson began work at the University of Cambridge, where he met British molecular biologist Francis Crick. The two joined forces to decipher the structure of DNA. Watson went to a seminar in November 1951 in which Franklin showed that DNA coiled in a spiral with at least two chains, with phosphate groups on the outside.

Watson did not understand the lecture and misremembered the structure's details, but he and Crick built a cardboard and wire model, like those they had seen Pauling make of other molecules. Their structure had three DNA chains twisted into a spiral, with sugar–phosphate backbones inside and the bases pointing outward. They

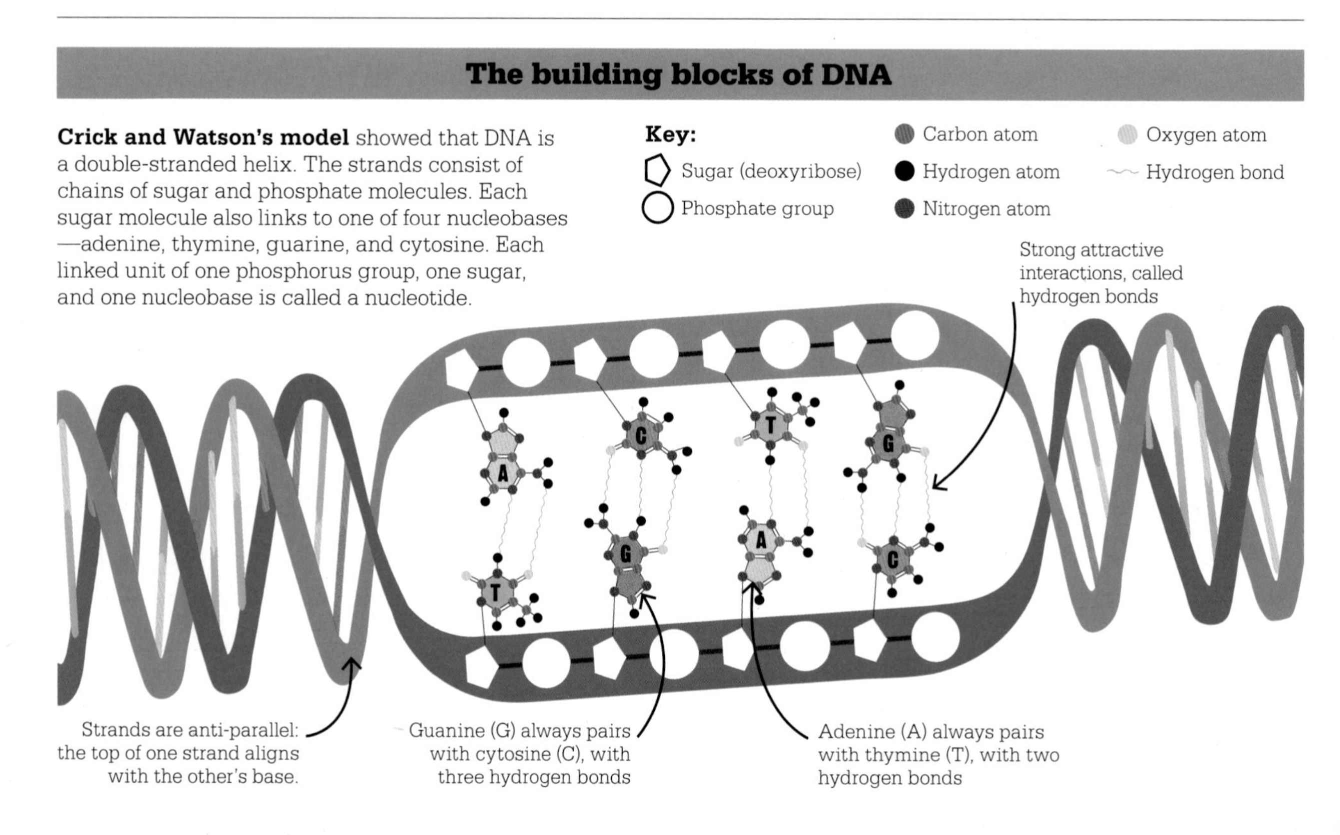
The building blocks of DNA
Crick and Watson's model showed that DNA is a double-stranded helix. The strands consist of chains of sugar and phosphate molecules. Each sugar molecule also links to one of four nucleobases —adenine, thymine, guarine, and cytosine. Each linked unit of one phosphorus group, one sugar, and one nucleobase is called a nucleotide.
Key:
Sugar (deoxyribose)
Phosphate group
Carbon atom
Hydrogen atom
Nitrogen atom
Oxygen atom
Hydrogen bond
Strong attractive interactions, called hydrogen bonds
A
C
T
G
T
G
A
C
Strands are anti-parallel: the top of one strand aligns with the other's base.
Guanine (G) always pairs with cytosine (C), with three hydrogen bonds
Adenine (A) always pairs with thymine (T), with two hydrogen bonds

invited Wilkins to see it, who brought Franklin with him. She said that their model did not fit her data, which showed the phosphate groups should be on the outside.

Franklin and Gosling took a key X-ray diffraction photo, photograph 51, in May 1952, showing that two chains coiled spirally to produce the now well-known double-helix structure. Shortly afterward, Crick and Watson met Austrian-American biochemist Erwin Chargaff, who had discovered another crucial structural clue. He had realized that in DNA the amounts of adenine and thymine are always equal, as are amounts of guanine and cytosine. The relative amounts of each pair of bases varies in different species, so DNA apparently differs between organisms, but there must be some underlying rule defining the relationship between bases.

Double-crossed helix

As 1952 ended, Crick and Watson heard that Pauling had figured out DNA's structure. But when the Cambridge researchers saw the model, they realized it was wrong: it was a three-chain helix with the phosphates inside and other flaws. Watson went to King's and suggested the groups collaborate to decipher DNA's structure, but Franklin declined. Wilkins agreed to help, however, and showed Watson photograph 51 without Franklin's permission.

Watson realized how the two complementary DNA spirals fitted in with Chargaff's findings of February 1953. Adenines on the inside of one DNA chain could pair with thymine molecules on the other, and the same could happen with guanine and cytosine. In this way, the bases could interact, locking together inside the helix through hydrogen bonds. Other molecules, including ones that copy DNA and read the instructions it carries in order to make proteins, could also easily engage with DNA via the hydrogen bonds.

Crick and Watson built another three-dimensional model. A double helix could spiral in two directions, right-handed or left-handed, but in their new model, DNA's double helix was only right-handed. It also showed that DNA chains are not symmetrical. One end could be viewed as the top and one as the base. The double helix was anti-parallel—the top of one strand aligned with the bottom of the other. This time, the structure convinced Wilkins and Franklin.

The Cambridge researchers published a short paper explaining the double-helix structure in April in the UK's *Nature* magazine, with a separate, supporting paper from the King's researchers. Yet Franklin was close to solving the structure of DNA on her own, as shown in a draft to *Nature* dated March 17, 1953, one day before she would see Watson and Crick's structure.

Wilkins, Crick, and Watson were awarded the Nobel Prize in Physiology or Medicine in 1962. Franklin's early death made her ineligible for nomination and lost her due recognition, although Crick wrote in a 1961 letter that the crucial data "was mainly obtained by Rosalind Franklin." James Watson's book *The Double Helix: A Personal Account of the Discovery of the Structure of DNA* (Atheneum, New York, 1968) belittles Franklin's role. It took Anne Sayre's *Rosalind Franklin and DNA* (Norton, New York, 1975) to firmly establish the credit Franklin deserved. The story often overshadows the double-helix discovery but makes the secret of the genetic code no less amazing. ■

Nucleic acids are basically simple. They are at the root of very fundamental biological processes, growth and inheritance.

Maurice Wilkins

Nobel Prize lecture (1962)

DNA's zippy copying system

In 1953, Crick and Watson hinted that DNA's structure related to how organisms copy their genetic material. Later, scientists revealed how this happens. DNA's unique A-to-G and C-to-T pairings in the double helix mean that each DNA strand binds to only one possible nucleobase sequence, like a zipper where each tooth needs a specifically shaped partner. When DNA "unzips," or uncoils, enzymes grab floating nucleobases—"loose teeth"—to build a new copy of the missing strand of the original double helix on each uncoiled strand, or side of the zipper. So two separated strands of one helix produce two new helices.

Crick also explained that base sequences in DNA form a code, biochemical assembly instructions for the order in which amino acids assemble to make proteins, and that the instructions may be translated via RNA (ribonucleic acid).

CHEMISTRY IN REVERSE

RETROSYNTHESIS

IN CONTEXT

KEY FIGURE
E.J. Corey (1928–)

BEFORE
1845 German chemist Hermann Kolbe makes acetic acid, showing for the first time that it is possible for chemical reactions to form bonds between carbon atoms.

1957 American chemist John Sheehan discovers a method for synthesizing penicillin.

AFTER
1994 Cypriot-American chemist Kyriacos Nicolaou and his team make the highly complex cancer drug taxol, using a retrosynthetic method that involves 51 steps.

2012 Polish-American chemist Bartosz Grzybowski and colleagues create Chematica, software that uses algorithms to predict pathways for the synthesis of molecules.

From the late 1940s, chemists had made great strides in synthesizing molecules. By the end of the 1950s, chemists had found many ways to assemble complex organic molecules—for use as agrochemicals, plastics, textiles, and medicinal drugs. But deciding which chemical reactions to use to synthesize the desired substance was largely a matter of intuition, trial, and error. Chemists picked a molecule that was commercially available and was also structurally similar to their target as a starting material. They then looked among the thousands of possible reactions for the transformations that they wanted.

> Organic substances … constitute the matter of all life on Earth, and their science at the molecular level defines a fundamental language of that life.
> **E.J. Corey (1990)**

Working backward

In 1957, E.J. Corey, professor of chemistry at the University of Illinois at Urbana-Champaign, had a simple idea that completely transformed the process of organic synthesis. He decided to develop a method for the theoretical deconstruction of target molecules, working backward toward a starting material. He called his method retrosynthetic analysis, or retrosynthesis.

Corey worked a target molecule through a series of hypothetical "transforms," each the reverse of a synthetic reaction. Each of the transforms separates the target into precursor structures—smaller, less complicated parts. The technique pinpoints strategic chemical bonds, often between two carbon atoms, and (theoretically) breaks them. The same process is then applied to the precursor structures. Gradually chemists build up a tree of molecular structures and reactions between them, representing possible synthetic routes to the target

See also: Functional groups 100–105 ▪ Structural formulae 126–127 ▪ Why reactions happen 144–147 ▪ Depicting reaction mechanisms 214–215 ▪ Chemical bonding 238–245

molecule. The tree ends with chemicals that are relatively cheap to buy or that can be made.

Corey's approach followed strict rules, so that each deconstruction step had to be the exact reverse of a chemical reaction. That way, he could be confident that the forward step would be successful in the laboratory. The retrosynthesis method could also help scientists identify entirely new chemical reactions for joining atoms together.

Lasting impact

One of the first molecules to which Corey applied the approach, in 1957, was longifolene, found in pine-tree resin. While not especially useful in itself, longifolene offered an important research challenge for chemists because its carbon atoms looped into rings that were difficult to synthesize. With retrosynthesis, Corey identified a bond that could be formed synthetically to connect the rings correctly.

In 1959, Corey moved to Harvard University, where he and his team went on to apply the retrosynthesis approach to more than a hundred natural products. For example, in 1967 scientists isolated the molecular structures of various types of insect juvenile hormone (JH) and wanted to explore the use of one of them, JH I, as an insecticide. It was impossible to acquire enough JH I from insects, but by 1968, Corey's team had figured out how to synthesize it. In the process, they invented four entirely new chemical reactions, three of which became widely used by chemists.

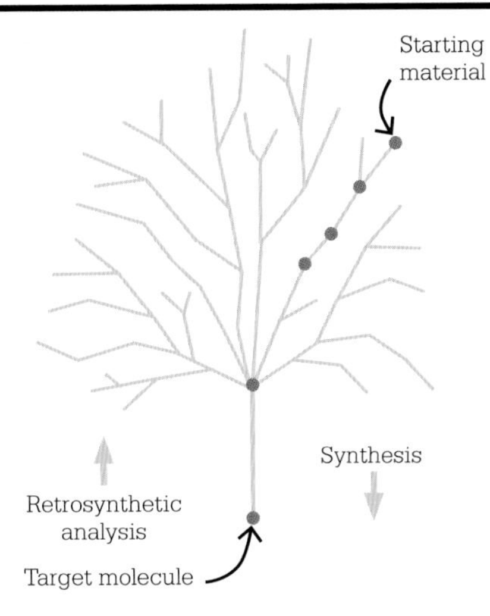

Retrosynthesis is likened to climbing a tree. From the target molecule—the root—the goal is to find the quickest, simplest, most reliable pathway to the best starting materials for synthesis.

Several important antibiotics, including the erythromycin family, have a complicated structure involving a large ring of molecules. In 1978, Corey's team achieved the landmark synthesis of erythronolide B, the precursor of erythromycin antibiotics. The new synthetic route they discovered enabled these drugs to be made easily.

Modern methodology

The retrosynthesis method itself remains Corey's most important contribution to science. It is an enormously powerful tool that chemists can use to determine how to build the molecules they want. Today, retrosynthesis is performed using computers, which suggest options from a multitude of different possible chemical reactions—but the choice of pathway relies on the expertise of the chemist. ▪

E.J. Corey

Born in 1928 to Lebanese parents in Methuen, Massachusetts, Elias James (E.J.) Corey was originally named William but was renamed Elias after his father, who died 18 months after the birth. At the age of 16, Corey entered MIT (the Massachusetts Institute of Technology), where he was captivated by the "intrinsic beauty" of organic chemistry. On finishing his PhD on synthetic penicillins, Corey joined the University of Illinois at Urbana-Champaign and became a professor in 1956. In 1959, he took up a professorship at Harvard University.

Corey remained at Harvard for the rest of his career and became famous for determining how to make incredibly complex natural molecules. He wrote more than 1,100 scientific papers and won more than 40 awards. His work on retrosynthesis resulted in the 1990 Nobel Prize in Chemistry.

Key works

1967 "General Methods for the Construction of Complex Molecules"
1995 *The Logic of Chemical Synthesis*

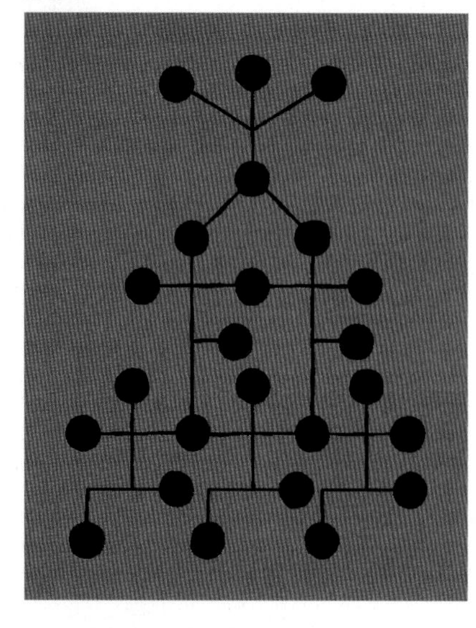

NEW COMPOUNDS FROM MOLECULAR ACROBATICS

THE CONTRACEPTIVE PILL

IN CONTEXT

KEY FIGURES
Gregory Pincus (1903–1967), **Min Chueh Chang** (1908–1991)

BEFORE
c. 1850 BCE Earliest account of birth control, in Ancient Egypt.

1905 British physiologist Ernest Starling first uses the word "hormone."

1929 American scientists Willard M. Allen and George W. Corner isolate progesterone and note how it affects ovaries.

AFTER
1971 Swedish pharmacologist Lars Terenius shows that drugs that block the hormone estrogen can fight breast cancer in rats.

1980 French scientist Georges Teutsch makes mifepristone, an emergency contraceptive that blocks progesterone receptors in the uterus.

2019 The Pill is used by 151 million women worldwide.

Making, or synthesizing, hormones that women can take in the form of an oral contraceptive pill to prevent pregnancy is one of most radical ways that chemistry has affected humanity. Campaigners for women to have more control over their fertility joined resourceful scientists to develop and produce the Pill.

American feminist Margaret Sanger followed and sometimes funded scientific research into birth control in the 1940s and 1950s. As a nurse, she knew of the harm to women's health due to poverty and large families. Her ally, American suffragist Katherine McCormick, inherited a large fortune in 1950 and pumped $2 million into contraceptive pill research, worth more than $18 million today.

In 1953, Sanger and McCormick teamed up with American biologist Gregory Pincus, who had set up the Worcester Foundation for Experimental Biology. There, with Chinese-American biologist Min

Progesterone prevents ovaries from releasing eggs during **pregnancy** and prevents the **womb** lining from forming.

Could women use a **progesterone-based** hormone drug as a **contraceptive**?

Hormones are **hard to extract** from natural sources and cannot be absorbed through the stomach and guts.

Synthetic hormones can be made more cheaply and absorbed more easily to work better as a contraceptive pill.

See also: Structural formulae 126–127 ▪ Antibiotics 222–229 ▪ Retrosynthesis 262–263 ▪ Rational drug design 270–271 ▪ Chemotherapy 276–277

Chueh Chang, Pincus had been using the hormone progesterone on animals. Progesterone helps prepare the body for conception, regulate the menstrual cycle, and maintain pregnancy. They hoped to use this hormone to produce signs of pregnancy in female animals so that the animals stopped ovulating.

Enter synthetic hormones

Until the 1940s, hormones made by scientists were laboriously extracted from animal organs. But American chemist Russell Marker discovered that diosgenin in the Mexican yam root vegetable had a hormonelike structure. By 1942, he could convert that chemical to progesterone. In 1944, Marker cofounded a company called Syntex to make progesterone. When Marker left in 1945, Syntex hired Mexican chemist George Rosenkranz, then Bulgarian-American chemist Carl Djerassi. Another Mexican chemist, Luis Ernesto Miramontes Cárdenas, became a vital part of the team.

Syntex's capabilities overcame a challenge that Pincus was facing. Progesterone itself did not work as a contraceptive when taken as a pill because it did not travel from the women's digestive system into their bloodstream. Once the Syntex researchers worked out how to modify hormones, in 1951, they made norethindrone, which worked in pill form.

> Birth control and the allied areas of sexual physiology and sexual behavior have long been battlegrounds of opinion-voicers.
>
> **Gregory Pincus**
> ***The Control of Fertility* (1965)**

Pincus and Chang used both norethindrone and noretynodrel, a similar progesterone modification first made by G. D. Searle and Company in 1952, in clinical trials of the Pill in Puerto Rico. The pills also unintentionally contained another hormone, mestranol, made during noretynodrel synthesis, which seemed to enhance the contraceptive effect.

After promising results, Pincus, Chang, Sanger, and McCormick turned to Searle to make the Pill. In 1960, the US approved Enovid, the contraceptive pill produced by Searle, which contained noretynodrel and mestranol. In 1964, Syntex launched its low-dose oral contraceptive pill Norinyl, containing norethindrone and mestranol. Many women used this formulation.

A mixed legacy

The Pill has attracted controversies beyond ideological opposition to contraception. Its widespread use means relatively rare side effects like blood clotting affect many women. Pincus ran his trials in Puerto Rico in an exploitative way. Sanger has been accused of being racist. Yet where it is available and affordable, the contraceptive pill has truly changed the world. ▪

Making hormones

Over billions of years, living creatures have evolved elegant ways to make hormones. By comparison, in the 1940s, synthetic chemistry struggled with their complex shapes. With carbon atoms looped together with four or more ring shapes, and more carbons pointing in various directions from those rings, hormones posed a huge challenge to replicate. Russell Marker's discovery that diosgenin had four rings that were arranged in a similar shape to those of progesterone gave him a valuable shortcut to produce progesterone using five chemical reactions.

When they took over from Marker at Syntex, George Rosenkranz, Carl Djerassi, and Luis Ernesto Miramontes Cárdenas built on his methods to improve on progesterone. They also incorporated findings from elsewhere on how to make a better alternative to progesterone that was effectively absorbed into women's bodies.

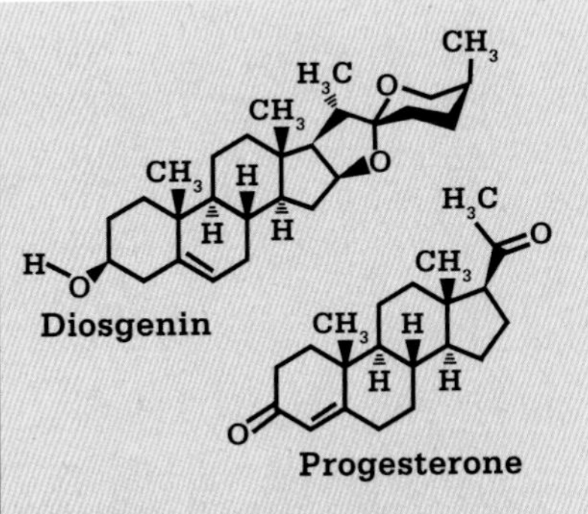

The chain of four carbon rings on the left of diosgenin's structure made it a good starting point for making hormones with similar structures, such as progesterone.

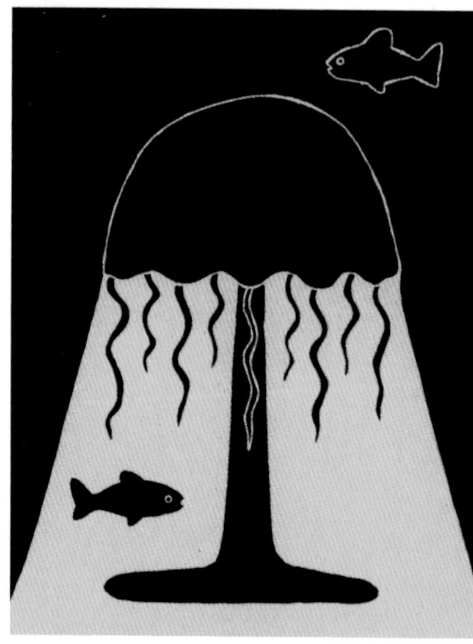

LIVING LIGHT

GREEN FLUORESCENT PROTEIN

IN CONTEXT

KEY FIGURE
Osamu Shimomura
(1928–2018)

BEFORE
1912 British scientists William and Lawrence Bragg, as well as others, develop X-ray crystallography, showing the positions of atoms in molecules.

1953 The structure of DNA—revealed by Rosalind Franklin, Maurice Wilkins, Francis Crick, and James Watson—enables Chalfie to genetically engineer organisms to make GFP.

AFTER
1997 American physicists Eric Betzig and William Moerner make a GFP they can switch on and off. It allows microscopes to view even smaller objects.

2007 American neuroscientists Joshua Sanes and Jeff Lichtman, with French colleague Jean Livet, devise the Brainbow. It uses fluorescent proteins to label brain cells.

Osamu Shimomura, a Japanese chemist at Princeton University, first isolated 5 milligrams of green fluorescent protein (GFP) from the glowing jellyfish *Aequorea victoria* in 1962. Shimomura then spent over 40 years working out how GFP functioned. He found that it is a relatively small protein made from 238 amino acid building blocks.

In the 1980s and 1990s, genetic engineers realized that they could clone the DNA sequence that encodes the instructions that tell jellyfish cells to make GFP. In 1994, Columbia University's Martin Chalfie used GFP's genetic instructions to color six cells in the transparent roundworm *Caenorhabditis elegans*. Genetic engineering could now modify an organism's DNA instructions to add GFP onto the end of any protein, showing scientists its location.

University of California, San Diego's Roger Tsien then spent many years changing the amino acids in GFP to produce different colored fluorescent proteins,

> Visualizing GFP was essentially noninvasive; the protein could be detected by simply shining blue light on to the specimen.
>
> **Martin Chalfie**
> **Nobel Prize Lecture (2008)**

allowing scientists to label and identify different proteins at the same time. GFP is now widely used in understanding how the cells of living creatures work. It also enables systems to detect other chemicals, such as metals and explosives. These detectors use other proteins to recognize the chemical and then trigger GFP's fluorescence—giving scientists an insight into otherwise undetectable processes. ■

See also: The structure of DNA 258–261 ▪ Protein crystallography 268–269 ▪ Editing the genome 302–303

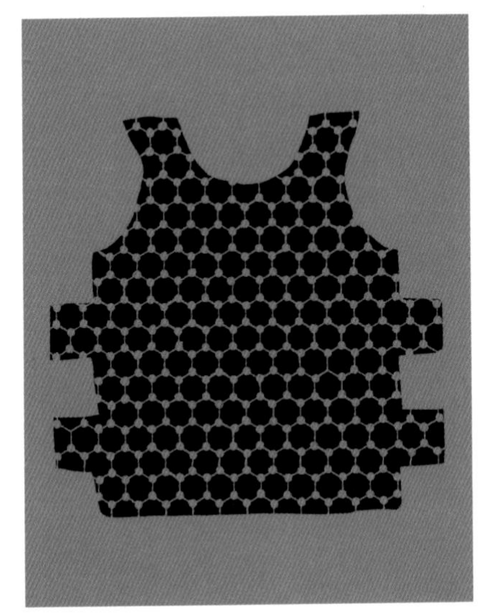

POLYMERS THAT STOP BULLETS

SUPER-STRONG POLYMERS

IN CONTEXT

KEY FIGURE
Stephanie Kwolek
(1934–2014)

BEFORE
1920 German organic chemist Hermann Staudinger finds that polymers such as starch, rubber, and proteins are chains of repeating units, enabling scientists to make new ones.

1935 DuPont scientists invent the polyamide polymer nylon.

1938 The discovery of Teflon by American chemist Roy Plunkett helps show the commercial and practical importance of novel polymers.

AFTER
1991 In the search for ever-stronger, tougher materials, Japanese physicist Sumio Iijima at NEC Corporation discovers carbon nanotubes.

2020 US House of Representatives passes PFAS Action Act to begin regulating use of "forever chemicals"..

In the 1960s, amid a forecast gasoline shortage in the US, the chemical company DuPont wanted to make car tires more durable and efficient. American chemist Stephanie Kwolek took on the project, exploring polymers known as polyamides that were similar to nylon.

Nylon is made from monomers with flexible carbon chains. Kwolek chose for her polymer to join rigid monomer building blocks, linking carbon atoms solidly into benzene rings. Carbon atoms in benzene rings share electrons more freely than carbon chains in nylon, and bond more strongly. The structure Kwolek made was a polyaromatic amide, or aramid. Intermolecular hydrogen bonding forces between amide groups on different chains hold the polymer together. Electron clouds around benzene rings provide another binding force.

When Kwolek spun it into fibers in 1964, the new polymer's tensile strength was five times greater than steel, yet it was as light as fiberglass. By the 1970s, DuPont sold the polymer as Kevlar. At first, this strong, lightweight material was used to toughen tires, but it was soon adopted for other purposes, most famously in body armor and bulletproof vests.

Kevlar is an example of polymers known as "forever chemicals", or PFASs, which accumulate in biological tissue and are linked to long-term adverse health effects such as cancer. They never break down and are found in human breast milk as well as ecosystems from Mount Everest to polar ice. ■

Woven Kevlar fibers form incredibly strong textiles that are lightweight and resist corrosion and heat. DuPont could easily make it into a range of fabrics that are as sturdy as armor.

See also: Synthetic plastic 183 ▪ Polymerization 204–211 ▪ Nonstick polymers 232–233

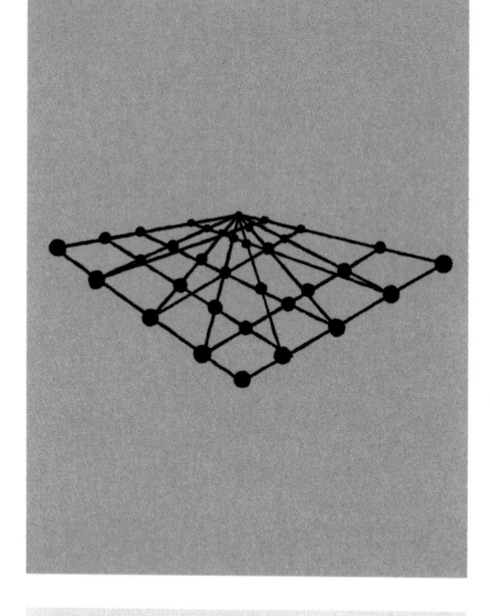

THE WHOLE STRUCTURE SPREAD OUT BEFORE ONE'S EYES

PROTEIN CRYSTALLOGRAPHY

IN CONTEXT

KEY FIGURE
Dorothy Hodgkin (1910–1994)

BEFORE
1913 British physicists William and Lawrence Bragg are the first to analyze crystal structures using X-rays.

1922 Canadian physician Frederick Banting administers insulin to a human patient.

AFTER
1985 American biophysicist James Hogle determines the 3D structure of poliovirus.

1985 German-American physicist Michael Rossmann is part of the group that publishes the structure of rhinovirus, responsible for the common cold.

2000 A team including Polish-American biochemist Krzysztof Palczewski publishes the first 3D structure of the family of G protein-coupled receptors that pass signals in living organisms.

In 1922, doctors gave the first life-saving injection of insulin to a 14-year-old diabetic at Toronto General Hospital, Canada. By the end of 1923, the hormone was treating around 25,000 patients in North America. However, the reason for its success was a mystery. British chemist Dorothy Crowfoot began to study the protein in 1934, in the hope of being able to unlock its structure and solve the puzzle.

Protein studies

In the 1930s, scientists knew a great deal about proteins in general, but they could not explain how they worked in detail. Scientists had determined that proteins seemed to be chain structures made of amino acids linked together, but it was difficult to obtain proteins that were pure enough to study.

At the start of the decade, British physicist William Astbury used X-ray diffraction to study fibers of two proteins that were of interest to the textile industry: keratin and collagen. In particular, he was interested in why wool, made of keratin, was more elastic than other textiles. Astbury passed X-ray beams through the fibers, deciphering structural clues from the patterns the beams made when interacting with the proteins'

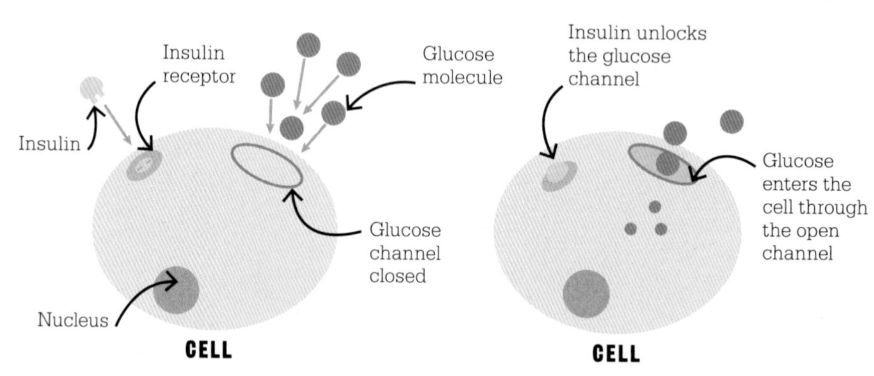

When insulin binds to the receptor on the surface of a body cell, it signals the cell to take in glucose from the bloodstream, which it then uses for energy.

See also: Benzene 128–129 ▪ X-ray crystallography 192–193 ▪ The structure of DNA 258–261 ▪ Atomic force microscopy 300–301

atoms. He found that the proteins coiled into a helix shape that could unfold when the fibers were stretched. This proved to have significance beyond textiles, and some have called this the beginning of molecular biology.

In 1934, Crowfoot worked with her mentor at the University of Cambridge, Irish chemist J. D. Bernal, to produce the first X-ray diffraction image of a crystallized protein, pepsin. Later that same year, Crowfoot set up her own research laboratory at Oxford University Museum of Natural History. Insulin, a small protein often known as a peptide, was one of the first substances she worked on. She passed X-ray beams through crystallized samples of insulin and analyzed the resulting diffraction patterns. This established the 2D structure of insulin in more detail than ever before, but the picture was far from complete.

A key advance came through a new mathematical Fourier transform technique developed by Arthur Lindo Patterson, a New Zealand–born crystallographer based in the US. It reveals electron density maps that show where atoms are located, and this allowed Patterson to work out molecular structures. In his experiments, he also included heavy atoms, such as those of metal elements. It was easier to establish the position of these heavy atoms because they deflected X-rays better. Finding their location helped him choose between options for structures.

Prized structures

At Oxford, Crowfoot—now working under her married name, Hodgkin—made Patterson maps of organic molecules using the heavy atom technique. In this way, she determined the 3D structure of penicillin, which she published in 1949. Her next challenge was the important dietary molecule vitamin B12. By assembling a large team of researchers and the necessary computational capacity, Hodgkin was able to map the 181 atoms of vitamin B12 in the early 1950s.

Meanwhile, in 1951, British biochemist Frederick Sanger had discovered insulin's amino acid sequence by using acids to break it down into smaller parts. Hodgkin returned to her search for the 3D structure of insulin in 1959, assembling another research team, which again used the heavy atom approach. Hodgkin and her team published a detailed 3D map of insulin's 788 atoms in 1969. This helped scientists work out how insulin engaged with receptors on body cells and identify the mutations in the insulin gene that cause diabetes. It also helped drug companies develop synthetic versions of human insulin that acted faster or lasted longer, for example, and were much less likely to cause allergic reactions. ■

I seem to have spent much more of my life not solving structures than solving them.
Dorothy Hodgkin

Dorothy Hodgkin

Born in 1910 in Cairo, Egypt, Dorothy Crowfoot went to school in England before going to the University of Oxford to study chemistry in 1928. In her final year, she asked X-ray crystallographer Herbert Powell to be her research supervisor. She then moved to J. Desmond Bernal's laboratory at the University of Cambridge in 1932 and co-authored 12 papers with him while working on her PhD. After returning to Oxford in 1934, she married historian Thomas Hodgkin in 1937. She remained at Oxford for the rest of her career and trained many students, including Margaret Thatcher. Hodgkin was awarded the 1964 Nobel Prize in Chemistry. She died in 1994.

Key works

1938 "The crystal structure of insulin I: The investigation of air-dried insulin crystals"
1969 "The structure of rhombohedral 2 zinc insulin crystals"

THE SIREN DRAW OF MIRACLE CURES AND MAGIC BULLETS

RATIONAL DRUG DESIGN

IN CONTEXT

KEY FIGURES
Gertrude Elion (1918–1999), **George Hitchings** (1905–1998)

BEFORE
1820 French chemists Pierre-Joseph Pelletier and Joseph-Bienaimé Caventou isolate quinine from the bark of trees in the *Cinchona* genus. The drug remains the main treatment for malaria until the 1940s.

1928 Alexander Fleming discovers that a substance produced by mold can kill bacteria; he names it penicillin.

AFTER
1975 German Georges Köhler and Argentinian César Milstein, both immunologists, synthesize forms of antibody molecules, paving the way for highly selective drugs.

1998 First clinical trial of imatinib—the first rationally designed selective cancer drug—led by American oncologist Brian Druker.

In 1906, German physician Paul Ehrlich conceived the "magic bullet"—a chemical drug that affects only the cause of a disease and does not harm the patient. He had been using synthetic dyes to view animal tissues and bacteria under a microscope. Ehrlich noted that some dyes stained specific tissues or bacteria, while others did not, and realized there was a link between the chemical structure of the dyes and the cells they colored.

Ehrlich reasoned he could use certain chemicals to target specific cells—such as disease-causing microbes—precisely, leaving others unharmed. He and his team used a dye to kill the malaria parasite and improved the toxic, arsenic-based dye arsanilic acid used to treat sleeping sickness. They made hundreds of similar compounds, randomly searching for less harmful options. In 1907, they created arsphenamine, which in 1909 they discovered killed the bacterium that causes syphilis.

Other scientists started to screen dyes as potential drugs, and German bacteriologist Gerhard Domagk tested more than 3,000. In 1935, he reported the first in the class of sulfonamide antibiotic drugs to be effective against a variety of bacterial infections.

> My greatest satisfaction has come from knowing that our efforts helped to save lives and relieve suffering.
> **George Hitchings**
> **Nobel Prize biography (1988)**

Rational thinking

Working at the Wellcome Research Laboratories in Tuckahoe, New York, biochemist George Hitchings wanted a more rational way of discovering drugs than sifting through large numbers of dyes. He focused on differences between how normal human cells and those causing disease—such as cancer cells, bacteria, and viruses—handle molecules such as DNA.

In 1944, Hitchings assigned Gertrude Elion to look at two of DNA's four main building blocks,

See also: Antibiotics 222–229 ▪ The structure of DNA 258–261 ▪ Chemotherapy 276–277 ▪ The polymerase chain reaction 284–285 ▪ Customizing enzymes 293

Disease-causing cells have specific and sometimes specialized **biochemical mechanisms.**

Selective drugs can target specific **mechanisms** in disease-causing cells.

Drugs can be designed to kill disease-causing cells but avoid healthy ones.

adenine and guanine. Bacteria need these molecules to make their DNA, and this gave Hitchings an idea. If they could use a chemical to stop the molecules from entering the biochemical mechanisms that disease-causing cells use to make DNA, that might stop them growing.

By 1950, Hitchings and Elion had made diaminopurine and thioguanine, which latched onto the enzymes that would normally attach to adenine and guanine, thus blocking DNA production. These drugs stopped leukemia cells from forming—the first treatments that could do so. However, they also affected stomach cells, making patients vomit badly.

Elion researched more than 100 compounds and created the more selective 6-mercaptopurine (6-MP). Today, 6-MP is part of a treatment that cures 80 percent of children with leukemia. Thioguanine is still used to treat acute myelocytic leukemia (AML) in adults.

Elion's work on nucleic acids led to allopurinol, a treatment for gout, and azathioprine, which suppresses the immune system and is used in organ transplants. In the 1960s, Hitchings and Elion developed the malaria drug pyrimethamine and the antibacterial trimethoprim, which is used to treat urinary and respiratory-tract infections, meningitis, and septicemia. However, most of these drugs still had bad side effects.

Magic bullets

In 1967, Elion's focus turned to viruses. She developed the herpes drug acycloguanosine, also called acyclovir and Zovirax. Zovirax proved that nucleic-acid drugs can be truly selective, making them Ehrlich's magic bullets. This helped Elion's colleagues later develop the HIV-AIDS drug azidothymidine (AZT). Similar selective molecules would become the COVID-19 drugs remdesivir and molnupiravir. ■

Gertrude Elion

Born in New York City in 1918, Gertrude Elion was motivated to fight cancer after it killed her grandfather when she was 15 years old. She studied chemistry at Hunter College, graduating in 1937. Science jobs for women were scarce, so Elion took on a poorly paid laboratory assistant role and also taught to cover graduate school costs at New York University. She earned a Master's degree in 1941, before World War II created vacancies for chemists in US industrial laboratories. After a brief stint at Johnson & Johnson in New Jersey, Elion became assistant to George Hitchings at Wellcome Research Laboratories in 1944, taking over from him in 1967. She published 225 papers before she retired in 1983. In 1988, Elion and Hitchings received the Nobel Prize in Physiology or Medicine. Elion died in 1999.

Key works

1949 *The effects of antagonists on the multiplication of vaccinia virus in vitro*
1953 *6-Mercaptopurine: effects in mouse sarcoma 180 and in normal animals*

THIS SHIELD IS FRAGILE

THE HOLE IN THE OZONE LAYER

IN CONTEXT

KEY FIGURES
Mario Molina (1943–2020), **F. Sherwood Rowland** (1927–2012)

BEFORE
1930 British mathematician Sydney Chapman devises the first photochemical theory for how sunlight forms ozone from atmospheric oxygen and how ozone breaks down again.

1958 James Lovelock invents the electron capture detector, capable of detecting extremely tiny amounts of substances, including gases in the air.

AFTER
2011 A team led by the American research scientist Gloria Manney discovers a rare ozone hole over the Arctic.

2021 American atmospheric scientist Stephen Montzka discovers that emissions of the banned chemical CFC-11 increased from 2011 to 2018 and then sharply declined.

In the early 1970s, Mexican chemist Mario Molina and American chemist F. Sherwood Rowland, from the University of California, Irvine, discovered that human-made chemicals threatened the protective layer of ozone that lies 12–18 miles (20–30 km) above Earth's surface. Their research included data gathered by British chemist James Lovelock, who had measured the abundance of chlorofluorocarbon (CFC) gases in the atmosphere using his electron capture device. Ozone (O_3) is a reactive gas, with three oxygen atoms instead of the usual two. It absorbs ultraviolet (UV) radiation from the sun, high levels of which can cause skin cancer.

CFCs and the ozone layer

In 1973, Molina and Rowland looked at how CFC gases affected the environment. CFCs were widely used as refrigerants, propellants in aerosol cans, and in plastic foams. However, Molina and Rowland figured out that CFCs gradually rise high into the atmosphere, where

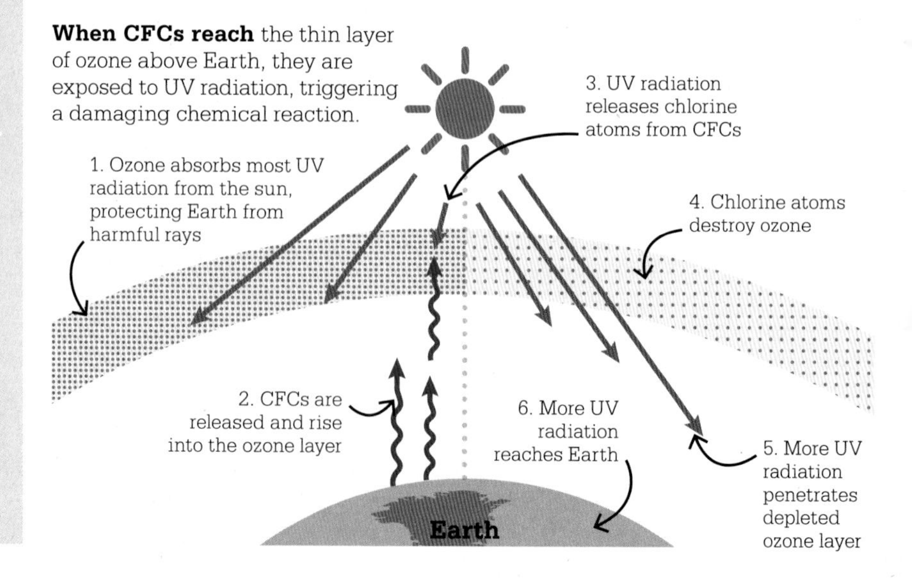
1. Ozone absorbs most UV radiation from the sun, protecting Earth from harmful rays
2. CFCs are released and rise into the ozone layer
3. UV radiation releases chlorine atoms from CFCs
4. Chlorine atoms destroy ozone
5. More UV radiation penetrates depleted ozone layer
6. More UV radiation reaches Earth
Earth

When CFCs reach the thin layer of ozone above Earth, they are exposed to UV radiation, triggering a damaging chemical reaction.

See also: Gases 46 ▪ Early photochemistry 60–61 ▪ The greenhouse effect 112–115 ▪ Leaded gasoline 212–213

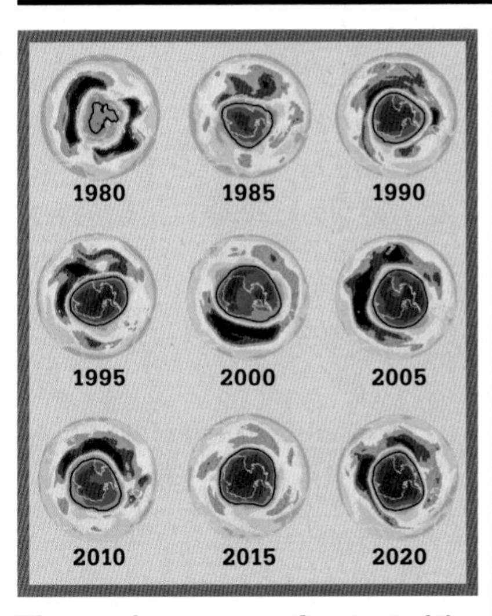

The maximum annual extent of the hole in the ozone layer over Antarctica is recorded by the EU's Copernicus Atmosphere Monitoring Service.

UV radiation is so strong it causes the molecules to break down, releasing chlorine atoms. These react with ozone, removing atoms to leave oxygen molecules (O_2).

In 1974, Molina and Rowland predicted that if CFC gas use continued, it would rapidly degrade the ozone layer. However, it took scientists nearly a decade to find any evidence. They expected chlorine from CFCs would affect the ozone layer most in the high atmosphere near the equator, but ozone levels there were steady. Instead, a combination of ground-based and satellite measurements revealed an "ozone hole" over Antarctica.

Looking into the hole

In 1983, British Antarctic Survey scientist Jon Shanklin was preparing for a public open day, drawing a graph intended to show unchanged ozone levels around the South Pole. In doing so, he realized that levels were falling dramatically each spring. Shanklin published the findings with two British colleagues, Joe Farman and Brian Gardiner, in 1985. The lowest values of ozone, seen in mid-October, had fallen by nearly half between 1975 and 1984. This discovery was so shocking that it caught the public's attention and motivated governments and scientists to act.

In 1986, Susan Solomon, an American researcher with the National Oceanic and Atmospheric Administration (NOAA), went to Antarctica to study the ozone hole. She showed that Antarctic clouds provide tiny, icy surfaces where a complex network of reactions release the ozone-depleting chlorine from CFCs when UV radiation from the sun reaches the Antarctic Circle in spring.

A thinning ozone layer weakens Earth's protection against dangerous radiation, which—for example—increases the risk of skin cancer and damages plants. Fortunately, the solution was clear, as chemical manufacturers could readily come up with alternatives to CFCs. In 1987, the Montreal Protocol, a global treaty to phase out CFCs and other ozone-depleting chemicals, was written, and came into effect in 1989.

Today, the size of the ozone hole varies each year, largely depending on temperatures high in the atmosphere above the South Pole. A small one formed in 2019, but a cold and stable Antarctic vortex helped the 12th largest ozone hole on record develop in 2020. Nevertheless, scientists expect ozone levels to fall back to pre-1980 levels by 2060, showing that given time, global action can reverse environmental damage. ■

Mario Molina

Born in Mexico City, Molina wanted to be a chemist from childhood. In 1960, he enrolled at the National Autonomous University of Mexico to study chemical engineering, before moving to the University of Freiburg, Germany, where he studied physical chemistry. He began his PhD on light-driven chemical reactions at the University of California, Berkeley, in 1968. In 1973, Molina moved to UC, Irvine, where he worked with Rowland on how CFCs affect the environment. In 1975, he became an assistant professor, and later joined the NASA Jet Propulsion Laboratory in Pasadena, California, where he helped investigate the Antarctic ozone hole.

Molina relocated to the Massachusetts Institute of Technology in 1989. In 1995, he jointly won the Nobel Prize in Chemistry for his work in atmospheric chemistry. He received the Presidential Medal of Freedom in 2013.

Key work

1974 "Stratospheric Sink for Chlorofluoromethanes: Chlorine Atom-Catalysed Destruction of Ozone."

POWER TO ALTER THE NATURE OF THE WORLD

PESTICIDES AND HERBICIDES

IN CONTEXT

KEY FIGURES
Paul Müller (1899–1965), **Rachel Carson** (1907–1964), **John E. Franz** (1929–)

BEFORE
c.2500 BCE Farmers use sulfur to kill insects and mites in Sumer (present-day Iraq).

1856 William Perkin sets the synthetic chemicals industry in motion by producing the first chemical dye, aniline purple.

AFTER
1985 German company Bayer patents the first neonicotinoid pesticide, imidacloprid, which blocks nerve signals in insects. Numerous studies have since linked neonicotinoids with a decline in bee numbers.

2015 A US Court of Appeal rules that the Environment Protection Agency violated federal law by approving sulfoxaflor—a pesticide similar to neonicotinoids—without reliable studies.

Chemical crop protection has helped farmers for millennia to produce better crop yields. From about 1800, farmers used poisonous salts of heavy metals such as arsenic and mercury to kill insects and bacteria, but these were later discovered to have huge health and environmental risks.

In 1974, glyphosate, a new type of synthetic herbicide, went on sale and became the world's most-sold farming chemical of all time.

From natural to synthetic

One of the world's oldest natural insecticides is pyrethrum. It was probably in use around 1000 BCE by Chinese farmers, who extracted it from pyrethrum daisies (*Tanacetum cinerariifolium*). From the 19th century, farmers in many countries also used nicotine from tobacco plants (*Nicotiana*). The first use of rotenone as an insecticide was in 1848. It occurs in the roots and stems of several tropical plants, including *Derris elliptica*, and was

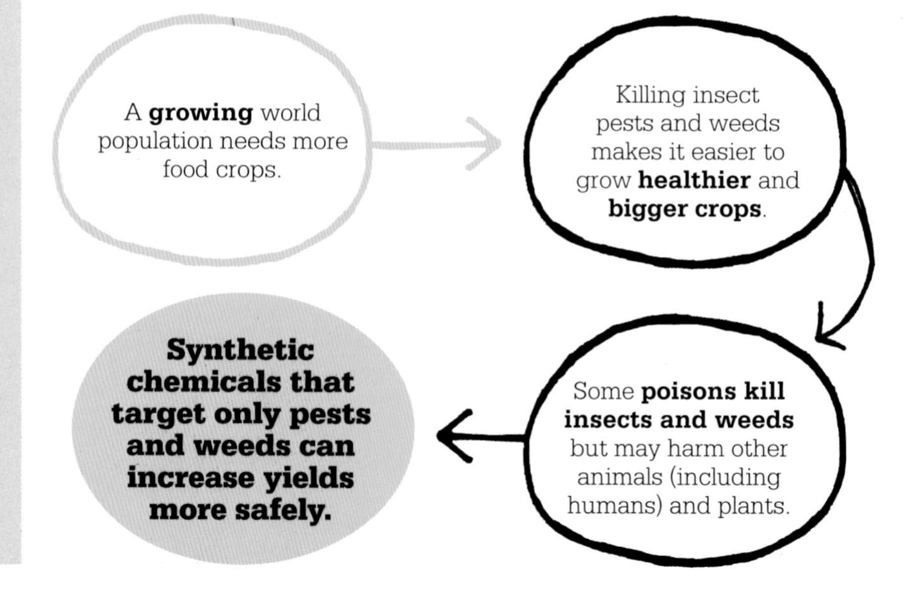

A **growing** world population needs more food crops.
Killing insect pests and weeds makes it easier to grow **healthier** and **bigger crops**.
Some **poisons kill insects and weeds** but may harm other animals (including humans) and plants.
Synthetic chemicals that target only pests and weeds can increase yields more safely.

See also: The synthesis of urea 88–89 ▪ Synthetic dyes and pigments 116–119 ▪ Enzymes 162–163 ▪ Fertilizers 190–191 ▪ Protein crystallography 268–269

Organic pesticides derive from natural sources, mainly plants, and are most often used as insecticides on food crops, such as this tomato crop. They can also be toxic to other animals.

also used to paralyze fish. These chemicals were costly, as they were laboriously extracted from natural sources. In the early 20th century, companies began to exploit growing knowledge of synthetic organic chemistry to make better, cheaper pesticides and herbicides.

In 1939, Swiss chemist Paul Müller, searching for a contact insecticide, discovered that dichlorodiphenyltrichloroethane, or DDT, was highly potent. A tank coated with DDT and cleaned to leave only trace amounts still killed flies. Müller won the Nobel Prize in Physiology or Medicine in 1948 for the use of DDT to kill insects that spread typhus, malaria, bubonic plague, and other diseases.

DDT was cheap and highly effective and became a wonder insecticide. It took a lot of DDT to kill larger animals, so scientists initially regarded it as safe to use. It could be sprayed over wide areas and did not wash away easily, so it did not need to be resprayed often. However, DDT persisted in the environment and did not break down when it was ingested. DDT became more concentrated as it passed from one animal to the next up the food chain until it reached lethal levels. In 1958, American biologist and author Rachel Carson received a letter about DDT's impact on birds. She researched the problems of the pesticide and published her findings in her famous book *Silent Spring* in 1962, arguing that humans were poisoning the environment. Chemical companies sought to discredit Carson, but respected scientific committees supported her conclusions, making pesticides a contentious issue.

Target selection

Since the 1970s, scientists have tried to develop safer pesticides and herbicides that work when spread in only tiny amounts and are highly selective in killing only their targets. The chemicals should also break down easily, so they do not leach into the environment.

John E. Franz, an American organic chemist at Monsanto in 1970, developed a new class of herbicides, including glyphosate. It inhibits a key plant enzyme, stopping plants from growing new cells. It binds tightly to soil, so it should not get into nearby crops or the wider environment, and breaks down into safer chemicals, which allows new plants to grow a month or two later. Monsanto soon started selling glyphosate under many names, most notably Roundup™.

In 2015, the International Agency for Research on Cancer (IARC) said glyphosate probably caused cancer in humans, but the US Food and Drug Administration (FDA) stated in 2018 that traces in food were too tiny to pose a risk to humans. Yet cancer patients in the US sued Bayer, which took over Monsanto in 2018. Many countries have banned some or all glyphosate products and others are reviewing its use. Chemists continue to search for more selective and safer products. ■

Unlocking the target

Pesticides and herbicides work by attaching their molecules to proteins, such as enzymes that are part of an organism's biochemical pathway, and disabling them or pushing them into action. Usually, the molecule interacts with a specific target in the pest's or weed's biochemical pathways. The target is the lock; the key is the pesticide or herbicide.

Often, pesticides seek to turn locks in insects' nerves. For example, pyrethrum and DDT stimulate insects' nerves constantly, making the insects spasm until they die. Rotenone blocks biochemical pathways in mitochondria, organelles in cells that make chemical energy, to stop them working. Glyphosate blocks an enzyme in microbes and plants used to make amino acids for protein molecules that are vital to life.

IF IT BLOCKS CELL DIVISION, THAT'S GOOD FOR CANCER

CHEMOTHERAPY

IN CONTEXT

KEY FIGURE
Barnett Rosenberg
(1926–2009)

BEFORE
1942 Two American pharmacologists, Alfred Gilman and Louis Goodman, administer the first chemotherapy treatment using mustard gas.

1947 American pathologist Sidney Farber and Indian biochemist Yellapragada Subba Rao develop methotrexate, one of the first approved chemotherapy drugs.

AFTER
1991 Scottish-born immunologist Ian Frazer and Chinese virologist Jian Zhou invent the first vaccine to protect against cervical cancer.

1994–1995 American immunologist James Allison realizes that drugs can stimulate the immune system to target cancer.

An accidental discovery in the 1960s is responsible for the chemotherapy drug cisplatin, referred to by some as "the most important anticancer drug ever developed." Chemotherapy drugs are those that destroy or inhibit the growth of cells. In cancer patients, they are used to shrink tumors and to stop the cancer from growing or spreading. Cisplatin was first approved for use as a treatment for testicular cancer in 1978 in the US.

After joining Michigan State University in 1961, American researcher Barnett Rosenberg decided to study cell division in bacteria by putting them in a solution with platinum electrodes. He borrowed some equipment and asked his laboratory technician, Loretta van Camp, to pass electrical current through the solution. When the power was on, van Camp observed the bacteria take on strange elongated shapes. The bacteria did not die, but they could not divide to form new cells.

> They had never seen a response like this, a total disappearance of so many cancers …
> **Barnett Rosenberg**

It seemed that electrical current was controlling cell division in the bacteria, but Rosenberg was careful to test that conclusion. He tried reusing a solution that had previously had current passed through it and saw the bacteria do the same thing. He realized that cell division was being blocked not by the electrical current, but by the platinum salts that had dissolved in the solution. Rosenberg published his results in 1965. He then tested several different platinum salts to reproduce the effect. The most powerful was cis-diamminedichloroplatinum (II) or cisplatin, first prepared by Italian chemist Michele Peyrone in 1844.

Life-saving structures

Cisplatin is a simple compound of platinum surrounded by four other chemical groups in a square shape. In each of two adjacent corners of the

See also: Coordination chemistry 152–153 ▪ The structure of DNA 258–261 ▪ Rational drug design 270–271 ▪ Editing the genome 302–303

Barnett Rosenberg inspects a vial of an anticancer drug derived from cisplatin at Michigan State University. Today, modern treatments cure more than 90 percent of testicular cancer patients.

square, there is a chlorine atom. In each of the other adjacent two, there is an ammine group. Each ammine comprises a nitrogen atom with three hydrogen atoms attached to it. This simple structure is perfectly shaped to interfere with cells' deoxyribonucleic acid (DNA), stopping them from multiplying. With cisplatin as an example, chemists started to understand the links between the molecular structures of drugs and their biological function.

In cancer, one or more mutations in a person's genes can cause a previously normal cell to divide and then multiply to create many more cells. Rosenberg quickly realized that a chemical such as cisplatin could be used as a cancer drug. He began to work with the US National Cancer Institute (NCI) in 1968, showing that cisplatin could stop cancer cells from dividing in mice. The NCI continued to invest in cisplatin, and clinical trials began in human cancer patients in 1972 at a specialist cancer hospital in New York. Cisplatin worked well in testicular, ovarian, head and neck, bladder, prostate, and lung cancers, but there were toxic side effects, including potential damage to a patient's hearing, nerves, and kidneys.

Despite this, in 1978, the US Food and Drug Administration (FDA) approved cisplatin for the treatment of testicular cancer, for which no effective anticancer drug was available at the time. Its use has since been improved, and cisplatin is now a key part of combination therapy—chemotherapy, surgery, and radiation therapy, for example—for many types of cancer.

Scientists continued to search for a less toxic alternative. The first came from researchers at two UK organizations—Johnson Matthey and the Institute for Cancer Research—which together discovered carboplatin. Replacing cisplatin's chlorine atoms with an organic group made the drug more stable in patients' bodies, and the FDA approved carboplatin for use in 1989. Similarly, Japanese research scientist Yoshinori Kidani discovered oxaliplatin in 1976, and it was approved by the FDA in 2002. Today, millions of patients have had their cancers cured by these treatments. ■

How platinum drugs fight cancer

Cisplatin's chemical structure is ideally shaped to react with DNA. Nitrogen atoms in DNA displace the two chlorine atoms that are attached to the platinum atom at cisplatin's center. With the DNA tied up in this way, cells cannot repair or copy it, and those cells will die. However, because cisplatin targets DNA, it affects all living cells. It particularly affects cells that are dividing rapidly, such as cancer. But there are other types of cells that divide rapidly, and these also stop multiplying and die. They include the stomach lining, the bone marrow that produces blood cells, and hair follicles. This leads to people feeling sick, getting more infections, and losing their hair, respectively. Cisplatin is effective for a number of cancers, such as some forms of leukemia and testicular cancer. In other cancers, either it does not work well or it is effective for a while but then stops working when cancer cells develop ways to expel the drug. Such resistant cancers need different drugs, such as carboplatin or oxaliplatin.

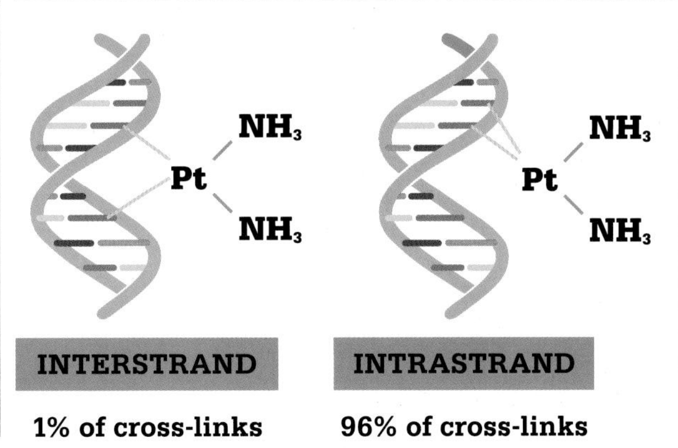

DNA strands displace cisplatin's chlorine atoms, leaving its platinum (Pt) atom and two ammine (NH_3) groups. Interstrand links, spanning DNA's two strands, or intrastrand links, within a single strand, kill cells.

THE HIDDEN WORKHORSES OF THE MOBILE ERA

LITHIUM ION BATTERIES

IN CONTEXT

KEY FIGURE
John Goodenough (1922–)

BEFORE
1800 Alessandro Volta invents the first battery, a pile of zinc and copper disks separated by cloth soaked in salty water.

1817 Swedish chemists August Arfwedson and Jöns Jacob Berzelius discover lithium, purifying it from a mineral sample.

1970 Swiss engineer Klaus-Dieter Beccu is the first to patent nickel-metal hydride rechargeable batteries, which are much more powerful than previous rechargeables.

AFTER
1995 American scientist K. M. Abraham reputedly discovers that a leak in a lithium-ion battery gives it a far higher energy content, leading to the invention of longer-lasting lithium-air batteries.

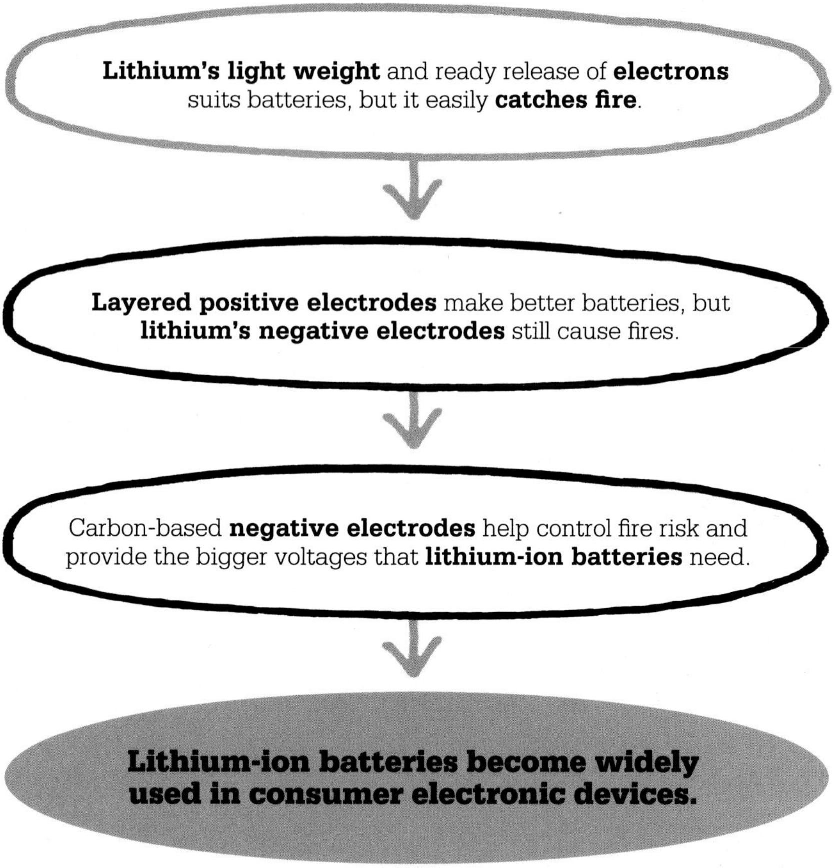

Lithium's light weight and ready release of **electrons** suits batteries, but it easily **catches fire**.

Layered positive electrodes make better batteries, but **lithium's negative electrodes** still cause fires.

Carbon-based **negative electrodes** help control fire risk and provide the bigger voltages that **lithium-ion batteries** need.

Lithium-ion batteries become widely used in consumer electronic devices.

Lithium-ion batteries are commonplace nowadays, powering almost every kind of portable electronic device. Mobile phones, laptops, earbuds, and cordless drills all operate thanks to a continuous trickle of chemical reactions. An American materials scientist named John Goodenough invented this type of battery in 1980.

Surprisingly, this relatively "green" energy technology owes its existence to one of the world's largest producers of fossil fuels. In 1973, Saudi Arabia triggered an oil crisis when it stopped exporting oil to several countries, including the US, Japan, the UK, Canada, and the Netherlands. Within a year, oil prices had quadrupled, and governments and scientists had also started to worry about how soon Earth's finite oil resources would be exhausted. As a result, interest in fossil fuel–free energy technologies grew.

Just before the start of the oil crisis, British chemist M. Stanley Whittingham had begun working in the research and engineering department of Exxon, one of the world's biggest oil companies, in New Jersey. He first worked on superconducting materials, whose electrical resistance is so low that they can supply electricity over long distances more efficiently. As part of this research, he studied conductive materials called layered sulfides. He had previously worked on batteries and realized that layered titanium sulfide was well suited to storing energy in batteries through a process called intercalation. The atoms in ionic compounds such as titanium sulfide arrange themselves in regular layered structures. In batteries, positively charged metal ions can slide—or intercalate—between those layers and back out again.

How batteries work

The basic workings of batteries have been the same for more than two centuries. They have a positively charged electrode and a negatively charged electrode—in traditional

See also: Rare-earth elements 64–67 ▪ Catalysis 69 ▪ The first battery 74–75 ▪ Isolating elements with electricity 76–79 ▪ Isomerism 84–87 ▪ Electrochemistry 92–93 ▪ The electron 164–165

How rechargeable batteries work

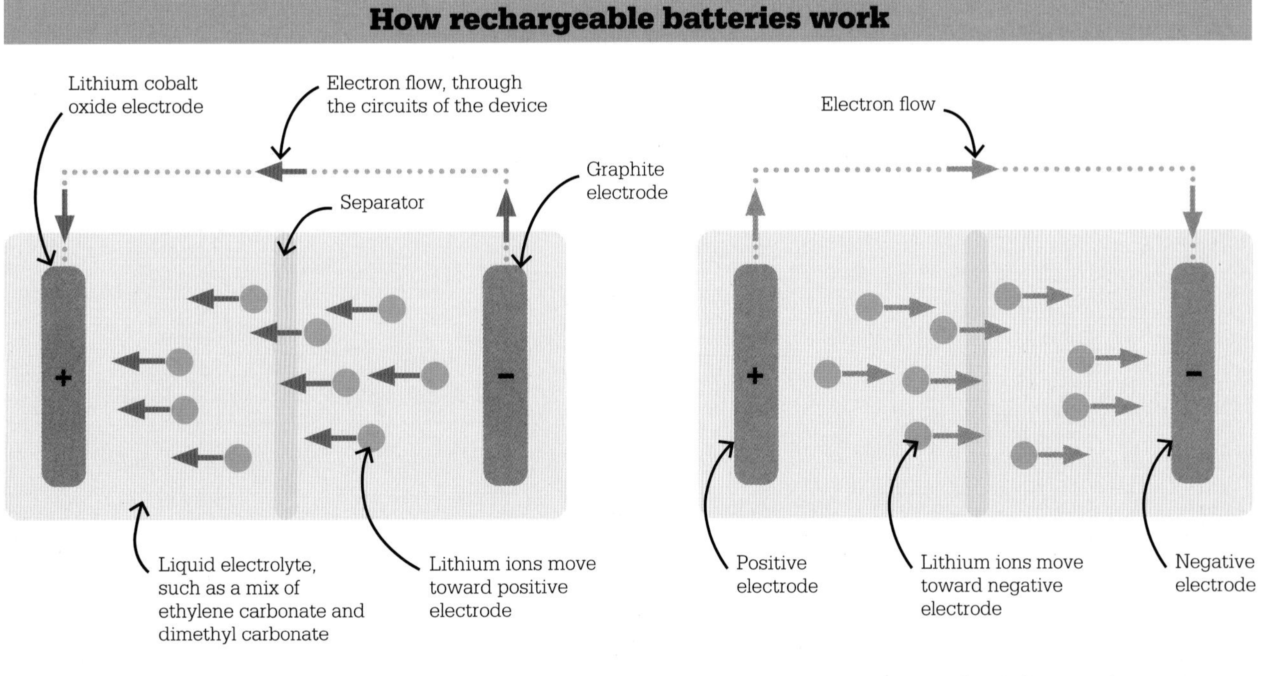

When an electronic device is in use, electrons and lithium ions flow from the negative electrode to the positive electrode. As electrons go through the gadget's circuits, lithium ions go through the electrolyte to balance them.

When you charge your gadget's battery, the reaction goes backward. However, many other unwanted reactions happen inside the battery at the same time, which is why old batteries work poorly.

batteries, these are respectively called the cathode and anode. Between the electrodes is a medium called an electrolyte, usually air or a liquid such as water, through which electrically charged ions can travel. When electrical charges move, they create an electric current. Usually, a battery also has a separator—a barrier material that prevents the electrodes from touching each other and short-circuiting; this would quickly and wastefully discharge stored energy. Instead, the battery electrodes connect to wires that take energy away, delivering it to the electrical circuits needed to power gadgets.

To release energy, a chemical process at the anode pushes electrons into the circuit. The chemicals formed travel through the electrolyte to the cathode, where they trigger another chemical process, which uses electrons coming into the battery from the circuit. The capacity of a battery is a product of its voltage, the current it can produce, and how long it can produce that current for. The voltage depends on how readily the chemicals at the anode give up electrons and how greedily those at the cathode suck them up. In rechargeable batteries, the process needs to be reversible. This means that battery designers must carefully choose the chemical processes to avoid causing any side reactions that would use up the materials in the battery and lower its voltage.

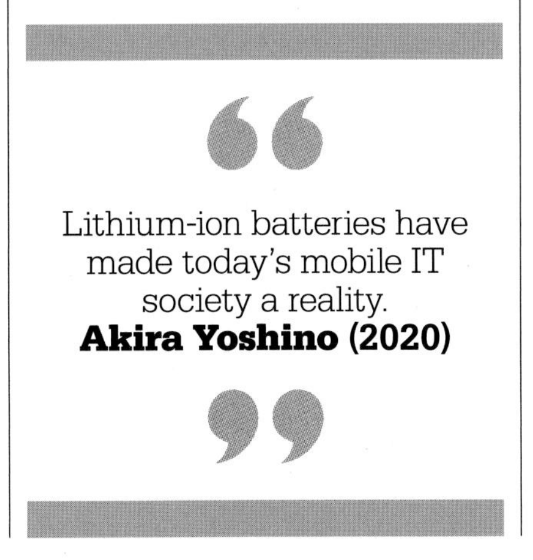

Lithium-ion batteries have made today's mobile IT society a reality.
Akira Yoshino (2020)

Disposable batteries

By the 1970s, disposable, or primary, batteries were in common usage. Rechargeable lead–acid batteries, such as those still found in many cars, had also been » around for decades but were based on chemical reactions that corroded the electrodes over time.

Seeking alternatives

Chemists had considered the potential of lithium for electrodes since at least the 1950s. Being the least dense metal, it had the potential to reduce the weight of batteries and so make them more practical for powering small devices. Although it does not occur as a metal in nature, lithium is present in small amounts in the minerals that make up many rocks.

Another of lithium's strengths is also a weakness, however. Lithium metal atoms give away electrons very easily, and that chemical process can release a lot of energy. This characteristic makes lithium metal very unstable, as it reacts vigorously in typical battery electrolytes of air and water, it forms lithium hydroxide and highly flammable hydrogen. For safety reasons, therefore, chemists store lithium in oil.

From the 1950s to the 1970s, scientists gradually worked out that some organic solvents would work well as the electrolyte in a lithium battery. They also confirmed that lithium metal worked as the anode despite its combustible tendencies, but the search for a suitable cathode continued. In 1973, American engineer Adam Heller devised a disposable lithium battery with a liquid cathode that gave the battery a long life, and this is still used widely today.

In the same year, while working on titanium sulfide, Whittingham realized that its conductive properties and the way lithium could slot between its layers would make it an excellent cathode in lithium batteries. Exxon moved

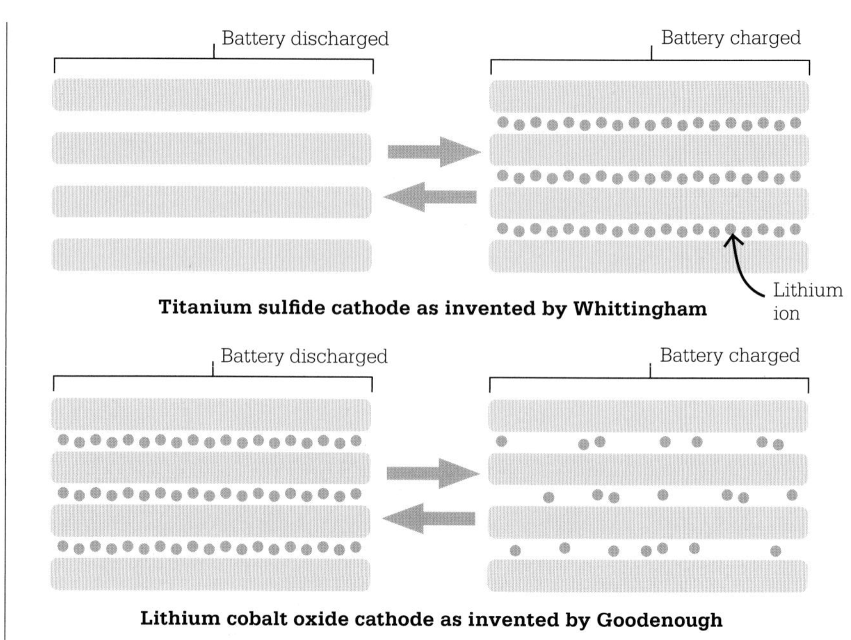

Batteries with titanium sulfide cathodes charge when lithium ions slide between the layers formed by the cathode's atoms. The layers in lithium cobalt oxide cathodes have fewer lithium ions when charged.

quickly to take advantage of this opportunity, demonstrating a 2.5-volt battery in 1976. However, side reactions caused long fingers of lithium metal, called dendrites, to grow between the electrodes, leading to short-circuits and fires. Meanwhile, oil prices fell, and around 1980, the corporation dropped its battery research.

Intercalation

Elsewhere, scientists looked toward intercalation at the anode, as well as the cathode, to avoid dendrites, but without much success. For example, lithium atoms could intercalate between the layers of the graphite form of carbon, but the solvent used in the battery electrolyte gradually broke graphite down into flakes. Goodenough and his team at the University of Oxford, UK, found a better cathode material, which they patented in 1979. Goodenough suggested that a battery could produce an even greater voltage if it was made using a metal oxide cathode instead of a metal sulfide. The smaller oxygen atom was greedier for electrons than the larger sulfur one, he reasoned, and could let lithium intercalate more

> We have to, in the near future, make a transition from our dependence on fossil fuels to a dependence on clean energy.
>
> **John Goodenough (2016)**

tightly. Exploring that avenue with cobalt oxide cathodes successfully produced a 4-volt battery, which is sufficient to power many devices.

Small is beautiful

Meanwhile, in Japan, corporations making portable electronic devices were becoming interested in lithium batteries to power their products. For example, at the Asahi Kasei Corporation, chemist Akira Yoshino tried various carbon-based materials as intercalating anodes. One, called petroleum coke, had some parts that were layered like graphite and some that were not. This arrangement prevented it from flaking and so made it strong enough to use. By 1985, Yoshino could therefore make an efficient, long-lasting 4-volt battery that did not cause fires and could be charged hundreds of times before wearing out.

Based on this design, the first commercial rechargeable lithium-ion batteries made by Japanese electronics firm Sony went on sale in 1991. Lithium-ion batteries have since enabled the widespread use of portable electronics, most notably computers, tablets, and mobile phones, becoming a market worth billions of dollars. And they have also continued to improve. Today's versions can use new cathode materials and electrolyte solvents that do not degrade graphite, finally allowing its use as an anode. In 2019, Whittingham, working at the US Department of Energy, declared that he wanted to double the energy density of lithium batteries. To achieve this, battery engineers are now replacing the cobalt in cathodes with nickel and searching for anode materials that can intercalate lithium at higher densities than graphite. Lithium extraction is water- and energy-intensive, contaminating waterways and soil; it is also linked to human rights abuses and serious health problems. Improved recycling of lithium batteries is an important goal. ■

Lithium-ion batteries have helped drive the boom in consumer products such as mobile phones, watches, toys, and cameras. However, mining lithium creates environmental problems.

John Goodenough

Born to American parents in Jena, Germany, in 1922, John Goodenough studied mathematics at Yale University before serving as a meteorologist in the US Army during World War II. He then studied at the University of Chicago with nuclear physics pioneer Enrico Fermi, earning his doctorate in physics in 1952.

Goodenough became a research scientist at the Massachusetts Institute of Technology (MIT) before moving to the University of Oxford, UK, in 1976, where he made his metal oxide cathode discovery. In 1986, he became a professor at the University of Texas at Austin.

In 2019, Goodenough was awarded the Nobel Prize in Chemistry—shared with M. Stanley Whittingham and Akira Yoshino—and became the oldest Nobel laureate in history.

Key work

1980 "Li_xCoO_2 ($0<x\leq1$): A New Cathode Material for Batteries of High Energy Density"

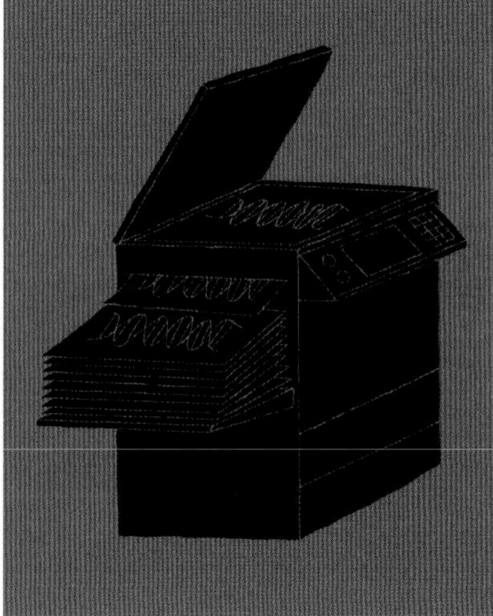

BEAUTIFULLY PRECISE COPYING MACHINES

THE POLYMERASE CHAIN REACTION

IN CONTEXT

KEY FIGURE
Kary Banks Mullis
(1944–2019)

BEFORE
1956 Arthur and Sylvy Kornberg discover DNA polymerase, the enzyme that builds new strands of DNA.

1961 Heinrich Matthaei and Marshall Nirenberg establish that the bases in the genetic code are read as three-letter "words," called codons.

AFTER
1997 Swedish geneticist Svante Pääbo and colleagues successfully sequence tiny amounts of Neanderthal DNA by amplifying the DNA to detectable levels with PCR.

1997 Hong Kong pathologist Dennis Lo and fellow scientists use PCR to extract a baby's DNA from its mother's blood.

2003 Human Genome Project declared complete, thanks in large part to PCR technology.

In 2020, the acronym PCR—polymerase chain reaction—became famous thanks to the COVID-19 pandemic, but it was already a scientific workhorse. Polymerase is an enzyme harnessed in chain reactions to make millions of copies of specific genetic molecules, such as a piece of DNA. In COVID-19 tests, the PCR samples tiny amounts of the virus's genetic material and makes an easily detectable number of copies.

PCR's origins date to the 1950s, with the growing acceptance that nucleic acids such as DNA and RNA

Basic steps of a polymerase chain reaction

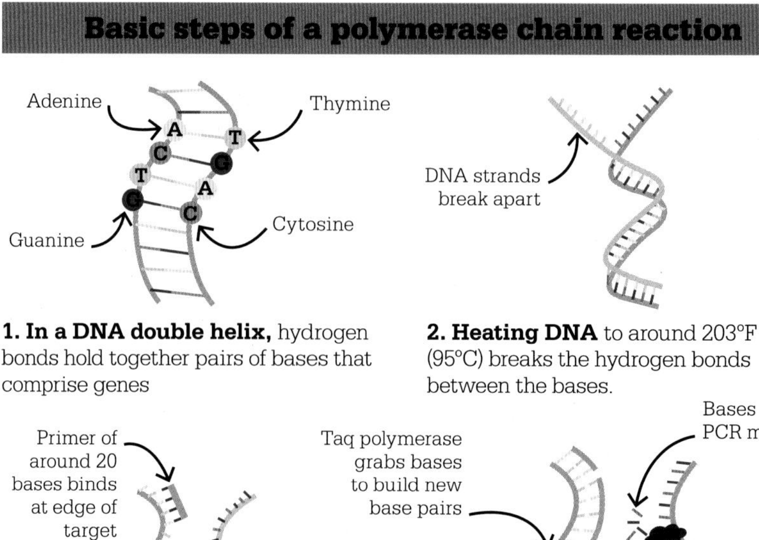

Adenine
Thymine
Guanine
Cytosine
DNA strands break apart
Primer of around 20 bases binds at edge of target region
Complementary strands of DNA
Taq polymerase grabs bases to build new base pairs

1. In a DNA double helix, hydrogen bonds hold together pairs of bases that comprise genes

2. Heating DNA to around 203°F (95°C) breaks the hydrogen bonds between the bases.

3. Cooling DNA to below 158°F (70°C) allows short, single primer DNA strands to bind to the unraveled strands.

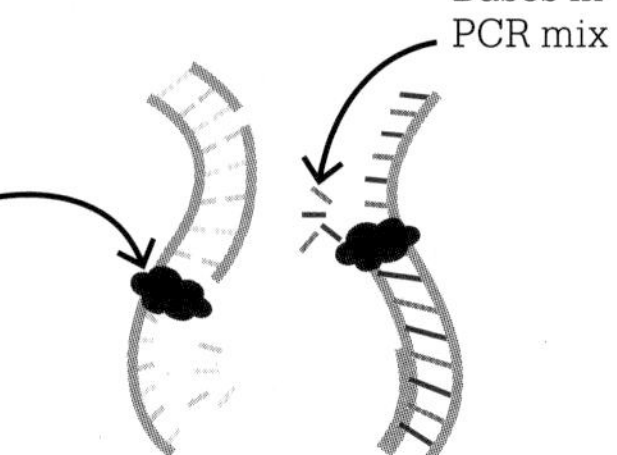

Bases in PCR mix

4. At 162°F (72°C) Taq polymerases extend each primer, leading to identical copies of the orignal DNA strands.

See also: Intermolecular forces 138–139 ▪ Enzymes 162–163 ▪ Polymerization 204–211 ▪ The structure of DNA 258–261 ▪ Customizing enzymes 293 ▪ Editing the genome 302–303

Primers

Primers are short pieces of DNA around 20 bases in length that focus Taq polymerase's effort on a specific genetic location in a PCR test. After a DNA double helix unravels, two selected primers flank the target region—the sequence of DNA to be copied—one on each strand. This is what happens in a COVID-19 PCR test. The SARS-CoV-2 virus's genetic material is RNA, so the COVID-19 PCR test first uses an enzyme called reverse transcriptase to copy its sequence into DNA. Then, the primers locate the gene encoding the protein that helps the virus enter human cells. Taq polymerase builds on that to create copies of both strands.

are the physical basis of genes. The discovery of DNA's double helix in 1953 suggested that the order of DNA's four chemical bases—represented by the letters A, C, T, and G—is a code. In 1961, American biochemist Marshall Nirenberg and his German colleague Heinrich Matthaei at the National Institutes of Health in the US established that the bases are read as three-letter "words" called codons. Each codon corresponds to a specific amino acid, the building blocks of proteins.

In 1956, American biochemists Arthur and Sylvy Kornberg had discovered DNA polymerase, the enzyme that is responsible for forming new copies of DNA. By the early 1960s, other scientists had shown that RNA polymerase uses DNA as a template to make RNA molecules with specific sequences.

In 1960, Indian American biochemist Har Gobind Khorana, working with a team in the US, made RNA chemically to decipher the code for how RNA molecules guide protein assembly in a cell. In 1970, he synthesized a short strand of DNA molecules—and created the world's first synthetic gene.

Living things naturally copy DNA by separating DNA's double helix into its two component strands, revealing the unique sequence of bases on each strand. This sequence acts as a template onto which new bases are matched. The DNA polymerase enzyme links the bases together, and two new identical strands of DNA are created. The process requires a primer—a short nucleic-acid sequence that attaches to one uncoiled DNA strand. Khorana's team brought these different components together but found that an enzyme called DNA ligase worked better in the laboratory than DNA polymerase.

Rapid copying machine

In 1971, Khorana's Norwegian team member Kjell Kleppe proposed using two suitable primers to copy both strands of a helix. Using DNA polymerase to copy short pieces of genetic material became routine, but no one had tried copying both strands at the same time.

In 1983, American biochemist Kary Mullis, who was working for the Cetus biotech corporation in California, which made primer sequences, had an idea for a process that would rapidly copy, or amplify, specific genes or strands of DNA. Mullis heated a mixture of the sample DNA, some DNA polymerase, and primer so the DNA separated into two strands. He cooled the mixture to allow the replication of strands into two copies and then started the process again. Within a few hours, 20–60 repeats of the procedure—which Mullis called PCR—could yield billions of copies of the DNA. All he needed was a test tube and some heat.

Cetus was searching for new ways to test for genetic conditions, such as sickle cell anemia, and PCR could rapidly increase the amount of the DNA required for testing. But heating each double helix to separate it destroyed the DNA polymerase, so Mullis had to add more after each heating step, which made the process expensive. After Mullis left the company in 1986, Cetus switched to the DNA polymerase from *Thermus aquaticus*, a species of bacteria that live in hot springs. Known as Taq polymerase, this enzyme survives each cycle of the PCR, greatly reducing costs. Although Mullis left Cetus two years before this innovation, in 1993, he alone won the Nobel Prize in Chemistry for the invention of PCR. ■

> We had just changed the rules in molecular biology.
> **Kary Banks Mullis**

60 CARBON ATOMS HIT US IN THE FACE

BUCKMINSTERFULLERENE

IN CONTEXT

KEY FIGURE
Harry Kroto (1939–2016)

BEFORE
1913 Francis Aston invents the mass spectrograph.

AFTER
1988 Scientists find C_{60} in soot from candle flames.

1991 Japanese physicist Sumio Iijima invents carbon nanotubes.

2004 Russian-born British physicists Andre Geim and Konstantin Novoselov confirm the existence of graphene, a carbon allotrope made of single flat sheets of atoms arranged into hexagons.

2010 Astronomers discover fullerene molecules in a planetary nebula.

2019 NASA's Hubble Space Telescope detects C_{60} in the interstellar medium—the gas and dust that lies in the space between stars.

Mystery molecules around red giant stars absorb **microwave radiation**.

Those molecules might be made in the **red giant's atmosphere**.

Trying to **simulate** the **formation conditions** reveals the existence of C_{60} molecules.

Evidence from space later confirms that C_{60} molecules absorb microwave radiation.

Carbon's chemical versatility is the basis for the processes of life on Earth, yet despite being so commonplace and much studied, it can still astound researchers. One of the biggest surprises in 20th-century science came in 1985, when British chemist Harry Kroto sought to understand a mystery in space.

At the University of Sussex, Kroto was trying to explain signals reaching Earth from carbon-rich stars known as red giants. The signals came in the form of microwave radiation, which sits between visible light and radio waves in the electromagnetic spectrum. All electromagnetic radiation is in the form of waves, for which the distance between wave peaks, known as wavelength, is a defining characteristic. In 1919, American astronomer Mary Lea Heger had first detected lines in the spectrum where very specific wavelengths were dimmer. Unknown chemicals in the interstellar clouds were absorbing microwave radiation from the stars, with their wavelength determined by their molecular structure.

Inspired by architecture

Around 1975, Kroto had found evidence that some lines detected in the atmospheres of red giants could be from long chain-shaped

See also: The synthesis of urea 88–89 ▪ Functional groups 100–105 ▪ Structural formulae 126–127 ▪ Benzene 128–129 ▪ Mass spectrometry 202–203 ▪ Carbon nanotubes 292 ▪ Renewable plastics 296–297 ▪ Two-dimensional materials 298–299

molecules of carbon and nitrogen called cyanopolyynes. He wanted to know how they could have formed. In 1985, Kroto visited American chemist Richard Smalley at Rice University, Texas. Smalley had built his own instrument, which used a laser to vaporize material to atoms, then stripped away their electrons to create a form of matter known as a plasma. Smalley's colleague Robert Curl studied the structures the vaporized atoms formed.

Over the course of 11 days, Kroto, Smalley, Curl, and two PhD students vaporized carbon in the form of graphite and allowed its atoms to form clusters. They used a spectrometer to analyze the carbon clusters to find out how many atoms had come together. The most abundant clusters had 60 atoms, labeled C_{60}, and were especially stable, while some had 70, labeled C_{70}. These forms (allotropes) of carbon had never been seen before.

Seeking to explain the stability of C_{60}, Kroto, Smalley, and Curl realized that 60 atoms were enough to form a very strong shape called a truncated icosahedron cage. They named their newly discovered structure buckminsterfullerene after the American architect R. Buckminster Fuller, who designed the iconic geodesic dome at Expo 67, held in Montreal, Canada.

Convincing evidence

The two forms of carbon, graphite and diamond, were well known, but no one expected carbon to take on this strange "buckyball" shape. When Kroto, Curl, and Smalley published their research, some scientists were critical, but others were enthusiastic. They continued to gather evidence, and by 1990, scientists could make buckyballs—also known as fullerenes—in quantities large enough to test. They also discovered other less stable carbon clusters, including C_{72}, C_{76}, C_{78}, and C_{84}.

The buckyball can withstand high temperatures and pressure, and can act as a semiconductor—and even a superconductor. It is also one of the largest known objects to exhibit the properties of both a particle and a wave.

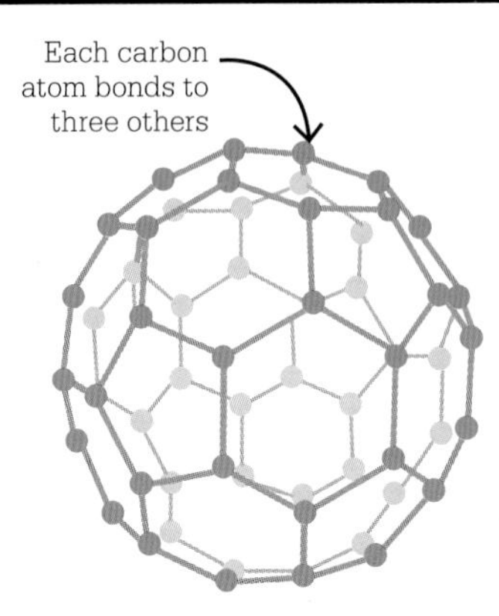
Each carbon atom bonds to three others

A C_{60} buckyball molecule has 12 pentagonal faces and 20 hexagonal faces—a shape used for soccer balls.

In 2010, a team led by Belgian astronomer Jan Cami at the University of Western Ontario, Canada, found fullerene molecules in space for the first time—C_{60} and C_{70}—in the planetary nebula Tc 1, 6,000 light-years from Earth. Kroto was elated, especially with how clear the evidence was, saying, "I thought I would never be as convinced as I am." ■

Harry Kroto

In October 1939, Harold Krotoschiner was born in Wisbech, UK, two years after his parents immigrated as refugees from Nazi Germany. His father shortened the family's name to Kroto in 1955, when he set up a balloon factory in Bolton. From 1958, Kroto studied chemistry at the University of Sheffield. After completing his PhD in 1964, he moved to the National Research Council in Ottawa, Canada, to study small molecules using microwave spectroscopy.

In 1966, Kroto moved to Bell Labs in the US before returning to the UK in 1967 to join the University of Sussex. There, he started using laser and microwave spectroscopy on carbon molecules, which led to the studies on buckminsterfullerene that would win him the Nobel Prize in Chemistry, shared with Curl and Smalley, and a knighthood in 1996. Kroto died in 2016.

Key works

1981 *The Spectra of Interstellar Molecules*
1985 *C_{60}: Buckminsterfullerene*

A CHANG
WORLD
1990 ONWARD

ING

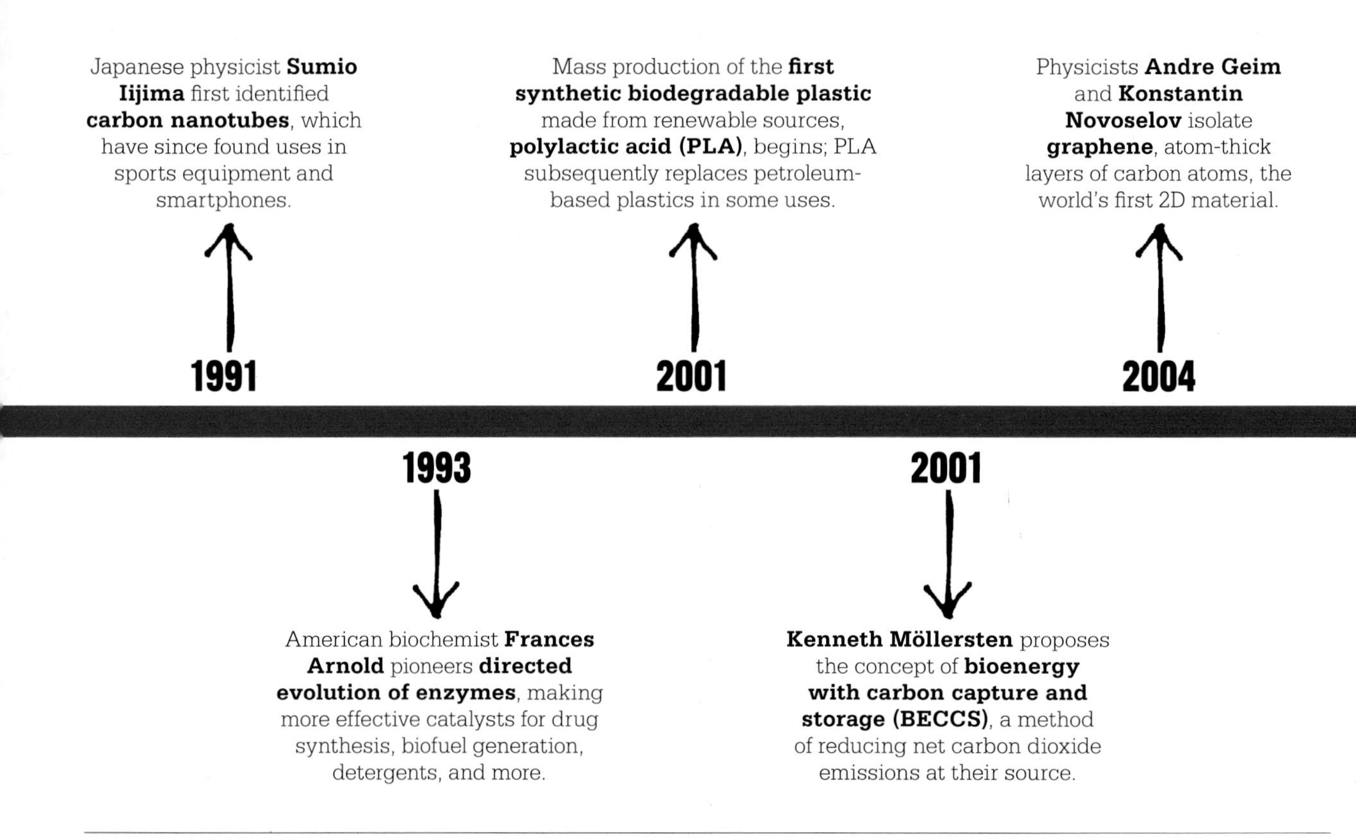

1991
Japanese physicist **Sumio Iijima** first identified **carbon nanotubes**, which have since found uses in sports equipment and smartphones.
1993
American biochemist **Frances Arnold** pioneers **directed evolution of enzymes**, making more effective catalysts for drug synthesis, biofuel generation, detergents, and more.
2001
Mass production of the **first synthetic biodegradable plastic** made from renewable sources, **polylactic acid (PLA)**, begins; PLA subsequently replaces petroleum-based plastics in some uses.
2001
Kenneth Möllersten proposes the concept of **bioenergy with carbon capture and storage (BECCS)**, a method of reducing net carbon dioxide emissions at their source.
2004
Physicists **Andre Geim** and **Konstantin Novoselov** isolate **graphene**, atom-thick layers of carbon atoms, the world's first 2D material.

Since the 1990s, the lines between the science disciplines have become increasingly blurred. There have always been areas of commonality, with aspects of science such as atomic structure straddling the border between chemistry and physics and the subdiscipline of biochemistry at the interface between chemistry and biology. Now, increasingly, advances in science are facilitated by a fusion of knowledge and techniques that defy these defined borders between the disciplines.

Chemistry meets biology

All living organisms are built up from carbon-based molecules, so organic chemistry has always had a close association with biology. Some of the most significant scientific advances in recent years have resulted from the application of chemistry to biological problems.

In the 1990s, biochemists pioneered the directed evolution of enzymes. This technique mimics the natural process of evolution. It has enabled scientists to tailor enzymes to catalyze reactions more effectively—thus enabling the creation of new biofuels or more environmentally friendly chemicals—or to make antibodies that are more selective for particular disease targets.

An even more powerful medical tool was developed in 2011: CRISPR-Cas9 gene editing. It is hoped that this technique may lead to treatment for some cancers and genetic diseases, and it has already been used in some testing methods to detect COVID-19 infections. The COVID-19 pandemic, which began at the end of 2019, offers the most significant example of the benefits of collaboration across biology and chemistry: the approved vaccines for the disease, which were produced in record-breaking time. The viral vector and RNA vaccines represented the first vaccines of their class but did not come out of the blue—they were the culmination of decades of work on the underlying concepts. While aspects of the vaccines are obviously biological in nature, chemistry played an important role in their formulation, ensuring that they function effectively.

Beyond the medical world, the sciences of chemistry and biology have been combined to address environmental hazards; examples include devising bioenergy-based

2009

Researchers use **atomic force microscopy (AFM)** to produce images of individual molecules for the first time.

2011

Jennifer Doudna and **Emmanuelle Charpentier** develop the CRISPR-Cas9 **gene editing technique**, which allows the "editing" of genomes with unprecedented accuracy.

2015

The discoveries of **elements 113**, **115**, **117**, **and 118** are verified; they are named and added to the periodic table the following year.

2020

A new vaccine type—**mRNA vaccines**—and new **viral vector vaccines** are approved for COVID-19. The process used to create the vaccines may be used for other diseases in the future.

methods to avoid carbon dioxide emissions from fossil fuels and creating biodegradable plastics from renewable sources.

Chemistry meets physics

The discovery of new elements, once the domain of the chemist, now requires cutting-edge physics. The discoveries of naturally occurring elements have now been exhausted; all of the chemical elements added to the periodic table since francium in 1939 were first discovered in a laboratory. Nor is element discovery a lone scientist's pursuit in the modern age. Vast teams of researchers work in dedicated facilities to smash elements into one another in the hope of fleetingly generating a few atoms of an element that is rarely—if ever—produced in nature. The challenge is one of exponential difficulty. The year 2015 was the last time that new elements were added to the periodic table, completing its seventh row; since then, we have seen the longest gap between discoveries since synthetic element discovery began. While there is still confidence that further elements will be created, there is also an increasing focus on understanding the exotic properties of the existing super-heavy synthetic elements.

Physicists have aided chemists not only in finding new elements, but also in uncovering new forms of existing ones. A number of elements can exist in different forms, or allotropes, with carbon the prime example. Diamond and graphite are commonly known substances that are allotropes of the element carbon, and in recent decades, further carbon allotropes—fullerenes, nanotubes, and graphene—have been identified. Scientists hope that the beneficial properties of these more recently discovered allotropes could have various uses; they have already found some actual and potential applications, from smartphones to energy storage, and systems for delivering drug therapies.

Finally, new physics-based techniques have allowed chemists to "see" molecules—a feat that would have astounded the early chemists who speculated about molecular structures. Notably, the use of atomic force microscopy has produced unprecedented images of individual molecules, making it possible to observe their behavior directly for the first time. ■

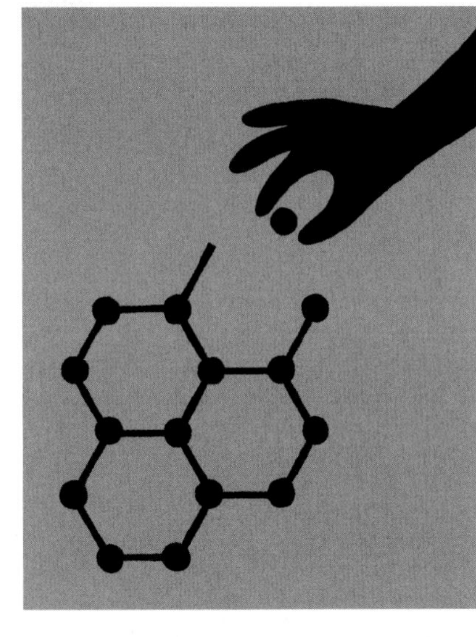

BUILD THINGS ONE ATOM AT A TIME

CARBON NANOTUBES

IN CONTEXT

KEY FIGURE
Sumio Iijima (1939–)

BEFORE
1865 Jacobus van 't Hoff and French chemist Joseph-Achille Le Bel confirm the tetrahedral structure of the carbon atom.

1955 American scientists L.J.E. Hofer, E. Sterling, and J.T. McCartney observe minute tubular carbon filaments.

1985 Richard Smalley and others discover fullerenes.

AFTER
1997 Swiss researchers demonstrate the special electrical properties of carbon nanotubes.

2004 Russian-born British physicists Andre Geim and Konstantin Novoselov confirm the existence of graphene.

After Richard Smalley and Harry Kroto's discovery of the buckyball, scientists wondered what other secrets lay in carbon molecules. In 1991, Japanese physicist Sumio Iijima discovered carbon nanotubes—fullerene molecules rolled into nanometer-sized tubes.

Minute carbon fibers had been observed as early as 1955, but it was Iijima who properly identified them. With an electron microscope, he could see the carbon fibers were tiny Swiss roll–like molecules (now called multiwall nanotubes). Two years later, he and American physicist Donald Bethune independently discovered even simpler single-layered nanotubes, or "buckytubes"—hollow cylinder-shaped molecules of carbon, bonded hexagonally and 100,000 times thinner than a human hair.

Carbon nanotubes are the strongest and stiffest material that has so far been discovered, in terms of tensile and elastic qualities.

Nanotubes conduct electricity and heat better than copper, and they are also very light, much stronger than steel, and resistant to corrosion. Using nanotubes to replace copper or steel in carbon-fiber materials makes the latter even stronger and more durable.

In 2004, scientists confirmed the existence of graphene, an allotrope of carbon consisting of a single layer of atoms. They realized that this could be the strongest of all nanotubes. If it becomes possible to produce thick layers of nanotubes, they could potentially be used instead of steel and other metals. They are already used in a wide range of products—from sports equipment to smartphones. ■

See also: Stereoisomerism 140–143 ▪ Super-strong polymers 267 ▪ Buckminsterfullerene 286–287 ▪ Two-dimensional materials 298–299

WHY NOT HARNESS THE EVOLUTIONARY PROCESS TO DESIGN PROTEINS?

CUSTOMIZING ENZYMES

IN CONTEXT

KEY FIGURE
Frances Arnold (1956–)

BEFORE
1926 American chemist James B. Sumner proves that enzymes are proteins.

1968 Researchers identify "restriction" enzymes that can cut short sections of DNA.

1985 American biochemist Alexander Klibanov shows how enzymes can function in organic solvents.

1985 American biochemist Kary Mullis invents the process of polymerase chain reaction (PCR), enabling scientists to copy short snippets of DNA.

AFTER
1998 Directed evolution is used to create lipases in detergents for the removal of fat stains.

2018 Arnold heads the creation of a mutant *E. coli* that converts sugars into isobutanol—a precursor for fuels and plastics.

In the 1980s, scientists learned how to engineer DNA so that organisms could produce chemicals to order. But it was laborious and not always effective. In 1993, American biochemist Frances Arnold pioneered the idea of nudging nature to get better results. Natural selection allows species to evolve, as favorable mutations thrive and unfavorable ones die out. Arnold's method, known as directed evolution, involves repeatedly introducing mutations in the gene of interest, then selecting each time only those heading in the desired direction.

Arnold's key experiment focused on the enzyme subtilisin, which breaks up the milk protein casein. Researching industrial applications for the enzyme, she sought a version that worked in the solvent dimethylformamide (DMF)—outside the watery environment of a cell. She provoked mutations in bacteria that produced subtilisin and grew them in culture dishes containing casein and DMF. The bacteria that were most successful in producing an enzyme to dissolve casein were then chosen for further mutations. In this way, Arnold created an enzyme that was 256 times more effective than the original in just three generations.

> Biology does chemistry very efficiently.
> **Frances Arnold (2018)**

Arnold's team used the same method to get enzymes to catalyze reactions that no enzyme had done before, and even to make substances with bonds that had never been made in nature, such as carbon and silicon. The technique is now used to make everything from new biofuels to synthesized drugs, and future possibilities are huge. ■

See also: Enzymes 162–163 ▪ Retrosynthesis 262–263 ▪ Rational drug design 270–271 ▪ The polymerase chain reaction 284–285

A NEGATIVE EMISSION IS GOOD

CARBON CAPTURE

IN CONTEXT

KEY FIGURE
Kenneth Möllersten (1966–)

BEFORE
1972 An early form of carbon capture technology is used for enhanced oil extraction.

1996 The Norwegian Sleipner facility begins to capture and store CO_2 in a saline reservoir in the North Sea.

2000 The Carbon Sequestration Initiative is launched at MIT.

AFTER
2014 SaskPower's Boundary Dam CCS project in Canada starts extracting CO_2 from power station emissions.

2020 The Shell Quest CCS project in Canada surpasses 5.5 million tons (5 million metric tons) of CO_2 stored in underground salt deposits.

2020 Around 44 million tons (40 million metric tons) of CO_2 is captured and stored globally, mainly to boost oil extraction.

To achieve the target of limiting global warming to 2.7°F (1.5°C) above pre-industrial levels, global carbon dioxide (CO_2) emissions from the burning of fossil fuels need to be reduced dramatically. This can be achieved by improving energy efficiency and switching from fossil fuels to renewable energy sources.

Research groups including some connected to fossil fuel companies have argued that these measures will not take place quickly enough and that CO_2 needs to be captured and stored. In 2001, Kenneth Möllersten, a Swedish PhD student, explained how bioenergy with carbon capture and storage (BECCS) could partly accomplish this. His idea involved planting agricultural crops or trees, which remove CO_2 from the atmosphere as they grow. The trees would be cut down and burned to produce bioenergy, and the resulting CO_2 emissions captured. However, the huge areas of land required to grow these crops and trees would increase pressure on strained natural resources.

> We have to learn how to do carbon capture and storage, and we have to learn how to do it quickly on a commercial scale.
> **Nicholas Stern**
> ***A Blueprint for a Safer Planet*** **(2009)**

CCS options

The concept of carbon capture and storage (CCS) was introduced in 1977, but Möllersten gave it a fresh impetus. CO_2 has to be separated from other emissions, compressed, and transported to storage locations, where it can be isolated from the atmosphere. One method of isolation is to pump the CO_2 underground, where it will dissolve in groundwater. In 2010, biological engineers at the Massachusetts Institute of Technology (MIT) found a way to combine CO_2 dissolved in water with mineral ions to form solid carbonates that could be used as a building material. A third option is to recycle captured CO_2 by reusing it in other industrial projects.

See also: Gases 46 ▪ Fixed air 54–55 ▪ The greenhouse effect 112–115 ▪ Cracking crude oil 194–195

SaskPower's Boundary Dam coal-fired power station in Saskatchewan, Canada, cost $1.1 billion to be converted to a facility that captures and stores carbon.

CCS facilities

In 1972, the Terrell methane processing plant in Texas piped CO_2 to a nearby oil field, where it was used to maximize oil extraction from underground. This technology was not "green"—because it was used to extract more fossil fuels—but it showed the potential of CCS.

In 1996, Norway's offshore Sleipner storage facility began to pump CO_2 derived from industry into undersea sandstone strata. The geology of an underground storage site is all-important: it must be beneath impermeable cap rocks, like those at Sleipner, to prevent the CO_2 from returning to the surface. By 2017, the facility had stored about 18.2 million tons (16.5 million metric tons) of CO_2.

In 2020, 26 commercial CCS projects were in operation globally, with others at various stages of development. Boundary Dam, a converted coal-fired power station in Canada, captures 90 percent of CO_2 emissions. The mixture of water vapor and gas produced when the coal burns bubbles up through an alkaline amine solution, which pulls the CO_2 molecules out of the air and into the solution. The liquid then flows to a heater, where the CO_2 molecules evaporate and are trapped and compressed.

Some scientists believe that with sufficient investment, carbon capture could achieve about 14 percent of the reduction in greenhouse gas emissions required by 2050. It could also be used to extract CO_2 already in the atmosphere via direct air capture (DAC). Alongside improving energy efficiency and using a far greater proportion of renewable energy for industry, homes, and transportation, CCS could play a vital role in carbon reduction. However, the high cost and uncertainties about where captured CO_2 can be stored need to be resolved. ■

Advantages and disadvantages of carbon capture

Advantages	Disadvantages
CCS is a tried-and-tested technique for reducing net emissions.	Long-term storage capacity is uncertain.
It is an effective method for reducing industrial emissions at source.	CCS (DAC) is poor at removing emissions of people, farming, heating, and transport
Other pollutants can be removed at the same time.	CCS is very expensive.

Direct air capture

Rather than capturing CO_2 as it leaves a source, such as a factory chimney, direct air capture (DAC) collects the gas that is already present in the atmosphere. A DAC plant has large fans that suck air into a compartment where a filter extracts the CO_2. This is then heated to 212°F (100°C) and dissolved in water. To store it underground, the dilute carbonic acid is pumped into reactive rock formations, such as basalt, where it mineralizes to become solid carbonates, such as calcite, over two years.

In 2020, there were 15 DAC plants globally; the largest, in Iceland, extracts about the equivalent of 400 people's annual emissions. This method of carbon capture is very expensive, although Iceland has the advantage of plentiful cheap geothermal energy to power its DAC plants. So far, DAC has captured only tiny amounts of CO_2, and scientists do not believe it offers a complete solution to the climate crisis. But with more investment in technology, and more plants, its contribution could improve.

Huge fans on the roof of a waste incineration plant in Switzerland capture CO_2 for recycling.

BIOBASED AND BIODEGRADABLE

RENEWABLE PLASTICS

IN CONTEXT

KEY FIGURE
Patrick Gruber (1961–)

BEFORE
1920s Wallace Carothers discovers polylactic acid (PLA), a biodegradable plastic that is made from renewable sources.

1990s Bioplastics made from renewable biomass sources are used to produce carrier bags, protective clothing, gloves, cups, and disposable cutlery.

AFTER
2010 Commercial production of bioplastics made from seaweed begins in France.

2018 Finnish oil company Neste starts producing bio-polypropylene for IKEA furniture.

2018 The first bioplastic, fully recyclable car prototype is assembled in Eindhoven, the Netherlands.

In 2001, a joint venture between two US corporations—Cargill and Dow Chemical—pioneered the mass production of a synthetic polymer made from renewable sources. Previously, petroleum-based polymers were the key ingredient in plastics manufacture, so this development had huge implications for reducing the industry's reliance on fossil fuels.

The era of bioplastics

The term "bioplastic" can be applied to any plastic that is primarily derived from renewable organic materials, such as cornstarch, vegetable fats, tapioca roots, or milk. The Cargill-Dow bioplastic used the polymer polylactic acid (PLA). This was not new—American chemist Wallace Carothers had discovered it in the 1920s—but its production was very expensive, so it was not manufactured commercially. Then in 1989, Patrick Gruber, a Cargill-Dow chemist, produced PLA from corn (maize) on his stove at home. By 2001, the plastic was being manufactured by a company now called NatureWorks, which makes PLA as a replacement for plastics such as polyethylene terephthalate (PET) and polystyrene in packaging and food service applications.

By 2019, more than 397 million tons (360 million metric tons) of plastics were being made annually, thereby creating a huge disposal problem. In 2020, around 2.3 million

Plastic-eaters

PET is used to make most plastic drinks bottles and many synthetic fibers. It decomposes very slowly, and accumulates in animals' stomachs and so may enter the human food chain. In 2016, Japanese researchers reported that the bacterium *Ideonella sakaiensis* had evolved the ability to use two enzymes—PETase and MHETase—to break down PET plastic into terephthalic acid and ethylene glycol, which it digests and uses as sources of carbon and energy. One problem is the slow speed: at 86°F (30°C), it takes the bacteria about six weeks to fully degrade a piece of PET plastic the size of a thumbnail. However, in 2020, a UK and US team reengineered the two enzymes into a "super-enzyme" that digests the plastic six times faster.

See also: The greenhouse effect 112–115 ▪ Enzymes 162–163 ▪ Synthetic plastic 183 ▪ Cracking crude oil 194–195 ▪ Polymerization 204–211 ▪ Customizing enzymes 293

tons (2.1 million metric tons) of bioplastics were manufactured, and this figure is projected to increase to just 3.1 million tons (2.8 million metric tons) by 2025. Around 60 percent of this is biodegradable, mostly PLA. Whereas traditional plastics are durable and degrade very slowly, PLA bioplastics are compostable and can be broken down into nutrient-rich biomass. However, this can only happen in specific conditions, which must include oxygen.

Usually derived from sugars, PLA has similar characteristics to polyethylene and polypropylene, but since it is a thermoplastic, it can be heated repeatedly to its melting point, cooled, and reheated again, without significant degradation. Today, PLA bioplastics are used for a variety of products, including plastic film, shrink wrap, bottles, 3D printing material, and medical implants intended to be absorbed by the body.

Alternatives to PLA

Polyhydroxyalkanoates (PHAs) are biodegradable plastics produced through the bacterial fermentation of sugars and lipids. Bacteria are deprived of the nutrients they need to function and instead are given high levels of carbon. The bacteria store the carbon in granules, which is then harvested by the manufacturer. PHAs have been used for agricultural and many medical applications, including sutures, bone plates, stents, and surgical mesh. Other bioplastics include cellulose-based types, manufactured using cellulose or derivatives of cellulose, and protein-based types derived from gluten, casein, and milk protein.

Bioplastics have the potential to replace petroleum-based plastics. However, critics argue that for this to happen, vast areas of farmland would need to be used for their production, which could create environmental problems—clearing forests for corn or sugar cane plantations, for example—and lead to higher food prices. There are other environmental factors, too. When bioplastics degrade, they emit the greenhouse gases carbon dioxide and methane, and they increase the acidity of soil and water.

Furthermore, a 2015 UN report expressed concern that people would recycle less if they thought that the plastic they used would harmlessly degrade when discarded. Bioplastics and recycling are only part of the solution to plastic pollution. Ultimately, humans need to significantly reduce plastic production and consumption. ■

Advantages and disadvantages of bioplastics

Advantages	Disadvantages
Reduce need for nondegradable single-use plastic	More expensive to produce
Fewer environmental problems	Require suitable moisture, acidity, and temperature to biodegrade
Reduce dependence on fossil fuels	Land needed to grow plants from which bioplastics are made
Production generates lower greenhouse gas emissions than traditional plastic	Release greenhouse gases when they biodegrade

How biodegradation works

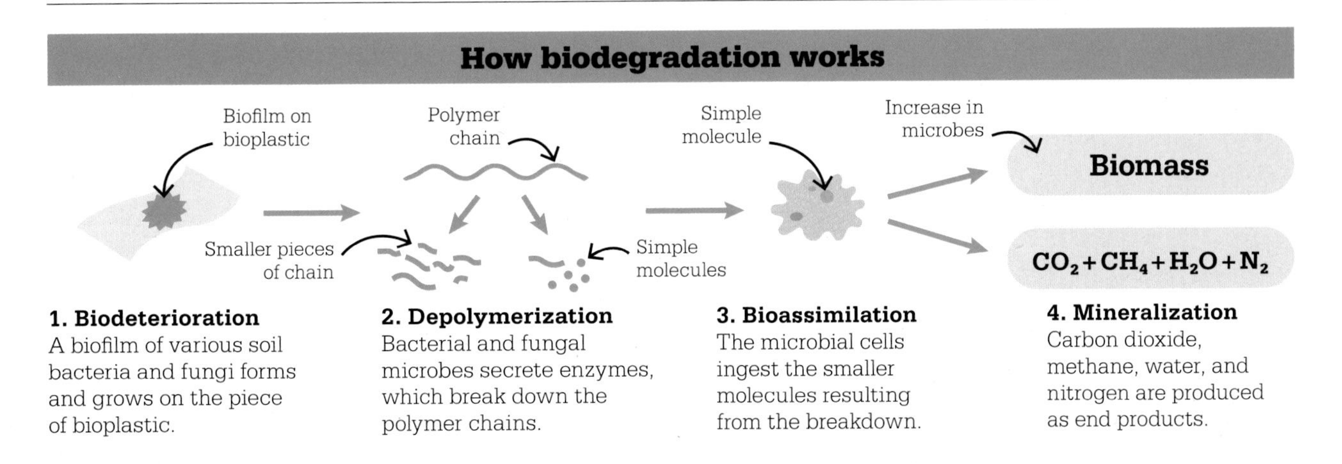

1. Biodeterioration A biofilm of various soil bacteria and fungi forms and grows on the piece of bioplastic.

2. Depolymerization Bacterial and fungal microbes secrete enzymes, which break down the polymer chains.

3. Bioassimilation The microbial cells ingest the smaller molecules resulting from the breakdown.

4. Mineralization Carbon dioxide, methane, water, and nitrogen are produced as end products.

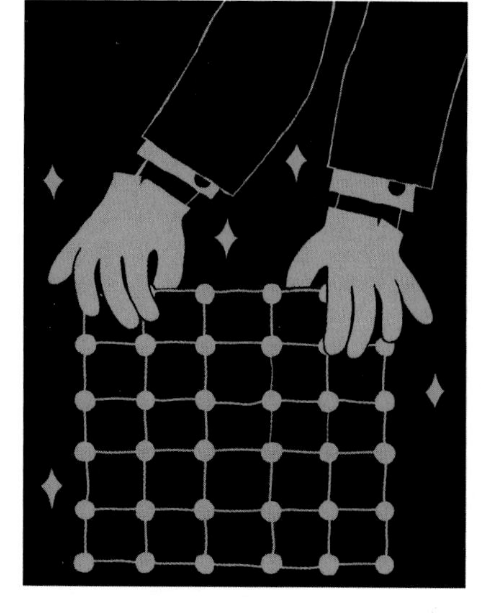

THE MAGIC OF FLAT CARBON

TWO-DIMENSIONAL MATERIALS

IN CONTEXT

KEY FIGURES
Andre Geim (1958–),
Konstantin Novoselov (1974–)

BEFORE
1859 British chemist Benjamin Brodie exposes graphite to strong acids, discovering what he called graphon, a new form of carbon with a molecular weight of 33.

1962 German chemists Ulrich Hofmann and Hanns-Peter Boehm produce single atom-thick layers from graphite oxide, but their discovery attracts little attention.

AFTER
2017 Samsung Electronics makes a transistor that harnesses electrons' full speed of movement in graphene.

2018 Swiss material scientist Nicola Marzari and his team discover that 1,825 different substances could have two-dimensional forms.

One of the most famous scientific discoveries of the 21st century relies on the same principles seen when applying a pencil to paper. Graphite consists of stacks of atom-thick layers of carbon held together by weak bonds. When a graphite pencil writes on a page, some layers slide off. The individual layers of graphite are known as graphene. Evidence that these flat graphene layers existed had been collected in the 19th and 20th centuries and then seemingly forgotten. Then, in 2002, Russian-born British physicists Andre Geim and Konstantin Novoselov began a series of experiments that firmly grabbed scientists' attention.

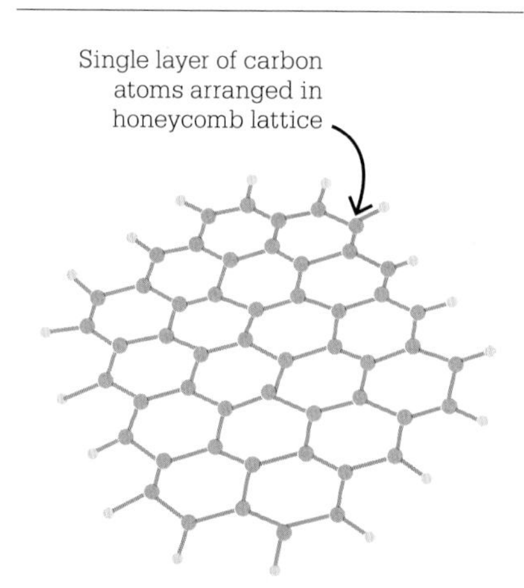

Made up of a single sheet of graphite, graphene is incredibly thin but very strong and resistant to tearing and can be rolled into fibers.

All done with sticky tape

Geim was leading a team that included Novoselov at the University of Manchester. He was interested in ultra-thin materials and their potential for electronics and thought that graphite was a good candidate. By placing a strip of adhesive tape on graphite, Geim and Novoselov could rip off some of its thin flakes. Folding the tape, then pulling it apart, split the flakes again. Repeating the process created thinner and thinner flakes. They dissolved the tape in solvent, dipped a silicon wafer into the solution, and found that flakes less than 10 nanometers thick stuck to it. Flakes this thin are transparent, but when viewed under a microscope, the thinnest layers looked dark blue against the silicon. They then took one of these ultra-thin flakes and used it to make a transistor—the tiny component at the heart of computer chips.

Geim and Novoselov had not yet made a 2D layer of graphene, which is less than 1nm thick, but they

See also: Benzene 128–129 ▪ Synthetic plastic 183 ▪ Super-strong polymers 267 ▪ Buckminsterfullerene 286–287 ▪ Carbon nanotubes 292 ▪ Renewable plastics 296–297

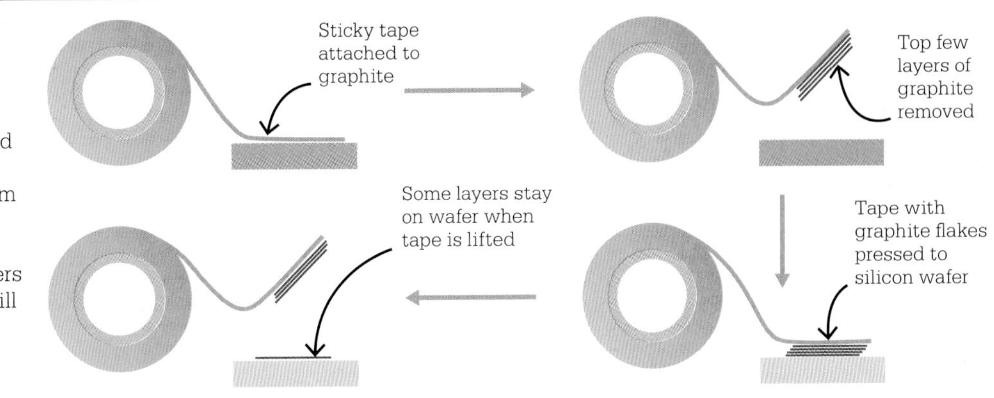
Sticky tape attached to graphite
Top few layers of graphite removed
Tape with graphite flakes pressed to silicon wafer
Some layers stay on wafer when tape is lifted

Officially known as micromechanical cleavage, the simplest method of producing adhesive tape, graphite, and a substrate such as silicon. If the adhesion of the bottom layer of graphene to the substrate is stronger than the bonds between the layers of graphene, some flakes will remain on the substrate.

continued their experiments and a year later had succeeded in making transistors using two layers of graphene, and one layer of graphene, and found that they behaved very differently. In 2004, they published their paper on the electrical conductivity of graphene and the scientific world took notice. By the time Geim and Novoselov were awarded the Nobel Prize in Physics in 2010, academics in many fields were exploring the possibilities of this "wonder material," as were several electronics firms.

Physically superior

In part because it is so hard to make, graphene became one of the most expensive materials on Earth. The high price is also due to its remarkable properties. Despite being extremely light, it has a breaking strength 200 times greater than steel, making it the strongest material ever tested. Today, graphene is added to some materials to reinforce them, such as the frames of tennis rackets. It is also extremely flexible, so it can be rolled into nanotubes.

As it is made of carbon, graphene retains the same movement of electrons across the positively charged hexagonal rings that makes graphite highly conductive. Geim and Novoselov measured graphene's electron mobility values—the speed at which electrons move through it—as about 100 times higher than in silicon used to make computer chips. And they were around 10 times higher than the fastest preexisting semiconductors, which stimulated interest in graphene's potential for ultra-fast advanced electronics.

This enhanced mobility comes because graphene is not like other materials, where the speed of electrons relates to their mass. Graphene electrons behave like photons—particles of light that move at a speed independent of their mass.

The discovery of graphene enabled researchers to explore what happens when they slim down other materials to single-atom layers. For example, phosphorene, equivalent to graphene but using phosphorus, has similar electron mobility but works better as a semiconductor. Today, the list of 2D materials is long and ever-growing and the possibilities seemingly endless. ■

Manufacturing graphene

The makers of silicon chips rely heavily on growing silicon crystals at high temperatures in a process known as epitaxy. This builds layers of material on wafers, depositing atoms one after another. A similar approach has produced graphene by reacting methane with hydrogen on copper films. Silicon carbide (SiC) is also a semiconductor material that is already manufactured as wafers and used industrially, albeit less widely than silicon. Heating high crystal quality silicon carbide wafers above 2,000°F (1,100°C) selectively evaporates silicon from the top surface and turns the silicon carbide there into a layer of graphene.

A more chemical approach might also be useful for exploiting graphene's physical properties. If graphene oxide paper is placed in a solution of pure hydrazine (N_2H_4)—a compound used in rocket fuel—it reacts and becomes a single layer of graphene.

ASTONISHING IMAGES OF MOLECULES

ATOMIC FORCE MICROSCOPY

IN CONTEXT

KEY FIGURES
Leo Gross (1973–),
Gerhard Meyer (1957–)

BEFORE
1965 American electrical engineer Harvey Nathanson invents a radio tuner that is the first micro-electro-mechanical system—a tiny electronic device with moving parts.

1979 Gerd Binnig and Heinrich Rohrer invent the scanning tunneling microscope (STM), which can map objects smaller than would be visible with an optical microscope.

AFTER
2016 Leo Gross uses AFM to visualize a reversible chemical reaction.

2020 American biophysicist Simon Scheuring increases the resolution of AFM by combining multiple images of the same area.

Although we cannot naturally see atoms in substances around us, some tools and techniques can help us better understand them and advance the field of chemistry.

X-ray crystallography is a method that works backward from the diffraction patterns that form when X-rays pass through crystals. Chemists may also use radio signals from nuclear magnetic resonance (NMR) to deduce how atoms are connected. In the 1980s, however, researchers found a way to produce much more striking atomic images: atomic force microscopy (AFM). This technique was taken further by German physicists Leo Gross and Gerhard Meyer, who were later able to produce images at atomic resolution.

AFM emerged from computing giant IBM's research lab in Zurich, Switzerland, closely following the related process of scanning tunneling microscopy (STM). STM was the first scanning probe microscopy method, and it works by scanning a sensor, or probe, across a surface. Once the probe has moved across the whole surface horizontally, it shifts down slightly and starts scanning another line. STM creates pictures by detecting tiny changes in the flow of electricity through the probe. It then maps the heights of very small features on a surface. German physicist Gerd Binnig and Swiss physicist Heinrich Rohrer both received the 1986 Nobel Prize in Physics for the invention of STM, but the technique only works with materials that conduct electricity.

Looking for improvements

Binnig wanted to open the method up to other substances. Binnig—along with fellow physicists,

> I was fascinated with the method, in part because the experimenter gets an immediate response.
> **Leo Gross**

See also: Intermolecular forces 138–139 ▪ X-ray crystallography 192–193 ▪ Nuclear magnetic resonance 254–255 ▪ Protein crystallography 268–269 ▪ Buckminsterfullerene 286–287 ▪ Carbon nanotubes 292 ▪ Two-dimensional materials 298–299

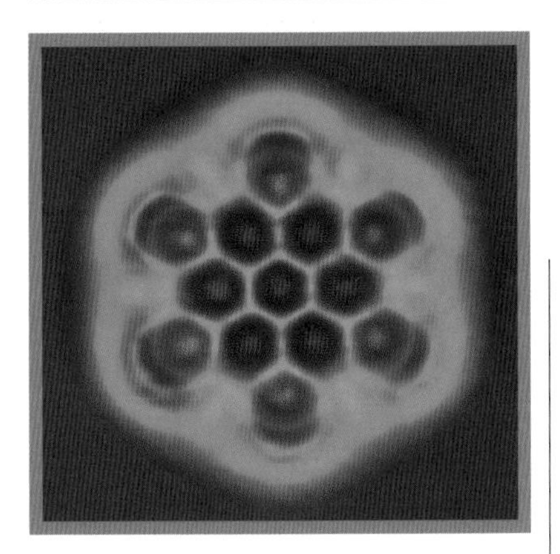

This image of a single molecule of nanographene, produced by IBM Zurich, shows carbon bonds of various lengths, proving that different bonds have different physical properties.

American Calvin Quate and Swiss Christoph Gerber—overcame the problem in 1985. As the name AFM suggests, they measured force, the fundamental phenomenon that moves things, rather than electrical current. AFM uses an extremely fine probe, just a few atoms wide at its point. This probe sits at the end of an arm called a cantilever, which can measure small changes in force.

Binnig, Gerber, and Quate made AFM cantilevers from gold foil with a diamond tip. Later cantilevers were usually made from silicon—the same material used to make microchips. However, silicon bends so much that it misses details smaller than 20 nanometers. While that is still a very high resolution, detecting individual atoms involves measuring details of 1 nanometer or even smaller.

German physicist Franz Giessibl found a solution to this problem during the 1990s, when he realized the quartz tuning forks that wristwatches use to count time were just the right stiffness for cantilevers. In 1996, he began making AFM sensors that relied on quartz cantilevers.

Fine details

The probe tip at the end of Giessibl's quartz cantilever was only a few atoms wide, but this was not sensitive enough to achieve atomic resolution. Trying to measure a single atom with a probe a few atoms across was like measuring a marble with a tennis ball. So, in 2009, an IBM team including Gross and Meyer affixed a single carbon monoxide (CO) molecule to the AFM probe. With just one carbon atom, which sits next to the probe, and one oxygen atom, which hangs underneath, carbon monoxide produces an atomically sharp tip. It makes a highly sensitive probe because it can detect tiny changes in the density of electrons around substances. Electron-dense areas—such as atoms or even the bonds between them—can deflect the carbon monoxide molecule, creating a force that the probe can detect.

Also in 2009, carbon monoxide helped the IBM researchers produce images of pentacene—a chain of five interlinked six-membered carbon rings. The AFM images showed bonds between atoms as clearly as if they were drawn on a piece of paper. This was groundbreaking. Their discovery captured the imaginations of chemists, opening the door to more direct observation of how chemicals behave. ■

How AFM works

Atomic force microscopes are a bit like record players. The equivalent to the record player's stylus is a very fine probe; this sits at the end of a cantilever, which is like a record player's arm. Much as the contours in a record's groove cause the stylus to vibrate and produce sound, the sample surface exerts a force on the probe that researchers can track. Electrons on the probe and surface can sometimes attract each other, locking the probe down. Scientists therefore vibrate the probe. As the probe nears the surface, the vibration frequency changes; researchers often measure this by shining a laser on the cantilever. Alternatively, cantilevers made from quartz produce an electric current as they vibrate, and scientists can track this to map the surface.

With such tiny vibrations involved, surrounding noise may be disruptive. One lab based in Vienna, Austria, has therefore suspended its AFM from the ceiling to protect it from the vibrations caused by traffic.

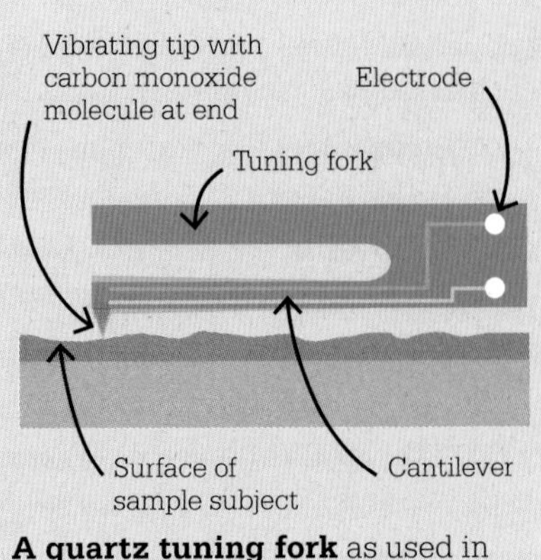

A quartz tuning fork as used in wristwatches was introduced to produce sharper images, making atomic-resolution AFM possible.

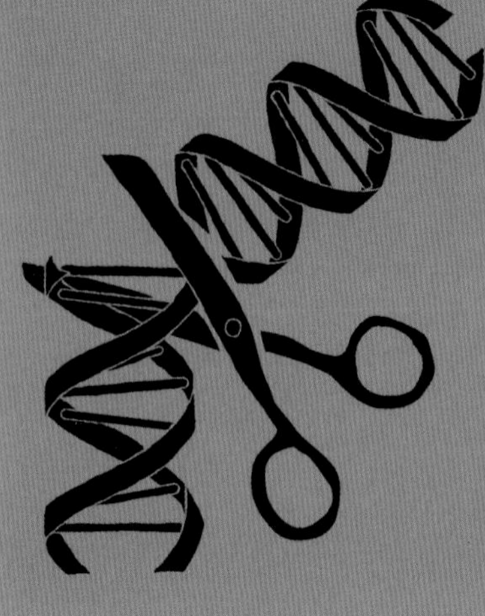

A BETTER TOOL TO MANIPULATE GENES

EDITING THE GENOME

IN CONTEXT

KEY FIGURES
Emmanuelle Charpentier (1968–), **Jennifer Doudna** (1964–)

BEFORE
1953 Francis Crick and James Watson describe the double helix structure of DNA.

1973 American biochemist Herb Boyer uses enzymes to splice DNA and insert genes into bacteria, widely seen as the start of genetic engineering.

AFTER
2016 American biochemist Douglas P. Anderson uses ancestral protein reconstruction to find the molecule that may be the reason why single-celled organisms evolved into multicelled organisms.

2017 American plant geneticist Zachary Lippman fine-tunes mutant genes to increase yields from tomato plants.

In 1987, a team of biologists at Osaka University in Japan discovered a strange pattern of DNA sequences in a gene belonging to *Escherichia coli*, a common bacterium. Five short repeating segments of DNA, each with identical sequences of 29 bases—the building blocks of DNA—were separated by short nonrepeating "spacer" sequences.

By the end of the 1990s, further scientific research had shown that this particular pattern was not peculiar to *E. coli* but could, in fact, be found in many different bacteria. Spanish biologist Francisco Mojica and Dutch microbiologist Ruud Jansen named the pattern "clustered regularly interspaced short palindromic repeats," or CRISPR for short, in 2002.

CRISPR sequences

Also in 2002, Jansen and his team observed that CRISPR was always accompanied by a second set of sequences, the CRISPR-associated, or Cas, genes. The Cas genes appeared to code for enzymes that cut DNA. By 2005, Mojica had determined that the "spacer" sequences between the CRISPR ones shared similarities with the DNA of viruses. This led Mojica to hypothesize that CRISPR acted as a bacterial immune system.

In 2005, Russian microbiologist Alexander Bolotin was studying *Streptococcus thermophilus*, a bacterium that contained some previously unknown Cas genes, including one that coded for an enzyme now known as Cas9. He noted that the spacers shared a

Jennifer Doudna and Emmanuelle Charpentier jointly received the Nobel Prize in Chemistry in 2020 for their groundbreaking work on editing and correcting DNA.

See also: The chemicals of life 256–257 ▪ The structure of DNA 258–261 ▪ Green fluorescent protein 266 ▪ Customizing enzymes 293

sequence at one end that seemed to be essential for target recognition of invading viruses.

Understanding Cas9

Scientists in Denmark, led by French microbiologist Philippe Horvath, revealed more about the function of the CRISPR-Cas9 system in 2007, when they infected *S. thermophilus* bacteria with two virus strains. While many bacteria were killed by the viruses, some survived. Further investigations revealed that the surviving bacteria were inserting DNA fragments from the viruses into their spacer sequences, enabling them to fight off subsequent attacks.

Confronted by an invader, such as a virus, the bacterium copies and incorporates viral DNA segments into its genome as spacers between the short DNA repeats in CRISPR. The spacers provide a template for the bacterium to recognize the DNA of an incoming virus in the future.

Charpentier and Doudna

In 2011, French microbiologist Emmanuelle Charpentier discovered another component of the CRISPR system, a molecule called crRNA that helps identify genetic sequences in invading viruses and works in tandem with a previously unknown molecule called tracrRNA to guide Cas9 to its targets.

That same year, Charpentier began working with American biochemist Jennifer Doudna. Together, they recreated the bacterium's genetic scissors—its ability to cut DNA—and simplified the molecular components of the scissors, so that they were easier to use. They reported their findings in a paper in 2012. Taking things further, Charpentier and Doudna reprogrammed the scissors so that they could be deployed to cut not just viral DNA but any DNA molecule at any point desired, making genome editing possible.

Doudna and Charpentier were not the only ones involved in developing CRISPR. In 2012, a team led by biochemist Virginijus Šikšnys in Lithuania independently showed how the Cas9 enzyme could be instructed to cut DNA sequences. And in the US, biochemist Feng Zhang also reported that he had modified the CRISPR-Cas9 system. Scientists now had a tool to allow them to modify and rewrite genomes with unprecedented accuracy. ■

CRISPR at work

CRISPR has the potential to revolutionize bioscience in medicine, agriculture, and other fields. The first trial of CRISPR cell therapy was in 2019, when it was used to treat patients with sickle cell disease by restoring fetal hemoglobin. It has also been used to engineer T-cells, part of the immune system, to make them more effective in destroying cancer cells. From 2019, during the COVID-19 pandemic, CRISPR was used as a diagnostic tool, utilizing Cas9's search function to target viral genetic material. It has been of benefit in stem-cell research, making it possible to reprogram and grow stem cells to produce the exact type of tissue desired.

In agriculture, it is predicted that CRISPR-modified foods from crops that have been made pest- and drought-resistant could be commercially available by 2030. CRISPR could also extend the shelf life of perishable foods. In the field of bioenergy, bacteria, yeasts, and algae have been modified to produce higher yields of biofuels.

CRISPR-Cas9 gene-editing technique

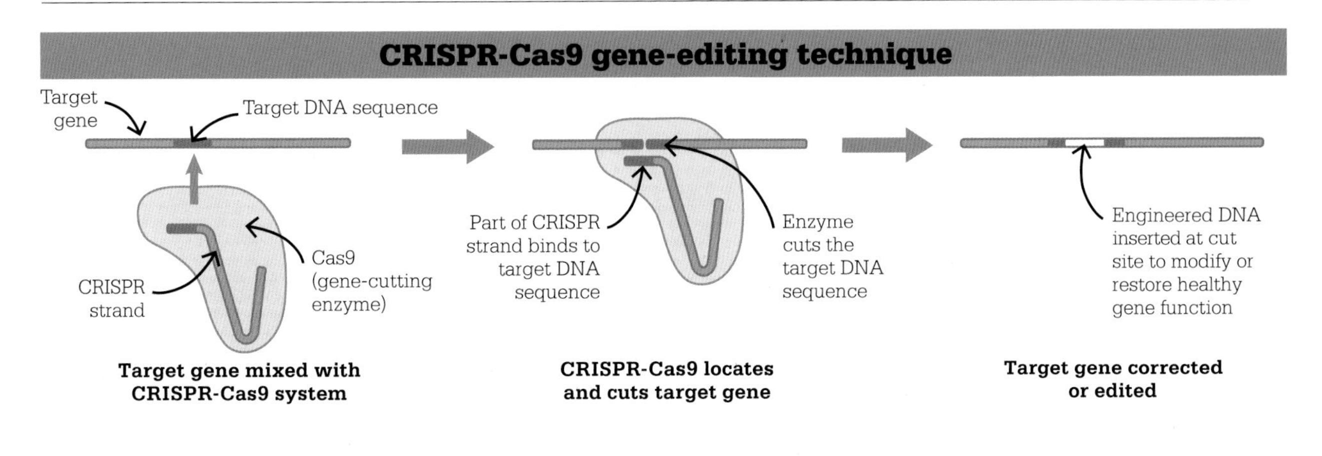
Target gene
Target DNA sequence
CRISPR strand
Cas9 (gene-cutting enzyme)
Target gene mixed with CRISPR-Cas9 system
Part of CRISPR strand binds to target DNA sequence
Enzyme cuts the target DNA sequence
CRISPR-Cas9 locates and cuts target gene
Engineered DNA inserted at cut site to modify or restore healthy gene function
Target gene corrected or edited

WE WILL KNOW WHERE MATTER CEASES TO BE

COMPLETING THE PERIODIC TABLE?

IN CONTEXT

KEY FIGURE
Yuri Oganessian (1933–)

BEFORE
1933 Maria Goeppert Mayer and Hans Jensen develop the shell model of atomic nucleus structure. Mayer proposes that certain numbers of protons or neutrons are more stable.

2002 Oganesson (element 118) is synthesized in Russia.

2010 Tennessine (element 117) is synthesized, filling one of the few remaining gaps in Mendeleev's periodic table.

2011 Finnish chemist Pekka Pyykkö publishes an extended periodic table—the Pyykkö model—showing the predicted properties of 54 elements up to atomic number 172.

AFTER
2016–2021 Unsuccessful attempts are made to synthesize elements with an atomic number above 118.

On December 30, 2015, the International Union of Pure and Applied Chemistry announced that it had verified the discoveries of four new chemical elements: numbers 113, 115, 117, and 118. The final missing elements in the periodic table's seventh row had been found, and six months later, they had confirmed names: nihonium, named after Nihon, one way to say "Japan" in Japanese; moscovium, named for the Moscow region where Russia's Joint Institute for Nuclear Research (JINR) is based; tennessine, named after the state of Tennessee, in which the Oak Ridge National Laboratory is based; and oganesson, named after Yuri Oganessian, the Russian nuclear physicist who played a major part in its discovery.

Superheavy elements

Nihonium, moscovium, tennessine, and oganesson are members of the family known as superheavy elements, with atomic numbers of 104 or more. They are also referred to as the transactinides because they have atomic numbers larger than the actinide elements, which run from 89 to 103. In the 1960s, the synthesis of the first transactinide was accompanied by protracted political posturing. American and Soviet scientists argued over who had first discovered elements 104, 105, and 106 and what names they should be given. These disputes became known as the transfermium wars, because they centered on the discoveries of the elements following fermium, element 100. A final resolution was agreed on in 1997.

Today, making superheavy elements is a much more collaborative process. While

The discovery of superheavy elements sometimes reminds me of the opening of Pandora's box.
Yuri Oganessian

Yuri Oganessian

Born in Rostov-on-Don in Russia in 1933, Yuri Oganessian spent most of his childhood in Yerevan, Armenia, but returned to Russia to study. In 1956, he graduated from the Moscow Engineering Physics Institute. He then moved to the Joint Institute for Nuclear Research, working under its then director Georgy Flerov. He became the laboratory's director after Flerov retired in 1989.

Oganessian invented two key methods for making superheavy elements. In 1974, he pioneered cold fusion, which was used to make elements 107 to 112. His later technique—hot fusion—was used to discover the elements from 113 to 118. The heaviest of these—oganesson, element 118—was named in his honor, which made Oganessian only the second person to have an element named after them during their lifetime.

Key work

1976 "Acceleration of 48Ca Ions and New Possibilities of Synthesizing Superheavy Elements"

See also: The periodic table 130–137 ▪ The electron 164–165 ▪ Radioactivity 176–181 ▪ Isotopes 200–201 ▪ Improved atomic models 216–221 ▪ Synthetic elements 230–231 ▪ The transuranic elements 250–253

the discovery of nihonium was credited solely to the Riken research institute in Japan, the discoveries of moscovium, tennessine, and oganesson were all jointly credited to teams from the United States and Russia, which worked together, sharing the necessary materials to produce the new elements.

Practical problems

On paper, making superheavy elements sounds relatively straightforward. Scientists need to combine atoms of two elements that together contain the number of protons in the new element being made. But it is not as simple as putting these two atoms next to each other. To get them to fuse together to form one larger nucleus, they have to be fired at each other at extraordinary speed so that they have sufficient energy to overcome the repulsive electrostatic forces between the positively charged protons. The discovery of elements has historically been the domain of chemists, but synthesizing superheavy elements requires the use of a key tool of physicists—the type of particle accelerator called a cyclotron.

In the lab, particle accelerators fire nuclei of one of the elements scientists are trying to combine at a target made of atoms of the other element. The projectile nuclei are fired at the target nuclei at a speed of around one tenth of the speed of light. The majority of these high-energy collisions end with both nuclei being smashed to pieces, but on rare occasions, two colliding nuclei fuse to form the nucleus of a new element.

Even this description makes the task sound much easier than it is. Sometimes, the new element undergoes radioactive decay so fast that it cannot even be detected, so identifying elusive atoms of new elements produced »

When an accelerated beam of ions is fired at a target in a cyclotron, unwanted by-products are separated, and—if a successful collision occurs—the newly generated element travels to detectors for identification.

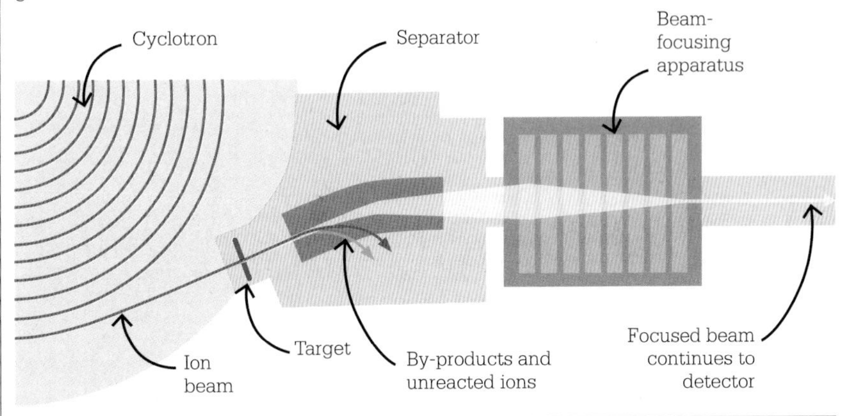

To create tennessine, scientists fired calcium-48 at a berkelium target. After fusion, the newly formed compound nucleus lost three neutrons to make the superheavy element tennessine, which has 117 protons and 177 neutrons.

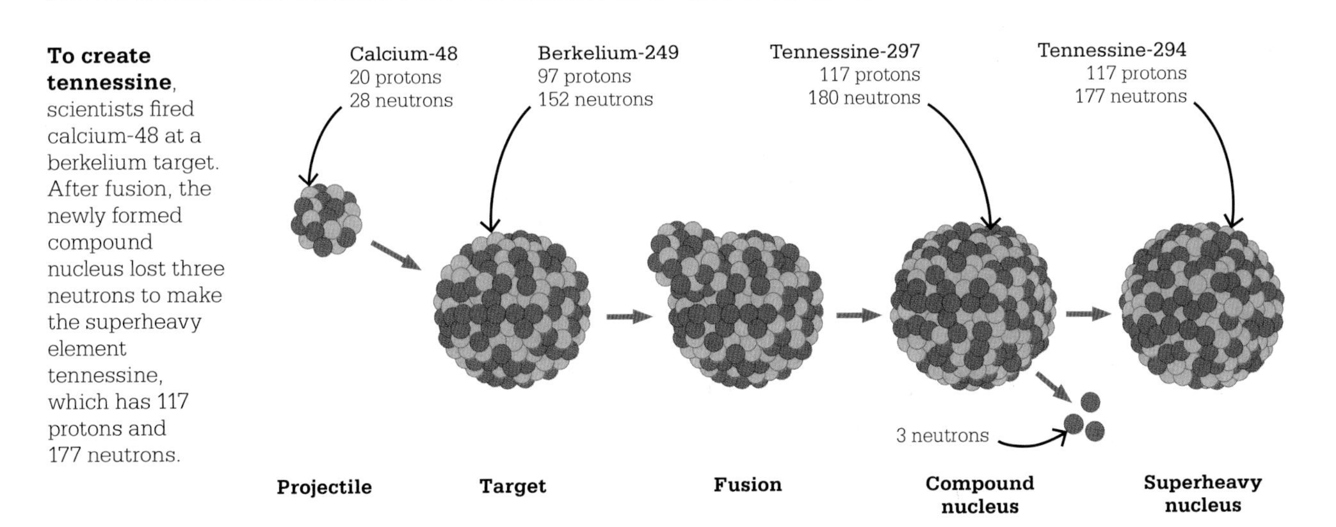

by these collisions is painfully slow. For the creation of element 118, oganesson, only around one atom was detected per month.

Neutron-rich isotopes

With the chances of creating and detecting an atom of a superheavy element so vanishingly rare, scientists working to make them have to tip the odds in their favor. One way of increasing the stability of the new atoms formed is to use neutron-rich isotopes as projectiles and targets. For example, the synthesis of three of the four elements confirmed in December 2015 used calcium-48 as a projectile. This is an isotope of calcium with 20 protons and 28 neutrons. After impact, the newly created atoms initially lose neutrons instead of instantly decaying, which makes detection easier. The technique isn't cheap though: in 2022, the projectile calcium-48 cost more than $250,000 per gram.

Finally, even detection poses challenges. Despite talk about detecting atoms of new elements, the reality is that these atoms are too fleeting to be picked up directly. Instead, they undergo radioactive decay, forming a distinct decay chain that scientists can trace back like a trail to determine the identity of the original decayed element.

The discoveries of the last elements of the periodic table's seventh row leave it looking complete, for now. But the work of those scientists creating superheavy elements is not finished—they are confident that elements beyond oganesson will be discovered and even believe that some may exist for longer than the short lifetimes of the currently known superheavies.

Finding superheavy elements

Superheavy elements are not directly detected, but their presence can be inferred from the evidence of characteristic radioactive decay chains. First, any superheavy element atoms produced by an experiment must be separated from other products, which is done using electric or magnetic fields. The superheavy elements commonly undergo alpha decay, which involves the nucleus losing an alpha particle (a helium nucleus) of two protons and two neutrons. The lighter resultant nucleus can again undergo alpha decay, or beta decay, or in some cases it may undergo fission and split into two smaller nuclei.

Detectors spot the decay products and fission products produced by superheavy element atoms. They record each individual event and allow researchers to follow the decay pathway all the way back to the original nucleus. The identification of the decay chain can be used as proof that a new element has been created.

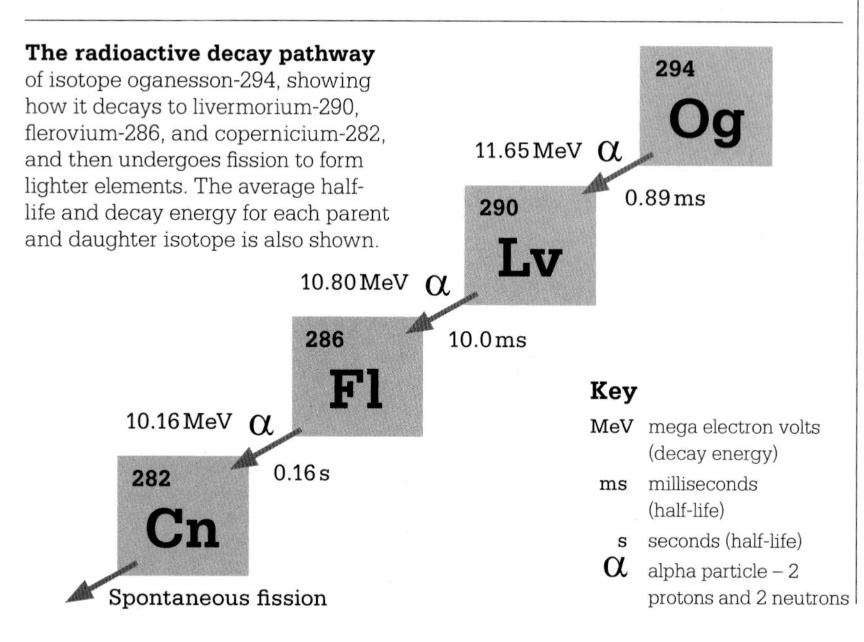

The radioactive decay pathway of isotope oganesson-294, showing how it decays to livermorium-290, flerovium-286, and copernicium-282, and then undergoes fission to form lighter elements. The average half-life and decay energy for each parent and daughter isotope is also shown.

Structure of the nucleus

The component parts of atoms influence their stability. Atoms with filled electron shells, such as the noble gases, are particularly stable. In school chemistry classes, a lot of focus is placed on how electrons arrange themselves in these electron shells, but the nucleus, which contains protons and neutrons, is often treated as a homogeneous blob. This was the case historically, too, until scientists noticed that, just as specific numbers of electrons gave more stable atoms, the same was true for atoms with specific numbers of protons and neutrons.

In 1949, German-born American theoretical physicist Maria Goeppert Mayer and German nuclear physicist Hans Jensen independently formulated a mathematical model of the structure of the nucleus. This was made up of individual nuclear shells, at different energy levels, in

Winning the prize wasn't half as exciting as doing the work itself.
Maria Goeppert Mayer
1963 Nobel Prize laureate

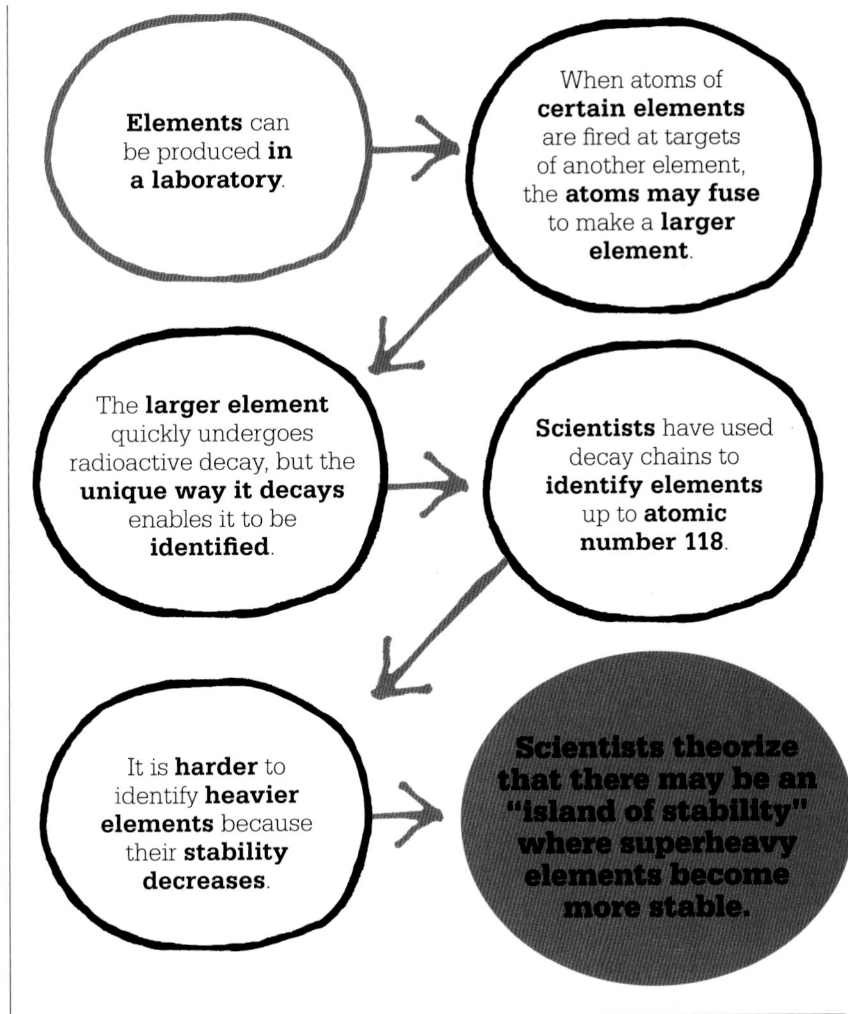

which pairs of neutrons and protons couple together. To explain this joining at a simple level, Mayer used a brilliant analogy of waltzing couples: "All the couples on the ballroom floor go one way, and that's your orbit. Then each couple is also circling in the dance step, and that's your spin." She continued the analogy by explaining that just as it is easier for waltzing couples to dance in one direction, so in an atom's nucleus, "each particle spins in the same direction that all are traveling in orbits." This is called spin-orbit coupling.

Elements can be produced **in a laboratory**.
When atoms of **certain elements** are fired at targets of another element, the **atoms may fuse** to make a **larger element**.
The **larger element** quickly undergoes radioactive decay, but the **unique way it decays** enables it to be **identified**.
Scientists have used decay chains to **identify elements** up to **atomic number 118**.
It is **harder** to identify **heavier elements** because their **stability decreases**.
Scientists theorize that there may be an "island of stability" where superheavy elements become more stable.

Mayer and Jensen's models explained why some configurations of protons and neutrons are more stable than others. Just as atoms with full electron shells are more stable, so too are atoms with full nuclear shells. The Hungarian-American physicist Eugene Wigner, one of Mayer's colleagues during the Manhattan Project, coined the term "magic number" to refer to these full shells—nuclei with 2, 8, 20, 28, 50, 82, or 126 protons or neutrons. Atomic nuclei with one of these magic numbers of protons or neutrons are more stable than any other nuclei. If the nuclei have magic numbers of both protons and neutrons, the nucleus is referred to as "doubly magic."

These magic numbers are a key driver behind superheavy element research. Part of the goal of making superheavy elements is the process of manufacture itself; most exist for only seconds or fractions of a second and will therefore never have applications outside of the laboratory. But the magic numbers also represent a tantalizing possibility, a so-called island of stability where isotopes of elements may exist for minutes, hours, or even longer.

Reaching the island of stability is the challenge. Part of the problem lies in knowing what the magic numbers are for the superheavy elements. We usually assume nuclei are spherical, but it is now thought that superheavy elements' nuclei are not spherical, and this may cause the positions of magic numbers to shift—or even lead to additional magic numbers. Scientists once predicted (and hoped) that element 114, flerovium, may contain a magic number of protons and consequently be more stable than its neighbors. However, subsequent research » ultimately dashed this hope. The next potential candidate as an island of stability is element 120, but this has yet to be synthesized.

Creating an element that has just two more protons than the current final element in the periodic table, oganesson, may not sound too difficult a task. However, superheavy element hunters have hit a roadblock with the synthesis methods currently available to them. Creating elements beyond oganesson requires a projectile nucleus with more protons than the calcium-48 that was used as a projectile in the discoveries of elements 115, 117, and 118. It also requires heavier target nuclei, which can be harder to obtain in sufficient quantities.

Thus far, scientists' efforts to create elements 119 and 120 have included slamming atoms of titanium (element 22) into targets of berkelium (element 97) or californium (element 98). Both of the targets can only be produced in milligram quantities in specialized nuclear reactors. While researchers are still confident that elements 119 and 120 will be made in the coming years, we are currently in the longest gap between discoveries of new elements since the era of synthetic element creation began.

The island of stability is believed to be about 112 protons and 184 neutrons. This diagram shows known and projected stabilities of superheavy elements; darker colors indicate greater stability.

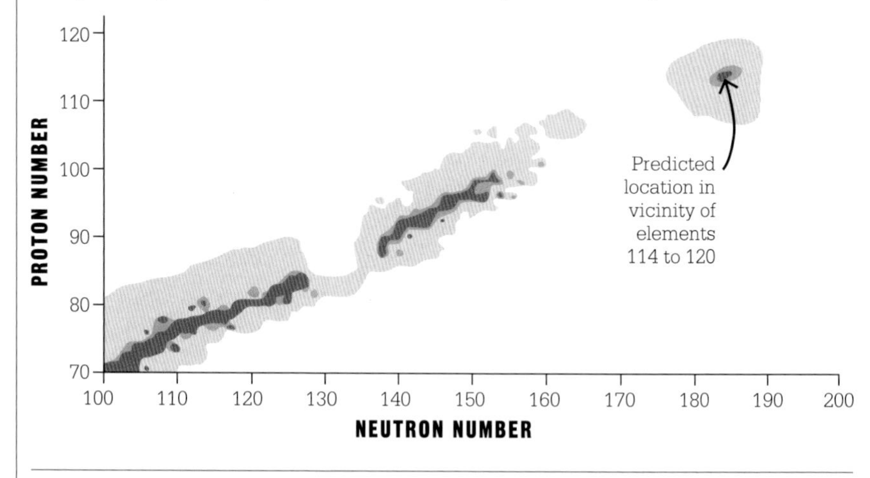

Change of direction

Only a small number of laboratories worldwide have the funding and equipment required to play a part in superheavy element synthesis, and in recent years, some of them have scaled down their attempts to produce further elements. Increasingly, the focus of superheavy element research is moving toward understanding the unusual properties of those already identified. Rather than extending the periodic table, this avenue of research could lead us to the point where the rules that govern its structure break down entirely.

Usually, atoms of elements behave in the way that our simple models of them predict they should. But occasionally, oddities appear. As atoms become heavier, their electrons start to move faster and faster, eventually reaching speeds that approach the speed of light, and scientists have to consider Einstein's theory of general relativity when dealing with them. Due to relativistic effects, the mass of the electrons moving at this speed is greater than that of an electron at rest, and this results in the contraction of the sizes of atomic orbitals. An everyday example is gold, where this contraction means that the energy difference between its two highest energy atomic orbitals is equivalent to that of blue light. Electrons in gold therefore absorb blue and violet light while reflecting red and orange wavelengths and appearing gold-colored.

Cold and hot fusion

Two methods have been used to make the elements from 107 to 118. The first of these is cold fusion (unrelated to the fantastical concept of nuclear fusion at room temperature). This involves using targets and projectiles that are similar in size, such as targets of lead or bismuth, and projectiles with a mass number of more than 40. Combining similar-sized nuclei reduces the amount of energy required for fusion. The resultant nucleus does not need to lose neutrons to become more stable. This in turn makes it easier to detect isotopes of new elements.

Hot fusion uses projectile and target nuclei that are less similar. Calcium-48, which has a magic number of protons and neutrons, is commonly used as a projectile, with the much heavier elements americium, berkelium, and californium used as targets. The large difference in mass increases the chance of the nuclei fusing. The recent use of more intense projectile beams has made this technique possible.

Relativistic effects can affect more than just an element's color; they can also affect its properties. The organization of the periodic table is governed by periodicity—the idea that elements' properties follow predictable and repeating patterns. Elements in a particular group of the periodic table behave in a particular way—for example, chemists expect all of the group 1 metals to readily react with water, while they expect the group 18 metals to refuse to react with almost anything. However, there is now evidence that the relativistic effects that superheavy elements are subject to could confound such expectations.

Surprising behaviors

Copernicium (element 112) is one of the longer-lived of the synthetic superheavy elements, so chemists have been able to probe its properties in more detail than for most, with surprising results. Copernicium sits below mercury (element 80) in the periodic table. Mercury is itself subject to relativistic effects, which is why it is the only metal in the periodic table that is a liquid at room temperature. Computer simulations suggest that, contrary to its position in the periodic table, copernicium should behave like a noble gas, but like mercury, it would be a liquid at room temperature.

Meanwhile, oganesson, which is a superheavy member of the noble gas class, is not expected to behave like a noble gas at all. Instead, calculations predict it to be a metallic semiconductor and that the relativistic effects it is subject to are so strong that its electron shell structure effectively disappears. This is significant. The number and arrangement of electrons in an electron shell determines an element's reactivity; if this structure blurs away into nothing for element 118 and beyond, the periodic table's power for predicting elements' properties is essentially useless.

Many questions to answer

In the case of oganesson, it seems unlikely that scientists will ever be able to experimentally confirm these predictions, given that its most stable isotope decays in less than a millisecond. But it may eventually be possible for copernicium, whose most stable isotope has a half-life of around 30 seconds. Intriguingly, the predictions generated by computer simulations raise interesting questions about whether the heavier regions of the periodic table may defy the criteria by which Mendeleev organized it more than 150 years ago.

In order to answer this question definitively, chemists require more than the limited number of superheavy element atoms that have been produced so far. Russian scientists have built a superheavy element factory with the purpose of producing superheavy isotopes in much greater quantities—up to 100 atoms a day, instead of as few as one a week. Given the strange chemistry that superheavy element researchers have uncovered with just the limited number of atoms at their disposal so far, who knows what oddities they will uncover in the coming years? ■

If you look backward over several decades, people have made roughly one new element maybe every three years—until now.
Pekka Pyykkö (2019)

Russian nuclear physicists Georgy Flerov (far left) and Yuri Oganessian had the superheavy elements 114 (flerovium) and 118 (oganesson) named after them.

HUMANITY AGAINST THE VIRUSES

NEW VACCINE TECHNOLOGIES

IN CONTEXT

KEY FIGURE
Katalin Karikó (1955–)

BEFORE
1796 British doctor Edward Jenner inoculates a 13-year-old boy with cowpox, which protects him from smallpox and begins the modern science of vaccinology.

1931 American virologists Alice Miles Woodruff and Ernest Goodpasture cultivate a virus in an embryonated chicken egg. The technique is later used to propagate other virus vaccinations.

1983 American biochemist Kary Mullis develops the polymerase chain reaction to produce large amounts of genetic material.

2005 Chinese virologist Shi Zhengli discovers that bats harbor coronaviruses such as SARS, which can infect people.

In early 2020, amid the urgency of dealing with the SARS-CoV-2 coronavirus pandemic, it became clear that the solution to the problem would mainly come through vaccines. That hope seemed distant because the fastest a vaccine had ever been developed previously—for mumps—had taken four years, from 1963 to 1967. However, by the beginning of December 2020, several vaccines were showing very good results in large-scale human trials in protecting against COVID-19, the disease caused by SARS-CoV-2.

Based on those results, drug regulators rapidly granted approval for three key innovative vaccines

See also: Intermolecular forces 138–139 ▪ Enzymes 162–163 ▪ Antibiotics 222–229 ▪ The chemicals of life 256–257 ▪ The structure of DNA 258–261 ▪ The polymerase chain reaction 284–285 ▪ Customizing enzymes 293

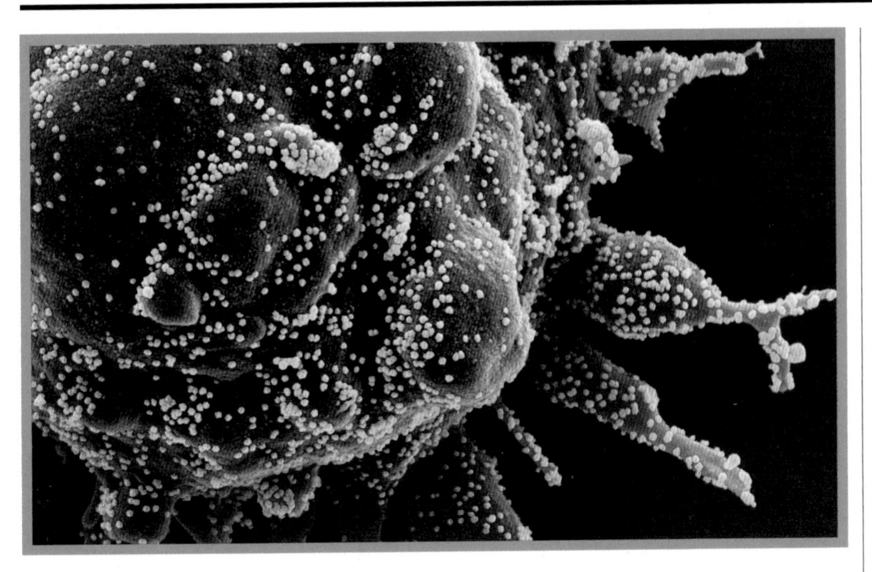

SARS-CoV-2 particles on a body cell are shown here in yellow. The virus causes COVID-19, a respiratory infection that can lead to fatal pneumonia.

before the year was out. Two directly used messenger RNA, or mRNA, to carry genetic information—one made by US-based company Moderna, the other made by German company BioNTech and US-based drug giant Pfizer. The third vaccine carried genetic information of a common cold virus, called an adenovirus, in a modified form. This was isolated from chimps and does not cause disease in humans. It was developed by the University of Oxford in the UK and made by UK/Sweden-based pharmaceutical company AstraZeneca.

One of the immune system's defenses involves white blood cells searching for recognizably infected cells. White blood cells can only do so if they have encountered the infecting bacterium or virus—known together as pathogens—before. Vaccination exposes our immune systems to weakened, dead, or partial versions of the pathogens. These safely teach our immune systems without causing any disease. Live attenuated vaccines, inactivated vaccines, and subunit and conjugate vaccines use disabled forms or inactive pieces of the pathogen they protect against.

By contrast, both the mRNA and adenovirus technologies are recombinant vaccines that work with the biological machinery that reads genes in our cells. They deliver genetic instructions for a vaccine protein into our cells, which produce the vaccine protein. This approach came to the fore during the COVID-19 pandemic because it is much faster to develop than more traditional vaccines.

Formulating a response

Scientists had to react urgently to the rapid spread of the SARS-CoV-2 virus. Chinese researchers published the sequence of its RNA genetic material on January 7, 2020. Modern nucleic acid synthesis chemistry enabled scientists across the world to start making copies for use in developing vaccines within days. Messenger RNA's journey to becoming a COVID-19 hero began in 1990, when Hungarian biochemist Katalin Karikó proposed using it as an alternative to DNA-based gene therapy. Gene therapy seeks to produce permanent changes that can tell cells to make new proteins and turn people into their own "drug factories." However, mRNA is the messenger taking instructions to our cells' protein-making machinery about what to create from DNA. So mRNA could build internal drug factories without permanently changing people's genes.

How mRNA works

When mRNA is injected into a person's body rather than letting the body make its own, virus-detecting immune sensors notice and cells shut down their protein production. In 2005, when working with her colleague, American immunologist Drew Weissman, Karikó found that it was surprisingly simple to avoid this shutdown. Like DNA, mRNA has four building blocks, each represented by a letter; mRNA »

> The total number of global deaths attributable to the COVID-19 pandemic in 2020 is at least 3 million …
> **World Health Organization**

uses one that DNA does not—uridine, which replaces thymine. Karikó and Weissman swapped uridine for pseudouridine. The mRNA made using pseudouridine could evade immune sensors.

To make a drug or vaccine, virologists also had to protect the mRNA from being broken down in patients' bodies before it could do its job. Their solution was to coat mRNA in fatty lipid molecules, forming tiny nanoparticle balls.

In COVID vaccines, mRNA encodes SARS-CoV-2's spike protein, which docks to human cell membranes and allows the coronavirus to invade. As Karikó foresaw, once the mRNA gets inside the cells, our bodies produce this spike protein. It acts as a foreign molecule that triggers an immune response but cannot infect the body with COVID-19.

There's a lot of vaccine development that we need to do now that we can do it.
Sarah Gilbert (2021)

Katalin Karikó is a specialist in mRNA therapy. Her expertise was key in creating the Pfizer/BioNTech and Moderna COVID-19 vaccines.

Lipid nanoparticles are still one of the biggest challenges for mRNA vaccines for two reasons. First, they are difficult to make, which slows down the process of making mRNA. Second, the lipid nanoparticles are unstable at room temperature. When the BioNTech/Pfizer vaccine became the first mRNA shot to be approved by the US Food and Drug Administration, it had to be stored at -94°F (-70°C). Both it and the Moderna vaccine can now be stored at -4°F (-20°C), but that still makes it hard to get them to remote, poor areas. Such challenges reflect the fact that this technology is only just mature enough to use widely. Making and distributing billions of doses is therefore an enormous achievement.

Using viral vectors

The Oxford/AstraZeneca vaccine carries instructions that tell our cells to make the SARS-CoV-2 spike protein. The chimpanzee adenovirus that it employs to carry those instructions into our bodies is called the viral vector. This technology emerged from years of research by vaccinologists at the University of Oxford's Jenner Institute, led by British scientist Sarah Gilbert and Irish researcher Adrian Hill. In 2014, the team used the technology to develop a potential vaccine intended as a rapid response to the Ebola virus disease in Africa that year. They were unable to complete their trials before the outbreak ended but gained valuable experience. Gilbert moved on to researching

mRNA vaccines

This type of vaccine harnesses how our cells naturally make proteins. The process starts with the DNA molecules on which our genes reside. In the cell nucleus, enzymes split apart the two strands that make up DNA's double helix. Other enzymes use unzipped DNA strands as templates to make matching strands of mRNA. This moves from the nucleus to the cells' ribosomes, biochemical factories that read the mRNA's genetic code for making proteins. In a laboratory, scientists can create mRNA that codes for other proteins and send it to cell ribosomes. COVID-19 mRNA vaccines contain genetic instructions for the spike protein with which the virus grabs and enters our cells. Ribosomes receive those instructions and make the spike protein, which attaches to the cells' surface. Next, the immune system learns to recognize and make antibodies for the spike protein without causing a COVID-19 infection. It also makes more white blood cells that can kill virus-infected cells.

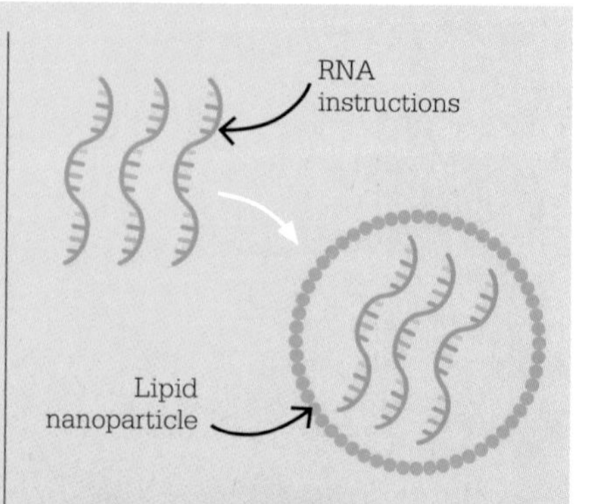

Synthetic mRNA, which codes for the virus spike protein, is protected by lipid nanoparticles that prevent the human body from breaking it down.

a coronavirus called MERS, first detected in 2012, in which half of the spike protein is the same as in SARS-CoV-2.

Safety and effectiveness

Two other viral vector vaccines entered wide use before that developed by Oxford/AstraZeneca. Sputnik V from the Gamaleya Research Institute, part of Russia's Ministry of Health, and Chinese company CanSino Biologics' Convidecia both used human adenoviruses. In each case, the countries where they were developed decided to use them before seeing the results of large-scale trials, whereas the Oxford/AstraZeneca vaccine—and those produced by Pfizer/BioNTech and Moderna—were first subjected to extensive trials. An additional problem was that people may have been infected previously by the human adenoviruses, so they had an immune response to the vaccine. That seems to have reduced the effectiveness of the CanSino vaccine in some tests.

Unlike live vaccines, none of the viral vectors used in COVID-19 vaccines can make new copies of themselves. This means lots of viral vector needs to be used for them to be effective, but it helps ensure that they are safe. They also get around the need to deliver vaccines in lipid nanoparticles, so the Oxford/AstraZeneca vaccine can be stored in a regular fridge.

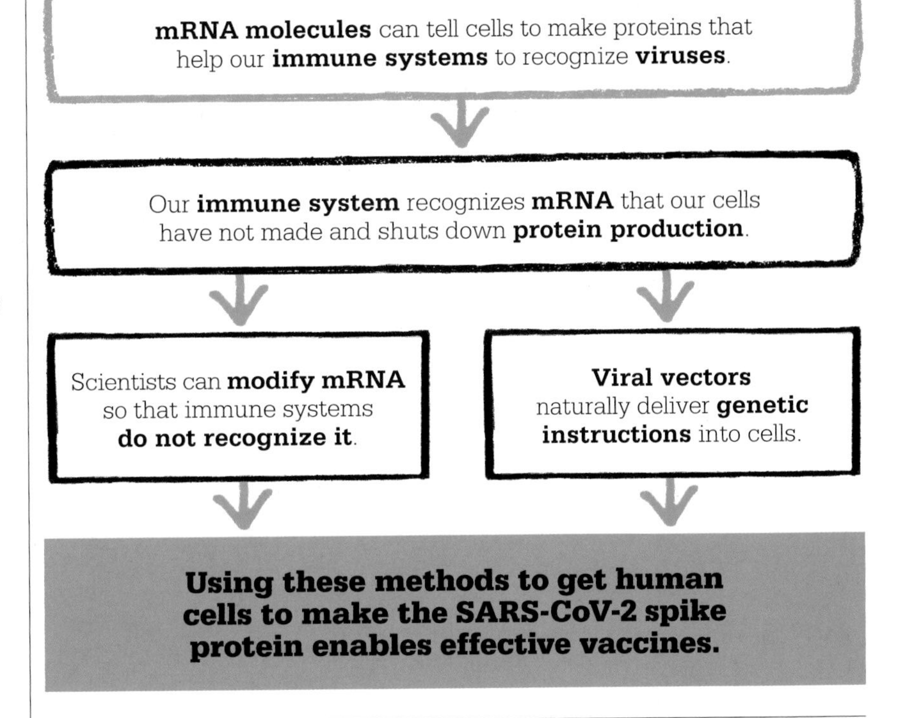

mRNA molecules can tell cells to make proteins that help our immune systems to recognize viruses.
Our immune system recognizes mRNA that our cells have not made and shuts down protein production.
Scientists can modify mRNA so that immune systems do not recognize it.
Viral vectors naturally deliver genetic instructions into cells.
Using these methods to get human cells to make the SARS-CoV-2 spike protein enables effective vaccines.

Agencies were only able to roll out COVID-19 vaccines quickly because researchers such as Gilbert, Hill, Karikó, and Weissman were already working on their innovative technologies. This shows how even research whose importance is unclear can change the world. ■

Viral-vector vaccines

Like mRNA vaccines, viral-vector vaccines exploit the way our bodies naturally make proteins. In this case, to get RNA into our cells, scientists add it to the genetic material of another virus, the viral vector. The new, altered virus is called a recombinant virus.

Once virus protein RNA enters the cells, the ribosomes—which follow instructions wherever they come from—follow the direction to make the virus protein. This then causes an immune response, training our body's immune system to recognize the virus. If we are later infected with this virus, the immune system will respond to it more quickly.

Viral-vector vaccines have not performed quite as well as mRNA vaccines against COVID-19. However, this may not be the case with future vaccines designed to combat viral infections. It is therefore important to have more than one type of vaccine available because virologists do not know what form the next pandemic might take.

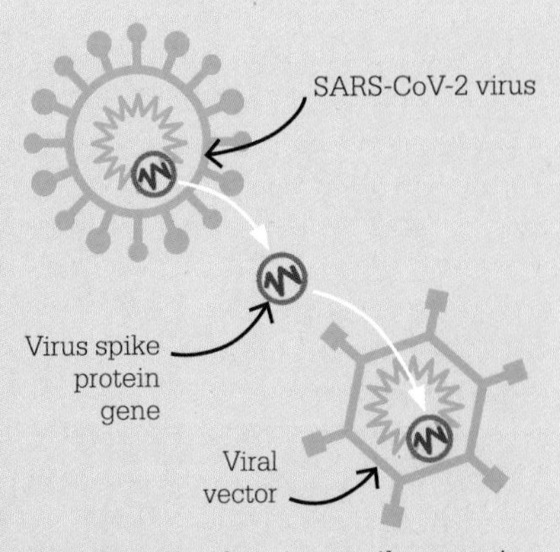

The gene for the virus spike protein is added to genetic material from a virus that has been genetically altered so it cannot cause disease.

DIRECTO

RY

DIRECTORY

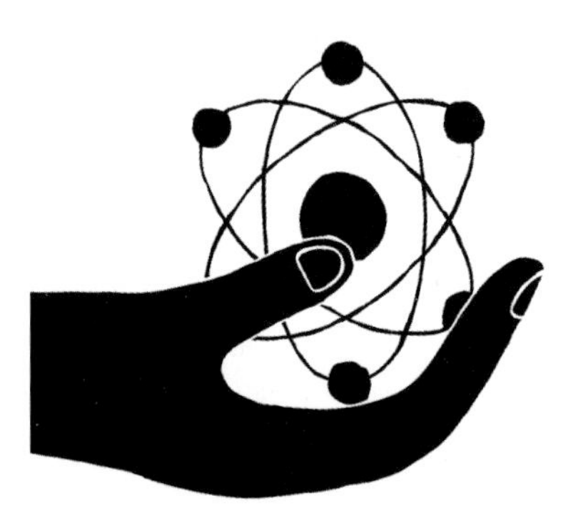

Key figures already featured in this book are but a few of many who played a part in the evolution of chemistry; this chronological directory lists a selection of other major contributors. It includes those who made great discoveries, from Hennig Brand, the first known individual to isolate a new element, to Akira Suzuki, who revolutionized industrial chemistry with a new technique to synthesize organic molecules. Others found ways to use chemistry to cure disease, such as Alice Ball's treatment for leprosy and Paul Ehrlich's drug for syphilis. Some were great educators: Jane Marcet made chemistry accessible to millions, and George Washington Carver's understanding of plant and soil chemistry helped countless American farmers.

HENNIG BRAND
c.1630–c.1710

Little is known about German alchemist Brand's early years; he was probably born in Hamburg, fought in the Thirty Years' War (1618–1648), and worked as a physician, albeit unqualified. Brand also pursued a quest for the philosopher's stone, a substance believed to turn common metals to gold. In an experiment in 1669, he boiled down a large volume of urine and produced a white, solid residue. This glowed in the dark, so Brand named it phosphorus—from the Greek *phos* (light) and *phoros* (bringing). In fact, Brand had discovered a new element, the first named individual to have done so.
See also: Attempts to make gold 36–41

JOHANN BECHER
1635–1682

One of the most influential alchemists of the 17th century, Becher was the son of a German Lutheran minister. In 1669, Becher published *Physica subterranea*, which investigates the nature of minerals and other substances. He proposed that materials are composed of three "earths"—vitrifiable, mercurial, and combustible—and that all inflammable materials contain *terra pinguis,* or a "fire-element," which is given off, or "liberated," during combustion. This substance was renamed phlogiston in the 18th century. Becher later attempted to transmute sand into gold.
See also: Attempts to make gold 36–41 ▪ Phlogiston 48–49 ▪ Oxygen and the demise of phlogiston 58–59

ÉTIENNE-FRANÇOIS GEOFFROY
1672–1731

French physician, chemist, and pharmacist Geoffroy was the first scientist to consider chemical affinities (*rapports*), or fixed attractions between different substances. His 1718 table of affinities was the first of many such tables created by him and other chemists. Each column in Geoffroy's table is headed by one element or compound; other chemical substances with which the main element or compound can react are listed below it in descending order of affinity. Affinity tables became authoritative references until the late 18th century.

Geoffroy's table of affinities coincided with a transition from alchemy to scientific chemistry as a distinct academic discipline, and some believe the table marked the start of the chemical revolution. Geoffroy became professor of chemistry at Paris's Jardin du Roi and dismissed some aspects of alchemy, such as the belief in the philosopher's stone.
See also: Attempts to make gold 36–41 ▪ Why reactions happen 144–147

DANIEL RUTHERFORD
1749–1819

While researching his doctoral thesis in 1772 at the University of Edinburgh, UK, Scottish physician

Rutherford used a flame to remove "good air" (oxygen) from a sealed container and passed the remaining gas through a solution to remove the "fixed air" (carbon dioxide). The remaining gas would neither support life nor burn. He called this "mephitic [poisonous] air," but we now know it as the gas nitrogen. It was Rutherford's only notable discovery, but he co-founded the Royal Society of Edinburgh in 1783 and went on to have a successful career in botany and medicine.
See also: Fixed air 54–55 ▪ Oxygen and the demise of phlogiston 58–59

JANE MARCET
1769–1858

Born Jane Haldimand to Swiss parents in London, Marcet became a science writer after attending a course given by English chemist Humphry Davy. In 1805, Marcet published—anonymously—*Conversations on Chemistry*, a series of fictional dialogues between a teacher and two female students that explored the basics of the science. Marcet intended it for girls, but its reach extended far wider. English chemist Michael Faraday, who had little formal education, was inspired by the book while working as a bookbinder.

Conversations on Chemistry was a bestseller in the UK and the US, where it became a standard text in girls' schools, and was translated into French and German. Marcet later went on to write *Conversations on Political Economy* (1816), as well as *Conversations on Vegetable Physiology* (1829).
See also: Inflammable air 56–57 ▪ Oxygen and the demise of phlogiston 58–59

JOHAN AUGUST ARFWEDSON
1792–1841

When Swedish mineralogist Arfwedson began work at the Royal Board of Mines, Stockholm, he met fellow Swede Jöns Jacob Berzelius, one of the most prominent chemists of the early 19th century. Berzelius gave Arfwedson access to his laboratory; here, in 1817, the latter analyzed the composition of the mineral petalite. In addition to aluminum, silicon, and oxygen, he found an element that formed salts similar to—but different from—sodium and potassium. He called the new element lithium, but he was unable to isolate it; that was achieved by English chemist William Brande four years later.
See also: Isolating elements with electricity 76–79 ▪ Lithium ion batteries 278–283

CARL GUSTAF MOSANDER
1797–1858

Originally, Swedish-born Mosander trained as a physician, but he went on to be curator of minerals at the Royal Swedish Academy of Sciences in Stockholm. In 1832, he became professor of chemistry and mineralogy at the city's Karolinska Institute, where his mentor was Jöns Jacob Berzelius.

In 1839, while investigating a sample of cerium oxide, Mosander realized that, while most of it was insoluble, some was soluble, and he deduced that it was the oxide of a new element. It was only the third rare-earth element to be discovered; Mosander called it lanthanum, from the Greek *lanthanein*, meaning "to lie hidden." Four years later, he isolated two more rare-earth elements—erbium and terbium—and thought he had found another, but that (didymium) was later found to be a mixture of oxides.
See also: The periodic table 130–137

CHARLES GOODYEAR
1800–1860

Self-taught chemist Goodyear started out in his father's hardware business in Connecticut. He became interested in developing a technique to make India rubber (latex) stronger, less adhesive, and less susceptible to extreme heat and cold. In 1839, during an experiment, he accidentally dropped some rubber mixed with sulfur onto a hot stove and came upon the process that made possible the manufacture of hard-wearing rubber, such as that used for car tires. He patented his invention (it was later named vulcanization) in 1844. Sadly, his patent was so widely infringed that while others grew wealthy from his discovery, Goodyear died heavily in debt from legal fees.
See also: Polymerization 204–211

CARL REMIGIUS FRESENIUS
1818–1897

A pioneer of chemical analysis, German chemist Fresenius devised a systematic method for identifying the constituents of a mixture of chemical substances. In 1841, he published the first edition of his *Manual of Qualitative Chemical Analysis*. He transferred to Giessen University to assist Justus von Liebig, a leading German chemist. In 1862, Fresenius founded the

Journal of Analytical Chemistry, probably the world's first specialist chemistry journal, which he edited until his death, by which time his *Manual* had passed through 16 editions and been translated into numerous languages.
See also: Chromatography 170–175

LOUIS PASTEUR
1822–1895

From humble beginnings in France, Pasteur became one of the greatest scientists of the 19th century. He is best known for his discoveries in microbiology, notably that bacteria cause disease—the core of germ theory. He also invented the pasteurization process and created vaccines for rabies and anthrax.

Pasteur's first research applied polarized light to the process of chemical analysis. Investigating fermentation in wine, he showed that two substances (tartaric acid and paratartaric acid) had the same chemical composition, but their atoms arranged differently in space to appear as mirror images of each other—they were optical isomers. He had demonstrated that you need to study a chemical's structure and its composition to understand how it behaves.
See also: Isomerism 84–87 ▪ Stereoisomerism 140–143 ▪ Antibiotics 222–229 ▪ New vaccine technologies 312–315

LOTHAR MEYER
1830–1895

German chemist Meyer began teaching chemistry in 1859. Five years later, he published *Modern Chemical Theory*, which proposed a periodic classification of the elements. He arranged 28 elements by atomic weight and investigated the relationship between the elements' weights and their properties. Meyer developed his ideas further in 1868 and 1870, but by this time Russian chemist Dmitri Mendeleev had published his own periodic table. Meyer was a professor of chemistry at the University of Tübingen from 1876 until his death.
See also: The periodic table 130–137

WILLIAM CROOKES
1832–1919

Best known for his work with vacuum tubes (the Crookes tube is named after him), British physicist and chemist Crookes was drawn toward optical physics, possibly after meeting Michael Faraday. Crookes was wealthy and worked tirelessly for decades in his own well-equipped laboratory. Much of his research involved flame spectroscopy, with which he discovered the post-transition metal thallium in 1861. He saw a green line in the spectrum of some impure sulfuric acid and realized that it signified a new element.
See also: Flame spectroscopy 122–125

HENRI MOISSAN
1852–1907

In 1884, French chemist Moissan, who was working at the School of Pharmacy in Paris, began to investigate fluorine compounds. A great unsolved problem in inorganic chemistry was isolating fluorine—the most reactive element in the periodic table. Fluorine is a highly toxic gas; Moissan was poisoned several times before achieving his aim in 1886 by electrolyzing a solution of potassium hydrogen fluoride and hydrofluoric acid. For isolating fluorine and inventing an electric furnace, he was awarded the 1906 Nobel Prize in Chemistry.
See also: Isolating elements with electricity 76–79 ▪ Electrochemistry 92–93 ▪ Why reactions happen 144–147

PAUL EHRLICH
1854–1915

German physician and chemist Ehrlich noticed the selective action of the aniline chemical dyes he used to stain cells. Ehrlich realized that they revealed different cell types and different chemical reactions and that some chemicals could treat disease. He developed a range of dyes to distinguish different blood cells and showed that cellular oxygen consumption varies throughout the body. Ehrlich won a half share of the 1908 Nobel Prize in Physiology or Medicine for his work on immunity, using antibodies in blood cells. In 1907, a chemical ("Compound 606") synthesized in Ehrlich's lab proved to be a very effective treatment for syphilis. Later called Salvarsan, it was the first major breakthrough in antibacterial chemotherapy.
See also: Antibiotics 222–229 ▪ Chemotherapy 276–277 ▪ New vaccine technologies 312–315

CARL AUER VON WELSBACH
1858–1929

Austrian chemist Welsbach used fractional crystallization in 1885 to separate the alloy didymium

(earlier thought to be an element) into its two constituent rare-earth elements: a green salt, which he called praseodymium, and a pink salt—neodymium. In the same year, he received a patent for a gas mantle that used lanthanum, another rare-earth element. In 1905, Welsbach was one of three chemists who independently discovered the rare-earth element lutetium; credit for the finding was eventually given to French chemist Georges Urbain.
See also: Rare-earth elements 64–67

GEORGE WASHINGTON CARVER
c. 1864–1943

Born into slavery a year before its abolition, African American Carver studied agricultural science at Iowa Agricultural College and Model Farm. In 1894, he was one of the first Black Americans to earn a science degree. He later taught at Iowa Agriculture and Home Economics Experiment Station and the Tuskegee Normal and Industrial Institute in Alabama. Although best known for developing multiple uses for peanuts, his most valuable work was on soil chemistry. He devised techniques for replenishing soil depleted by years of cotton monoculture by, for example, cultivating nitrogen-fixing peanuts, soybeans, and sweet potatoes.
See also: Fertilizers 190–191

ALICE BALL
1892–1916

Born in Seattle, Washington, Ball gained degrees in pharmaceutical chemistry and pharmacy. Then, at the College of Hawaii, she became the first African American to earn a Master's degree in chemistry and, at age 23, the institution's first woman chemistry teacher. Here, Ball developed the first useful treatment for leprosy—an injection of a solution of oil extracted from the chaulmoogra tree (*Hydnocarpus* spp.). The "Ball method" was used successfully for more than 30 years until sulfonamide drugs were brought in. Ball never saw the true impact of her treatment: she died the following year after accidentally inhaling chlorine gas. The college's dean initially claimed her discovery for himself before Ball was duly credited in 1922.
See also: Functional groups 100–105 ▪ Antibiotics 222–229

IRÈNE JOLIOT-CURIE
1897–1956

The daughter of French physicists Marie and Pierre Curie, Joliot-Curie operated X-ray machines with her mother in mobile field hospitals during World War I. She then studied chemistry at her parents' Radium Institute in Paris, writing her doctoral thesis on radiation emitted by polonium. Together with husband Frédéric Joliot, she researched radioactivity and the transmutation of elements. In 1935, they shared the Nobel Prize in Chemistry for their discovery that new radioactive elements could be synthesized from stable elements. In 1938, Joliot-Curie's work on the action of neutrons on heavy elements was a major step in the development of uranium fission. Like her mother, she died from leukemia contracted in the course of her work.
See also: Radioactivity 176–181 ▪ Synthetic elements 230–231 ▪ Nuclear fission 234–237

PERCY JULIAN
1899–1975

African American chemist Julian pioneered chemical synthesis of medical drugs from plants. Racist attitudes in the US of the 1920s meant Julian could not get a teaching post in a major university despite a scholarship to Harvard University. In 1929, however, he received a Rockefeller Foundation fellowship to study at the university in Vienna, gaining a PhD in 1931.

Returning to the Glidden Company in the US, Julian devised his own technique to synthesize the sex hormones progesterone and testosterone from chemicals isolated from soybean oil. He also developed a cheap, soy-based substitute for cortisone, which was used for pain relief. In 1953, he established his own research company, Julian Laboratories. He still had to contend with racism; in the 1950s, his Chicago house was attacked on at least two occasions.
See also: Functional groups 100–105 ▪ The contraceptive pill 264–265

SEVERO OCHOA
1905–1993

The Spanish Civil War and, later, the outbreak of World War II limited research opportunities in Europe so, in 1941, Spanish physiologist and biochemist Ochoa moved to the US. He later became an American citizen. Ochoa could now focus his research on his passions: enzymes and protein synthesis. In 1955, at the New York University College of Medicine, he and his co-worker Marianne Grunberg-Manago found an enzyme that can link nucleotides—the building

blocks of RNA (ribonucleic acid) and DNA (dioxyribonucleic acid). Their discovery enabled a better understanding of how genetic information is translated. For this, he received a half share in the 1959 Nobel Prize in Physiology or Medicine—he was the first Hispanic American recipient.
See also: Enzymes 162–163 ▪ The structure of DNA 258–261

FREDERICK SANGER
1918–2013

Sometimes known as "the father of genomics," British biochemist Sanger is the only scientist to have been awarded two Nobel Prizes in Chemistry. At the University of Cambridge, UK, in 1943, he began researching the protein insulin. In 1955, he identified the unique sequence of amino acids that make up the insulin molecule, for which he won his first Nobel Prize in 1958. This work provided a key to understanding the way DNA codes for making proteins in the cell.

In 1977, Sanger developed a method for mapping an organism's genome by establishing the order of the nucleotides within its DNA molecules. For this work, he was awarded a quarter share of the 1980 Nobel Prize in Chemistry.
See also: The structure of DNA 258–261 ▪ The polymerase chain reaction 284–285 ▪ Editing the genome 302–303

AKIRA SUZUKI
1930–

In 1979, while professor in applied chemistry at Hokkaido University, Japan, and working with fellow chemist Norio Miyaura, Suzuki used palladium as a catalyst to synthesize large organic molecules by bonding carbon atoms. This cross-coupling reaction, which became known as the Suzuki (or Suzuki-Miyaura) reaction, had a dramatic impact on organic chemistry, enabling the production of biaryls, alkenes, and styrenes, which are important in industrial chemistry and the pharmaceuticals industry. Products of the Suzuki reaction may become important in nanotechnology. For this work, Suzuki shared the 2010 Nobel Prize in Chemistry with Japanese chemist Ei-ichi Negishi and American chemist Richard Heck.
See also: Catalysis 69 ▪ Functional groups 100–105

YOUYOU TU
1930–

After studying pharmacology at Beijing Medical College, China, Tu spent her career at the Academy of Traditional Chinese Medicine. In 1969, Tu was tasked with finding a new cure for malaria because the malarial parasite (*Plasmodium* spp.) had become resistant to the antimalarial drug chloroquine.

Wormwood (*Artemisia annua*) was already utilized in Chinese traditional medicine: in 1972, Tu's team isolated a key chemical within it—the lactone she called artemisinin. It kills the parasite by blocking protein synthesis. The following year, her team isolated dihydroartemisinin. These compounds led to a new generation of antimalarial drugs and saved millions of lives. In 2015, Tu was awarded a half share of the Nobel Prize in Physiology or Medicine for what has been described as "arguably the most important pharmaceutical intervention of the last half-century."
See also: Functional groups 100–105

JAMES ANDREW HARRIS
1932–2000

Despite being a well-qualified graduate in 1953, African American chemist Harris struggled to find work in scientific research. In 1960, however, he was employed in the Lawrence Radiation Laboratory at the University of California, Berkeley, and tasked with finding transuranic (super-heavy) elements. He was part of a team that isolated element 104 (rutherfordium) in 1969 and, in 1970, element 105 (dubnium), so he was the first Black American to contribute to the discovery of new elements. Harris devoted much of his free time and later career in administration encouraging young African American scientists.
See also: The transuranic elements 250–253

GERHARD ERTL
1936–

At the suggestion of leading electrochemist Heinz Gerischer at the University of Munich, German chemist Ertl began in the 1960s to research the emerging discipline of surface science, especially the solid–gas interface. Over a number of years, he investigated why chemical reactions accelerate at surfaces and also developed ultrapure vacuum technology to help study surface chemical reactions. Applications of his research have included improving the Haber–Bosch process to synthesize ammonia and making

hydrogen-fuel cells perform more efficiently. Ertl was director of the Fritz Haber Institute in Berlin from 1986–2004. He received the 2007 Nobel Prize in Chemistry for his work on chemical processes on solid surfaces.
See also: Fertilizers 190–191

MARGARITA SALAS
1938–2019

For three years from 1964, Spanish biochemist Salas was part of the team at Severo Ochoa's laboratory in New York University, looking at the mechanisms of replication, transcription, and translation that convey genetic information from DNA to protein. Back in Spain in 1977, at the Center for Molecular Biology "Severo Ochoa" (CBMSO) in Madrid, Salas and biochemist Luis Blanco developed a new mechanism for the replication of DNA. Multiple displacement amplification enabled faster and more accurate DNA testing than had been possible with the polymerase chain reaction (PCR). Salas's method needed only tiny DNA fragments to generate many copies of whole genomes, so it is ideal for forensic analysis, identification of mutations in tumors, and genetic analysis of fossils.
See also: The structure of DNA 258–261 ▪ The polymerase chain reaction 284–285

ADA YONATH
1939–

Born to a poor Israeli family, Yonath studied chemistry, biochemistry, and biophysics at the Hebrew University of Jerusalem, and X-ray crystallography (XRC) at the Weizmann Institute of Science. Yonath employed XRC to examine the atomic structure and function of ribosomes (the protein factories in cells). She also developed the cryocrystallography technique to limit radiation damage to proteins during XRC, and later investigated the atomic structure of antibiotics. For her work on ribosome structure, she was awarded a third share of the 2009 Nobel Prize in Chemistry.
See also: X-ray crystallography 192–193 ▪ The chemicals of life 256–257 ▪ Protein crystallography 268–269

DAN SCHECHTMAN
1941–

While a visiting researcher at the National Institute of Standards and Technology (NIST), Maryland, in 1982, Israeli materials scientist Schechtman noticed strange diffraction patterns in an aluminum–manganese alloy. The patterns were ordered but not periodic or repeating, indicating a previously unknown way of stacking atoms and molecules in crystals. Since then, hundreds of quasicrystals, as they were later named, have been discovered. Their applications range from nonstick frying pans to materials that convert heat to electricity. For his discovery, Schechtman received the 2011 Nobel Prize in Chemistry.
See also: X-ray crystallography 192–193 ▪ Chemical bonding 238–245

AHMED ZEWAIL
1946–2016

In 1976, Egyptian American chemist Zewail began work at the California Institute of Technology (CalTech). His team caused and observed chemical reactions by using ultrafast lasers to create light pulses lasting a quadrillionth of a second—rather than a picosecond (trillionth of a second), the timescale of reactions at molecular level. The new femtosecond spectroscopy, as Zewail named it, allowed scientists to observe reaction dynamics and molecular pathways during reactions. For his pioneering ultrafast chemistry, Zewail won the 1999 Nobel Prize in Chemistry and was the first Nobel winner in science from the Arabic-speaking world.
See also: Chemical bonding 238–245

DAVID MACMILLAN
1968–

Until the 21st century, only metals and enzymes could be used as catalysts for chemical reactions. In 2000, while at the University of California, Berkeley, Scottish American organic chemist MacMillan invented the organocatalysis technique. It uses a small, carbon-based molecule as a catalyst to produce special molecules called enantiomers (nonsuperimposable, mirror-image structures). This breakthrough enabled manufacture of new drugs and materials; organocatalysts also are biodegradable and cheaper than traditional, sometimes toxic catalysts. MacMillan developed the technique further at Princeton University. In 2021, he shared the Nobel Prize in Chemistry with German chemist Benjamin List, who had also progressed asymmetric organocatalysis.
See also: Catalysis 69 ▪ Functional groups 100–105

GLOSSARY

Aerosol A dispersion of tiny liquid or solid particles in air.

Akamptisomer A stereoisomer discovered in 2018 involving normally flexible chemical bonds being hindered from rotating by the molecule's structure.

Alkali See *base*.

Alloy A metal that is a blend or mixture of more than one metallic element, sometimes including nonmetals as well.

Alveoli (singular, alveolus) Microscopic air sacs in the lungs where exchange of oxygen and carbon dioxide takes place between air and blood.

Amino acid Small nitrogen-containing compounds found in all living things. Protein molecules are long chains of up to 20 different types of amino acids arranged in a unique order for each protein.

Atom The smallest unit of any chemical element, made up of a central heavy nucleus with electrons orbiting around it.

Atomic bomb A bomb whose energy comes mainly from the fission of U-235, a uranium isotope, or Pu-239, a plutonium isotope. See also *isotope*, *nuclear fission*.

Atomic mass (more fully "relative atomic mass"; also called "atomic weight") The average mass of the atoms of an element expressed relative to the isotope carbon-12. Atomic masses are not usually whole numbers because most elements consist of a mixture of isotopes of different masses. See also *isotope*.

Atomic number The number of protons in an atom's nucleus. Each chemical element has the same number of protons in all its atoms, which is a different number from any other element. In a neutral atom, the number of electrons orbiting an atom's nucleus, which determines an element's chemical properties, is the same as the number of protons.

Atomic weight See *atomic mass*.

Atropisomer A stereoisomer involving restricted rotation of single bonds, which in most other molecules are free to rotate.

Base (1) The chemical opposite of an acid. A soluble base is called an alkali. Acids and bases react together to form salts. (2) In biology, a nitrogen-containing carbon compound of which there are four types in DNA and four (one different from DNA) in RNA. The order of different bases along DNA and RNA molecules "spells out" the genetic instructions that they code for.

Biochemical pathway An organized sequence of chemical reactions in living things. Biochemical pathways are controlled by enzymes.

Blast furnace The main type of furnace for smelting iron. Iron ore, coke, and other materials are introduced at the top of its tall structure and molten iron is tapped off at the bottom. Air is blasted in through nozzles to keep the smelting process going. See also *smelting*.

Bloomery furnace A small-scale furnace for smelting iron, in use since ancient times, in which the iron does not become liquid.

Bond, chemical A link between two atoms, binding them together to form a molecule or compound. Strong bonds are formed either by atoms sharing electrons (covalent bonding) or by transferring electrons and being held together by electrical attraction (ionic or electrovalent bonding), or by bonds intermediate between these types. There are weaker bonds that do not belong to the above categories. See also *hydrogen bond*.

Borosilicate glass Glass that includes boron trioxide among its ingredients. It expands very little on heating, making it useful in applications such as kitchenware. See also *soda-lime glass*.

Calotype The earliest photographic process to use negative images from which many positive images could be obtained.

Calx Crumbly or powdery residue left when a mineral or metal has been roasted or burned.

Camera obscura A device that uses a lens to project an image of a view onto a surface inside a darkened box. It was used by artists and is an ancestor of the photographic camera.

Cantilever A rigid projecting structure attached only at one end.

Cast iron A hard, brittle form of iron formed by remelting metal produced by a blast furnace. It has a high carbon content.

Catalyst A substance that speeds up a chemical reaction without being permanently altered or decomposed by the end of the reaction.

Cathode ray A stream of electrons emerging from a negative electrode (cathode). See *electrode*.

Cathode ray tube A sealed high-vacuum device in which a beam of electrons is directed onto a screen, as in a traditional television receiver.

CFCs Short for chlorofluorocarbons, artificially produced halocarbons formerly used industrially; they were later found to damage the ozone layer. See also *halocarbon*, *ozone layer*.

Chirality See *handedness*.

Collimator A device for producing a parallel beam of radiation.

Compound A substance made up of atoms of more than one element bonded together in a definite ratio. For covalent compounds, the smallest part of the compound is a single molecule.

Conservation of mass The principle that the total mass of substances neither increases nor decreases in a chemical reaction.

Contact insecticide A substance that is toxic to insects upon contact.

Contact process The main modern industrial process for making sulfuric acid.

Corona, solar The thin outer atmosphere of the Sun, extending for millions of miles (km).

Covalent bond See *bond*.

Crucible A container for melting metals or other substances at high temperatures.

Crystal Any solid consisting of a regular array of atoms or molecules arranged in a repeated 3D geometrical pattern.

D lines Dark lines of particular wavelengths in the spectrum of the Sun that indicate the presence of sodium, which absorbs those wavelengths.

Daguerreotype An early type of photograph, obtained by a process that results in single positive images.

Density functional theory A method of calculating the distribution of electrons in molecules or solids.

Deoxyribose A type of sugar molecule that forms part of the structure of DNA.

Dephlogisticated air An obsolete name for oxygen, dating from when it was believed that oxygen consisted of air with phlogiston removed. See also *phlogiston*.

Diatomic Of a molecule: made up of two atoms.

Diorama A mobile device for showing giant scenes.

DNA Short for deoxyribonucleic acid, a long-stranded molecule made up of small individual units. The genes of living things are recorded in the DNA of their cells: the order of the different bases in the DNA (see *base*) "spells out" each gene. The genes of some viruses occur as RNA, not DNA. See also *nucleic acid*.

Dry-cell battery A battery in which the electrolyte is a paste, not a free liquid.

Dry distillation Heating a solid to collect gases without completely vaporizing or burning the solid.

Dynamic equilibrium The state of affairs in a chemical reaction when reactants are changing into products and products into reactants at the same rate, resulting in no net change.

Electrochemistry The branch of chemistry that studies the relationship between chemistry and electricity—for example, how compounds are decomposed into elements using electrolysis.

Electrode A conductor that gives out or takes in electrons as part of an electric circuit.

Electrolysis The breaking down of chemical substances using electricity.

Electrolyte A liquid or paste that can conduct electricity as a result of the movement of positive and negative ions within it.

Electromagnetic radiation Radiation that transmits energy as waves of fluctuating electric and magnetic fields traveling at "the speed of light." See also *electromagnetic spectrum*.

Electromagnetic spectrum The complete range of electromagnetic radiation from radio waves (the lowest frequencies and longest wavelengths) through microwaves, infrared radiation, visible light, and ultraviolet radiation, to X-rays and gamma rays, which have the highest frequencies and energy.

Electromagnetism Magnetic forces produced by electricity; more generally, any phenomena involving the connection between electricity and magnetism.

Electron A tiny negatively charged particle. In an atom, electrons orbit around the much heavier central nucleus, balancing out the positive charge of the atom's protons. Electrons can flow freely through metals and can also occur as beams of radiation. See also *cathode ray*.

Electron orbital The path that an electron follows when orbiting around the nucleus of an atom. Orbitals are not exact routes but represent the probability of finding the electron at any particular point. Orbitals can be different shapes and are fundamental to forming chemical bonds. See also *hybrid orbital*.

Electronegativity The tendency of an atom of a particular element to attract electrons when it is covalently bonded with the atom of another element. Fluorine is the most electronegative element.

Element, chemical A substance made up of atoms all with the same atomic number. See *atomic number*.

Emission lines Bright lines at particular positions in a spectrum that indicate the presence of chemical elements whose atoms emit light at those particular frequencies when heated.

Enzyme Any of several thousand types of large molecules in living things, each of which catalyzes (promotes) a particular type of chemical reaction. Nearly all enzymes are proteins.

Ester An organic compound formed by the reaction of an acid with an alcohol.

Fission See *nuclear fission*.

Flame spectra Spectra observed when substances are heated up in a flame. Different elements emit light of characteristic frequencies and can be identified via their spectra.

Fungicide A substance used to kill fungi.

Greenhouse effect The tendency of some gases in the atmosphere, such as carbon dioxide, to trap heat energy radiating from Earth and make the atmosphere warmer.

Haber's Law A law used to calculate the effect of exposure to toxins.

Half-life The time by which half of a sample of a particular radioactive substance will have decayed into other substances. The term is used in other contexts, such as the length of time a drug stays in the body.

Halocarbon Organic chemical similar to hydrocarbon, where some or all of the hydrogen atoms have been replaced by halogen atoms.

Halogen Any of the group of similar elements that includes fluorine, chlorine, bromine, iodine, and astatine. (Halogen literally means "salt former.")

Handedness Term used in relation to molecules that exist in two mirror-image forms, in a similar way to right- and left-handed gloves. Also called chirality. See also *stereoisomerism*.

Heavy metal A term used for any metallic element other than those with light atomic masses. For example, copper and lead are heavy metals, but aluminum and magnesium are not.

Hybrid orbital A type of electron orbital found in bonds between atoms, where the orbitals of individual atoms overlap and combine. See also *electron orbital*.

Hybridization In the context of chemical bonding, the forming of hybrid orbitals.

Hydrocarbon An organic compound consisting only of carbon and hydrogen atoms, such as methane or octane.

Hydrogen bond A type of attraction, weaker than a covalent bond, between a hydrogen atom and certain other atoms such as oxygen and nitrogen. Hydrogen bonds help stabilize the shapes of many biological molecules such as proteins and nucleic acids.

Ice core A long cylindrical column of ice obtained from a glacier or ice cap by drilling. Ice cores provide evidence of changing climatic conditions and pollution levels over millennia.

Incandescence Light given off when a substance is heated sufficiently.

Inert Unreactive.

Infrared radiation Electromagnetic radiation having wavelengths longer than visible light, experienced as heat radiation in everyday life.

Ion An atom that has gained or lost one or more electrons and so has an overall negative or positive charge. The process of this happening is called ionization.

Ionizing radiation Any form of radiation that causes atoms to ionize. Such radiation is often dangerous to the human body.

Isomer A molecule that contains the same number and type of atoms as another molecule but in a different arrangement.

Isomerism Where a substance occurs as more than one isomer.

Isotope An atom of a given element that contains a particular number of neutrons. Many elements have several different isotopes, which normally have the same chemistry but may differ in physical characteristics, e.g. some are radioactive.

Kaliapparat A 19th-century glass apparatus for measuring the carbon content of different substances.

Keeling curve A graph showing the increase of world atmospheric carbon dioxide levels since 1958, recorded at Mauna Loa Observatory, Hawai'i, in a program begun by geochemist Charles Keeling.

Latent image A photographic image whose details exist in the form of chemical differences on the photographic surface but which are not visible to the eye without further processing.

Law of Definite Proportions The chemical law that pure substances form compounds with each other only in fixed ratios measured by weight.

Lead chamber process An older method for making sulfuric acid, now largely replaced by the *contact process*.

Leyden jar An apparatus invented in the 18th century for storing static electricity. It can give an electric shock when discharged.

Mass The amount of matter in a particle or substance.

Meniscus The curved upper surface of a liquid in a tube.

Mitochondria (singular, mitochondrion) Organelles in living cells that make energy available to the cells. See also *organelle*.

Molecule A combination of two or more atoms joined together by (usually covalent) chemical bonds. The atoms may be of the same element (for example, the hydrogen molecule, which contains two hydrogen atoms) or of different elements.

Negative image A photographic image where the light areas of the original scene are visible as dark areas and vice versa.

Neonicotinoids Modern synthetic insecticides, chemically similar to nicotine. Less toxic to mammals and birds than many insecticides, it can be very harmful to bees and other "beneficial" insects.

Neutron A particle with no electric charge, found in the nuclei of all atoms except the main isotope of hydrogen, whose nucleus is a single proton. A neutron has almost the same mass as a proton. Free neutrons are released by nuclear fission.

Nuclear fission The splitting of certain kinds of atomic nuclei into two or more smaller nuclei of other elements. This can occur if the nuclei are hit by neutrons, as happens in a nuclear reactor or atomic bomb.

Nucleic acid Either of the two long-stranded molecules in living things, DNA (deoxyribonucleic acid) and RNA (ribonucleic acid). Nucleic acids consist of long chains of individual units called nucleotides; in turn, each nucleotide contains a sugar molecule (ribose or deoxyribose), a phosphate group, and a base (see *base*, sense 2).

Nucleobase See *base*.

Nucleus (plural, nuclei) (1) The central part of an atom, containing its protons and neutrons. Nearly all of an atom's mass is in the nucleus. (2) The organelle in most biological cells that contains the cell's genetic information (DNA).

Orbital See *electron orbital*.

Ore A commercially important mineral source of a metal.

Organelle A small structure, usually surrounded by a membrane, that carries out particular functions within a cell. Examples include mitochondria and cell nuclei.

Organic chemical Any compound of carbon. (A few very simple compounds such as carbon dioxide are usually excluded.) Carbon can form chains and rings with itself at the same time as bonding with other elements, so a huge number of different organic chemicals exist, including major biological molecules such as carbohydrates and proteins. See also *hydrocarbon*.

Osmosis Movement of water or another solvent from a more dilute to a more concentrated solution through a semipermeable membrane.

Oxidation, oxidize A substance is oxidized when oxygen is combined with it, or more generally when electrons are removed from it, during a chemical reaction. See also *reduction*.

Oxide A compound of oxygen with one other element.

Ozone layer The layer containing gaseous ozone (a three-atom form of oxygen) high in the atmosphere that protects Earth from harmful ultraviolet radiation.

Pathogen A bacterium or other microscopic living thing that causes disease.

Peptidoglycan A large molecule that forms part of the cell wall in bacteria.

Petri dish A shallow flat-bottomed circular dish with a lid, usually transparent and made of glass or plastic.

Phlogiston In 18th-century chemical theory, a supposed substance that was given off when all substances burned. The theory of phlogiston was later discredited.

Phosphorescence Light given off by a substance other than from heating, especially when it persists for a time.

Plum pudding model An outdated model of the structure of an atom in which negatively charged electrons were imagined as embedded in a positively charged matrix, like plums in a plum pudding.

Polymer A long molecule made of smaller repeating units joined together.

Proton A positively charged particle found in the nucleus of all atoms. Each chemical element has a different number of protons in its atoms. See also *atomic number*.

Radiation (1) See *electromagnetic radiation*. (2) Beams of subatomic particles such as electrons or neutrons.

Radioactivity Phenomenon where some types of atomic nuclei give off radiation while transforming into other kinds of nuclei. See also *radiation, transmutation*.

Reactant A substance that takes part in a chemical reaction and decomposes or changes in the process. See also *catalyst*.

Reagent A substance that causes a chemical reaction, especially one that is used to detect another substance present.

Reducing agent A substance that causes another substance to be reduced. See *reduction*.

Reduction, reduce In a chemical reaction, a substance is reduced when oxygen is removed from it, or more generally when electrons are added to it. For example, iron oxide is reduced to iron during smelting. See also *oxidation*.

Resonance In chemistry, a term used to indicate a pattern of bonding intermediate between two more easily described states.

Reversible reaction A chemical reaction that can take place in both directions.

RNA Short for ribonucleic acid, a long-stranded molecule similar to DNA. Genetic instructions need to be copied from DNA into RNA in cells to have an effect. RNA molecules also play other roles in cells.

Salt Any substance formed by the reaction of an acid with a base. See *base*.

Silver salts Compounds of silver, in particular salts such as silver bromide that are used in photography.

Smelting Extracting a metal from its ore using heat and a reducing agent. See *reducing agent*.

Soda-lime glass The most common type of glass, whose ingredients include sodium carbonate (soda) and calcium oxide (lime), as well as silicon dioxide (silica).

Spectrum The pattern produced when different wavelengths of light or other electromagnetic radiation are spread out using a prism or other device. The term can also denote a range of wavelengths—for example, "the infrared spectrum."

Static electricity Phenomena where positive and negative electric charges become separated via friction between different materials or naturally in a thundercloud. The energy built up may discharge suddenly, as in a lightning flash.

Stereoisomerism A type of isomerism where two or more isomers contain the same chemical subgroups, but the groups are arranged differently in space. Mirror-image isomers are an example. See *handedness, isomerism*.

Tetraethyl lead (TEL) A lead compound formerly added to gasoline to improve its performance but now banned worldwide because of the lead pollution it creates.

Tetrahedron A 3D geometrical shape with four triangular faces and four vertices (corners). In a carbon atom bonded to four other atoms, the carbon's bonds are normally arranged in a tetrahedral shape—that is, extending from the atom to the corners of an imaginary tetrahedron.

Thalidomide A drug that was formerly used to treat morning sickness in pregnant women but that turned out to cause major birth defects in their offspring.

Thermodynamics The theory and study of the relationship between heat and other forms of energy. These principles are vital to an understanding of chemical reactions.

Transmutation The change of one chemical element or isotope into another, attempted unsuccessfully by alchemists but achieved today in nuclear reactors.

Transuranic elements Elements with a higher atomic number than uranium in the periodic table.

Ultraviolet radiation Electromagnetic radiation with wavelengths shorter than visible light but longer than X-rays.

Valence, valency The "combining power" of an atom, i.e. the number of different bonds it can form with other atoms or molecules.

Voltaic pile The earliest electric battery, invented by Alessandro Volta and first demonstrated publicly in 1800.

Wave A form of regular motion that transmits energy. The wavelength of a water or light wave is the distance between the crests of successive waves.

Wrought iron A low-carbon form of iron created by hammering or rolling more impure iron. It is much less brittle than cast iron but has largely been replaced by steel today. See also *cast iron*.

X-ray diffraction The technique of investigating the structure of crystals by beaming X-rays at them and studying the way the rays are affected by being passed through the crystal.

INDEX

Page numbers in **bold** refer to main entries.

B

C

E

F

G

H

I

J

L

M

N

P

Q

S

T

U

V

W

QUOTE ATTRIBUTIONS

The primary quotations below are attributed to people who are not the key figure for the topic.

PRACTICAL CHEMISTRY

18 He who does not know beer, does not know what is good
Sumerian proverb

20 Sweet oil, the fragrance of the gods
"The debate between sheep and grain," Sumerian creation myth

24 Dusky iron sleeps in dark abodes
Erasmus Darwin, *The Botanic Garden* (1791)

26 If it were not so breakable, I should prefer it to gold
Gaius Petronius, *Satyricon*, (1st century CE)

27 Money is by nature gold and silver
Karl Marx, *Das Kapital* vol 1 (1867)

28 Atoms and the vacuum were the beginning of the universe
Diogenes Laetius, *Democritus* vol IX

THE AGE OF ALCHEMY

42 The whole house burned down
Zhenyuan miaodao yaolüe (Taoist text)

ENLIGHTENMENT CHEMISTRY

60 I have seized the light
Louis Daguerre, letter to his friend Charles Chevalier, 1839

69 Chemistry without catalysis would be a sword without a handle
Alwin Mittasch, German chemist (1948)

THE CHEMICAL REVOLUTION

84 The same but different
Jöns Jacob Berzelius

90 The instantaneous union of sulfurous acid gas with oxygen
The Manufacture of Sulphuric Acid, Patent No 6096 (1831)

94 Air reduc'd to half its wonted extent, obtained twice as forcible a spring
Robert Boyle, *A Defense of the Doctrine Touching the Spring and Weight of the Air* (1662)

98 Any object may be copied by it
"Remarkable invention," *Boston Daily Advertiser* (February 23, 1839)

106 O, excellent air-bag!
Robert Southey, letter to his brother Tom (1800)

THE INDUSTRIAL AGE

116 Coal-derived blues
August von Hofmann, German chemist

120 Powerful explosives have enabled wonderful work
Pierre Curie, Nobel lecture (1905)

122 Bright lines when brought into the flame
Robert Bunsen, "Chemical Analysis by Observation of Spectra" (1860)

144 The entropy of the universe tends to a maximum
Rudolf Clausius, German physicist

150 Change prompts an opposing reaction
John Gall, *The Systems Bible* (2002)

154 A glorious yellow effulgence
Joseph Lockyer, British astronomer

162 Proteins responsible for the chemistry of life
Randy Schekman, American cell biologist

THE MACHINE AGE

182 Molecules, like guitar strings, vibrate at specific frequencies
Infrared: Interpretation (2020); https://chem.libretexts.org/@go/page/1846

183 This material of a thousand purposes
TIME Magazine (September 22, 1924)

184 The most measured chemical parameter
Maria Filomena Camões, "A Century of pH Measurements" (2010)

190 Bread from air
Contemporary advertising slogan

192 The power to show unexpected and surprising structures
Dorothy Hodgkin

196 The throat seized as by a strangler
A. T. Hunter, Canadian soldier, Ypres (1915)

202 Each line corresponds to a certain atomic weight
Professor H. G. Söderbaum, Nobel presentation speech (1922)

204 The biggest thing chemistry has done
Lord Todd, president of the Royal Society of London (1980)

212 Development of motor fuels is essential
Frank Howard, Vice President Ethyl Gasoline Corporation (1925)

214 Curly arrows are a convenient electron accounting tool
Yu Liu, Philip Kilby, Terry J. Frankcombe, and Timothy W. Schmidt, "Calculating curly arrows from ab initio wavefunctions" (2018)

230 Out of the atom smasher
Nature (January 28, 2019)

THE NUCLEAR AGE

254 Delicate motion that resides in ordinary things
Edward Purcell, American physicist

258 The language of the genes has a simple alphabet
Steve Jones, British geneticist

264 New compounds from molecular acrobatics
George Rosenkranz, Mexican chemist

266 Living light
Jules Verne, *Twenty Thousand Leagues Under the Sea* (1872)

267 Polymers that stop bullets
Dan Samorodnitsky, *Massive Science* (2019)

270 The siren draw of miracle cures and magic bullets
Angie Wan, Amy Zhang, *The History of Magic Bullets: Magical Thinking in Times of Infectious Diseases* (2020)

274 Power to alter the nature of the world
Rachel Carson

278 The hidden workhorses of the mobile era
Paul Coxon, *Quanta Magazine* (2019)

284 Beautifully precise copying machines
Har Gobind Khorana, Nobel lecture (1968)

A CHANGING WORLD

292 Build things one atom at a time
Richard E. Smalley, American chemist

296 Biobased and biodegradable
X. Chen, N. Yan, "A brief overview of renewable plastics," *Materials Today Sustainability* (2020)

304 We will know where matter ceases to be
Dawn Shaughnessy, American radiochemist

312 Humanity against the viruses
Sarah Gilbert, British vaccinologist

ACKNOWLEDGMENTS

Dorling Kindersley would like to thank: Aparajita Kumar and Hannah Westlake for editorial assistance; Nobina Chakravorty for design assistance; Bimlesh Tiwary for CTS assistance; Ahmad Bilal Khan for picture research assistance; Ann Baggaley for proofreading; and Helen Peters for indexing.

PICTURE CREDITS

The publisher would like to thank the following for their kind permission to reproduce their photographs:

(Key: a-above; b-below/bottom; c-center; f-far; l-left; r-right; t-top)

18 Alamy Stock Photo: Adam Ján Figeľ (bc). **19 Hop Growers of America**. **21 akg-images:** Roland & Sabrina Michaud (cr). **24 Alamy Stock Photo:** Science History Images (b). **27 Alamy Stock Photo:** Zev Radovan / www.BibleLandPictures.com (cr). **29 Alamy Stock Photo:** Chronicle (bl). **30 Alamy Stock Photo:** The History Collection (cb). **38 Dreamstime.com:** Rob Van Hees (t). **39 Alamy Stock Photo:** Heritage Image Partnership Ltd. (crb). **41 Alamy Stock Photo:** Realy Easy Star (bc). **Getty Images:** Hulton Archive / Apic (tl). **42 akg-images:** Erich Lessing (b). **44 Alamy Stock Photo:** PvE (cb). **47 Alamy Stock Photo:** Granger Historical Picture Archive (cr). **49 Science Photo Library. 54 Science & Society Picture Library:** Science Museum (bc). **55 Science Photo Library:** Sheila Terry (bl). **57 Science Photo Library:** Sheila Terry (tr). **59 Alamy Stock Photo:** Prisma Archivo (bl). **61 Alamy Stock Photo:** Granger Historical Picture Archive (tr). **Getty Images:** mashuk / DigitalVision Vectors (tl). **63 Alamy Stock Photo:** Chronicle (bl). **Science Photo Library:** (t). **65 Alamy Stock Photo:** ART Collection (bl); SBS Eclectic Images (tr). **68 Alamy Stock Photo:** Panagiotis Kotsovolos (crb). **74 Getty Images:** Universal Images Group Editorial / Leemage (crb). **75 Dreamstime.com:** Nicku (bl). **77 Alamy Stock Photo:** Science History Images (bc). **78 Science Photo Library:** Paul D. Stewart (tl). **79 Science Photo Library:** Alexandre Dotta (tl). **80 Science Photo Library:** Cordelia Molloy (cb). **81 Science Photo Library:** Library Of Congress (bl). **83 Depositphotos Inc.:** georgios (bl). **Getty Images:** Hulton Archive / Apic (cr). **85 Alamy Stock Photo:** The Print Collector / Oxford Science Archive / Heritage Images (tr). **86 Dreamstime.com:** Sergey Tsvirov (b). **87 Getty Images:** DigitalVision Vectors / ZU 09 (clb). **89 Alamy Stock Photo:** Library Book Collection (bl). **90 Alamy Stock Photo:** World History Archive (br). **93 Alamy Stock Photo:** GL Archive (bl). **Science Photo Library:** Sheila Terry (tr). **95 Science Photo Library:** Sheila Terry (crb). **96 Alamy Stock Photo:** Chronicle (b). **97 Alamy Stock Photo:** FLHC 96 (tr). **99 Alamy Stock Photo:** Pictorial Press Ltd. (bl). **Getty Images:** Hulton Archive / V&A Images (t). **103 Alamy Stock Photo:** Hirarchivum Press (bl). **Dreamstime.com:** Tashka2000 (tc). **104 Dreamstime.com:** Tezzstock (b). **106 Science & Society Picture Library:** Science Museum (bc). **107 Alamy Stock Photo:** Everett Collection Inc. (b); Science History Images (tr). **113 Alamy Stock Photo:** Pictorial Press Ltd. (tr); Nikolay Staykov (b). **114 Our World in Data | https://ourworldindata.org/:** Global Carbon Budget—Global Carbon Project (2021) (Graph's visual representation). **115 Getty Images:** Josh Edelson / AFP (t). **117 Alamy Stock Photo:** Zuri Swimmer (tr). **Science Photo Library. 119 Science Photo Library. 120 Getty Images:** GraphicaArtis / Archive Photos (crb). **123 Science Photo Library:** Charles D. Winters (tr). **124 Alamy Stock Photo:** Chronicle (bl). **Science Photo Library:** (tr); Sheila Terry (cl). **125 Dreamstime.com:** Reese Ferrier (crb). **127 Getty Images:** Apic / Hulton Archive (bl). **129 Alamy Stock Photo:** Pictorial Press Ltd. (bl); Science History Images (tr). **132 Getty Images:** Sovfoto / Universal Images Group (bl). **133 Alamy Stock Photo:** Photo Researchers / Science History Images (ca). **134 Science & Society Picture Library:** Science Museum. **135 Getty Images:** Science & Society Picture Library / SSPL. **137 Alamy Stock Photo:** Everett Collection Historical (br). **139 Alamy Stock Photo:** Historic Collection (tr). **141 Science Photo Library:** Alfred Pasieka (t). **142 Alamy Stock Photo:** History and Art Collection (bl). **145 Getty Images:** Hulton Archive / Stringer (tr). **148 Science Photo Library:** Charles D. Winters (bc). **149 Alamy Stock Photo:** Pictorial Press Ltd. (bl). **151 Dreamstime.com:** Egortetiushev (cr). **153 Alamy Stock Photo:** Science History Images (tr). **156 Alamy Stock Photo:** The History Collection (bl). **157 Internet Archive:** *The Gases of the Atmosphere: The History of Their Discovery* by William Ramsay, 1896. **158 Shutterstock.com:** Kim Christensen (bl); Kim Christensen (bc/Helium); Kim Christensen (bc); Kim Christensen (bl/Crypton); Kim Christensen (fbl). **159 Alamy Stock Photo:** Aardvark (tr). **161 Getty Images:** Nicola Perscheid / ullstein bild (bl). **162 Science Photo Library:** Laguna Design (br). **163 Alamy Stock Photo:** Chronicle (tr). **164 Science Photo Library:** Science Source / Charles D. Winters (b). **165 Alamy Stock Photo:** World History Archive (bl). **172 Science Photo Library:** Charles D. Winters (crb). **174 Getty Images:** Mondadori Portfolio Editorial (tl). **Science Photo Library:** Cordelia Molloy (tr). **178 Getty Images:** Archiv Gerstenberg / ullstein bild (crb). **179 Alamy Stock Photo:** GL Archive (bl). **180 Dorling Kindersley:** USGS. **181 Alamy Stock Photo:** Yogi Black (b). **183 Alamy Stock Photo:** Christopher Jones (cr). **187 Carlsberg Archives:** (bl). **188 Alamy Stock Photo:** Aleksandr Dyskin (clb). **189 Dreamstime.com:** Pramote Soongkitboon (b). **191 Alamy Stock Photo:** GL Archive (tr). **Dorling Kindersley:** Data: Erisman, J., Sutton, M., Galloway, J., et al. "How a century of ammonia synthesis changed the world." Nature Geosci 1, 636–639 (2008). https://doi.org/10.1038/ngeo325 / Our World in Data | https://ourworldindata.org/ (bl). **193 © The University of Manchester 2022. All rights reserved**. **194 Getty Images / iStock:** blueringmedia (b). **195 Getty Images:** Photo12 / Universal Images Group (ca). **197 Alamy Stock Photo:** Interfoto / Personalities (tr). **198 Archiv der Max-Planck-Gesellschaft, Berlin**. **201 Austrian Central Library for Physics:** (tc). **203 Alamy Stock Photo:** Yogi Black (tr). **207 Alamy Stock Photo:** John Davidson Photos (ca). **Getty Images:** Fritz Eschen / ullstein bild (bl). **209 Alamy Stock Photo:** World History Archive (t). **211 Alamy Stock Photo:** Trinity Mirror / Mirrorpix (t). **Getty Images:** Meinrad Riedo (br). **213 Alamy Stock Photo:** Historic Images (tr); Suzanne Viner / Retro AdArchives (cb). **220 Alamy Stock Photo:** Pictorial Press Ltd. (bl). **224 Getty Images:** Baron / Hulton Archive (bl). **225 Science Photo Library:** St. Mary's Hospital Medical School (cla). **226 Alamy Stock Photo:** Retro AdArchives (b). **228 Dorling Kindersley:** Data: 1369-5274/© 2020 The Authors. Published by Elsevier Ltd. This is an open access article under the CC BY license (http://creativecommons.org/licenses/by/4.0/). **229 Science Photo Library:** Stephanie Schuller (b). **231 Alamy Stock Photo:** Science History Images (bl). **Science Photo Library:** ISM (cra). **232 Shutterstock.com:** AP (br). **235 Bridgeman Images:** © Estate of Lotte Meitner-Graf (tr). **237 Getty Images:** Universal History Archive / Universal Images Group. **241 Alamy Stock Photo:** Sueddeutsche Zeitung Photo (bl). **242 Oregon State University Special Collections and Archives Research Center, Corvallis, Oregon:** The Ava Helen and Linus Pauling Papers (MSS Pauling). **245 Science Photo Library:** Ramon Andrade 3dciencia. **251 Alamy Stock Photo:** Alpha Stock (tr). **252 Science Photo Library:** National Archives (t); US Department Of Energy (clb). **255 akg-images:** Bruni Meya (tr). **xkcd.com:** © Andrew Hall 2016 (b). **257 Science Photo Library. 259 Alamy Stock Photo:** Pictorial Press Ltd. (bl). **260 Alamy Stock Photo:** Science History Images (tl). **263 Getty Images:** Pam Berry / The Boston Globe (bl). **267 Science Photo Library:** Sinclair Stammers (crb). **269 Alamy Stock Photo:** Keystone Press (bl). **271 Shutterstock.com:** Sipa (tr). **273 Getty Images:** Brooks Kraft / Sygma (tr). **© The European Centre for Medium-Range Weather Forecasts (ECMWF):** (tl). **275 Getty Images:** Pramote Polyamate / Moment (t). **277 Getty Images:** James L. Amos / Corbis Historical (cla). **283 Alamy Stock Photo:** Alex Segre (t); University of Texas at Austin via Sipa USA (bl). **287 Alamy Stock Photo:** Andrew Hasson (bl). **292 Science Photo Library:** Laguna Design (cb). **295 Alamy Stock Photo:** Ken Gillespie / First Light / Design Pics Inc. (ca). **Shutterstock.com:** Walter Bieri / EPA (crb). **301 Reprint Courtesy of IBM Corporation ©:** (tl). **302 Alamy Stock Photo:** picture alliance / dpa (bc). **306 Alamy Stock Photo:** ITAR-TASS News Agency (bl). **310 Dorling Kindersley:** IOP Science: Yuri Oganessian 2012 J. Phys.: Conf. Ser. 337 012005. **311 TopFoto:** Sputnik. **313 Science Photo Library:** NIAID / National Institutes Of Health (t). **314 Shutterstock.com:** Csilla Cseke / EPA-EFE (tr)

All other images © Dorling Kindersley
For further information see: www.dkimages.com